METAL IONS IN SOLUTION

METAL IONS
IN SOLUTION

by

Dr. JOHN BURGESS, M.A., Ph.D.
Lecturer in Chemistry
University of Leicester

ELLIS HORWOOD LIMITED
Publisher   Chichester

Halsted Press: a division of
JOHN WILEY & SONS
New York · London · Sydney · Toronto

The publisher's colophon is reproduced from James Gillison's drawing of the ancient Market Cross, Chichester

First published in 1978 by

ELLIS HORWOOD LIMITED

Market Cross House, 1 Cooper Street, Chichester, Sussex, England

Distributors:

Australia, New Zealand, South-east Asia:
Jacaranda-Wiley Ltd., Jacaranda Press,
JOHN WILEY & SONS INC.,
G.P.O. Box 859, Brisbane, Queensland 4001, Australia.

Canada:
JOHN WILEY & SONS CANADA LIMITED
22 Worcester Road, Rexdale, Ontario, Canada.

Europe, Africa:
JOHN WILEY & SONS LIMITED
Baffins Lane, Chichester, Sussex, England.

North and South America and the rest of the world:
HALSTED PRESS, a division of
JOHN WILEY & SONS
605 Third Avenue, New York, N.Y. 10016, U.S.A.

© 1978 J. Burgess/Ellis Horwood Limited

British Library Cataloguing in Publication Data

Burgess, John
Metal ions in solution.
1. Metal ions 2. Solution (Chemistry)
I. Title
541'.372 QD561 78-040122
ISBN 0-85312-027-7 (Ellis Horwood Ltd., Publishers)
ISBN 0-470-26293-1 (Halsted)

Typeset in Press Roman by Coll House Press, Chichester, Sussex.
Printed in Great Britain by Cox and Wyman Ltd., Fakenham.

Table of Contents

3 METAL ION SOLVATION: OTHER SPECTROSCOPIC TECHNIQUES

4 METAL ION SOLVATION: NON-SPECTROSCOPIC METHODS OF INVESTIGATION

14 KINETICS AND MECHANISMS: REACTIONS OF COORDINATED SOLVENTS

15 ENVOI

Author's Preface

This book presents an inorganic chemist's view of metal ions in solution, a subject which spans inorganic and physical chemistry. There is a copious literature concerned with this subject, ranging from elementary textbooks through to the most detailed and erudite reviews, among which the present book is intended to occupy an intermediate position. In other words, it is hoped that it will appeal to senior undergraduates and also to specialised readers engaged in research work. The level of the discussion is aimed at the former group, but the wide coverage of topics and generous provision of examples and data should prove of value to both classes of reader. Additionally, some sections may prove useful to amplify and illustrate topics normally presented to undergraduates at all levels. The book is fully referenced, both in order that interested readers can readily trace details of experimental conditions and methods, and that those who wish to pursue matters in greater depth can find their own way into the voluminous literature. In addition to numerous references in the text, 'Further Reading' lists of selected books and reviews of general relevance are provided at the end of most chapters.

The scope of the book can be gauged from the Table of Contents. Solvents covered include water, a variety of non-aqueous solvents, and binary aqueous mixtures. Aqueous systems predominate, of course. The book deals with metal ions from all appropriate areas of the Periodic Table. Metal oxoanions are considered to be complexes rather than simple solvated metal ions and are therefore not covered. The present volume is not concerned with the chemistry of metal complexes, a topic fully covered elsewhere. The subject of the kinetics and mechanism of formation of complexes has been included because this is an important aspect of the chemistry of metal ions in solution. Stability constants of complexes, and the kinetics and mechanisms of solvolysis of complexes, are not covered.

Some knowledge of spectroscopy, thermodynamics, kinetics, and descriptive inorganic chemistry is assumed. Experimental techniques are not described, though often the derivation of results from experimental data is discussed. Illustrative examples are for the most part drawn from the normal realms of inorganic and physical chemistry, though some examples from bio-inorganic and from industrial chemistry are added where possible.

The preliminary draft of this book was made in 1971; the final manuscript was prepared during 1976. Literature coverage was intended to be complete for books, reviews, and journals received by the University of Leicester library up to the end of 1975, but some of the more important relevant articles published during 1976 have also been included, although often the need to incorporate these into the manuscript with the minimum of disruption has limited mention to a brief citation or relegated them to a footnote. While efforts have thus been made to keep references as up to date as possible, important early work has not been entirely neglected. Several historically important and seminal references have been included, and certain sections devoted to areas which were formerly considered more significant than they are now.

I am grateful to many colleagues for advice, assistance, and encouragement during the prolonged period of writing. In particular I wish to thank Dr. David Rosseinsky, Professor Alex McAuley, and Professor Martin Tobe for detailed and critical reading of large sections of the manuscript. Thanks to them the printed text is both more accurate and more readable than the original, while the responsibility for errors and infelicities which remain is of course entirely mine. Numerous discussions with undergraduates, research students, and colleagues at the University of Leicester have contributed to the development of the text. Dr. Mike Blandamer's advice on various aspects of physical chemistry, especially those relating to the thermodynamics of ions in solution, has been of great value over the past few years. I would also like to record my gratitude to Drs. Valda and Jack McRae and to Professor Don Stranks for their most generous hospitality during a sabbatical in Melbourne, where the agreeable environment greatly encouraged the process of organising the raw material for this book into some sort of coherent shape. Finally it is my pleasure to thank my publisher Ellis Horwood for his patience, help, and encouragement during the protracted writing of this book.

J. BURGESS
University of Leicester

June 1977

Chapter **1**

INTRODUCTION

1.1   GENERAL

Ionic salts have large lattice energies. Therefore considerable expenditure of energy is required in separating their constituent ions when they are dissolved so as to give solutions containing discrete cations and anions dispersed throughout the solvent. The fact that innumerable salts dissolve in water, and that many also dissolve in other polar non-aqueous solvents, is a reflection of the favourable nature of the interactions between the solvent molecules and the dissolved ions. The opposite, compensatory features of the situation can be succinctly summarised in the basic thermochemical cycle of Fig. 1.1.

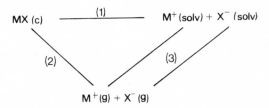

Fig. 1.1 The relation between dissolution, lattice, and ion solvation thermodynamic parameters.

The energy (free energy, enthalpy, ...) for the overall process of the salt dissolving in the solvent, (1), is the balance or difference between two processes. These are the unfavourable process of separating the ions from the lattice to a state of non-interacting or 'infinite' separation, (2), and the favourable process of solvating the ions, (3). Generally (2) and (3) are large, their difference small. The observed solubility characteristics of salts are thus the reflection of small differences between large opposing contributions. For example, the enthalpy of solution of potassium chloride is $4.2 \text{ kcal mol}^{-1}$, whereas the lattice enthalpy and the sum of the cation and anion hydration enthalpies both have magnitudes of about $170 \text{ kcal mol}^{-1}$.

1.1.1   Metal Ions

The ionisation energies

$$M(g) \rightarrow M^+(g) + e^-$$

$$M(g) \rightarrow M^{2+}(g) + 2e^- \quad \text{etc.}$$

of the elements are all positive, the formation of gaseous cations is always an energetically unfavourable process. The ionisation energies of the elements cover a wide range [1]. First ionisation energies range from $567 \text{ kcal mol}^{-1}$ for helium down to 90 kcal mol^{-1} for caesium. In general the lower the ionisation energy, the more likely the element to form positive ions and to exhibit metallic character in its chemistry. The unfavourability of the gas-phase ionisation process can be offset to a greater or lesser extent by solvation of the metal ion M^{n+} when it is transferred into solution (see below, and Chapters 7 and 8).

Many properties of metal ions in solution depend on the charge and size of the ion. The charge on an isolated metal ion is integral and obvious, but the size of a metal ion is more difficult to establish. The estimation of crystal radii for ions has been the subject of much discussion over several decades. Two well-established series of values are those of Pauling and Goldschmidt [2], which agree within a few per cent for most cations. However, agreement is worst for Li^+, where the difference is over 20%. A particularly extensive compilation of crystal radii for ions is available in Adams's recent book on inorganic solids [3].

1.1.2   Solvents

The deposition of ions into a polar, solvating, solvent is advantageous from the point of view of the ions, but may not be so advantageous from the point of view of the solvent. Solvents which are good at solvating ions are in general liquids with pronounced intermolecular forces, such as dipolar interactions and hydrogen bonding. In the solvation of an ion there will obviously be strong perturbation of solvent–solvent interactions in the vicinity of the ion. In other words, the structure of the solvent will be greatly modified near an ion.

Water. This is the commonest, most studied, and most discussed solvent. The early background to the chemistry of water can be found in the interesting book by Dorsey, published in 1940 [4]. More recent books devoted to water and its properties include those written by Eisenberg and Kauzmann [5] and edited by Horne [6] and by Franks [7]. An exhaustive bibliography, running to 119 pages, of references to water from the period 1969 to 1974 is available [8].

The water molecule is bent, with $\angle H-O-H = 104.523°$. The oxygen-hydrogen distances are 0.9571 Å, and the dipole moment of water is 1.84 D. The O–H bond energy is estimated to be $110 \text{ kcal mol}^{-1}$ [5], while the angle subtended by the two lone pairs at the oxygen atom has been calculated to be close to $120°$ [9]. In liquid water there is extensive hydrogen bonding between adjacent molecules. It is generally assumed that the O–H $\cdots$ O arrangement is linear in normal circumstances, with an oxygen–oxygen distance of between 2.5 and 3.0 Å. The energy of dimerisation of water is about $-4\ \text{kcal mol}^{-1}$. A unique property of water is that it can hydrogen bond *from* two points, the two hydrogen atoms, in each molecule. Further, each molecule can bond *to* one of two points, the two lone pairs, in another molecule. This balance permits uniformity of growth of hydrogen-bonded aggregates in three dimensions [10]; the growth of aggregates is energetically favourable. A popular description [11] of liquid water is in terms of a mixture of low density hydrogen-bonded clusters with non-hydrogen-bonded molecules, in the ratio of about 4 to 1. Rearrangements within the system are very rapid; the half-life for a water molecule in a particular environment is thought to be only about 10^{-11} sec. X-ray studies of liquid water suggest that the number of nearest neighbours to a given water molecule is 4.4 rather than 4.0, and are consistent with such a picture of open hydrogen-bonded frameworks with non-hydrogen-bonded molecules in cavities. The interpretation of infrared and Raman spectra of liquid water should provide further support for, and information on, this model but it remains a matter of controversy.

Non-aqueous Solvents. These [12] range from dipolar hydrogen-bonded solvents much like water, such as alcohols and liquid ammonia, through dipolar aprotic solvents such as dimethyl sulphoxide and acetonitrile, to non-polar solvents such as alkanes and benzene. The last type of solvent is of little concern to the present book, but the hydrogen-bonded and dipolar aprotic categories are both important with respect to the chemistry of metal ions in solution.

A variety of physical properties, including molecular geometry, bond strengths, dipole moments, densities, viscosities, and dielectric constants, for a range of inorganic and organic solvents can be found in the invaluable *Chemical Rubber Handbook* [13]. More comprehensive tabulations of densities, viscosities, and dielectric constants of organic solvents are included in Timmermans's book [14], while Horvath's up-to-date compilation of such physical properties for inorganic solvents has recently appeared [15].

Hydrogen-bonded solvents such as the alcohols are, like water, extensively associated in the liquid state, but the same uniformity of growth of aggregates is not possible in the alcohols. This is because each alcohol molecule has two lone pairs of electrons on the oxygen, but only one hydrogen atom available to take part in hydrogen bonding [10]. These solvents are thus less structured than water, as is liquid ammonia for a similar reason. Dipolar aprotic solvents can be accommodated by a simpler model, for the main intermolecular interactions are simply dipole–dipole in nature. These solvents are non-structured.

Several extreme cases of non-aqueous solvents have not been included in the present book. These include molten salts [16], salts of the tetra-alkylammonium tetra-alkylboronate type which are liquid at room temperature

[17], and liquid metals [18].

Mixed Solvents. There are several reasons for the use of mixed solvents, for instance the important but technical need for solubility. This may dictate the use of mixed aqueous media for studying the reaction between a water-soluble inorganic salt (or of one of its constituent ions) and a water-insoluble reactant, for example an uncharged hydrophobic organic compound. Such a situation can arise in studying kinetics of formation of complexes (Chapter 12). Use can also be made of a range of mixtures of two components taken in varying proportions so as to permit the continuous adjustment of a property of the medium. An appropriate illustration would be the variation of the dielectric constant of the solvent, for example over the wide range from 2 to 80 by the use of dioxan + water mixtures. Similarly, if one is concentrating on solvent structure and its bearing on solute properties, the addition of cosolvents to water can have a marked effect on the structure of the solvent. A key reference to the properties of binary aqueous mixtures is the 1966 review by Franks and Ives, which is devoted to alcohol + water mixtures [18]. Much succeeding work in the area of binary aqueous mixed solvents has drawn inspiration from this article.

In any discussion of chemistry in mixed solvents it is necessary to consider both intra-component and inter-component interactions. This point can be illustrated most clearly by reference to two organic systems, acetone + chloroform and methanol + carbon tetrachloride. In the former case, non-ideality of solvent mixtures stems from intercomponent interactions, from hydrogen bonding between the oxygen of the acetone and the hydrogen of the chloroform. Neither component separately is markedly associated. In the latter case, the carbon tetrachloride breaks up the hydrogen-bonded structure of the methanol without itself interacting with methanol molecules. The nett additional bonding interactions in the acetone + chloroform mixtures are reflected in negative excess free energies (G^E) of mixing, whereas the nett bonding disruption in the methanol + carbon tetrachloride mixtures is reflected in positive excess free energies of mixing. The situation with respect to aqueous mixtures is usually more complicated. Often both components are themselves associated, and there are thus both inter- and intra-component interactions to consider in the solvent mixtures. This situation obtains in, for instance, alcohol + water mixtures. It is convenient to classify binary aqueous mixtures according to their thermodynamic excess functions of mixing (Table 1.1). This classification is based on whether G^E is positive or negative, and whether G^E is dominated by H^E or by $T.S^E$

In 'typically aqueous' mixtures there is usually some enhancement of water structure (cf. clathrate hydrates) when small amounts of the organic cosolvent are added to water. On the other hand, addition of greater quantities of the organic component eventually leads to disruption of solvent structure. The changeover from structure enhancement to structure disruption is reflected in a variety of spectroscopic and thermodynamic properties [20]. The composition corresponding to this changeover is characteristic of the cosolvent; it occurs at a

Table 1.1 Classification of binary aqueous mixtures according to their thermo-dynamic excess functions of mixing.

Classification	Excess functions	Cosolvent examples
Typically aqueous (TA)	G^E positive $|T.S^E| > |H^E|$	monohydric alcohols acetone tetrahydrofuran dioxan
Typically non-aqueous positive (TNAP)	G^E positive $|H^E| > |T.S^E|$	acetonitrile sulpholane propylene carbonate
Typically non-aqueous negative (TNAN)	G^E negative $|H^E| > |T.S^E|$	hydrogen peroxide dimethyl sulphoxide hexamethylphosphoramide glycol [†]

† Glycol is a borderline TNAN–TA case.

successively higher mole fraction for methanol, ethanol, and t–butyl alcohol. The larger the hydrophobic portion of the organic cosolvent, the bigger the effect on water structure and on physical properties which are affected thereby. These points are well illustrated by the kinetic pattern for complex formation in mixed aqueous solvents described in Chapter 12.9. On the other hand, there are no extrema of properties in TNAP mixtures. A cosolvent such as acetonitrile merely disrupts water structure over the whole range of compositions. Hydrogen peroxide and dimethyl sulphoxide (TNAN mixtures) are both structure breakers because they compete with the water for hydrogen bonding. Hydrogen peroxide distorts and disrupts three-dimensional water frameworks as it is a different size from water and less symmetrical. The extensive intercomponent interactions (G^E is negative) characteristic of TNAN mixtures are reflected in a variety of spectroscopic and dielectric properties.

A variety of physical properties of mixed solvents are detailed in Volume 4 of Timmermans's treatise on binary systems [21]. A useful brief selection of data are included in the first chapter of Covington and Dickinson's book on solvent systems [22].

1.1.3 Ions in Solution

The environment of a cation in aqueous solution is depicted in Fig. 1.2 [11, 23]. The regions shown are:

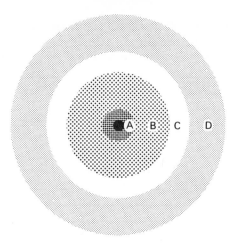

Fig. 1.2   The environment of a cation in solution. • Metal cation; (A) primary solvent; (B) secondary solvent; (C) disordered solvent; (D) bulk solvent.

(A) The primary solvation shell of the cation. This contains water molecules, often six in number, directly interacting or bonding with the cation.

(B) A region of secondary solvation. Here the cation has an influence over the water molecules, and its electron-withdrawing effect on the primary waters of hydration encourages them to hydrogen-bond to secondary waters of hydration. The volume of this secondary shell varies greatly with the nature of the cation, being large for small ions of high charge and small or negligible for large monopositive cations.

(C) A disordered region, or 'fault zone', necessarily separating the ordered region of (A) and (B) from the differently ordered region (D). Again the extent of this region depends on the properties of the metal ion involved.

(D) Bulk solvent.

This picture only applies fully in very dilute solutions, where the mole ratio water:cations is large enough to provide sufficient water molecules so as to surround each cation to the extent indicated. The average separations of ions in a 1:1 electrolyte solution of concentration 10 M, 1 M, or 0.1 M are 4.4, 9.4, and 20 Å respectively [24]. Hence the model of Fig. 1.2 can only be employed legitimately at salt concentrations of lower than about 0.1 M for a 1:1 electrolyte, and at considerably lower concentrations for electrolytes containing ions of higher charge than one.

Regions (B), (C) and (D) of Fig. 1.2 merge imperceptibly into each other. The same may be true of regions (A) and (B), but a distinction between these two regions can sometimes by made and demonstrated meaningfully. For some cations, of high charge and relatively small radius, the rate of exchange of water molecules (or of other solvent molecules) between region (A) and the other regions is sufficiently slow for a real physical distinction to be possible, for example by the use of n.m.r. spectroscopy (see Chapter 2.2.1). In the extreme case of chromium(III), water exchange between the primary solvation (coordination) shell and the other regions is so slow that solvated cations containing various solvent ratios can be isolated and characterised (Chapter 6.2.1). Kinetic aspects of solvent exchange in aqueous and in non- and mixed aqueous media are dealt with in Chapter 11.

In non-aqueous solvents similar to water, for example methanol, ethanol, or formamide, the general picture of Fig. 1.2 still applies. In non-aqueous solvents much less structured than water, the differences between the properties of the bulk solvent, region (D), and those of the solvating solvent, regions (A) and (B), will be less, and region (C) will be less important. This situation holds for, for instance, acetonitrile or dimethylformamide.

Phrased in terms of structural properties of solvents, (A) and (B) are regions where the ion makes its presence felt in a structure-making capacity, region (C) in a structure-breaking capacity. Ions with a low charge density (small charge, low radius) will have a relatively small structure-making region and act as nett structure-breakers, whereas for ions of high charge density (large charge, small radius) structure-forming will dominate. For example, region (C) is thought to be practically absent from the surroundings of the very small Li^+ cation. These ideas of structure-breaking and structure-forming by ions in aqueous solution date back at least as far as 1933 [25]. Since Bernal and Fowler's use of these terms in that year, the concepts have been developed, particularly by Frank and Evans [26] and in Gurney's book [27]. The use of the second differential of partial molar volume with respect to temperature, $\partial^2 \bar{V}^{\ominus}/\partial T^2 |_P$, has been suggested as a criterion for nett structure-breaking (negative) and structure-making (positive) [28]. The concept of structure-breaking is an alternative description of so-called **negative hydration** [29].

A similar picture will apply to ions in mixed solvents, though here there will be an extra variable, the ratio of solvent molecules of different types in the various regions (A) to (D). In particular the composition of the primary solvation shell may differ greatly from that of the bulk solvent. Taking water and acetone as an extreme case of a solvent mixture consisting of a 'good' and a 'bad' solvent, the primary solvation shell may well contain only the one type of solvent molecule, in this example water. Altogether there are a large number of interactions to bear in mind in this type of system. The interactions of the dissolved ions with each solvent must be considered, as well as the effects of the dissolved ions on inter- and intra-solvent interactions.

The information needed to characterise ionic solvation includes the solvation number of the ion, the distance between the ion and adjacent solvent molecules, the strength of ion-solvent interactions, and the delocalisation of charge from the ion onto adjacent solvent molecules. The most important point to make at this stage with respect to cation solvation numbers is that they vary according to the experimental method used in their estimation. For cations at which solvent exchange is sufficiently slow (Chapter 2.2.1) n.m.r. spectroscopy gives the number of solvent molecules in the primary coordination shell, region (A) of Fig. 1.2. On the other hand, methods of estimation based on the motion of solvated ions through the liquid (Chapter 4.2) give an indication of the number of solvent molecules which actually move with the ion. Such estimates include some secondary solvent molecules, region (B), as well as those adjacent to the ion in the primary solvation shell. This problem, and solvation numbers determined for cations, will be dealt with in Chapter 5.

The second requirement, information about cation-solvent distances, can often be obtained from X-ray diffraction experiments (Chapter 4.4.1). It is more difficult to estimate the strength of cation-solvent interactions, though for transition metal cations of electronic configuration d^1 to d^9 inclusive, relative cation-solvent interactions can readily be obtained from ultraviolet-visible spectroscopy (Chapter 3.1). Thermodynamic measures of cation-solvent interaction are discussed in Chapter 7. There is currently some activity in calculating strengths of cation-solvent interactions from appropriate models, especially using molecular orbital theory [30]. Such calculations are giving reasonable estimates for strengths of cation-solvent interactions, but often poor agreement with cation-solvent distances established by X-ray diffraction.

The presence of the adjacent cation has a marked effect on the chemistry of coordinated solvent molecules, and indeed on coordinated ligands in general (Chapter 14). Especially important is the increased ease of loss of a proton from, for instance, a coordinated water molecule (Chapter 9). The acidity of water coordinated to a unipositive cation or to a dipositive alkaline earth or transition metal cation is very small, but when coordinated to a tri- or tetra-postive cation water is markedly acidic.

The foregoing discussion has been concerned primarily with the environment of the cation in solution, and has ignored the presence of the anion simultaneously and necessarily introduced when a salt dissolves. In sufficiently dilute solutions it is possible to treat the cation in isolation, but as the concentration of the solution increases there comes a time when cation–anion interactions can no longer be ignored. In the limit, the anion may enter the primary coordination sphere of the cation and thereby form a complex. Kinetic aspects of this process are reviewed in Chapter 12. Lesser degrees of interaction are grouped together under the generic title of **ion-pairing**. The extent of ion-pair formation depends very much on the solvent. It is particularly great in solvents of low dielectric constant, where the formation of ion-triples and higher aggregates is sometimes evident. When the tendency to form ion-pairs and related species is strong, it may prove difficult to satisfactorily extrapolate experimental results to infinite dilution so as to obtain single ion properties. To give an idea of the propensity of ions to pair in solvents of low dielectric constant, the situation in liquid sulphur dioxide ($D = 15$ at $0°C$) can be cited. Here ion-pairing is essentially complete at concentrations as low as $10^{-2}M^{\dagger}$.

The situation where several anions (or neutral molecules) bond to one cation to give a multiligand complex is extremely familiar. The converse situation, where two or more cations bond to one anion to give a bi- or poly-nuclear complex is rather less commonly encountered. Such cases as $.Ag_3I^{2+}$, mentioned in 1900, and cations bonded via hydroxo-bridges have long been known (Chapter 10). More recent descriptions deal with extensions to these species such as $Ag_2(en)^{2+}$ and Ag_8Te^{6+}, complexes of the $Fe(CN)_6$. $Hg(CN)_2^{4-}$ type, bi- and tri-nuclear complexes of the decadentate ligand triethylenetetra-minehexaacetate, and alkali metal complexes of polydentate cyclic ethers [32].

Even for very dilute solutions where cation-anion interactions are negligible it is often not possible to determine cation and anion parameters separately. Physical measurements of such quantities as enthalpies of solution, conductivi-ties, or viscosities are necessarily made for whole salts. To separate such whole salt values into ionic components requires the making of at least one assumption. This can sometimes be done in a satisfactory manner, sometimes not. Examples will be encountered many times later in this book, particularly in Chapters 4 and 7. On the other hand, certain physical techniques do give single ion parameters directly. These include n.m.r. spectroscopy in slow solvent exchange situations, ultraviolet-visible spectroscopy, and X-ray diffraction.

Table 1.2 Distribution of the elements which give one or more well-established solvated cation species; such elements are indicated in bold type.

H																	
Li	**Be**											B	C	N	O	F	Ne
Na	**Mg**											Al	Si	P	S	Cl	Ar
K	**Ca**	**Sc**	**Ti**	**V**	**Cr**	**Mn**	**Fe**	**Co**	**Ni**	**Cu**	**Zn**	**Ga**	Ge	As	Se	Br	Kr
Rb	**Sr**	**Y**	Zr	Nb	Mo	Tc	**Ru**	Rh	Pd	**Ag**	**Cd**	**In**	**Sn**	Sb	Te	I	Xe
Cs	**Ba**	**La** **Ln†**	Hf	Ta	W	Re	Os	Ir	Pt	Au	**Hg**	**Tl**	**Pb**	**Bi**	Po	At	Rn
Fr	**Ra**	**Ac** **Act††**															

† Lanthanide cations (Chapter 1.2.4)
†† Actinide cations (Chapter 1.2.5)

†For discussions of very concentrated solutions, for instance 'aqueous melts' of concentra-tions in the region of 20 to 30M, where there is insufficient solvent to solvate the ions adequately, the reader is referred to relevant articles [31].

1.2   DISTRIBUTION OF AQUO-CATIONS

When discussing ionic solvation from an inorganic point of view, some classification of known solvento-cations in terms of the periodic arrangement of the elements would seem both obvious and desirable. A comprehensive tabulation of the distribution of all known solvento-cations would be impossibly bulky, though it would serve to emphasise the random and sketchy state of present knowledge of solvated cations in non-aqueous media. Therefore the incidence of aquo-cations only will be discussed in this present section, as information is abundant, though still not entirely complete, about this area.

1.2.1   The sp-Elements

The behaviour of the alkali metal and alkaline earth elements of Groups IA and IIA is simple and straightforward. These elements give aquo-cations M^+aq and $M^{2+}aq$ respectively. There are no complications from redox reactions with solvent water, and complex formation is easy to avoid. Hydrolysis and polymerisation are only significant for the very small Be^{2+} cation. There is some evidence to support the transient existence of the Mg^+aq cation [33].

The Group IIB elements zinc, cadmium, and mercury also give $M^{2+}aq$ species. However mercury(II) often forms fairly strong complexes in solution, so only salts of anions which are weakly complexing (e.g. perchlorate, nitrate) can be used to generate reasonable concentrations of $Hg^{2+}aq$. Moreover such solutions have to be kept acidic to prevent the formation of hydroxo- and polynuclear species (cf. Chapters 9 and 10). Mercury also gives the dimeric cation Hg_2^{2+}, which is unique in its stability in aqueous solution. Cations Hg_3^{2+}, Hg_4^{2+} are also known, but only in such media as liquid sulphur dioxide containing arsenic pentafluoride (Chapter 10.6.1). The Cd_2^{2+} cation has been detected in melts and in the solid state, but not in aqueous solution. The monomeric mercury(I) cation Hg^+ may have a transient existence in solution, but this a matter of controversy [34]. Ephemeral Zn^+aq and Cd^+aq cations can be generated by pulse radiolysis of aqueous solutions of salts of the respective elements in their normal 2+ state [35].

The Group III elements aluminium to thallium all form $M^{3+}aq$ cations. A low pH is required to prevent the formation of hydroxo-species; Tl_2O_3aq precipitates from aqueous solutions at a pH of 1 or above. The 1+ oxidation state is of increasing stability and importance as Group III is descended. In^+aq persists in solution just long enough for its physical and kinetic characteristics to be established. Tl^+aq is perfectly stable, with physical properties similar to hydrated alkali metal ions. The thallium(II) ion $Tl^{2+}aq$ can be generated by pulse radiolysis of mixed thallium(I)–thallium(III) solutions in perchloric acid. Like In^+aq, $Tl^{2+}aq$ has only a transient existence, but persists for long enough for its ultraviolet absorption spectrum and its pK_a to be determined, and for rate constants for some of its reactions to be estimated [36].

With Group IV, the predominant character of the elements has changed to non-metallic. There is no report of the existence of simple aquo-cations of carbon or silicon of course, though mixed complexes $SiF_4(OH_2)_2$ and $SiF_5(OH_2)^-$ have been proposed [37]. The suggestion of a Ge^{4+}aq cation in strongly acid media must be considered tentative [38], and the similar suggestion of Sn^{4+}aq at a pH of less than -1.5 is hardly less tentative [39]. Pb^{4+}aq seems even less plausible, especially in the light of the strong oxidising powers of lead(IV) compounds. The situation is more promising for the 2+ oxidation state, for compounds of tin(II) and lead(II) are known, and some of these are probably ionic. However tin(II) and lead(II) tend to exist as hydroxo- or polynuclear species rather than as the simple aquo-cations in aqueous solution.

By the time Group V is reached, evidence for simple aquo-cations is almost non-existent. Bismuth(V) is normally anionic; bismuth(III) in aqueous media is generally found in the form of oxo- or hydroxo- species or polymeric units. However the existence of Bi^{3+}aq in aqueous perchloric acid of strength greater than 0.6M has been claimed [40]. Antimony(III) appears to exist as the antimonyl, SbO^+, entity in potassium antimonyl tartrate, for example, but no such simple species has been detected in solutions prepared from antimonyl compounds. The $Sb_2(OH)_2^{4+}$ cation has been said to exist in dilute perchloric acid [41].

In Group VI a claim for the existence of Te^{4+}aq in strong perchloric acid has recently been made [42]. There is a possibility that polonium at the foot of this Group may give a Po^{2+}aq cation, though this seems unlikely. Here the situation is complicated by the chemical consequences of the strong radioactivity of this element.

1.2.2 Transition Elements: First Row

All the first row transition elements from titanium through to copper form M^{2+}aq cations in aqueous solution. The early members of this series are powerful reducing agents, which complicates the study of their aqueous solution chemistry. In particular Ti^{2+}aq reduces water within a very short time of its generation. This and the next cations V^{2+}aq and Cr^{2+}aq are all air-sensitive.

All the elements from scandium through to cobalt give M^{3+}aq ions, at least in acid solution. Ti^{3+}aq and V^{3+}aq are reductants, but some of the latter members oxidise water, specifically Mn^{3+}aq and Co^{3+}aq. Copper(III) exists only transiently in aqueous media, in anionic form in saturated sodium hydroxide solution [43].

Simple, stable M^{4+}aq cations are unknown for this series of elements, even in strong acid. Solutions of titanium(IV) are extensively hydrolysed, with $Ti(OH)_2^{2+}$ and $Ti(OH)_3^+$ the suggested species present in dilute perchloric acid. The TiO^{2+}aq ion seems not to exist in aqueous media, whereas VO^{2+}aq is the normal, stable form of vanadium(IV) in solution (though vanadium(IV) occurs in the form of $VO(OH)_3^-$ in strong aqueous alkali [44]). Several stable chromium(IV) compounds are known in the solid state, ranging from $Cr(C_6H_5)_4$ (reported in 1926, [45]) and several alkoxides and alkylamido-compounds [46] to salts of the CrO_4^{4-} and CrF_6^{2-} anions [47], but chromium(IV) species in aqueous solution are too fugitive for their chemistry to have been established. There is strong kinetic evidence for their intermediacy in many, if not all, reductions of chromium(VI). In these kinetic studies it is assumed that the chromium(IV) is in a cationic form [48]. Manganese(IV) is strongly oxidising in acid solution, and precipates as manganese dioxide from alkaline solution.

Oxidation states of 5+ and higher occur as anionic species in aqueous media, with only one exception. That exception is vanadium(V), which can exist under appropriate conditions as the VO_2^+aq cation.

Stable cations M^+aq are not known for this series of elements, but evanescent Co^+aq and Ni^+aq have been generated by pulse radiolysis [35]. Copper(I) is stable in several non-aqueous media, but unstable with respect to disproportionation in aqueous solution. However solutions of Cu^+aq in dilute perchloric acid can be generated, for example by chromium(II) reduction of copper(II) in perchlorate medium [49]. These solutions have been kept long enough for kinetic studies of some of their reductions to have been carried out (Chapter 13.4.4).

1.2.3 Transition Elements: Second and Third Rows

The number of characterised aquo-cations of these elements is small. Their rarity is consistent with the low stability or non-existence of compounds of these elements in oxidation states 1+ to 3+, those oxidation states most favourable for the existence of simple M^{n+}aq cations. The first pair of elements yttrium and lanthanum (and the lanthanide elements, see below) do form M^{3+}aq cations. However zirconium(IV) exists as polynuclear hydroxo-species with no evidence for significant concentrations of Zr^{4+}aq or of ZrO^{2+}aq even in strong perchloric acid. Only recently has evidence for the existence of $Zr(OH)^{3+}$ and of $Hf(OH)^{3+}$ in aqueous perchloric acid been published [50]. A pK value has been estimated for 'Hf^{4+}aq' (see Chapter 9), but the hafnium(IV) species present were not unequivocally characterised. The presence of polymeric species, like the hafnium(IV)–hydroxide–sulphate polynuclear species postulated in hafnium(IV) sulphate solutions [51], cannot be ruled out. Niobium and tantalum have 'virtually no cation chemistry' [52], though the existence of $Nb(OH)_4^+$ in dilute sulphuric acid has been assumed by some authors [53].

The aqueous solution chemistry of molybdenum has been studied in some detail in the past few years. Molybdenum(II) exists in aqueous solution as the dimer Mo_2^{4+}aq, with a postulated molybdenum–molybdenum bond order of four [54]. After some argument, the preparation and properties of Mo^{3+}aq have been clearly established [55]. Molybdenum(IV) in aqueous solution has been variously said to exist as MoO^{2+}aq, $MoO(OH)^+$aq, $Mo(OH)_2^{2+}$aq [56], and as a

dimer in strongly acidic solution as shown in Fig. 1.3(a) [57]. Recent investigators believe that molybdenum(IV) exists as a monomer, which seems more
likely to be MoO^{2+}aq than $Mo(OH)_2^{2+}$aq [58], but this problem of monomer
versus dimer remains to be settled [59]. Molybdenum(V) can be obtained either
by reduction of molybdenum(VI) or by oxidation of lower oxidation states. In
dilute hydrochloric or perchloric acid the predominant species is binuclear with
a nett charge of 1+ per molybdenum atom, which suggests the structure shown
in Fig. 1.3(b) [60]. There is also evidence for polynuclear cationic species in
weakly acid solution [61]. The oxidation states of molybdenum in aqueous
solution have been the subject of a short review [62]. The situation for tungsten
is less clear. Lead amalgam is said [63] to reduce tungsten to the 3+ state
in hydrochloric acid. Both for molybdenum and tungsten the lower oxidation states usually exist in the form of complexes, dimers, and other non-simple
forms, and are air-sensitive.

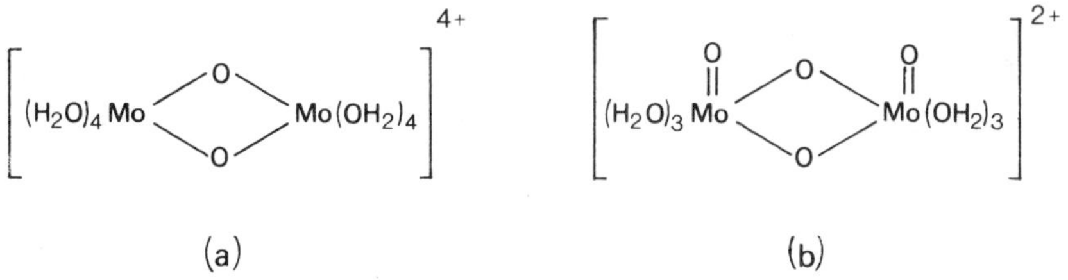

Fig. 1.3 Structures proposed for (a) molybdenum(IV) and (b) molybdenum(V) in
aqueous solution.

No simple aquo-cations of rhenium or of technetium have been reported;
the existence of $Re(OH)_3(OH_2)_3^+$ has been claimed [64] and disputed [65].
By the time Group VIII is reached, the stability of lower (2+, 3+) oxidation
states has increased sufficiently to permit the existence of several well-
established aquo-cations. Ru^{2+}aq, Ru^{3+}aq, and Ru^{4+}aq have all been characterized.
The former can be generated from solutions containing higher oxidation states
of ruthenium, provided that the anions present are non-complexing. No
Os^{n+}aq cations have been established.

Rh^{3+}aq is a stable species which, in contrast to Co^{3+}aq, does not oxidise
water. There is evidence for dimeric aquorhodium(II) in aqueous solution [66],
and a tentative hint of RhO^{2+}aq in a thermodynamic compilation [67]. The
generation of Ir^{3+}aq has recently been described in detail [68], while vague
intimations of Ir^{4+}aq or of its hydroxo-derivatives exist [69]. The Pd^{2+}aq cation
is well established, but the generation of Pt^{2+}aq was reported only late in 1976.
The latter cation is prepared by the reaction of $PtCl_4^{2-}$ with silver(I) or
mercury(II) perchlorate in aqueous solution [70]. Previously the nearest approaches to $Pt(OH_2)_4^{2+}$ were analogues such as $Pt(MeCN)_4^{2+}$ [71] and mixed

ligand complexes such as $Pt(OH_2)_2(NH_3)_2^{2+}$, $Pt(OH_2)_3Cl^+$, and $Pt(OH_2)_2(dmso)$ Cl^+ [72]. For platinum(IV) there is the suggestion that the $PtMe_2Br(OH_2)_3^+$ cation is generated by attack of Ag^+ at $[PtMe_2Br_2]_n$ [73], and an indication from n.m.r. spectroscopy of a $PtMe_3(OH_2)_3^+$ cation [74].

Group IB is similar in behaviour to the Group VIII pairs just discussed. Ag^+aq is a very familiar species, but Au^+aq (and $Au^{3+}aq$) are unknown. Neither $Ag^{2+}aq$ nor $Ag^{3+}aq$ have yet been characterised, though one or both may well exist. Complexes are known of both silver(II) and silver(III), including the $Ag(OH)_4^-$ anion [75], which may be regarded as a distant relative of $Ag^{3+}aq$.

1.2.4 Lanthanide Elements

The 3+ oxidation state of these elements gives $M^{3+}aq$ cations, at least in modestly acid solutions preventing any hydrolysis or polymerisation. Samarium(II), europium(II), and ytterbium(II), which presumably exist as $M^{2+}aq$ ions in aqueous solution, can be generated without too much trouble but are strongly reducing. Cerium(IV) is the only 4+ lanthanide cation to have an established aqueous solution chemistry, though $Ce^{4+}aq$ hydrolyses to give $Ce(OH)^{3+}$ and subsequently polynuclear species [76]. It also has a strong tendency to form complexes, even with such anions as nitrate or sulphate. So $Ce^{4+}aq$ itself is only to be found in relatively strong solutions of acids with non-complexing anions. Some kinetic results are available for reactions of $Pr^{4+}aq$, generated by pulse radiolysis (cf Chapter 12.6.5) [77].

1.2.5 Actinide Elements

Only the 3+ state is to be found for all these elements. Probably all the actinides, except thorium whose 3+ state is confined to the solid state, form $M^{3+}aq$ in solutions acidic enough to prevent the formation of $M(OH)^{2+}aq$. The 3+ oxidation state has been little studied for protoactinium, is strongly reducing for uranium, modestly reducing for neptunium, and the 'normal' oxidation state from americium onwards (except for nobelium, whose 2+ state appears to be the most stable). However one should add here that from plutonium onwards the highly radioactive nature of the nuclei has a profound effect on the study of their chemistry.

The 4+ oxidation state is known in aqueous solution for the elements thorium to berkelium, though by the time berkelium is reached it is strongly oxidising. $Th^{4+}aq$ exists as such, rather than as $Th(OH)^{3+}aq$, at pHs below about 0.5 in solutions 0.5M in thorium(IV). The dominant species in acid aqueous solutions of protoactinium(IV) is said to be $PaO^{2+}aq$ (or $Pa(OH)_2^{2+}aq$) [78]. Uranium(IV) is claimed to be stable in aqueous solution at pH $\lesssim 2.9$, but to give a colloid at higher pHs [79]. The 5+ and 6+ oxidation states of the elements uranium to americium can given cationic species in aqueous solution, but these are of the form MO_2^+aq and $MO_2^{2+}aq$ respectively. A claim for cationic neptunium(VII) in the form of NpO_2^{3+}, isoelectronic with the well-known UO_2^{2+}, has been made. But this seems to be in the solid state; in solution derived anionic forms such as $NpO_2(OH)_6^{3-}$ and $NpO_4(OH)_2^{3-}$ seem more likely [80]. The stable oxidation states of the actinides are summarised in Table 1.3.

Table 1.3 Oxidation states of the actinides in (aqueous) solution; the most stable state is in bold type. Most of the entries with footnotes refer to transient species.

Ac	Th	Pa	U	Np	Pu	Am	Cm	Bk	Cf	Es	Fm	Md	No	Lw
				2^a	2^a	$2^{a,b}$	2^a		$2^{c,d}$	2^c	2^e	2^f	2^f	
3	**3**	**3**	**3**	**3**	**3**	**3**	**3**	**3**	**3**	**3**	**3**	**3**	**3**	**3**
	4	4	**4**	4	**4**	4	4	4						
		5	5^g	**5**	5^g	5								
			6	6	6	6								
				7^h										

a Generated in aqueous ethanol [81]; *b* in acetonitrile [82]; *c* in aqueous ethanol [83]; *d* in acetonitrile [84]; *e* in aqueous ethanol [85]; *f* fairly easy to obtain in aqueous solution [82]; *g* disproportionate readily; *h* ref. [80].

1.2.6 Extensions

There is no reason to believe that there are no more aquo-cations to discover. The relatively recent work on molybdenum (Chapter 1.2.3) shows how suitable investigation can still lead to the characterisation of new aquo-species. It seems likely that further experimental work under carefully chosen conditions using weakly complexing anions, such as perchlorate or trifluoromethyl sulphonate, will reveal hitherto unknown aquo-cations, perhaps of rhenium, tungsten, or other members of the currently under-represented third row transition elements. Fast working may also be needed in the characterisation of ephemeral entities such as the likely Cr^{4+}aq. Doubtless pulse radiolysis, mentioned several times already in this chapter, will produce further new low oxidation state aquo-cations.

One should perhaps mention, and dismiss, sundry fringe and special cases of metal ions in solution. The chemistry — solubilities, chemical potentials, spectra, and reactions — of several metals in the special situation of M^{n+} with $n = 0$ are known, for example Hg^0 or Ag^0. It is also possible to generate metals in anionic form under certain exceptional conditions, for instance Na^- in cryptate complexes [86, 87] and both Na^- and K^- in amine or ether solutions [88]. Gold is reckoned to be Au^- in (solid) caesium auride. These exceptional cases will be considered no further.

Presumably ranges of solvento-cations similar to those aquo-cations discussed in Chapters 1.2.1 to 1.2.5 will exist for many non-aqueous solvents, particularly those of a water-like nature such as the lower alcohols. Likely species can be guessed by judicious and cautious extrapolation from known situations in water, bearing in mind the different redox properties and nucleophilicities of the solvents in question. Thus H_2S is a more difficult solvent to work with than H_2O in respect of solvento-cations, both in view of the diminished range of redox stability of this solvent (Chapter 8.4.2) and of the higher nucleophilicities of S^{2-} and SH^- compared with O^{2-} and OH^-.

1.3   SOLVATING PROPERTIES OF SOLVENTS

There is no single solvent property which uniquely determines its ability to solvate ions. Indeed a solvent which effectively solvates cations may solvate anions only poorly, and vice versa. Hexamethylphosphoramide and dimethyl sulphoxide are both good examples of the former category.

The dielectric constant (D) is obviously an important factor in determining the ease of dissolution of a salt. A high dielectric constant seems to be a necessary requirement, but by no means the only one. Formamide, N–methylformamide, sulphuric acid, hydrogen peroxide, liquid hydrogen cyanide, and liquid hydrogen fluoride all have higher dielectric constants than water, but in many respects are inferior solvents for inorganic salts. This has long been realised. 48 years ago it was pointed out [89] that salts dissolve less readily in liquid hydrogen cyanide ($D = 115$ at $20°C$) than in water ($D = 80$ at $20°C$). It is only recently that a series of salts containing $M(HCN)_6^{2+}$ cations have been characterised [90].

Properties other than the dielectric constant, especially the ability of the solvent to donate an electron pair to a cation, are of major importance. Most common solvents for inorganic salts contain potentially ligating nitrogen and oxygen atoms available for favourable interaction with the cation of the salt. However the lone pair must be readily accessible to the cation. Hence nitromethane is not effective at solvating cations because the negative end of its dipole is spread over the nitrogen and two oxygen atoms. Again t-butyl alcohol is hampered by its bulk, preventing the close approach of several molecules to one cation needed to solvate that cation fully [91].

Yet other factors have to be invoked to rationalise the poor cation-solvating powers of other solvents. These have recently been summarised [23] as the Lewis acidity of the ion, the extent of solvent structure, and dispersion forces and hydrogen bonding interactions between ion and solvent. Of these, the last two are of greater relevance to complex ions than to metal ions, but the first two are of general applicability.

Despite the obvious difficulties and complications involved, much effort has been devoted to the construction of scales of solvating powers of solvents.

Such scales are based either on an intrinsic property of the solvent (dielectric constant, ionisation potential) or on the respective interactions of a series of solvents with a reference solute (the remaining parameters mentioned below). Values of some of these parameters for a range of solvents are listed in Table 1.4 [92-96]. In this Table the solvents are listed in order of their E_T (Reichardt's solvent parameter) values. This choice is fairly arbitrary, being dictated by the availability of these values for the great majority of solvents of interest.

The unsuitability of dielectric constant as a solvation parameter has already been stated, and examination of dielectric constants listed in Table 1.4 reveals several difficulties. For instance solvents such as acetone and nitromethane, which are very poor solvators of metal cations, come in the same region as the monohydric alcohols. Various functions based on dielectric constants have also been used in correlating chemical properties in series of solvent, among them the functions $1/D$ and $(D-1)/(2D+1)$. These variants obviously rank solvents in the same sequence as D itself and thus offer no advantages in the present context. Indeed the function $(D-1)/(2D+1)$ only shows a marked variation between solvents of very low dielectric constant, and we are little concerned with such solvents here. Solvent 'electrostatic factors', equal to the product of dielectric constant and dipole moment, have found use in classifying solvents in respect of their dissolving powers and other properties [97]. The other intrinsic solvent property included in Table 1.4, the ionisation potential, is also a poor guide to cation-solvating propensities. Pairs of solvents of such different solvating powers as pyridine and benzene, or liquid ammonia and hexane, or water and carbon tetrachloride, have similar ionisation potentials.

Thus quantitative characterisation of a solvent's solvating abilities in terms of any of its intrinsic properties was found to be impossible. This led to attempts to express these abilities in terms of empirical scales based on a variety of manifestations of solvent-solute interaction. Solvent Y values, proposed by Grunwald and Winstein in 1948 [98], are based on the rates of solvolysis of t-butyl chloride. They are largely determined by solvation of the leaving chloride ion in this S_N1 reaction. The range of Y values is obviously limited to those protic solvents with which t-butyl chloride reacts. The Z value scale of Kosower [99] is based on charge-transfer frequencies for 1-ethyl-4-carbomethoxypyridinium iodide, Fig. 1.4(a). This scale is also limited in range, this time by the solubility of the ionic substrate.

A much more extensive scale of values is provided by Reichardt's E_T parameter [100]. Once again the scale is related to a charge-transfer process, but now the charge-transfer is intramolecular. The betaine used as shown in Fig. 1.4(b) is uncharged, and soluble in a conveniently large number of solvents. Brownstein has attempted to provide a more broadly based scale in his S values [101]. These values represent the average solvating powers of solvents as estimated from a variety of phenomena, including reactivity, acid-base properties,

Table 1.4 Solvent parameters. All values refer to 25°C unless stated otherwise; values from reference [92], except for Gutmann donor numbers (DN) and ionisation potentials (IP) [93].

Solvent	E_T /kcal mol^{-1}	Z /kcal mol^{-1}	Y	S	D	$\frac{D-1}{2D+1}$	IP /eV	DN
					Dielectric parameters			
Water	63.1	94.6	3.493	0.1540	78.5	0.491	12.6	18.0
Formic acid	56.6		2.054	0.1139	57.9a	0.487a		~17e
Formamide	56.3	83.3	0.604	0.0463	109.5	0.494		24.7e
Ethylene glycol	55.5	85.1		0.0679	37.7	0.480		
Methanol	54.1	83.6	−1.090	0.0499	32.6	0.477		23.5e
N-Methylformamide	51.9			0			10.5	30.0e
Ethanol		79.6	−2.033	0.0050	24.3	0.470	10.4	
Acetic acid	50.7	79.2	−1.639	−0.0158	6.2	0.388		
n-Propanol	50.2	78.3		−0.0240	19.7	0.463		
n-Butanol	48.6	77.7	−2.73	−0.0413	17.7	0.459		
i-Propanol	46.6	76.3			18.3	0.460		
Propylene carbonate	46.3			−0.134	65.1	0.489		15.1
Nitromethane	46.0	71.3		−0.1039	38.6a	0.481a		2.7
Acetonitrile	45.0	71.1			37.5a	0.480a		14.1
Dimethyl sulphoxide	44.0				48.9a	0.485a		29.8
Sulpholane	43.9b	71.3		−0.1047	44.0	0.483		14.8
t-Butyl alcohol	43.8		−3.26	−0.1416	12.2	0.441		
NN-Dimethylformamide	42.2	68.5		−0.1748	36.7	0.480		26.6
Acetone	41.1	65.7		−0.1890	20.5	0.464	9.7	17.0
Dichloromethane	40.9f	64.2			8.9	0.420	11.3	
Hexamethylphosphoramide	40.2	62.8c		−0.1970	29.6d	0.475		38.8
Pyridine	39.1	64.0		−0.2000	12.3	0.441	7.9	33.1
Chloroform	38.1	63.2		−0.210	4.70	0.356	11.4	
Ethyl acetate	37.4				6.03	0.385	10.1	17.1
Tetrahydrofuran	36.0				7.39	0.405		20.1
Dioxan	34.6			−0.179	2.21	0.223		
Diethyl ether	34.5			−0.277	4.22	0.341	9.5	19.2
Benzene	32.6			−0.215	2.27	0.229	9.2	
Carbon disulphide	32.5			−0.240	2.64	0.261	10.1	
Carbon tetrachloride	30.9			−0.245	2.23	0.225	11.5	
n-Hexane				−0.337	1.90	0.188	10.4	

a At 20°C; b at 30°C; c ref. [94]; d ref. [95]; e ref. [96]; f see Part I of ref. [97].

and several forms of spectroscopy. Gutmann concentrated on the electron pair donor properties of solvents in establishing his scale of donicities or donor numbers, DN [93]. These are related to the enthalpy of interaction of the various solvents with antimony pentachloride in 1,2-dichloroethane solution.

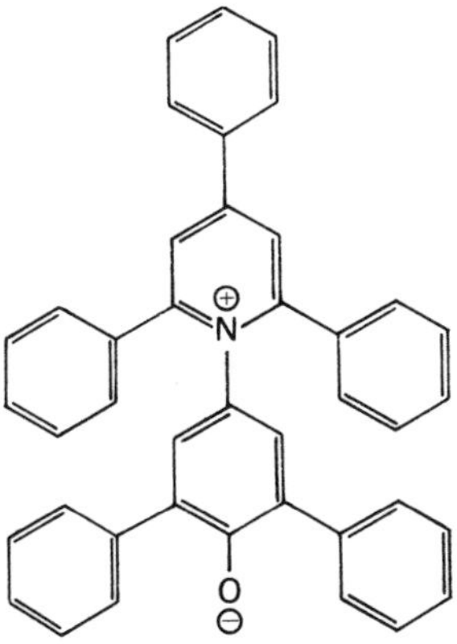

Fig. 1.4 (a) 1-Ethyl-4-carbomethoxypyridinium iodide; (b) the pyridinium N-phenolbetaine on whose charge-transfer spectrum the Reichardt scale of solvent E_T values is based.

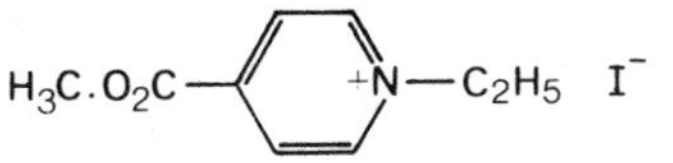

Values for these parameters are listed in Table 1.4. There is a reasonable correlation between the two spectroscopically based scales, those of E_T and Z values [92]. There is also a reasonable correlation between these and S values [92]; correlation of Y values with E_T, Z, and S is not quite as good. Correlation of these scales with Gutmann's donor numbers is not satisfactory, as indicated by the irregular variations down the DN column in Table 1.4. None of the parameters included in Table 1.4 gives a satisfactory ranking of the relative powers of solvents to solvate cations. Sufficient reason for this can be found in the great difference in nature between metal cations and the respective organic substrates of the Y and Z scales.

The most closely related parameter is Gutmann's DN, for this reflects the ease of donation of the solvent molecule's lone pair. However, an inspection of Table 1.5 shows that DN values [93, 102] are a fallible guide [103] in this area. It is true that good solvents such as pyridine, dimethyl sulphoxide, and dimethylformamide have high donor numbers. It is also true that moderate solvents such as propylene carbonate and acetonitrile come lower in the scale, and that a poor solvent such as nitrobenzene has a low donor number. Thus far the *DN* values parallel Parker's qualitative scale [104] of cation solvating abilities. However the donor numbers of acetone, ethyl acetate, and diethyl ether are near equality to that for water, and this is utterly at variance with the cation solving abilities of these organic compounds, which are all very different from that of water.

Differences in acceptor properties between metal cations and antimony pentachloride, the substrate for the *DN* scale, are obviously significant [103]. Two such differences may be mentioned. First, steric hindrance is less likely to be of overriding importance in formation of the 1:1 solvent:SbCl$_5$ complex via

Table 1.5   Ranking of solvents according to their Gutmann donor numbers, *DN* [93].

DN	Solvent
47.2	NN'N"-tris(tetramethylene)phosphoramide [a]
38.8	hexamethylphosphoramide
33.1	pyridine
30.0	ethanol [b]
29.8	dimethyl sulphoxide
26.6	dimethylformamide
24.7	formamide [b]
23.5	methanol [b]
20.0	tetrahydrofuran
19.2	diethyl ether
18.0	water
17.1	ethyl acetate
17.0	acetone
~ 17	formic acid [b]
15.1	propylene carbonate
14.8	sulpholane
14.1	acetonitrile
4.4	nitrobenzene
2.7	nitromethane

a DN value from ref. [102]; *b DN* values from ref. [96].

whose enthalpy of formation the solvent's position in the DN scale is assessed. However because the solvation of a cation involves several solvent molecules approaching it closely, inter-solvent molecule steric hindrance will be significant in the case of bulky molecules. Hexamethylphosphoramide is a reasonably good solvent [105] for some inorganic salts [106]. With its high DN value (the highest for a common solvent), it is potentially a good solvator of cations. Yet it is prevented from being exceptionally good in this respect by its bulk. A second difference between metal cations and the reference acceptor antimony pentachloride is in the relative σ- and π-contributions to the interaction. An ion such as Ag^+, with its strong π-interaction with suitable molecules, will be much more favourably solvated by such solvents as acetonitrile and even benzene than would be suggested by solvent DN values. Recently Gutmann has proposed a complementary scale of solvent 'acceptor numbers', AN. These AN values correlate with such solvent parameters as Y and Z values [107].

Other empirical solvent parameters are described and discussed in Reichardt's comprehensive but succinct review [92]. These include Gielen and Nasielski's X values, based on rates of reaction of tetramethyltin with bromine, and Berson, Hamlet, and Mueller's Ω values. These are also kinetically based, this time on rates of Diels-Alder reactions. Further empirical parameters have been proposed since the appearance of Reichardt's review. F values derived from solvent effects on $n \rightarrow \pi^*$ transitions of ketones [95, 108], and $g(S)$ values based on solubilities of a range of organic compounds in non- and mixed aqueous solvents [109] are among these. Though based on properties of solutions of nonelectrolytes, Hildebrand and Scott's long-established solubility parameter δ_s [110] may also turn out to be of relevance to some aspects of cation solvation. The parameters mentioned in this paragraph correlate variously amongst themselves and with some of the parameters cited earlier; none of them provides a good index of cation solvating ability.

It is impossible to avoid a pessimistic conclusion at this stage. Several instances of the success of limited correlations between cation solvation and one or more of the empirical solvent parameters have been noted (see Chapter 2.5), but such cases remain isolated examples. An overall correlation has yet to be established. Indeed, for the range and variety of metal ions known to exist in solution, it seems unlikely that anything less than an unwieldy multi-parameter scale would suffice.

The situation in mixed solvents will be even more complicated, with selective solvation of the cation and effects of the cation on intersolvent interactions to be considered. Nonetheless tabulations of E_T [111] and of Y [98, 112] values in several series of mixed aqueous and mixed non-aqueous solvents exist for use in appropriate circumstances.

FURTHER READING

Cotton, F. A. and Wilkinson, G., *Advanced Inorganic Chemistry, 3rd Edn.*, Wiley Interscience (1972).
Comprehensive Inorganic Chemistry, Volumes 1–5, ed. Bailar, J. C., Emeléus, H. J., Nyholm, R. S. and Trotman-Dickenson, A. F., Pergamon (1973).

REFERENCES

[1] E.g. Dasent, W. E., *Inorganic Energetics*, Penguin (1970) pp. 56–58.
[2] See pp. 81–82 of reference [1].
[3] Adams, D. M., *Inorganic Solids*, Wiley (1974).
[4] Dorsey, N. E., *Properties of Ordinary Water Substance*, Reinhold (1940).
[5] Eisenberg, D. and Kauzmann, W., *The Structure and Properties of Water*, Clarendon Press (1969).
[6] *Water and Aqueous Solutions*, ed. Horne, R. A., Wiley Interscience (1972).
[7] *Water – A Comprehensive Treatise*, ed. Franks, F., Plenum Press (1973).
[8] Hawkins, D. T., *J. Solution Chem.*, **4**, 625 (1975).
[9] Duncan, A. B. F. and Pople, J. A., *Trans. Faraday Soc.*, **49**, 217 (1953).
[10] Symons, M. D. R., *Proc. R. Soc. B.*, **272**, 13 (1975).
[11] Frank, H. S. and Wen, W. -Y., *Disc. Faraday Soc.*, **24**, 133 (1957); Nemethy, G. and Scheraga, H. A., *J. Chem. Phys.*, **36**, 3382 (1962).
[12] *Non-aqueous Solvent Systems*, ed. Waddington, T. C., Academic Press, (1965); Waddington, T. C., *Non-aqueous Solvents*, Nelson, (1969); Sisler, H. H., *Chemistry in Non-aqueous Solvents*, Reinhold (1961).
[13] *Handbook of Chemistry and Physics, 55th edn.*, ed. Weast, R. C., Chemical Rubber Company (1975).
[14] Timmermans, J., *Physico-chemical Constants of Pure Organic Compounds*, Elsevier (1950, 1965).
[15] Horvath, A. L., *Physical Properties of Inorganic Compounds*, Edward Arnold (1975).
[16] E. g. Sundheim, B. R., *Fused Salts*, McGraw-Hill, New York (1964); Clarke, J. H. R. and Hills, G. J., *Chem. Brit.*, **9**, 12 (1973).
[17] Ford W. T., *Analyt. Chem.*, **47**, 1125 (1975).
[18] E.g. Bredig, M. A., in *Molten Salt Chemistry*, ed. Blander, M. Interscience (1964) p. 367.
[19] Franks, F. and Ives, D. J. G., *Q. Rev. Chem. Soc.*, **20**, 1 (1966).
[20] E.g. Blandamer, M. J., Clarke, D. E., Claxton, T. A., Fox, M. F., Hidden, N. J., Oakes, J., Symons, M. C. R., Verma, G. S. P., and Wootten, M. J., *Chem. Commun.*, 273 (1967); and references therein.
[21] Timmermans, J., *The Physico-chemical Constants of Binary Systems in Concentrated Solutions, Volume 4, Systems with Inorganic + Organic or Inorganic Compounds*, Wiley Interscience (1960).
[22] *Physical Chemistry of Organic Solvent Systems*, ed. Covington, A. K. and

Dickinson, T., Plenum Press (1973).

[23] Cox, B. G., Hedwig, G. R., Parker, A. J., and Watts, D. W., *Aust. J. Chem.*, **27**, 477 (1974).

[24] Robinson, R. A. and Stokes, R. H., *Electrolyte Solutions, 2nd edn., revised*, Butterworth (1968) p. 15.

[25] Bernal, J. D. and Fowler, R. H., *J. Chem. Phys.*, **1**, 515 (1933).

[26] Frank, H. S. and Evans, M. W., *J. Chem. Phys.*, **13**, 507 (1945).

[27] Gurney, R. W., *Ionic Processes in Solution*, McGraw-Hill (1953).

[28] Hepler, L. G., *Can. J. Chem.*, **47**, 4613 (1969).

[29] Samoilov, O. Ya., *Disc. Faraday Soc.*, **24**, 141 (1957).

[30] E.g. Clack, D. W. and Farrimond, M. S., *J. Chem. Soc. A*, 299 (1971); Clementi, E. and Popkie, H., *J. Chem. Phys.*, **57**, 1077 (1972); Eliezer, I. and Krindel, P., *J. Chem. Phys.*, **57**, 1884 (1972); Kistenmacher, H., Popkie, H., and Clementi, E., *J. Chem. Phys.*, **58**, 1689 (1973).

[31] E.g. Nikolić, R. M. and Gal, I. J., *J. Chem. Soc. Dalton.*, 162 (1972); Angell, C. A. and Bressel, R. D., *J. Phys. Chem.*, **76**, 3244 (1972); Ambrus, J. H., Moynihan, C. T., and Macedo, P. B., *J. Phys. Chem.*, **76**, 3287 (1972); Stokes, R. A. and Robinson, R. H., *J. Solution Chem.*, **2**, 173 (1973); Franck, E. U., *J. Solution Chem.* **2**, 339 (1973).

[32] E.g. Bohigian, T. A. and Martell, A. E., *Inorg. Chem.*, **4**, 1264 (1965); Beck, M. T. in *Coordination Chemistry in Solution*, ed. Högfeldt, E. Swedish National Science Research Council (1972) p. 241; Ohtaki, H. and Ito, Y., *J. Coord. Chem.*, **3**, 131 (1973); Anderson, K. P., Butler, E. A. and Wooley, E. M., *J. Phys. Chem.*, **78**, 2244 (1974); Touche, M. L. D. and Williams, D. R., *J. Chem. Soc. Dalton*, 1355 (1976).

[33] Rausch, M. D., McEwen, W. E. and Kleinberg, J., *Chem. Rev.*, **57**, 417 (1957).

[34] Fujita, S., Horii, H. and Taniguchi, S., *J. Phys. Chem.*, **77**, 2868 (1973); Fujita, S., Horii, H., Mori, T. and Taniguchi, S., *J. Phys. Chem.*, **79**, 960 (1975).

[35] Basco, N., Vidyarthi, S. K. and Walker, D. C., *Can. J. Chem.*, **52**, 343 (1974); Buxton, G. V. and Sellers, R. M., *J. Chem. Soc. Faraday I*, **71**, 558 (1975); and references therein.

[36] Schwarz, H. A., Comstock, D., Yandell, J. K. and Dodson, R. W., *J. Phys. Chem.*, **78**, 488 (1974); O'Neill, P. and Schulte-Frohlinde, D., *J. Chem. Soc. Chem. Commun.*, 387 (1975).

[37] Borodin, P. M. and Nguen Kim Zao, *Russ. J. Morg. Chem.*, **17**, 959 (1972); Plakhotnik, V. N., *Russ. J.phys. Chem.*, **48**, 1651 (1974).

[38] Alekseeva, I. I. and Nemzer, I. I., *Russ. J. Inorg. Chem.*, **16**, 987 (1971).

[39] Nazarenko, V. A., Antonovich, V. P. and Nevskaya, E. M., *Russ. J. Inorg. Chem.*, **16**, 980 (1971).

[40] Baron, A. F. M. and Wright, G. A. in *Proceedings of the First Australian Conference on Electrochemistry*, ed. Friend, J. A. and Gutmann, F.,

Pergamon Press (1965) p. 124.

[41] Bovin, J-O., *Acta Chem. Scand.*, **28A**, 723 (1974).

[42] Nabivanets, B. I., Kapantsyan, E. E., and Oganesyan, E. N., *Russ. J. Inorg. Chem.*, **19**, 394 (1974).

[43] Anbar, M. in *Advances in Chemistry Series, No. 49*, ed. Gould, R. F., American Chemical Society (1965) Ch. 6.

[44] Iannuzzi, M. M. and Rieger, P. H., *Inorg. Chem.*, **14**, 2895 (1975).

[45] Hein, F. and Eissner, W., *Chem. Ber.*, **59**, 362 (1926).

[46] Dyrkacz, G. and Roček, J., *J. Amer. Chem. Soc.*, **95**, 4756 (1973); and references therein.

[47] Scholder, R. and Klemm, W., *Angew. Chem.*, **66**, 461 (1954).

[48] E.g. Beattie, J. K. and Haight, G. P. in *Progress in Inorganic Chemistry*, ed. Lippard, S. J.; *Volume 17: Inorganic Reaction Mechanisms, Part II*, ed. Edwards, J. O., Interscience, New York (1972) p. 93; Espenson, J. H., *Accts. Chem. Res.*, **3**, 347 (1970); Mitewa, M., Malinovski, A., Bontchev, P. R. and Kabassanov, K., *Inorg. Chim. Acta*, **8**, 17 (1974); Newton, T. W., *Inorg. Chem.*, **14**, 2394 (1975); and references therein.

[49] Dockal, E. R., Everhart, E. T. and Gould, E. S., *J. Amer. Chem. Soc.*, **93**, 5661 (1971).

[50] Tribalat, S. and Schriver, L., *C. R. Hebd. Séanc. Acad. Sci., Paris*, **279C**, 443. (1974).

[51] Codneva, M. M., Motov, D. L. and Okhrimenko, R. F., *Russ. J. Inorg. Chem.*, **18**, 767 (1973).

[52] Cotton, F. A. and Wilkinson, G., *Advanced Inorganic Chemistry, 3rd edn.*, Wiley Interscience (1972) p. 934.

[53] Myers. O. E. and Brady, A. P., *J. Phys. Chem.*, **64**, 591 (1960).

[54] Cotton, F. A., Frenz, B. A. and Webb, T. R., *J. Amer. Chem. Soc.*, **95**, 4431 (1973).

[55] Hartman, H. and Schmidt, H. J., *Z. Phys. Chem. Frankf. Ausg.*, **11**, 234 (1957); Bowen, A. R. and Taube, H., *J. Amer. Chem. Soc.*, **93**, 3287 (1971), Andruchow, W. and DiLiddo, J., *Inorg. Nucl. Chem. Lett.*, **8**, 689 (1972); Kustin, K. and Toppen, D., *Inorg. Chem.*, **11**, 2851 (1972); Sasaki, Y. and Sykes, A. G., *J. Chem. Soc. Chem. Commun.*, 767 (1973); idem *J. Less-common Metals*, **36**, 125 (1974); idem *J. Chem. Soc. Dalton*, 1048 (1975).

[56] Souchay, P., Cadiot, M. and Duhameaux, M., *C. R. Hebd. Séanc. Acad. Sci., Paris*, **262C**, 1524 (1966); Souchay, P., Cadiot, M. and Viossat, B., *Bull. Soc. Chim. France*, 592 (1970); Souchay, P., *J. Morg. Nucl. Chem.*, **37**, 1307 (1975); Lamache, M., *J. Less-common Metals*, **39**, 179 (1975).

[57] Ardon, M. and Pernick, A., *J. Amer. Chem. Soc.*, **95**, 6871 (1973).

[58] Ramasami, T., Taylor, R. S., and Sykes, A. G., *J. Amer. Chem. Soc.*, **97**, 5918 (1975).

[59] Ardon, M., Bino, A., and Yahav, G., *J. Amer. Chem. Soc.*, **98**, 2338

(1976).

[60] Ardon, M. and Pernick, A., *Inorg. Chem.*, **12**, 2484 (1973).

[61] Viossat, B. and Lamache, M., *Bull. Soc. Chim. France A*, 1570 (1975).

[62] Haight, G. P. and Boston, D. R., *J. Less-common Metals*, **36**, 95 (1974).

[63] Grindley, D. N., *An Advanced Course in Practical Inorganic Chemistry*, Butterworth (1964) p. 63.

[64] Rulfs, C. L. and Meyer, R. J., *J. Amer. Chem. Soc.*, **77**, 4505 (1955).

[65] Pavolova, M., Jordanov, N. and Popova, N., *J. Inorg. Nucl. Chem.*, **36**, 3845 (1974).

[66] Maspero, F. and Taube, H., *J. Amer. Chem. Soc.*, **90**, 7361 (1968); Wilson, C. R. and Taube, H., *Inorg. Chem.*, **14**, 2276 (1975).

[67] Goldberg, R. N. and Hepler, L. G., *Chem. Rev.*, **68**, 229 (1968).

[68] Beutler, P. and Gamsjäger, H., *Chimia*, **29**, 525 (1975); *J. Chem. Soc. Chem. Commun.*, 554 (1976).

[69] Pshenitsyn, N. K., Ginzburg, S. I. and Sal'skaya, L. G., *Russ. J. Inorg. Chem.*, **5**, 399 (1960); Pilipenko, A. T., Falendysh, N. F., and Parkhomenko, E. P., *ibid.*, **20**, 1683 (1975).

[70] Elding, L. I., *Inorg. Chim. Acta*, **20**, 65 (1976).

[71] De Renzi, A., Panunzi, A., Vitagliano, A. and Paiaro, G., *J. Chem. Soc. Chem. Commun.*, 47 (1976).

[72] Elding, L. I., *Acta Chem. Scand.*, **24**, 1331 (1970); Kukushkin, Yu. N., *Inorg. Chim. Acta*, **9**, 117 (1974).

[73] Hall, J. R. and Swile, G. A., *J. Organomet. Chem.*, **96**, C61 (1975).

[74] Glass, G. E., Schwabacher, W. B. and Tobias, R. S., *Inorg. Chem.*, **7**, 2471 (1968).

[75] Cohen, G. L. and Atkinson, G., *J. Electrochem. Soc.*, **115**, 1236 (1968).

[76] E.g. Adamson, M. G., Dainton, F. S. and Glentworth, P., *Trans. Faraday Soc.*, **61**, 689 (1965).

[77] Faraggi, M. and Feder, A., *J. Chem. Phys.*, **56**, 3294 (1972).

[78] Lundqvist, R., *Proc. Int. Solvent Extraction Conf.*, **1**, 469 (1974) (*Chem. Abs.*, **83**, 153451m (1975)).

[79] Davidov, Yu. P. and Efremenkov, V. M., *Radiokhimiya*, **17**, 155 (1975).

[80] Burns, J. H., Baldwin, W. H. and Stokely, J. R., *Inorg. Chem.*, **12**, 466 (1973).

[81] Mikheev, N. B., Kamenskaya, A. N., Dyachkova, R. A., Rozenkevitch, N. A., Rumer, I. A. and Auerman, L. N., *Inorg. Nucl. Chem. Lett.*, **8**, 523 (1972).

[82] Baybarz, R. D., Asprey, L. B., Strouse, C. E. and Fukushima, E., *J. Inorg. Nucl. Chem.*, **34**, 3427 (1972); and references therein.

[83] Mikheev, N. B., Spitsyn, V. I., Kamenskaya, A. N., Rozenkevitch, N. A., Rumer. I. A. and Auerman, L. N., *Inorg. Nucl. Chem. Lett.*, **8**, 869 (1972).

[84] Friedman, H. A., Stokely, J. R. and Baybarz, R. D., *Inorg. Nucl. Chem. Lett.*, **8**, 433 (1972).

[85] Mikheev, N. B., Spitsyn, V. I., Kamenskaya, A. N., Gvozdev, B. A., Druin, V. A., Rumer, I. A., Dyachkova, R. A., Rozenkevitch, N. A., and Auerman, L. N., *Inorg. Nucl. Chem. Lett.*, **8**, 929 (1972).

[86] Dye. J. L., Ceraso, J. M., Lok, M. T., Barnett, B. L., and Tehran, F. J., *J. Amer. Chem. Soc.*, **96**, 608 (1974).

[87] Tehan, F J., Barnett, B. L., and Dye, J. L., *J. Amer. Chem. Soc.*, **96**, 7203 (1974); Ceraso, J. M. and Dye, J. L., *J. Chem. Phys.*, **61**, 1585 (1974).

[88] Dye, J. L., DeBacker, M. G., Eyre, J. A. and Dorfman, L. M., *J. Phys. Chem.*, **76**, 839 (1972), Mei Tak Lok, Tehran, F. J., and Dye, J. L., *Phys. Chem.*, **76**, 2975 (1972).

[89] Fredenhagen, K. and Dahmols, J., *Z. Anorg. Allg. Chem.*, **179**, 77 (1929).

[90] Everstein, P. L. A., Zuur, A. P. and Driessen, W. L., *Inorg. Nucl. Chem. Lett.*, **12**, 277 (1976).

[91] Price, E. in *The Chemistry of Non-aqueous Solvents, Volume 1*, ed. Lagowski, J. J., Academic Press (1966) p. 67.

[92] Reichardt, C., *Angew. Chem., Int. Ed. Engl.*, **4**, 29 (1965).

[93] Gutmann, V., *Fortschr. Chem. Forsch.*, **27**, 59 (1972).

[94] Dubois, J. E. and Bienvenüe, A., *Tetrahedron Lett.*, 1809 (1966).

[95] Dubois, J. E. and Viellard, H., *J. Chim. Phys.*, **62**, 699 (1965).

[96] DeWitte, W. J. and Popov, A. I., *J. Solution Chem.*, **5**, 231 (1976).

[97] Dack, M. R. J. in *Techniques of Chemistry*, ed. Weissberger, A., *Volume VIII: Solutions and Solubilities, Part II,* ed. Dack, M. R. J., Wiley (1976) Ch. 11 (see especially pp. 96–102).

[98] Grunwald, E. and Winstein, S., *J. Amer. Chem. Soc.*, **70**, 846 (1948).

[99] Kosower, E. M., *J. Amer. Chem. Soc.*, **80**, 3253 (1958); *J. Chim. Phys.*, **61**, 230 (1964).

[100] Dimroth, K., Reichardt, C., Siepmann, T., and Bohlmann, F., *Justus Liebigs Annln Chem.*, **661**, 1 (1963).

[101] Brownstein, S. *Can. J. Chem.*, **38**, 1590 (1960).

[102] Ozari, Y. and Jagur-Grodzinski, J., *J. Chem. Soc. Chem. Commun.*, 295 (1974).

[103] Lim, Y. Y. and Drago, R. S., *Inorg. Chem.*, **11**, 202 (1972).

[104] Parker, A. J., *Q. Rev. Chem. Soc.*, **16**, 163 (1962); *Chem. Rev.*, **69**, 1 (1969).

[105] Normant, H., *Angew. Chem., Int. Ed. Engl.*, **6**, 1046 (1967).

[106] Robert, L., *Chim. Ind.*, **97**, 337 (1967).

[107] Mayer, W., Gutmann, V., and Gerger, W., *Mh. Chem.*, **106**, 1235 (1975).

[108] Dubois, J. E., Goetz, E., and Bienvenüe, A., *Spectrochim. Acta*, **20**, 1815 (1964).

[109] Letellier, P. and Gaboriaud, R., *J. Chim. Phys.*, **70**, 941 (1973).

[110] E.g. Hildebrand, J. H., Prausnitz, J. M., and Scott, R. L., *Regular and Related Solutions,* Van Nostrand Reinhold (1970); Herbrandson, H.F. and Neufeld, F. R., *J. Org. Chem.*, **31**, 1140 (1966).

[111] Reichardt, C., *Lösungsmitteleffekte in der Organischen Chemie,* Verlag Chemie (1968).

[112] Wells, P. R., *Chem. Rev.,* **63**, 171 (1963).

Chapter 2

METAL ION SOLVATION: N. M. R. SPECTROSCOPY

2.1 INTRODUCTION

N.m.r. spectroscopy is widely applicable in the study of ionic solutions. Chemical shifts, peak areas, linewidths, and coupling constants all reflect ion-solvent interactions. In a few favourable cases n.m.r. spectroscopy gives the fundamentally important primary solvation number directly from peak ·area integration (see Chapter 2.2.1, 2.3.1 and 2.3.2; also Fig. 2.1). This number is not otherwise obtainable directly, except for the few cases where it can be determined by the laborious isotopic dilution method (Chapter 4.5.1). Solvation numbers can also be estimated from chemical shifts at times. In these and similar instances n.m.r. spectra provide valuable information about ion-solvent interactions, both from a static and from a dynamic (kinetic) point of view. On the other hand previously established models of ionic solvation must often be used to explain features of observed n.m.r. spectra of electrolyte solutions. This latter situation commonly obtains for linewidths and coupling constants, and the present Chapter deals with both of these aspects of n.m.r. studies of ions in solution.

N.m.r. spectroscopy may be used to look, separately and individually, at the solvent and at dissolved ions. The majority of solvents contain hydrogen atoms, whose resonances can be readily monitored. Many solvents contain oxygen or nitrogen as well, and here the isotopes ^{17}O or ^{14}N can also be monitored by n.m.r., albeit with a little more technical difficulty.

Table 2.1 [1–23] gives examples of nuclei whose cationic species in solution can be studied by n.m.r. spectroscopy. For instance the whole series of alkali metal cations have at least one isotope suitable for n.m.r. investigations. Other suitable nuclei are scattered throughout the Periodic Table, the suitability of a given nucleus depending on its abundance, its sensitivity, and its nuclear spin (I). Thus ^{27}Al is a good nucleus to monitor, for its abundance is 100% and its sensitivity high. It does have the disadvantage of $I > \frac{1}{2}$ (in fact $I = 5/2$) but this is not important when dealing with symmetrical species such as octahedral hexa-solvento-species [5].

Until recently the majority of studies used proton n.m.r. spectroscopy, but it is interesting that the possibility of monitoring cation resonances in solution was recognised fairly early in the development of n.m.r. spectroscopy. For thallium(I) salts, the concentration dependence of the thallium chemical shifts was investigated about two decades ago [14]. Nonetheless the possibilities of monitoring the potentially wide range of cations indicated by Table 2.1 have only been realised extensively in the past few years. Although the present volume deals with cations in solution, complementary studies of anion nuclei, for example of the halides, are both possible and informative.

Table 2.1 Examples of solute nuclei whose resonances have been monitored by n.m.r. spectroscopy [a].

^{7}Li	b	^{9}Be [1]		
^{23}Na	b	^{25}Mg [2]	^{27}Al [4,5]	
^{39}K	b		^{45}Sc [6]	
^{85}Rb, ^{87}Rb	b	^{87}Sr [3]		
^{133}Cs	b		^{139}La [7,8]	
			^{69}Ga [11], ^{71}Ga [12]	
	^{113}Cd [9]		^{115}In [13]	^{119}Sn [18]
	^{199}Hg [10]		^{203}Tl, ^{205}Tl [14-17]	^{207}Pb [19]
^{47}Ti, ^{49}Ti [20]	^{51}V [21]	^{59}Co [22]	^{63}Cu [23]	
			^{109}Ag [c]	

a Details of experimental techniques, and whether chemical shifts or linewidths have been monitored, will be found in the references cited; b numerous references to n.m.r. studies of these nuclei will be found in Chapter 2.5.1 and 2.5.2; c ^{109}Ag n.m.r. studies of silver (I) salts in solution have recently been described by A. K. Rahimi and A. I. Popov, *Inorg.nucl.chem. Lett.*, 12, 703 (1976).

N.m.r. spectroscopy is most straightforward in the study of diamagnetic systems, but its use in systems containing paramagnetic species is by no means ruled out. Indeed the large chemical shifts and line-broadening produced by paramagnetic species can sometimes be turned to considerable advantage. Solvent resonances, for example ^{1}H or ^{17}O, can be monitored satisfactorily in systems containing paramagnetic ions with short electronic relaxation times. Thus solvent-nucleus n.m.r. spectra of solutions containing salts of cobalt(II) or of nickel(II), have often been reported, as will be seen in subsequent sections and chapters dealing with solvation numbers and kinetic characteristics. It has even proved possible to monitor solute resonances, for example ^{59}Co (Table 2.1).

Recently the superiority of deuterium n.m.r. over protium n.m.r. for solutions containing paramagnetic ions has been demonstrated. For example solutions containing Ti^{3+}, Cr^{3+}, or Fe^{3+} which proved difficult or impossible to probe with ^1H n.m.r. spectroscopy have yielded useful ^2H n.m.r. spectra [24].

The chemical shift of a given nucleus is determined by the environment of that nucleus. Except in very unusual circumstances a cation or an anion nucleus will give one signal, but solvent nuclei may give more complicated patterns. In a system containing a solvated metal cation, a solvated anion, and bulk solvent (Fig. 1.2) several resonances for a given solvent nucleus are possible in principle, these resonances corresponding with the various types of solvent molecule. In fact the amount of detail distinguishable in the n.m.r. spectrum of such a system is always much less than this, because rapid solvent molecule exchange occurs between some or all of the different environments. To permit the observation of a separate n.m.r signal, a nucleus must remain in a particular environment for a time appreciably greater than the reciprocal of the frequency of the n.m.r. signal. This time depends on the specific system in question, but in general is of the order of milliseconds.

The exchange of solvent molecules between a cation or anion's secondary solvation shell and bulk solvent is always fast on the n.m.r. time-scale, with the possible exception of chromium(III) solvates (Chapter 5.4). Hence separate signals from nuclei in these three situations will not be observed. Similarly, exchange of molecules between the primary solvation shell of anions and their surrounds is also fast, and separate n.m.r. signals therefore not observed. Solvent exchange between the primary solvation shell of the cation and the remaining solvent is often also too fast for the observation of separate signals, in which case only one resonance will be observed for a given solvent nucleus. The chemical shift of this single line will be determined by the various environments and the average residence of the solvent molecules in each. However, the case is different with solutions containing certain cations of relatively small size and relatively high charge, for example Be^{2+}, Mg^{2+}, Co^{2+}; Al^{3+}, Ga^{3+}, Cr^{3+}. Here the rate of exchange of solvent molecules between the primary solvation shell of the cation and the remaining ('bulk'[†]) solvent is sufficiently slow for two separate resonances to be observed for each type of solvent nucleus (Fig. 2.1). It is often necessary to cool the solution since this slows the exchange sufficiently for satisfactory resolution of the two signals.

2.2   SLOW SOLVENT EXCHANGE: PROTON N.M.R.

The two resonances expected for water protons, from water in the cation primary solvation shell and from water in the remaining environments, are illus-

---

† Here and subsequently the term 'bulk' solvent will be used in apposition to cation primary solvation solvent, to include the anion primary solvation shell and the two secondary solvation shells as well as the bulk solvent.

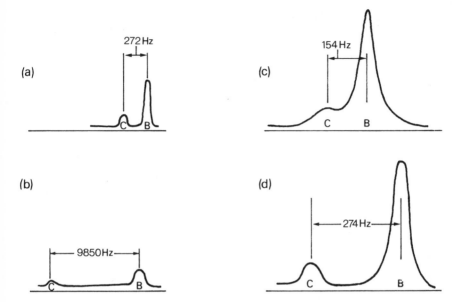

Fig. 2.1   Proton n.m.r. spectra of aqueous solutions of (a) gallium(III) perchlorate (1.29M, -62°C [13]), (b) cobalt(II) perchlorate (3.2M, -60°C [25]),
and (c) indium(III) chloride (3.81M, -60°C [26]); (d) proton n.m.r. spectrum of
1:5 aqueous acetone solution of indium(III) perchlorate (0.11M, -100°C [26]).
The signals marked C arise from cation primary hydration, those marked B from
the remaining water molecules in the system.

trated for the case of the gallium(III) cation in water in Fig. 2.1(a). Although
the environments of these two types of water molecule are fairly similar, the
chemical shift difference between the two proton resonances compared with the
line widths is large enough for two distinct peaks to be observed under the conditions of the experiment. If the metal cation is paramagnetic, then the coordinated water peak will be shifted greatly downfield from the 'bulk' water peak.
This is because all the water in the latter environments is shielded from the paramagnetic centre by the thickness of at least one water molecule, as shown in
Fig. 2.1(b). Many paramagnetic cations cause much line-broadening as well as a
large shift and in such cases the primary cation water peak may no longer be
observable. The favourable cases, where slow solvent exchange permits the
detection of two peaks, furnish two useful pieces of information: from the peak
areas, the primary solvation number of the cation, and from the chemical shifts
some idea of cation-solvent interaction.

2.2.1   Determination of Cation Solvation Numbers

If the solvent resonances can be split into two signals, then determining solvation numbers by peak area integration is a straightforward matter. In order to

get a cation primary solvation peak of reasonable size, it is usual to work with strong solutions. Water itself is 55.5M, so in a molar solution of an aluminium(III) salt where the cation hydration number is six, there would still be nearly ten times as much water in the secondary+bulk+anion environments ('bulk solvent') as in the cation hydration shell. Thus even in such a strong solution as this, the cation hydration peak is only about one tenth of the area of the 'bulk' water peak.

However working with such concentrated solutions means that conditions are favourable for the formation of ion-pairs or of complexes, though this disadvantage can be minimised by using perchlorates or trifluoromethylsulphonates. In the case of aqueous solutions, one advantage of working with such strong solutions is that it is easy to work at temperatures well below $O^{u}C$. Therefore a greater chance of slowing down water exchange sufficiently for the detection of the two separate signals exists. If the dissolved salt does not depress the freezing point enough, then solution n.m.r. spectra can be obtained at yet lower temperatures by using solutions made up in acetone–water mixtures. As discussed in Chapter 6, acetone is a poorly coordinating solvent which does not compete with water for a place in the primary solvation shell of cations. This trick of adding acetone permits the attainment of a temperature sufficiently low for resolution of two signals from aqueous solutions of indium(III) (Fig. 2.1(c) and (d) shows this) and of thorium(IV) salts. However even this technique is not enough in the cases of such cations as Sc^{3+} and Y^{3+}. When working with acetone–water mixtures it is even more important to work with salts of perchloric acid or other non-complexing species. The apparent hydration number of Th^{4+} is 9 in aqueous acetone perchlorate medium, but only 2.9 in nitrate medium [27].

Solvation numbers determined by this method are included in Table 2.3 below, which lists cation solvation numbers determined from n.m.r. spectroscopy of other nuclei as well as [1]H. Solvation numbers obtained by this n.m.r. peak area method will be compared with those obtained by other methods in Chapter 5.2.

This peak area method can also be used for mixed ligand-solvent complexes which exchange coordinated solvent with bulk solvent sufficiently slowly, for example cis-$Pt(NH_3)_2(MeCN)_2^{2+}$ [28].

2.2.2 Chemical Shifts

For diamagnetic cations the ^{1}H chemical shifts of the primary cation water and 'bulk' water are of the order of 100 Hz apart. For paramagnetic cations the difference is measured in thousands of hertz (see Fig. 2.1).

Diamagnetic cations. Cnemical shifts for the primary coordinated water protons in a number of slow-exchange systems [29] are listed in Table 2.2. The main factor controlling these shifts is the electric field of the cation. This polarises coordinated water molecules, withdrawing electron density from the vicinity of the hydrogen atoms. As a result there is a decrease in the diamagnetic shielding by the bonding electrons around these hydrogen atoms so that the resonances for these protons occur at lower field than that for bulk solvent protons. The magnitude of the shift between coordinated and bulk water protons will depend on the size and charge of the cation. The same applies to other solvents, though for a given cation the size of the shift between coordinated and bulk signals will depend on the distance between the cation and the nucleus in question. The bulk to coordinated shift difference will be less for the methyl protons of ethanol, say, than for water protons.

Although the chemical shifts listed in Table 2.2 do correlate approximately with cation charge and size, there are significant irregularities, especially in the aluminium–gallium–indium series. This suggests that some other effect, of smaller importance, is superimposed on the main field effect of the cation, causing an irregularity. This has been ascribed to a change in geometry of the hydrated cation. The four atoms in an $M-OH_2$ unit may be coplanar or pyramidal (Fig. 2.2), two geometries which give different M-H distances. Investigations of the crystal structures of metal hydrates have shown that $M-OH_2$ units are close to planarity in salts containing $Mg(OH_2)_6^{2+}$ or $Al(OH_2)_6^{3+}$, but are pyramidal in $Be(OH_2)_4^{2+}$ and (probably) $Ga(OH_2)_6^{3+}$.[†]

Fig. 2.2   The pyramidal and planar geometries possible for $M-OH_2$ units in the solid state and in solution.

Subtle differences in secondary solvation may also be reflected in the chemical shifts for the protons of water molecules in the primary hydration shell. One particularly thorough discussion of ^1H chemical shifts in aqueous systems containing $Al^{3+}aq$ [30] deals with contributions from primary and secondary hydration shells, and from disrupted bulk water structure. The most recent suggestion is that a significant contribution to the chemical shift of nuclei in the coordinated solvent molecules is made by interactions between the primary coordination shell molecules and neighbouring solvent molecules in the secondary solvation shell. This proposal is based on an examination of the solvent composition and temperature dependences of water and methanol hydroxyl–proton shifts in aqueous methanol containing Mg^{2+} or Al^{3+} after inert cosolvents such as acetone or acetonitrile have been added to affect the various inter- and intra-component interactions [31].

[†] The latest discussion of the trigonal and tetrahedral geometries found, or postulated, in crystal hydrates can be found in Friedman, H. L. and Lewis, L., *J.solution Chem.*, 5, 445 (1976).

Table 2.2. Proton chemical shifts for water in the primary hydration shell of cations, from ref. [29].

	Chemical shift /p.p.m.[a]
water (25°C)	− 4.13
$Be(OH_2)_4^{2+}$	− 8.04
$Mg(OH_2)_6^{2+}$	− 5.55
$Al(OH_2)_6^{3+}$	− 8.74
$Ga(OH_2)_6^{3+}$	− 8.98
$In(OH_2)_6^{3+}$	− 7.22
$Sn(OH_2)_x^{4+}$	− 10.1[b]

a Chemical shifts measured at low temperatures, 90 MHz, and relative to ethane gas; b this value is derived from solutions containing chloride or sulphate. It is therefore in some doubt due to the real possibility of ion-pair or complex formation, not to mention the likelihood of significant hydrolysis to species of the type $SnOH^{3+}$.

Paramagnetic cations. The paramagnetic (isotropic) shift due to the unpaired electrons in a paramagnetic cation is commonly divided into two components. Thus the isotropic shift results from the contact shift added to the pseudocontact (dipolar) shift. The contact shift is ascribed to electron delocalisation via chemical bonds. The pseudocontact shift is ascribed to the through-space effects of the electron(s) on the nuclei in question [32, 33]. Interpretation of these shifts in terms of cation-solvent bonding has often been attempted, but it is still a matter of lively controversy at the time of writing [34].

2.2.3 Proton Exchange

A complication commonly encountered in probing ionic solvation by proton n.m.r. spectroscopy is the possibility that, although complete solvent molecules may be exchanging slowly, proton exchange may still be fast. Processes of the type

$$Al(OH_2)_6^{3+} + {}^*H^+ \rightleftharpoons Al(OH_2)_5(OH^*H)^{3+} + H^+$$

average the environment of all the protons in the system. In many cases, including the aluminium(III) case used as an example here, proton exchange rates[†] can be reduced sufficiently to resolve a combined n.m.r. signal into coordinated

† Kinetics of proton exchange with aquocations are discussed in Chapter 14.

and 'bulk' components by means of cooling the system. However it is better to avoid such complications by investigating the n.m.r. spectrum of the ligating nucleus, for instance ^{17}O for water or ^{14}N for liquid ammonia. With solutions in alcohols fast proton exchange effects can be avoided by monitoring the methyl and methylene proton resonances rather than the hydroxy proton signal (Chapter 11.2.2).

2.2.4   Coupling Constants

Coupling constants (A/h) for ^1H n.m.r. spectra of solutions containing paramagnetic cations will be mentioned in Chapter 2.3.3, together with A/h values for other solvent nuclei.

2.3 SLOW SOLVENT EXCHANGE: N.M.R. OF OTHER SOLVENT NUCLEI

There are considerable chemical advantages in monitoring the n.m.r. signals from the solvent atom actually bonded to the cation, as has been noted in Chapter 2.2.3. To offset these, there are also certain practical difficulties. In the case of aqueous systems, the natural isotopic abundance of the appropriate nucleus is low (0.037% ^{17}O), so that isotopically enriched water or an extremely sensitive instrument and technique are obligatory. The sensitivity is lower and the resonance broader ($I = 5/2$) for ^{17}O than for ^1H. Indeed for diamagnetic cations the chemical shift between coordinated-^{17}O and 'bulk'-^{17}O is **small** relative to the natural linewidth, so that only one broad ^{17}O resonance is usually observed even for slow-exchange cations. For paramagnetic cations there is a very large shift of the coordinated solvent resonance. This is sufficient to separate the two peaks for slow exchange cases, though the coordinated solvent peak may be so broadened as to be undetectable. The use of paramagnetic ions to achieve peak separation for slow-exchange diamagnetic systems in a neat and efficient manner is described in Chapter 2.3.2.

2.3.1   Paramagnetic Cations

At temperatures below $0°C$, exchange of water and of alcohols at Co^{2+} becomes slow on the n.m.r. time-scale. There are two well separated peaks for the solvation shell of the cation and for 'bulk' solvent in the ^{17}O n.m.r. spectra of such solutions (Fig. 2.3) [35]. Ni^{2+} and Fe^{2+} cause somewhat more line broadening than Co^{2+}, but similar two peak ^{17}O n.m.r. spectra have been obtained from solutions of their respective perchlorates in ^{17}O-enriched water [36–38]. Solvation numbers can be determined from the ratio of the peak areas and the known composition of the solution, as described for the analogous ^1H case in Chapter 2.2.1 above. With ^{14}N monitored, the method for liquid ammonia as solvent is similar.

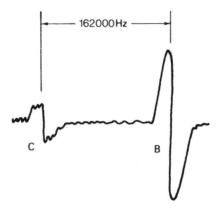

Fig. 2.3 ^{17}O–N.m.r. spectrum of 3.6M cobalt(II) perchlorate solution in ^{17}O-enriched water at $-10°$C [35].

This method has also been used for complexes, for instance to confirm the number of water ligands in the cis-Pt(NH$_3$)$_2$(OH$_2$)$_2^{2+}$ cation [39]. Sometimes separate bulk and coordinated resonances can still be detected despite their significant linewidth even for a diamagnetic cation, as in the early-studied case of aqueous solutions containing Co(NH$_3$)$_5$(OH$_2$)$^{3+}$ [40]. Here the two resonances are about 1 gauss apart, just enough for them to appear as separate peaks. Recently pulsed Fourier transform techniques of high sensitivity have been developed. These have permitted the observation of ^{13}C n.m.r. spectra showing cation-coordinated and 'bulk' alcohol molecules for solutions of (diamagnetic) Mg^{2+} and Al^{3+} in alcohols containing ^{13}C at natural abundance [41].

The difference in chemical shifts between the cation-coordinated and 'bulk' solvent peaks is, as one would expect, sensitive to the nature of the cation. Thus the shift difference for the similar cations Fe^{2+} and Ni^{2+} in water (^{17}O shifts) are 165,800 and 133,500 Hz respectively, at 8.034 MHz, $-29°$C [37]. As in proton n.m.r. spectroscopy, for a given cation the shift difference between the cation-coordinated and 'bulk' solvent signals is not particularly sensitive to the nature of the solvent. Witness the differences of 162,000 and 154,000 Hz for Co^{2+} in water and in methanol respectively, in 3.6 and 2.2M solutions of cobalt(II) perchlorate at $-10°$C, 8.133 MHz [35].

2.3.2 Diamagnetic Cations

Except in the rare cases mentioned in the previous section, ^{17}O, ^{14}N, and ^{13}C n.m.r. spectra of aqueous and non-aqueous solutions containing diamagnetic cations consist of one broad line. This is so even when solvent exchange between the cation coordination sphere and the rest of the solvent is known to be slow. The lack of resolution of two peaks arises because the chemical shift difference between the two potential peaks is small compared with their natural linewidths. However the two peaks can be separated by the addition of a suitable paramagnetic species, an ingenious trick which has since been taken up and developed by organic chemists ('shift reagents') [42] and biochemists [43].

The application of this paramagnetic shift method to the determination of solvation numbers of slow-exchange cations will now be described for a particular case. Figure 2.4 shows the spectrum of an aqueous solution of an aluminium(III) salt to which a small quantity of a cobalt(II) salt has been added [40]. At temperatures around or a little below 0°C, water exchange at Al^{3+} is slow, at Co^{2+} fast, by n.m.r. standards. Thus two sorts of water can be distinguished, those

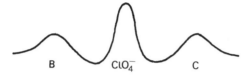

$$\text{B} \qquad ClO_4^- \qquad \text{C}$$

Fig. 2.4   Diagrammatic representation of the ^{17}O n.m.r. spectrum of an aqueous solution of aluminium(III) perchlorate containing cobalt(II) perchlorate and perchloric acid. The peak due to water coordinated to the aluminium(III) is labelled C, that due to the 'bulk' water is labelled D.

in the cation solvation shell of the aluminium(III), and the rest. The latter molecules are exchanging rapidly between bulk solvent, anion solvation shells, the secondary solvation shell of the aluminium(III), and the solvation shell of the cobalt(II). For such time as they spend in the cobalt(II) environment, the water ^{17}O atoms will experience the full effect of the unpaired electrons on the cobalt(II), and these nuclei will thus experience a large paramagnetic shift. On time-averaging the environments of these water molecules, they will suffer a marked downfield shift of a magnitude depending on the cobalt(II) concentration. On the other hand, the primary solvation shell of water molecules around the aluminium(III) cation will not come into contact with the paramagnetic cobalt(II) centre and will not have their ^{17}O resonance shifted significantly. The overall result of adding the cobalt(II) to the aqueous solution containing aluminium(III) is thus to separate the single broad ^{17}O resonance into separate primary cation and 'bulk' peaks by inducing a large shift in the latter as shown in Fig. 2.5(a) to (b). Peak areas may now be integrated to obtain an estimate of the cation solvation number. This method can also be used for beryllium(II), magnesium(II), and gallium(III) [40].

Cobalt(II) and dysprosium(III) [39] were the preferred cations for use in this type of experiment, because they induce large shifts but cause little extra broadening. More recently it has become apparent that dysprosium(III) may produce bigger shifts than europium(III) or praseodymium(III), say, but that the latter cations give less line-broadening [44]. In fact this latter pair are useful because they cause shifts in opposite directions. Holmium(III) gives rise to rather more line-broadening, and gadolinium(III) to yet more; these nuclei are there-

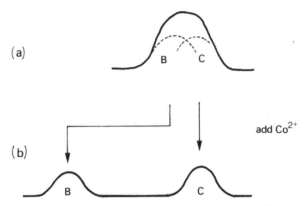

Fig. 2.5 Diagrammatic representation of the peak separation effect of adding a paramagnetic cation, e.g. Co^{2+}, on the ^{17}O n.m.r. spectrum of the water molecules in an aqueous solution containing a slow-exchange cation, e.g. Al^{3+}; C: water coordinated to Al^{3+}, B: 'bulk' water.

fore valuable in relaxation studies rather than in the shift application described here. Amongst transition metal cations manganese(II) also produces much line-broadening, often to the point of making the 'bulk' solvent signal too broad to detect. Copper(II) has been used for analogous experiments in liquid ammonia [45]. Solvation numbers of diamagnetic cations determined by this method are included in Table 2.3.

2.3.3 Coupling Constants

In a solvento-complex of a paramagnetic metal ion, the unpaired electron(s) of the metal ion exert a strong influence on the resonances of the nuclei in the coordinated solvent. The scalar couping constant A/h indicates how much the spin density from the unpaired electrons(s) is delocalised from the metal towards a given solvent nucleus. The coupling constant, derived from the contact shift (Chapter 2.2.2) [65], is determined by the resultant of delocalisation by σ and by π routes. In favourable cases these two contributions can be separated fairly satisfactorily to give some idea of the extent of σ- and π- bonding and of spin densities at various coordinated solvent nuclei.

A selection of coupling constants for various solvento-complexes of transition metal cations is given in Table 2.4 [66–76]. Obviously the ^{17}O and ^{14}N coupling constants give information about ion-solvent interaction more directly than ^{1}H coupling constants, because the former are bonded directly to the cation whereas the protons are two or more bonds removed from the paramagnetic centre. Contribution to the coupling constants are of opposite sign for σ and for π delocalisation[†]. Values for solvation by acetonitrile provide a good example of

[†] Signs for ^{17}O coupling constants will be the opposite from those for ^{1}H or ^{14}N, because the magnetogyric ratio [77] is opposite in sign for ^{17}O as compared with ^{1}H and ^{14}N.

Table 2.3 Cation solvation numbers determined by n.m.r. peak area techniques. Each entry indicates the solvation number, the nucleus monitored, and the appropriate literature reference.

	Water	Ammonia	Methanol	Ethanol	Acetonitrile	Dimethyl-formamide	Dimethyl sulphoxide
sp-Elements							
Be²⁺	4 ¹H [13] 4 ¹⁷O [46]ᵃ					4 ¹H [59]	
Mg²⁺	6 ¹H [13]	6 ¹H [50,51]	6 ¹H [52]				
Zn²⁺	6 ¹H [27]		6 ¹H [53]	4ᵈ ¹³C [41]			
Al³⁺	6 ¹H [13] 6 ¹⁷O [46]ᵃ	6 ¹⁴N [45]ᶜ	6 ¹H [54]	5.5ᵈ ¹³C [41]	2.8 ¹H [57]	6 ¹H [60]	6 ¹H [63]
Ga³⁺	6 ¹H [12,13] 6 ¹⁷O [47]ᵃ [48]ᵇ 6 ⁶⁹Ga [11]		6 ¹H [54]		6 ⁷¹Ga [13]	6 ¹H [61]	
In³⁺	6 ¹H [13]						
Transition metal cations							
Fe²⁺	6 ¹⁷O [37]						
Co²⁺	6 ¹H [25]		6 ¹H [55]		6 ¹H [58]	6 ¹H [62]	6 ¹H [64]
Ni²⁺	6 ¹H [49] 6 ¹⁷O [37,38]		~5 ¹H [56]				6 ¹H [64]
Actinide cations							
Th⁴⁺	9 ¹H [27]						

a Cobalt(II) added; *b* dysprosium(III) added; *c* copper(II) added; *d* ±1, as ¹³C peak areas were difficult to measure accurately.

this. When the delocalisation of spin density is exclusively through σ orbitals, as with solvates of Ni^{2+} or Cu^{2+}, then the ^{14}N coupling constant is positive. When delocalisation occurs primarily through π orbitals, as with Ti^{3+}, then the ^{14}N coupling constant is negative. In cases where both σ and π delocalisation is important, as with Mn^{2+}, then the coupling constant is the small resultant of the two opposite contributions. The ratio of coupling constants to different nuclei in the coordinated solvent may also give some idea of the changing balance between σ and π delocalisation. Thus, still for acetonitrile as solvating molecule, $A(^{14}N)/A(^{1}H)$ is 72 and 80 for Cu^{2+} and Ni^{2+} respectively, where σ delocalisation predominates. However this ratio has a value of 332 for Co^{2+}, where both σ and π delocalisation is important.

The deduction of the relative importance of σ and π bonding from coupling constants (see the case of hydrated transition metal cations [66]) needs to be done with care and circumspection. One must both feed in assumptions in order to make the σ vs. π split, and also assume that the coupling constants are correct in the first place. Reference [78] gives several examples of situations in which plots of chemical shifts against temperature do not have the zero intercept required for the use of the standard equation for calculating coupling constants [65]. Secondary solvation is another factor usually ignored in these bonding estimates which may have a small though not negligible effect [38].

As mentioned above, it is possible to estimate spin densities at solvent nuclei from the observed coupling constants. In the example of V^{3+}aq, the spin densities on the hydrogen and oxygen of the coordinated water are -0.0157 and $+0.0014$ electrons/(atomic unit)^3 respectively [69]. Spin densities on co-ordinated water atoms are also available for other aquo-cation, e.g. Ni^{2+}aq [71]. Relative degrees of covalency in cation–solvent bonds can also be estimated. Comfortingly, this method of estimation gives, at least in the case of the comparison of Fe^{3+} with Mn^{2+}, a very similar result to that derived from Racah parameters (Chapter 3.1.2) [70]. There is another useful connection between coupling constants and other chemical properties. This is the correlation between A/h for acetonitrile solvates of metal(II) cations with activation enthalpies $(\Delta H^{\ddagger})$ for acetonitrile exchange at these cations (Chapter 11.3). Such a correlation would be expected for a dissociative solvent exchange mechanism, where $\Delta H^{\ddagger}$ would be determined by cation–solvent bond strengths [79].

2.4   FAST SOLVENT EXCHANGE: N.M.R. OF SOLVENT NUCLEI

2.4.1   Chemical Shifts

Consider a situation where solvent exchange between all the environments listed earlier, that is primary and secondary solvation shells of the cation and the anion and bulk solvent, is rapid by n.m.r. standards. Here each solvent nucleus may experience all environments within a time corresponding to the frequency of

Table 2.4 Coupling constants for solvento-complexes of some paramagnetic metal cations ($10^{-5} A/h/Hz$)[a].

	Water		Methanol			Acetonitrile		Dimethyl-formamide formyl-^1H	Dimethyl sulphoxide ^1H
	^{17}O	^1H	^{17}O	hydroxyl-^1H	methyl-^1H	^{14}N	^1H		
Mn^{2+}		5.9 [70]		7.5 [73]	2.1 [73]	+32 [74]			+2.0 [75]
Fe^{2+}	−110 [37]	5.0 [70]				+50 [74]			+0.49 [75]
Co^{2+}	−174 [35]	4.2 [70]		8.0 [73]	4.1 [73]	+83 [74] +95 [66]	0.23 [58]	3.88 [62]	+1.02 [75] +0.86 [76]
Ni^{2+}	−274[b]	1.1 [70] 1.3[c] [71]				+150 [74] +270 [66]	−3.94 [74] −2.63 [58] 3.34 [66]	6.81 [62]	+2.2 [75]
Cu^{2+}	−550 [67]	1.5 [70]	290 [72]	8.8 [72]	10.6 [72]	+300 [57]	4.7 [66]		+6.65 [75]
Ti^{3+}	+44 [68]					−ve [66]			
V^{3+}	50 [69]	31 [69]							
Cr^{3+}				41.5 [73]					−1.2 [75]
Fe^{3+}	7.7 [70]								−0.04 [75]

a Signs are quoted only when specifically indicated by authors; b this value is a mean of published values of −240 × 10⁵ Hz [38], −282 × 10⁵ Hz [41], and −300 × 10⁵ Hz [15,66] (an early estimate of −193 × 10⁵ Hz [36] has been disregarded); c A/h for deuterium is 2.0 × 10⁴ Hz, giving a ratio of A/h values for H and for D which is consistent with their gyromagnetic ratios.

the n.m.r. frequency. Thus each solvent nucleus will give rise to one n.m.r. signal only, and the position of this resonance, in other words its chemical shift, will depend on the effects of the cation and of the anion. The chemical shift will therefore depend on the **nature** of the cation and the anion. Moreover the observed chemical shift will also depend on the **concentration** of the salt solution, as the time each solvent molecule spends under the influence of the cation or of the anion will depend on the ratio of bulk to solvating solvent. This situation may be contrasted with that for slow-exchange conditions. Here the cation primary solvent resonance will be at a constant position (chemical shift), but its area will change according to the cation concentration.

Therefore one would expect that the chemical shift for the solitary solvent resonance in a fast exchange system would be a linear function of salt concentration, at least in dilute solution. Many aqueous and non-aqueous solutions do conform to this pattern, and Fig. 2.6 shows the dependence of shift on concentration for one such well-behaved system [80]. Where the dependence of shift on concentration is not linear one may reasonably assume that the simple solvation model of separate solvated ions is no longer valid. Ion–ion interactions, to give ion-pairs or complexes, will result in non-linearity of the shift vs. concentration plot. Such deviations are more likely in concentrated solutions and in mixed and non-aqueous systems. Nevertheless, it is usually possible to obtain the slope or the tangent at the origin, and so characterise the shift behaviour of a given electrolyte solution by a value for $(d\delta/dc)_{c=0}$, where δ is the chemical shift.

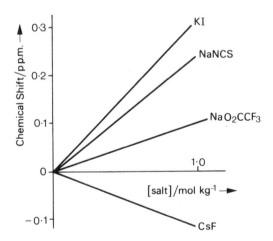

Fig. 2.6 The dependence of hydroxyl-proton chemical shifts on salt concentration for some alkali metal salts in methanol, at $-69°$C [80].

The next step is to attempt to split the electrolyte shift into its constituent ionic components:

$$\delta_{obs} = c(n\delta_{n+} + m\delta_{m-}) \, .$$

This split can be achieved in an internally consistent manner, but, as so often, the choice of reference point is arbitrary. No matter what assumption is made in the separation of cation and anion contributions, differences between δ_{n+} values for different cations of the same charge will always be correct and meaningful. Likewise differences between anions of the same charge will also be real and unaffected by the choice of reference point. However the actual absolute values of each individual ion shift do depend on the choice of zero. Thus the degree of meaning in any comparison between cation and anion shifts (or even in the sign of an ion shift) is determined by whether or not the method for separating the electrolyte shift $(d\delta/dc)_{c=0}$ into its single ion components is wisely chosen.

Different authors have made several different assumptions for the separation of electrolyte shifts into their components. An early assumption set the chemical shift for the hydrated ammonium cation at zero [81], because the ammonium ion can be accommodated into the structure of liquid water with the minimum distortion of solvent structure. There are disadvantages to this choice, due to the existence of hydrogen-bonding between the ammonium ion and water, and the ready exchange of protons between these two species. Alternatively, the chemical shift for a cation NR_4^+ with large alkyl groups R may be set at zero, on the assumption of relatively small ion-solvent interaction for such a large cation. Another possibility is to set the shift for K^+aq equal to that of Cl^-aq, on the grounds of the similarity of their ionic radii. Finally, one may make an intuitive guess or a detailed theoretical calculation for one given ion (Chapter 2.4.3).

2.4.2 Cation Shifts: Individual Assignments

Aqueous Solutions. Table 2.5 lists some representative values of individually assigned molar chemical shifts for water protons and water oxygen-17 nuclei derived from n.m.r. spectra of solutions of alkali and alkaline earth salts [81–83]. Such ionic shifts are commonly considered in terms of a balance between two opposing contributions [84]. The first is polarisation of the water molecules, which results in a high-field shift. The second arises from hydrogen-bond breaking, or structure breaking, in the solvent as a consequence of the introduction of the cation. This results in a low-field shift. The latter assignment is supported by the shifts observed, both for ^1H and ^{17}O resonances, when water itself is heated or diluted with an inert (non-hydrogen-bonding) solvent. It is possible, perhaps necessary when considering such a strange sequence of shifts as the ^{17}O set in Table 2.5, to break down these two opposing contributions further [81].

Table 2.5 Water-^{1}H and water-^{17}O chemical shifts (p.p.m. M^{-1}) for aqueous solutions containing cations.[a]

	^{1}H		^{17}O
	[82]	[81]	[83]
Li$^+$	− 0.022	− 0.016	+ 1.03
Na$^+$	+ 0.044	+ 0.048	+ 1.00
K$^+$	+ 0.044	+ 0.048	+ 0.85
Rb$^+$	+ 0.047	+ 0.037	+ 0.92
Cs$^+$	+ 0.040	+ 0.027	+ 0.47
Mg^{2+}	− 0.165		
Ca^{2+}	− 0.059		
Sr^{2+}	− 0.008		
Ba^{2+}	+ 0.007		

a The assumptions used in dividing the observed salt shifts into single ion components are zero shift for NH$_4^+$ [81, 83] or equal shifts for K$^+$ and Cl$^-$ [82]. The tolerable agreement between the two sets of ^{1}H shifts above shows that these two assumptions are reasonably compatible.

It has proved difficult to establish precisely the relative importance of the polarisation and structure-breaking contributions to cation solvent shifts. A general picture emerges of 1:1 electrolytes acting primarily as structure-breakers, but of 2:2 (and higher) electrolytes acting predominantly through the polarisation effects of their constituent ions. Thus the M^{2+} shifts in Table 2.5 are nearly all downfield from the M$^+$ shifts, with the largest downfield shift caused by the smallest and so most polarising cation in the series, Mg^{2+}. Unfortunately no clear parallels emerge between ^{1}H and ^{17}O shifts. Cation–water interactions and polarisation may well have a greater effect in determining ^{17}O shifts, while co-ordinated − bulk solvent interactions may be of greater relative importance in determining ^{1}H chemical shifts.

Non-aqueous solutions. Single ion cation shifts for several metal(I) cations in some non-aqueous solvents [85–87] are listed in Table 2.6. The results may be discussed in terms similar to those used for aqueous systems above. With non-aqueous solvents the degree of solvent structure or ordering will be important in determining the importance or otherwise of this factor in relative cation shifts. Thus solvent structural changes are likely to be important in the case of ordered solvents such as liquid ammonia or hydrogen fluoride, less important for dimethyl sulphoxide or acetonitrile, and of small significance for a solvent as little structured as sulpholane [87]. Polarisation will still be important, for instance the greater polarisability of ammonia compared with water is echoed in the cation shifts in the two solvents. The cation shifts determined in liquid hydrogen fluoride correlate with the cation radii [86].

Table 2.6   Solvent-^{1}H chemical shifts (p.p.m. M^{-1}; all positive) for non-aqueous solutions containing cations.

Solvent	Liquid ammonia	Acetonitrile	Dimethyl sulphoxide	Sulpholane	
				α-H	β-H
Assumption[a]	b	c	d	d	d
Reference	[85]	[87]	[87]	[87]	
Li$^+$	0.14	3.0	0.9	3.9	1.4
Na$^+$	0.21	2.0	0.5	3.1	0.2
K$^+$	0.23	1.5			
Ag$^+$					1.0

a The assumptions used in splitting observed salt shifts into single ion components are: b $\delta(NH_4^+) = 0$; c $\delta(Et_4N^+) = \delta(Cl^-) = 1.0$; d $\delta(Et_4N^+) = 0$.

2.4.3   Cation Shifts: Absolute

For the purpose of probing the nature and extent of cation-solvent inter-action, the cation shifts obtained via the arbitrary ionic contribution splitting methods mentioned in the previous section are of limited value. What is needed is some way of obtaining at least one single ion shift of established magnitude and sign. The key to the solution of this problem lies in the intercomparison of results from fast and slow exchange, that is to say high and low temperature, spectra of a cation exhibiting both types of behaviour within the accessible temperature range. This approach has been tried only relatively recently and is, at the time of writing, not yet fully worked out. In particular, the role of secondary solvation around the cation is a matter of controversy.

The first step in fixing absolute cation shifts is to start with the slow-exchange state. Then either some formula to accommodate all the slow-exchange cation shifts if found, and extrapolated to the fast-exchange cases [29], or the anion plus bulk solvent shift is extracted and in turn used to split the observed shift for a fast exchange case into its components. The latter approach was first demonstrated with magnesium perchlorate solution in methanol [88] and sub-sequently developed for methanol solutions of aluminium and zinc perchlorates [89] and the corresponding aqueous solutions of these and other cations [90]. In this method care is needed to ensure that the slow exchange cation solvent shift is measured at a temperature sufficiently low to rule out any question of this peak having started to move towards the 'bulk' solvent peak as a prelude to coalescence. The two approaches mentioned here cross-check satisfactorily, so probably do produce good approximations to absolute cation solvent shifts for fast exchange cations.

2.4.4   Solvation Number Estimation from Shifts

There are several empirical ways of estimating cation solvation numbers from chemical shift information. The method most used is applicable to cations dissolved in solvent mixtures consisting of one solvating and one non-solvating component. Measured shifts for one of the nuclei of the solvating solvent are plotted against the mole ratio of cation to solvating solvent (Fig. 2.7). When such a plot shows a marked discontinuity the appropriate solvent:cation ratio is taken as the solvation number of the cation. The method seems to work best for

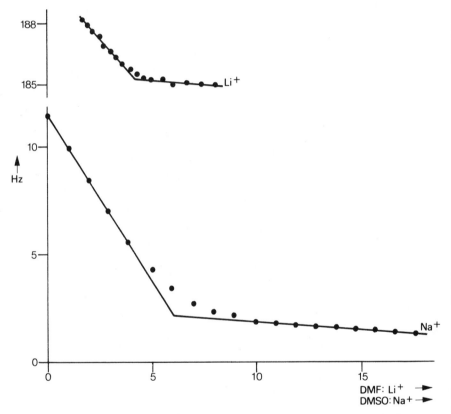

Fig. 2.7 The dependence of solvent chemical shift on the cation to solvating solvent mole ratio for (a) Li^+ in dimethylformamide+dioxan [91] and (b) Na^+ in dimethyl sulphoxide+dioxan [94].

lithium salts. Dimethyl sulphoxide, dimethylformamide, and 1-methyl-2-pyrrolidinone (Fig. 2.8) have been examined as solvating solvents, while dioxan and pentan-1-ol have been used as non-solvating components. The plot of dimethylformamide-^{1}H shifts against the Li^+:DMF mole ratio shown in Fig. 2.7(a) suggests a solvation number of 4 for Li^+ here, from both lithium iodide and lithium perchlorate solutions [91]. Similar plots for Li^+ in dimethyl sulphoxide+pentan-1-ol mixtures [92] and in 1-methyl-2-pyrrolidinone+dioxan mixtures [93] suggest solvation numbers of two and four respectively for the Li^+ cation. All these estimates seem eminently reasonable when compared with solvation numbers for Li^+ in the respective solvents obtained by other methods. When this method is tried for sodium salts, the plots have less well marked discontinuities, as Fig. 2.7(b) shows. However estimates are still possible and solvation numbers for Na^+ of six (with respect to dimethyl sulphoxide) and of four (with respect to 1-methyl-2-pyrrolidinone) have been obtained this way [94].

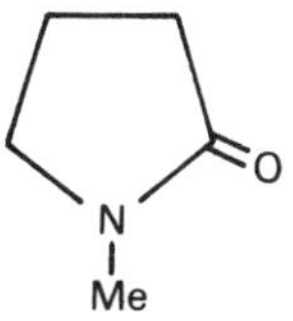

Fig. 2.8   1-Methyl-2-pyrrolidinone.

Another method for estimating solvation numbers from solvent chemical shift data makes use of the fact that when a paramagnetic species is added to a slow-exchange cation it causes a large shift to the 'bulk' solvent peak(s), but has little effect on the signals of the solvent coordinated to the cation. From the observed shifts it is possible to estimate how much solvent remains bonded to the slow exchange cation, beyond the full influence of the paramagnetic centre [40, 95]. This method works satisfactorily for Al^{3+} and Be^{2+}, but is not so successful for Cr^{3+} where two paramagnetic species are involved. This method seems less direct and attractive than the peak area method, though recent extensions to complexes and to mixed solvent systems may prove useful [96].

A third way of estimating solvation numbers from chemical shifts is based on examination and interpretation of their temperature dependence. The original application of this method indicated a value of 13.4 for the hydration number of the Al^{3+} cation [97], but after reinterpretation it was suggested that the expected value of 6 could be obtained for the primary hydration number here [98]. Nonetheless there are still serious reservations about this method [31, 99], and its recent extension to the estimation of solvation numbers in methanol [100] should be regarded with caution.

2.4.5   Solvent Coordination Site

Certain solvents contain two or more atoms in each molecule which could coordinate to a cation. For instance, dimethyl sulphoxide could coordinate to a cation either through the sulphur or through the oxygen atom, and dimethylformamide through nitrogen or through oxygen. There are two established ways for deciding which is the ligating atom, n.m.r. spectroscopy (see below) and infrared spectroscopy (Chapter 3.2.3).

The principle of the n.m.r. method can most simply be demonstrated by using pyridine as an example. Figure 2.9(a) shows the shift differences for the various sites on the pyridine ring between the signals in pyridine (50% aqueous solution) and in a solution containing cobalt(II) chloride [101]. A marked decrease in the difference between the free and coordinated shifts is seen as the proton in question recedes from the ligating nitrogen. Now take the cases of N-methylformamide and of NN-dimethylformamide, where there is a choice of ligating atoms. When a paramagnetic cation is added to either of these solvent, the carbonyl-proton resonances are shifted more than the N-methyl resonances, as shown in Fig. 2.9(b) and (c). This indicates that these solvents bond to cations such as Co^{2+} through oxygen [102].

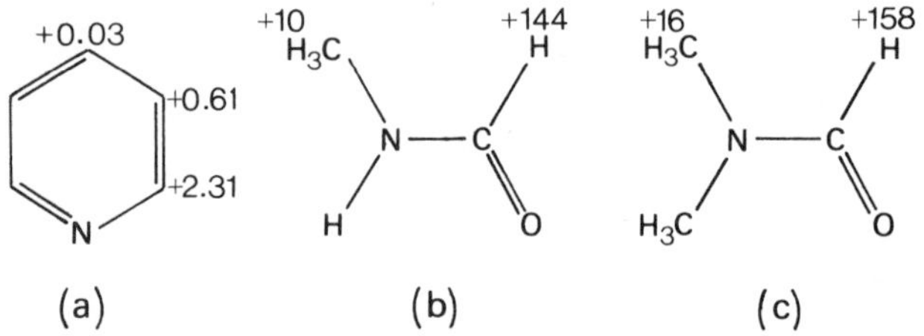

(a)   **(b)**   **(c)**

Fig. 2.9 Differences in proton chemical shifts, in p.p.m. relative to the shifts in salt-free solvents, for pyridine, N-methylformamide, and NN-dimethylformamide in the presence of added cobalt(II). In (a) the pyridine is in 50% aqueous solution containing 0.5M $CoCl_2$; in (b) and (c) the solutions are 0.3M $CoCl_2$ in 1:10 mole ratio cosolvent: water [101, 102].

2.4.6 Linewidths

The linewidth of the n.m.r. absorption provides an alternative source of information concerning cation-solvent interactions [32]. Linewidths are determined by the so-called spin-lattice or longitudinal relaxation time T_1 and the spin-spin or transverse relaxation time T_2. Linewidth measurements for resonances from solvent nuclei, and comparisons of these with corresponding natural linewidths for the nuclei in the pure solvent, give some idea of the effect of dissolved ions on the solvent. T_1 is of the order of a second for water in most (liquid) environments, only differing greatly in the vicinity of paramagnetic species ($\sim 10^{-4}$ sec). So in an aqueous solution of a salt, except for a salt of a slow-exchange cation, the environment of an individual water molecule changes many times within the time interval corresponding to T_1. Observed T_1 values are thus weighted mean quantities. Proton T_1 values increase as the structuredness of the solvent increases, in other words as the mobility of the water decreases. They are therefore much used in probing the structure-making and structure-breaking effects of solutes, including metal cations. Perhaps the most useful application here has arisen from investigations into the state of hydration of large ions of low charge, such as Rb^+ and Cs^+, where T_1 measurements suggest enhanced solvent fluidity right up to the cation. In other words these cations act as structure breakers without, on this evidence, having a firmly bonded primary hydration shell [103]. Support for this picture comes from observations of the effects of temperature and pressure [104] on T_1 values for aqueous solutions of salts of these cations. This method of investigation has also proved successful in probing ion-pairing and complex formation in aqueous solution [103] and in examining the effects of ions on the structural properties of the solvent in some non-aqueous solvents [105].

There is an empirical method for determining solvation numbers for fast-exchange cations. This is based on their effect on the width of the water proton resonance when this has been broadened by the addition of a little paramagnetic manganese(II) [106]. The effect of the added fast-exchange cation on this linewidth is thought to be due to its decreasing the amount of free 'bulk' water (note the similar shift method in Chapter 2.4.4 above). The method was calibrated by using Al^{3+} and Be^{2+}, under fast exchange conditions, so as to establish the points for hydration numbers of six and four respectively. Once it had been confirmed that Ga^{3+} also gave the expected hydration number of six, the estimation of hydration numbers for fast exchange cations was undertaken. Estimated hydration numbers are shown in Table 2.7. These numbers vary in a consistent way with cation charge and radius. However the trend to an increase in hydration number with increasing cation radius, while being geometrically reasonable, is the opposite to that normally proposed. In addition the hydration number for Mg^{2+} is unusually low. Doubts have been expressed as to the physical validity of this method, but the originators subsequently provided a stout defence [107].

Table 2.7 Hydration numbers of cations estimated from the linewidth method [106].

Mg^{2+}	3.8						
Ca^{2+}	4.3	Zn^{2+}	3.9				
Sr^{2+}	5.0	Cd^{2+}	4.6				
Ba^{2+}	5.7	Hg^{2+}	4.9	Pb^{2+}	5.7	Th^{4+}	10

2.5   FAST SOLVENT EXCHANGE: N.M.R. OF SOLUTE NUCLEI

Table 2.1 has already indicated the range and variety of solute nuclei with resonances which can be monitored by n.m.r. spectroscopy. Of the nuclei there listed, the series of alkali metal cations have been the most thoroughly studied. A few investigations have dealt with the linewidths (Chapter 2.5.2), but the majority have been concerned with chemical shifts (Chapter 2.5.1). These n.m.r. results yield information on cation-solvent interactions, on ion-pairing, and on solvation in mixed solvents. They can also shed some light on molecular motion, for example of water coordinated to Li^+ [108]. Similarly ^{139}La n.m.r. spectra have helped in the elucidation of the dynamic behaviour of bovine serum albumin in solution [8]. Cation solute n.m.r. spectroscopy is currently proving useful in several areas of biochemistry. ^{205}Tl is a good probe, for its sensitivity is high and its chemical shift changes enormously on varying the environment [17].

2.5.1 Chemical Shifts

The magnitudes of chemical shifts for alkali metal resonances increase markedly from lithium to caesium, and are a further order of magnitude larger for thallium [17]. However the sensitivity of the alkali metal chemical shifts to environment, here solvation variation specifically, increases in the opposite direction. Chemical shifts depend both on the nucleus and the nature of the solvent. They also, for a given alkali metal cation, depend on the nature of the gegenion and the concentration of the salt. This is illustrated for ^{23}Na [109, 110] and for ^{39}K [111] resonances, for aqueous solutions of a variety of salts, in Fig. 2.10. From observations such as those illustrated in this Figure, orders of increasingly negative shift producers can be constructed:

$$Cr_2O_7^{2-} > NO_3^- > CrO_4^{2-} > SO_4^{2-} > NO_2^- > N_3^- > F^- > CH_3CO_2^- > NCS^- > CO_3^{2-} >$$

$$NCO^- > Cl^- > OH^- > CN^- > Br^- > I^- \quad [111]$$

$$ClO_4^- > BPh_4^- > Cl^- > OH^- \quad [109].$$

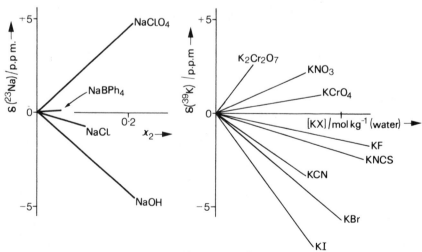

Fig. 2.10 The dependence of (a) ^{23}Na and (b) ^{39}K chemical shifts on the nature and concentration of salt in aqueous solutions.

Chemical shift dependence on anion nature and on concentration is ascribed to interactions during random collisions [16, 112]. This dependence has been variously correlated with anion basicity [109] and with the nephelauxetic series [111].

If cation-solvent interactions are to be assessed, shift (δ) versus salt concentration plots must obviously be extrapolated carefully to infinite dilution so as to get δ_0 values. Chemical shifts at infinite dilution δ_0 for the nuclei ^{7}Li, ^{23}Na, and ^{205}Tl from solutions of their salts in a range of solvents are given in Table 2.8. The values in this Table illustrate the great sensitivity of ^{205}Tl resonances to solvent environment. Correlations between these sets of shifts are not good, and neither are correlations of these shifts with such empirical solvent. parameters (Chapter 1.3) as Z, E_T, or S. The ^{23}Na shifts have been variously claimed to correlate tolerably well with solvent basicities [110] and with.

Table 2.8 ^{7}Li, ^{23}Na, and ^{205}Tl chemical shifts (p.p.m) for Li$^+$, Na$^+$, and Tl$^+$ in non-aqueous solutions; solvents are ordered according to their E_T values (Chapter 1.3) as far as possible.

	$\delta(^7\text{Li})$	$\delta(^{23}\text{Na})$		$\delta(^{205}\text{Tl})$	E_T
	[113][a]	[110][b]	[114][c]	[16][d]	[116]
Water	0	0		−233	63.1
Methanol	+0.85		+ 72		55.5
Acetic acid		+5.4			51.2
Acetonitrile	+2.90	+6.4	+127		46.0
Dimethyl sulphoxide	+1.29	0.0	+ 37		45.0
Dimethylformamide	−0.32	+4.5		− 29	43.8
Dimethylacetamide	−0.25				43.7
Acetone	−1.23	+5.8	+136		42.2
Pyridine	−2.26	− 6.6	− 20	+432	40.2
1,2-Dimethoxyethane	+1.66				38.2
Ethyl acetate	+0.43	−3.5	+147		38.1
Tetrahydrofuran	+1.06				37.4
Triethylamine	+1.00				33.3
n-Butylamine				+1662	
Formamide				− 39	
Hexamethylphosphoramide		− 41			

a Shifts for 7.5 × 10^{-2} M LiClO$_4$ solutions relative to 23% aqueous LiBr solution; b for sodium iodide solutions at infinite dilution; c shifts, in Hz, for 0.5M NaBPh$_4$ solutions; d for thallium(I) perchlorate solutions at infinite dilution, relative to 5M aqueous thallium(I) formate solution.

Gutmann donor numbers (Chapter 1.3) [114]. The latter correlation is satisfactory for ten non-aqueous solvents, but does not include water. Moreover,
although this correlation with Gutmann donor numbers is satisfactory for
^{23}Na and for ^{133}Cs shifts [114], it is not for ^{7}Li shifts [115]. However, the ^{7}Li
shifts correlate with Z values [113]! Worst of all, ^{205}Tl shifts in a dozen solvents
correlate neither with Z values, Gutmann donor numbers, nor dielectric constants [17]. It is difficult to avoid the pessimistic conclusion that there are no
wide-ranging correlations between cation chemical shifts and solvent properties.
Even for such restricted ranges as sodium salts in strongly basic nitrogenous
solvents [117] and in tetrahydrofuran and chelating diethers [118], the
observed variation of δ_0 values could not be rationalised in a simple manner.

In some of these studies the observed dependence of chemical shift on salt
concentration is not linear, and this requires the invocation of ion-pairing in
explanation. The consequences on chemical shift trends of this ion-pairing are
illustrated in Fig. 2.11, where the curvature of the plots in such 'poor'
solvents as acetone or sulpholane is immediately apparent. Here the extent of
ion-pairing in the various solvents correlates with their Gutmann donor numbers
[119].

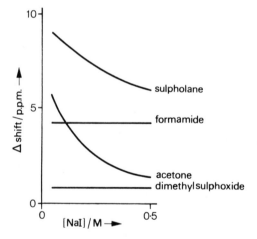

Fig. 2.11   The effects of ion-pairing on ^{23}Na chemical shifts for solutions of
sodium iodide in a range of non-aqueous solvents [119].

Selective solvation of alkali metal ions in mixed solvents can be monitored
from the extent to which the chemical shift of the alkali metal nucleus varies
with solvent composition. This is illustrated in Fig. 2.12 for the typical case of
Na$^+$ in acetonitrile+dimethyl sulphoxide mixtures. The dependence of the ^{23}Na
chemical shift on solvent composition reflects strong selective solvation by the
latter component [120].

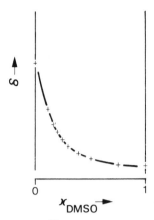

Fig. 2.12 The dependence of ^{23}Na chemical shifts on solvent composition for acetonitrile+dimethyl sulphoxide solutions of sodium tetraphenylboronate [120].

A rather different aspect of solvent effects on solute cation n.m.r. spectra is provided by studies of micellar effects on alkali metal, for instance ^{23}Na and ^{133}Cs, resonances [121]. Here the behaviour of the cation shift reflects the attachment of these ions to the macromolecules and macromolecular aggregates.

2.5.2 Linewidths

There have been several investigations of alkali metal resonance linewidths in solution, but the amount of chemical information pertaining to cation-solvent interactions so far extracted is small. A systematic and extensive study of such linewidths has recently been started, and the first results show that the dependence of linewidth on salt concentration is not simple (Fig. 2.13). At this early stage discussion is concerned primarily with attempting to explain the observed patterns in terms of known or reasonable amounts of ion-solvent and ion–ion interactions [122, 123].

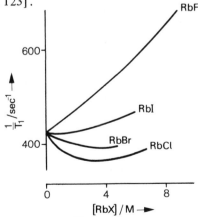

Fig. 2.13 The dependence of ^{85}Rb linewidths $(1/T_1)$ on salt concentration for aqueous solutions of rubidium halides [123].

Studies of alkali metal resonances have also been undertaken in several non-aqueous media. ^{7}Li, ^{23}Na, ^{87}Rb, and ^{133}Cs linewidths have been measured in methanolic solutions of appropriate salts, while the ^{87}Rb nucleus has been monitored in ethanol, glycol, and formamide as well. The ^{87}Rb linewidths and relaxation times parallel the extent of cation solvation, with a long T_1 for highly solvating methanol and a considerably smaller value for less well solvating glycol [124].

There are indications that linewidth measurements may give some idea of solvation in mixed solvents. The dependence of the ^{23}Na resonance linewidth on solvent composition in some mixtures containing one good and one poor solvent is outlined in Fig. 2.14. The difference in pattern between dimethyl sulphoxide or dimethylformamide as good solvent, on the one hand, and methanol or water, on the other, is striking. The difference is attributed to relatively long-lived cation-solvent complexes forming for the former solvents, but not for methanol or water. Such a difference is behaviour is reasonable. It is consistent with Parker's qualitative ranking of solvents in order of their cation solvating abilities, and with their respective Gutmann donor numbers (Chapter 1.3) [125]. The effects of micelle formation on cation solvation have been probed via linewidth measurements for several alkali metal resonances (see shifts, in the previous section) [126].

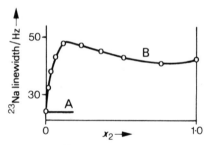

Fig. 2.14 The variation of ^{23}Na linewidth with solvent composition in mixed solvents. Curve A applies to acetonitrile+water and to acetonitrile+methanol; curve B applies to acetonitrile+dimethylformamide. In each case the latter-named solvent is component 2, at mole fraction x_2. The solute is sodium perchlorate. [125].

This area, where cation nucleus n.m.r. spectroscopy is applied to the study of ion-solvent interactions, is still in its infancy, but its potential value seems large.

FURTHER READING

Hertz, H. G., *Ber. Bunsenges.phys. Chem.*, **67**, 311 (1963); *Prog.nucl.magn. Reson. Spectrosc.*, **3**, 159 (1967).

Hinton, J. F. and Amis, E. S., *Chem. Rev.*, **67**, 367 (1967).

Burgess, J. and Symons, M. C. R., *Q. Rev. Chem. Soc.*, **22**, 276 (1968).
Deverell, C. in *Progress in Nuclear Magnetic Resonance Spectroscopy, Volume 4*, ed. Emsley, J. W., Feeney, J., and Sutcliffe, L. H., Pergamon (1969) pp. 235-334.
Lincoln, S. F., *Coord. Chem. Rev.*, **6**, 309 (1971).
McLauchlan, K. A., *Magnetic Resonance*, Oxford University Press (1972).
Fratiello, A., in *Progress in Inorganic Chemistry*, ed. Lippard, S. J., *Volume 17: Inorganic Reaction Mechanisms, Part II*, ed. Edwards, J. O., Wiley Interscience (1972) p. 57.
Hertz, H. G. in *Water: A Comprehensive Treatise – Volume 3: Aqueous Solutions of Simple Electrolytes*, ed. Franks, F., Plenum Press (1973) Ch. 7.
Harris, R. K., *Chem. Soc. Rev.*, **5**, 1 (1976).
Popov, A. E. in *Solute-Solvent Interactions, Volume 2*, ed. Coetzee, J. F. and Ritchie, C. D., Marcel Dekker (1976).

REFERENCES

[1] Kotz, J. C., Schaeffer, R. and Clouse, A., *Inorg. Chem.*, **6**, 620 (1967).
[2] Lurio, A., *Phys. Rev.*, **126**, 1768 (1962).
[3] Bucka, H., Kopfermann, H. and Putlitz, G.zu, *Z. Phys.*, **168**, 542.(1962); Putlitz, G.zu, *Z. Phys.*, **175**, 543 (1963).
[4] E.g. O'Reilly, D. E., *J. Chem. Phys.*, **32**, 1007 (1960).
[5] Delpuech, J. -J., Khaddar, M. R., Peguy, A. A. and Rubini, P. R., *J. Amer. Chem. Soc.*, **97**, 3373 (1975).
[6] Melson, G. A., Olszanski, D. J. and Roach, E. T., *J. Chem. Soc. Chem. Commun.*, 229 (1974); Buslaev, Yu. A., Petrosyants, S. P., Tarasov, V. P. and Chagin, V. I., *Russ. J. Inorg. Chem.*, **19**, 975 (1974).
[7] Nakamura, K. and Kawamura, K., *Bull. Chem. Soc. Japan*, **44**, 330 (1971); Reuben, J., *J. Phys. Chem.*, **79**, 2154 (1975).
[8] Reuben, J., *J. Amer. Chem. Soc.*, **97**, 3823 (1975).
[9] Maciel, G. E. and Borzo, M., *J. Chem. Soc. Chem. Commun.*, 394 (1973); Cardin, A. D., Ellis, P. D., Odom, J. D. and Howard, J. W., *J. Amer. Chem. Soc.*, **97**, 1672 (1975).
[10] Godfrey, P. D., Heffernan, M. L., and Kerr, D. F., *Aust. J. Chem.*, **17**, 701 (1964).
[11] Fratiello, A., Lee, R. E. and Schuster, R. E., *Inorg. Chem.*, **9**, 82 (1970).
[12] Lincoln, S. F., *Aust. J. Chem.*, **25**, 2705 (1972).
[13] Fratiello, A., Lee, R. E., Nishida, V. M. and Schuster, R. E., *J. Chem. Phys.*, **48**, 3705 (1968); Fratiello, A., Davis, D. D., Peak, S. and Schuster, R. E., *Inorg. Chem.*, **10**, 1627 (1971).
[14] Gutowsky, H. S. and McGarvey, B. R., *Phys. Rev.*, **91**, 81 (1953).
[15] Chan, S.·O. and Reeves, L. W., *J. Amer. Chem. Soc.*, **96**, 404 (1974).
[16] Dechter, J. J. and Zink, J. I., *J. Chem. Soc. Chem. Commun.*, 96 (1974).

[17] Dechter, J. J. and Zink, J. I., *J. Amer. Chem. Soc.*, **97**, 2937 (1975).

[18] Burke, J. J. and Lauterbur, P. C., *J. Amer. Chem. Soc.*, **83**, 326 (1961).

[19] Hawk, R. M. and Sharp, R. R., *J. Chem. Phys.*, **60**, 1522 (1974).

[20] Jeffries, C. D., *Phys. Rev.*, **92**, 1262 (1953).

[21] Howarth, O. W. and Richards, R. E., *J. Chem. Soc.*, 864 (1965).

[22] Procter, W. G. and Yu, F. C., *Phys. Rev.*, **81**, 20 (1951); Freeman, R., Murrary, G. R. and Richards, R. E., *Proc. R. Soc.*, **242A**, 455 (1957).

[23] McConnell, H. M. and Weaver, H. E., *J. Chem. Phys.*, **25**, 307 (1956).

[24] Johnson, A. and Everett, G. W., *J. Amer. Chem. Soc.*, **94**, 1419 (1972); Achlama (Chmelnik), A. M. and Fiat, D., *J. Chem. Phys.*, **59**, 5197 (1973).

[25] Matwiyoff, N. A. and Darley, P. R., *J. Phys. Chem.*, **72**, 2659 (1968).

[26] Fratiello, A. in *Progress in Inorganic Chemistry*, ed. Lippard, S. J., *Volume 17: Inorganic Reaction Mechanisms Part II,* ed. Edwards, J. O., Wiley/Interscience (1972) p. 57.

[27] Fratiello, A., Lee, R. E. and Schuster, R. E., *Inorg. Chem.*, **9**, 391 (1970).

[28] O'Brien, J. F., Glass, G. E. and Reynolds, W. L., *Inorg. Chem.*, **7**, 1664 (1968).

[29] Akitt, J. W., *J. Chem. Soc. Dalton Trans.*, 42 (1973).

[30] Akitt, J. W., *J. Chem. Soc. Dalton Trans.*, 1177 (1973).

[31] Symons, M. C. R., *Spectrochim. Acta*, **31A**, 1105 (1975).

[32] McLauchlan, K. A., *Magnetic Resonance*, Oxford University Press (1972) p. 55.

[33] Kurland, R. J. and McGarvey, B. R., *J. Magnetic Resonance*, **2**, 286 (1970).

[34] Bertini, I. and Gatteschi, D., *Inorg. Chem.*, **12**, 2740 (1973).

[35] Fiat, D., Luz, Z. and Silver, B. L., *J. Chem. Phys.*, **49**, 1376 (1968).

[36] Connick, R. E. and Fiat, D., *J. Chem. Phys.*, **44**, 4103 (1966).

[37] Chmelnik, A. M. and Fiat, D., *J. Amer. Chem. Soc.*, **93**, 2875 (1971).

[38] Neely, J. W. and Connick, R. E., *J. Amer. Chem. Soc.*, **94**, 3419 (1972).

[39] Glass, G. E., Schwabacher, W. B. and Tobias, R. S., *Inorg. Chem.*, **7**, 2471 (1968).

[40] Jackson, J. A., Lemons, J. F. and Taube, H., *J. Chem. Phys.*, **32**, 553 (1960).

[41] Stockton, G. W. and Martin, J. S., *Can. J. Chem.*, **52**, 744 (1974).

[42] See pp. 56-58 of reference [32].

[43] *Metal Ions in Biological Systems – Volume 4: Metal Ions as Probes,* ed. Sigel, H., Marcel Dekker (1974).

[44] Jones, C. R. and Kearns, D. R., *J. Amer. Chem. Soc.*, **96**, 3651 (1974).

[45] Glaeser, H. H., Dodgen, H. W., and Hunt, J. P., *J. Amer. Chem. Soc.*, **89**, 3065 (1967).

[46] Connick, R. E. and Fiat, D. N., *J. Chem. Phys.*, **39**, 1349 (1963).

[47] Fiat, D. N. and Connick, R. E., *J. Amer. Chem. Soc.*, **88**, 4754 (1966).

[48] Swift, T. J., Fritz, O. G., and Stephenson, T. A., *J. Chem. Phys.*, **46**, 406 (1967).

[49] Swift, T. J. and Weinberger, G. P., *J. Amer. Chem. Soc.*, **90**, 2023 (1968).

[50] Harrison, L. W. and Swift, T. J., *J. Amer. Chem. Soc.*, **92**, 1963 (1970).

[51] Swift, T. J. and Lo, H.H., *J. Amer. Chem. Soc.*, **89**, 3988 (1967).

[52] Nakamura, S. and Meiboom, S., *J. Amer. Chem. Soc.*, **89**, 1765 (1967).

[53] Butler, R. N. and Symons, M. C. R., *Trans. Faraday Soc.*, **65**, 945 (1969); Al-Baldawi, S. A. and Gough, T. E., *Can. J. Chem.*, **47**, 1417 (1969); Al-Baldawi, S. A., Brooker, M. H., Gough, T. E., and Irish, D. E., *Can. J. Chem.*, **48**, 1202 (1970).

[54] Richardson, D. and Alger, T. D., *J. Phys. Chem.*, **79**, 1733 (1975).

[55] Luz, Z. and Meiboom, S., *J. Chem. Phys.*, **40**, 1058 (1964).

[56] Luz, Z. and Meiboom, S., *J. Chem. Phys.*, **40**, 1066 (1964).

[57] Supran, L. D. and Sheppard, N., *Chem. Commun.*, 832 (1967).

[58] Matwiyoff, N. A. and Hooker, S. V., *Inorg. Chem.*, **6**, 1127 (1967).

[59] Matwiyoff, N. A. and Movius, W. G., *J. Amer. Chem. Soc.*, **89**, 6077 (1967).

[60] Fratiello, A. and Schuster, R. E., *J. Phys. Chem.*, **71**, 1948 (1967); Movius, W. G. and Matwiyoff, N. A., *Inorg. Chem.*, **6**, 847 (1967).

[61] Movius, W. G. and Matwiyoff, N. A., *Inorg. Chem.*, **8**, 925 (1969).

[62] Matwiyoff, N. A., *Inorg. Chem.*, **5**, 788 (1966).

[63] Thomas, S. and Reynolds, W. L., *J. Chem. Phys.*, **44**, 3148 (1966).

[64] Frankel, L. S., *Chem. Commun.*, 1254 (1969).

[65] Bloembergen, N., *J. Chem. Phys.*, **27**, 595 (1957); and references therein.

[66] Kapur, V. K. and Wayland, B. B., *J. Phys. Chem.*, **77**, 634 (1973); and references therein.

[67] Lewis, W. B., Alei, M., and Morgan, L. O., *J. Chem. Phys.*, **44**, 2409 (1966).

[68] Chmelnik, A. M. and Fiat, D., *J. Chem. Phys.*, **51**, 4238 (1969).

[69] Chmelnik, A. M. and Fiat, D., *J. Magnetic Resonance*, **8**, 325 (1972).

[70] Wayland, B. B. and Rice, W. L., *Inorg. Chem.*, **5**, 54 (1966).

[71] Granot, J., Achlama (Chmelnik), A. M., and Fiat, D., *J. Chem. Phys.*, **61**, 3043 (1974).

[72] Poupko, R. and Luz, Z., *J. Chem. Phys.*, **57**, 3311 (1972).

[73] Luz, Z., *Israel J. Chem.*, **9**, 293 (1971).

[74] Lincoln, S. F. and West, R. J., *Aust. J. Chem.*, **26**, 255 (1973).

[75] Vigee, G. S. and Ng, P., *J. Inorg. Nucl. Chem.*, **33**, 2477 (1971).

[76] Beech, G. and Miller, K., *J. Chem. Soc. Dalton*, 801 (1972).

[77] Pople, J. A., Schneider, W. G. and Bernstein, H. J., *High-resolution Nuclear Magnetic Resonance*, McGraw-Hill (1959) p. 4 and Appendix A.

[78] Perry, W. D. and Drago, R. S., *J. Amer. Chem. Soc.*, **93**, 2183 (1971).

[79] West, R. J. and Lincoln, S. F., *Aust. J. Chem.*, **24**, 1169 (1971).

[80] Butler, R. N. and Symons, M. C. R., *Trans. Faraday Soc.*, **65**, 2559 (1969).

[81] Hindman, J. C., *J. Chem. Phys.*, **36**, 1000 (1962).
[82] Hertz, H. G. and Spalthoff, W., *Z. Elektrochem.*, **63**, 1096 (1959).
[83] Luz, Z. and Yagil, G., *J. Phys. Chem.*, **70**, 554 (1966).
[84] Shoolery, J. N. and Alder, B. J., *J. Chem. Phys.*, **23**, 805 (1955).
[85] Allred, A. L. and Wendricks, R. N., *J. Chem. Soc. A*, 778 (1966).
[86] Shamir, J. and Netzer, A., *Can. J. Chem.*, **51**, 2676 (1973).
[87] Coetzee, J. F. and Sharpe, W. R., *J. Solution Chem.*, **1**, 77 (1972).
[88] Butler, R. N., Phillpott, E. A., and Symons, M. C. R., *Chem. Commun.*, 371 (1968).
[89] Butler, R. N. and Symons, M. C. R., *Chem. Commun.*, 71 (1969).
[90] Davies, J., Ormondroyd, S., and Symons, M. C. R., *Trans. Faraday Soc.*, **67**, 3465 (1971); Brown, R. D. and Symons, M. C. R., *J. Chem. Soc. Dalton,* 426 (1976).
[91] Lassigne, C. and Baine, P., *J. Phys. Chem.*, **75**, 3188 (1971).
[92] Maxey, B. W. and Popov, A. I., *J. Amer. Chem. Soc.*, **90**, 4470 (1968).
[93] Wuepper, J. L. and Popov, A. I., *J. Amer. Chem. Soc.*, **91**, 4352 (1969).
[94] Wuepper, J. L. and Popov, A. I., *J; Amer. Chem. Soc.*, **92**, 1493 (1970).
[95] Alei, M. and Jackson, J. A., *J. Chem. Phys.*, **41**, 3402 (1964).
[96] Ablov, A. V., Gulya, A. P., Vdovenko, V. M., Shcherbakov, V. A., Morkovin, N. V. and Chepurnykh, I. I., *Russ. J. Inorg. Chem.*, **18**, 381 (1973).
[97] Malinowski, E. R. and Knapp, P. S., *J. Chem. Phys.*, **48**, 4989 (1968).
[98] Akitt, J. W., *J. Chem. Soc. A*, 2865 (1971).
[99] Sare, E. J., Moynihan, C. T. and Angell, C. A., *J. Phys. Chem.*, **77**, 1869 (1973).
[100] Vogrin, F. J. and Malinowski, E. R., *J. Amer. Chem. Soc.*, **97**, 4876 (1975).
[101] Fratiello, A. and Christie, E. G., *Trans. Faraday Soc.*, **61**, 306 (1965).
[102] Fratiello, A. and Miller, D. P., *Molec. Phys.*, **11**, 37 (1966).
[103] Hertz, H. G. and Keller, G., in *Nuclear Magnetic Resonance in Chemistry*, ed. Pesce, B., Academic Press (1965) p. 199.
[104] Lee, Y. and Jonas, J. *J. Magnetic Resonance*, **5**, 267 (1971).
[105] Engel, G. and Hertz, H. G., *Ber. Bunsenges. Phys. Chem.*, **72**, 808 (1968).
[106] Swift, T. J. and Sayre, W. G., *J. Chem. Phys.*, **44**, 3567 (1966).
[107] Meiboom, S., *J. Chem. Phys.*, **46**, 410 (1967); Swift, T. J. and Sayre, W. G., *J. Chem. Phys.*, **46**, 410 (1967).
[108] Hertz, H. G., Tutsch, R., and Versmold, H., *Ber. Bunsenges. Phys. Chem.*, **75**, 1177 (1971).
[109] Templeman, G. J. and Geet, A. L. van, *J. Amer. Chem. Soc.*, **94**, 5578 (1972).
[110] Bloor, E. G. and Kidd, R. G., *Can. J. Chem.*, **46**, 3425 (1968).
[111] Bloor, E. G. and Kidd, R. G., *Can. J. Chem.*, **50**, 3926 (1972).
[112] Kondo, J. and Yamashita, J., *Physics Chem. Solids*, **10**, 245 (1959); Ikenberry, D. and Das, T. P., *Phys. Rev.*, **138A**, 822 (1965); Sears, R. E.

J., *J. Chem. Phys.*, **61**, 4368 (1974).

[113] Maciel, G. E., Hancock, J. K., Lafferty, L. F., Mueller, P. A., and Musker, W. K., *Inorg. Chem.*, **5**, 554 (1966).

[114] Erlich, R. H., Roach, E., and Popov, A. I., *J. Amer. Chem. Soc.*, **92**, 4989 (1970), Ehrlich, R. H. and Popov, A. I., *J. Amer. Chem. Soc.*, **93**, 5620 (1971); Erlich, R. H., Greenberg, M. S., and Popov, A. I., *Spectrochim. Acta*, **29A**, 543 (1973); DeWitte, W. J., Schoening, R. C. and Popov, A. I., *Inorg.nucl. Chem. Lett.*, **12**, 251 (1976).

[115] Cahen, Y. M., Handy, P. R., Roach, E. T., and Popov, A. I., *J. Phys. Chem.* **79**, 80 (1975).

[116] Dimroth, K., Reichardt, C., Siepmann, T., and Bohlmann, F., *Justus Liebigs Annln. Chem.*, **661**, 1 (1963).

[117] Herlem, M. and Popov, A. I., *J. Amer. Chem. Soc.*, **94**, 1431 (1972).

[118] Canters, G. W., *J. Amer. Chem. Soc.*, **94**, 5230 (1972).

[119] Greenberg, M. S., Bodner, R. L., and Popov, A. I., *J. Phys. Chem.*, **77**, 2449 (1973).

[120] Greenberg, M. S. and Popov, A. I., *Spectrochim. Acta*, **31A**, 697 (1975).

[121] Gustavsson, H. and Lindman, D., *J. Chem. Soc. Chem. Commun.*, **93**, (1973).

[122] Hertz, H. G., Holz, M., Klute, R., Stalidis, G., and Versmold, H., *Ber. Bunsenges. Phys. Chem.*, **78**, 24 (1974); Hertz, H. G. and Rädle, C., *Ber. Bunsenges. Phys. Chem.*, **78**, 509 (1974).

[123] Hertz, H. G., Holz, M., Keller, G., Versmold, H., and Yoon, C., *Ber. Bunsenges. Phys. Chem.*, **78**, 493 (1974).

[124] Melendres, C. A. and Hertz, H. G., *J. Chem. Phys.*, **61**, 4156 (1974).

[125] Green, R. D. and Martin, J. S., *Can. J. Chem.*, **50**, 3935 (1972).

[126] Robb, I. D. and Smith, R., *J. Chem. Soc. Faraday I*, **70**, 287 (1974): Klink, J. J. van der, Zuiderweg, L. H., and Leyte, J. C., *J. Chem. Phys.*, **60**, 2391 (1974); Manning, G. S., *J. Chem. Phys.*, **62**, 748 (1975); Leyte, J. C. and Klink, J. J. van der, *J. Chem. Phys.*, **62**, 749 (1975).

Chapter 3

METAL ION SOLVATION:
OTHER SPECTROSCOPIC TECHNIQUES

3.1 ULTRAVIOLET-VISIBLE SPECTROSCOPY

Ultraviolet-visible (electronic) spectroscopy provides information germane to metal ion–solvent interaction in one main area, that of the transition metal cations. The five d orbitals, which are degenerate in the free ion, are differently affected by coordinated ligands. Thus in an octahedral field, as in a hexa-solvento-cation, the three d orbitals pointing between the axes and the solvent molecules are much less affected than the two d orbitals pointing along the axes and so interacting strongly with the coordinated solvent (Fig. 3.1). For any d electron configuration between d^1 and d^9 inclusive, there will be one or more absorption bands in the visible or near-ultraviolet region of the spectrum (though these weak bands may be obscured by intense allowed transitions, e.g. charge-transfer). The energies of these bands, reflected in the observed frequencies of maximum absorption, reflect the strength of the interaction between the metal ion and the coordinated solvent. Information concerning ion-solvent interactions can be extracted from the crystal field splitting parameter (Dq, Chapter 3.1.1) or from the Racah parameter (B', Chapter 3.1.2). From the former, one can insert solvents into their appropriate positions in the standard spectrochemical series [1] of ligands. From the latter, one can estimate the positions of solvents in the nephelauxetic series [1], and adduce an approximate degree of cation-solvent covalent interaction.

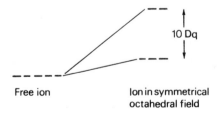

Free ion   Ion in symmetrical
octahedral field

Fig. 3.1 Effect of ligand (solvent) coordination on the d levels of a transition metal cation.

The usefulness of ultraviolet-visible spectroscopy in probing ion-solvent interactions is effectively limited to transition metal cations. The absence of $d \to d$ transitions for cations of the sp-block elements obviously rules these out. One might expect to be able to extract some information about ion-solvent interactions from the spectra of lanthanide and actinide cations, since co-ordinated solvent molecules should have some effect on f-orbitals. Unfortunately crystal field effects are very small because the f-orbitals are too well shielded, and so their frequencies are not sufficiently sensitive to coordinated ligand effects to be of much value even though $f \to f$ transitions cause very sharp spectroscopic lines [2]. The situation is further complicated by the large number of possible spectroscopic states, and therefore of possible $f \to f$ transitions, in a ligand field [3]. An investigation of lanthanide cations in melts recently indicated that the intensity of $f \to f$ transitions is somewhat more sensitive to environment than their frequencies [4]. It remains to be seen whether intensity measurements for solutions of lanthanide cations will provide some useful insights into cation-solvent interactions.

3.1.1 Ligand Field Effects

Solvent Dq values can readily be determined from the spectra of d^1 and d^9, in practice usually titanium(III) and copper(II), cations. In these ions the ^2D term† for the free ion splits into just ^2E_g and $^2T_{2g}$ terms in an octahedral field, as Fig. 3.2(a) shows. The value of $10Dq$ is thus obtained directly from the single observed transition illustrated in Fig. 3.2(b); $E = h\nu$. There is the complication that hexasolvento-complexes both of d^1 and of d^9 configurations will be tetragonally distorted (Jahn–Teller) from octahedral symmetry. Indeed the asymmetry actually observed for these absorption bands (see Fig. 3.2(b)) is strong evidence supporting the postulate of such distortion. It is easy to estimate an average value for $10\,Dq$ from titanium(III) spectra with a small degree of uncertainty only. However copper(II) systems are often more markedly tetragonal and therefore more complicated.

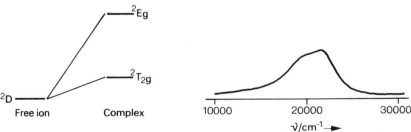

Fig. 3.2   (a) Effect of an octahedral ligand field on a ^2D term (d^1, e.g. Ti^{3+}); (b) visible absorption spectrum of Ti^{3+}aq [5].

The derivation of Dq values for cations with electron configurations between d^2 and d^8 is slightly less straightforward. The commonly studied

†For explanations of term symbols, see Chapter 6 of reference [5].

cation nickel(II), which is d^8, is like the much less investigated complementary d^2 cation vanadium(III) in having three $d \rightarrow d$ bands in an octahedral ligand (solvent) field (Fig. 3.3). Dq can be estimated directly from the $^3T_{2g} \leftarrow {}^3A_{2g}$ band, whose wavenumber gives 10 Dq, but a more involved method of estimation using all three bands gives a better estimate. Similarly a value approximating to 8 Dq for the d^7 cation cobalt(II) can be obtained from the $^4T_{2g} \leftarrow {}^4T_{1g}$ transition, and 10 Dq comes from the $^5E_g \leftarrow {}^5T_{2g}$ band for high-spin iron(II). For further details of the estimation of Dq values from visible absorption spectra, the reader is referred to the comprehensive account in Chapter 9 of reference [5].

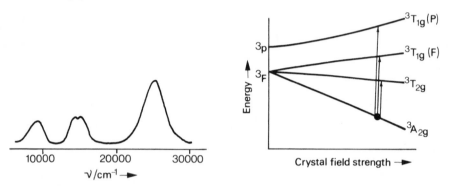

Fig. 3.3 (a) Absorption spectrum of Ni^{2+}aq [5]; (b) the three transitions involved in the absorption spectrum of a d^8 cation, e.g. Ni^{2+}aq, in an octahedral ligand field.

Table 3.1 Values of the crystal field splitting parameter Dq (cm^{-1}) for transition metal cations in aqueous solution; from ref. [5] except where otherwise indicated.

d^1			Ti^{3+}	2030^a		
d^2			V^{3+}	1860		
d^3	V^{2+}	1230	Cr^{3+}	1700	Mo^{3+}	2630^e
d^4	Cr^{2+}	1410^b	Mn^{3+}	2000^c		
d^5	Mn^{2+}	850	Fe^{3+}	$\sim 1400^d$		
d^6	Fe^{2+}	1000	Co^{3+}	1820	Rh^{3+}	~ 2700
d^7	Co^{2+}	930				
d^8	Ni^{2+}	890				
d^9	Cu^{2+}	1200				

a See also ref. [6]; *b* or 1390 [7]; *c* or 1860 [8], 2100 [9]; *d* difficult to estimate accurately because of hydrolysis; *e* from ref [10].

Cations in aqueous solution. Values of Dq for transition metal cations in aqueous solution are listed in Table 3.1 [5–10]. The variation in values from different sources for the Mn^{3+} and Cr^{2+} cations indicates how precise the figures listed are. A spectrochemical series of metal cations can be constructed from these Dq values, and similar spectrochemical series for cations can be constructed from these Dq values for other solvents or ligands. The order of the cations in these series is determined by numerous factors, including charge, size, electronegativity, π-bonding propensity, and polarisability. It seems wiser to follow Figgis [5] and accept this order as a fact of nature, rather than try to explain it in terms of so many variables, especially as some of these are of a nebulous nature.

Cations in non-aqueous media. There are many reports of Dq values for hexa-solvento-complexes of transition metal cations, especially Co^{2+}, Ni^{2+}, Cu^{2+}, and Cr^{3+}, in a variety of solvents. Unfortunately these values are derived from a multitude of different sources, and Dq values for a given cation-solvent pair can vary by an uncomfortably large amount from source to source. Values of Dq for Co^{2+}, Ni^{2+}, and Cr^{3+} in selected solvents (and from selected sources) are listed in Table 3.2. The Co^{2+} and Ni^{2+} values are derived perforce from a variety of references, [11] to [20] inclusive, but fortunately a single extensive, and hopefully internally consistent, tabulation of Dq values for Cr^{3+} exists [21]. Chromium(III) is perhaps the most suitable cation for estimating solvent Dq values, because it has a symmetrical (d^3) distribution of electrons. Cu^{2+} is rather unsuitable as a probe for cation-solvent interactions from ultraviolet-visible spectroscopy, as its environment tends to be tetragonal rather than octahedral. The effect of this tetragonal distortion, together with the question of how ultraviolet-visible absorption spectra of solvento-copper(II) species should be interpreted, have recently been fully illustrated, referenced and discussed [22]. Values of Dq are also available for Ti^{3+} [23] and for V^{3+} [24] in a few non-aqueous solvents.

All the cations mentioned so far have come from the first row of transition elements, because the majority of simple aquo-cations of transition elements are derived from this series. Such solvento-cations of the second and third row transition elements as exist will have bigger Dq values than those of the analogous first row elements. For a given ligand (solvent), Dq is about 1.3 times larger for a second row transition metal cation than for its first row analogue. The third row metal value is again higher than that for the second row metal.

From experimentally derived Dq values, solvents can be inserted into their appropriate positions in the spectrochemical series of ligands. This is illustrated diagrammatically for Ni^{2+} and for Cr^{3+} in Fig. 3.4. The diagram for Cr^{3+}, by including a wide range of non-solvent ligands, shows how closely most solvents cluster around water in the spectrochemical series. The Ni^{2+} diagram is drawn to a larger scale so that individual solvent positions can be shown. As for the spectrochemical series of cations (*v.s.*), the ordering of solvents within the spectrochemical series of ligands is determined by so many factors that simple explanation of the observed order is not possible.

Table 3.2   Values of the crystal field splitting parameter Dq (cm^{-1}) for selected transition metal cations in octahedral solvento-environments.

	Co^{2+}	Ni^{2+}	Cr^{3+}
N-Methylacetamide		752 [17]	
NN-Dimethylacetamide	805 [11]	769 [17,18]	1538 [21]
Tetramethylene sulphoxide	830 [12]	775 [12]	1575 [21]
Dimethyl sulphoxide	840 [12]	777 [12], 781 [14], 773 [17]	1587 [14,21]
Trimethyl phosphate	975 [13]	796 [19]	1628 [21]
N-Methylformamide		838 [17, 18]	
NN-Diethylformamide		840 [18]	
Acetamide			1645 [21]
Ethanol			1668 [21]
Tetrahydrofuran			1680 [21]
Methanol		850 [17]	1695 [21]
NN-Dimethylformamide		850 [17,18]	1710 [21]
Water	930–1020 [5,13–16]	860 [17], 890 [5]	1700–1740 [5, 21]
Benzonitrile		970 [17]	
Ethylamine		987 [17]	
Methylamine		993 [17]	
Pyridine		1000 [17]	
Acetonitrile	1022 [13]	1026 [17], 1041 [19]	
Ammonia	1010 [16]	1070 [16,20], 1080 [17]	2155 [21]

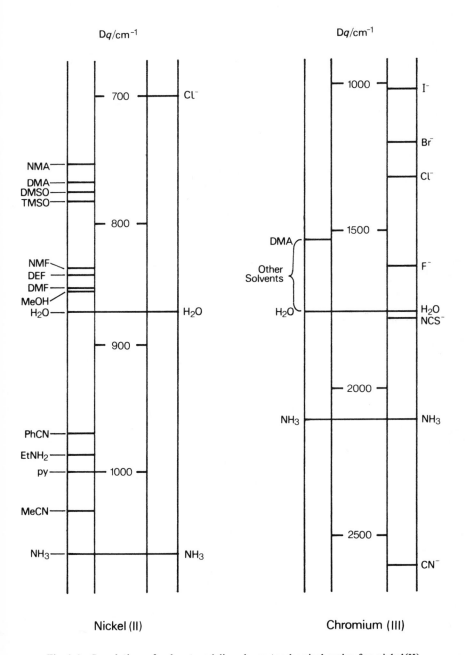

Fig. 3.4 Correlation of solvent and ligand spectrochemical series for nickel(II) and for chromium(III).

3.1.2 Racah Parameters

The Racah parameters A, B, and C [25] are electron repulsion integrals for free ions, of which B is of prime concern here. This interelectronic repulsion parameter has a different value, B', in a complex from that (B) in the free ion since the 'metal' electrons are delocalised to a greater extent in the complex [26]. The relative degree of covalent character in metal–ligand bonding is reflected in the value of the quotient B'/B, represented by β. Lower values of β correspond to greater covalent interaction with more polarisable ligands. The order of β values corresponds to the so-called **nephelauxetic series.**

Some values of β for transition metal cations in various solvents are listed in Table 3.3. Apart from the obvious differences between di- and tri- positive cations, it is rather difficult to pick out and discuss trends in series of solvents or of cations. Differences are small, of the same order as the uncertainties in individual values. For instance, estimated values of β for Cr^{3+} in water range from 0.68 to 0.79, and for Co^{2+} in water from 0.74 to 0.86. Given more precise estimates of β values it would be possible to arrange the cations and the solvents in satisfactory nephelauxetic series. However this desirable systematisation of results still lies in the future.

3.1.3 Solvation Numbers

The parameters Dq and β may throw some light on the nature of cation-solvent bonding, but they do not provide measures of the solvation numbers of cations. Indeed the cation solvation number really must be established or assumed before Dq and β can be calculated and interpreted.

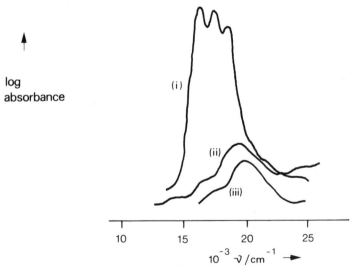

Fig. 3.5   Ultraviolet-visible spectra of Co^{2+} in (i) tetrahedral, (ii) octahedral, and (iii) aqueous environments [29].

Table 3.3 Values of β, the ratio of the Racah parameters B and B', for transition metal cations in octahedral hexasolvento-environments. The solvents are arranged in order of Dq values (Table 3.2).

	Dimethyl acetamide	Tetramethylene sulphoxide	Dimethyl sulphoxide	i-Propanol	Methanol	Water	Liquid ammonia
Mn^{2+}			0.81 [14]				
Co^{2+}	0.86 [11]	0.85 [12]	0.85 [12, 14]	0.79 [15]	0.75^a [27]	b	
Ni^{2+}		0.87 [12]	0.88 [12, 14]		0.87^a [27]		0.84 [16]
Cr^{3+}		0.75 [12]	0.71 [12, 14]			b	0.71 [16]
Fe^{3+}						0.75 [16]	
Co^{3+}						0.61 [16]	0.56 [16]
Rh^{3+}						0.73 [16]	0.60 [16]

a These values are derived from observations on *solid* $M(MeOH)_6(ClO_4)_2$ salts; b range of values, see text.

Ultraviolet-visible spectra do sometimes yield an indication of the solvation number of a cation. Large differences in both wavenumbers of maximum absorption and molar extinction coefficients exist between cations in, say, octahedral and tetrahedral environments. These differences permit deduction of the stereochemistry of a cation in solution by comparing its spectrum with those of the same cation in known environments in the solid state [28]. The method can be illustrated by reference to cobalt(II) [29]. Figure 3.5 shows spectra of, firstly cobalt(II) in the tetrahedral four-oxygen environment in $Co_{0.1}Zn_{1.9}SiO_4$, secondly cobalt(II) in the octahedral six-oxygen environment in $Co_{0.1}Mg_{0.9}O$, and thirdly cobalt(II) in aqueous solution. The similarity between the third and second spectra and the difference between the third and first strongly suggest an octahedral environment of six oxygens around cobalt(II) in aqueous solution. Table 3.4 illustrates the same approach, this time to demonstrate that the nickel(II) cation in aqueous solution is much more likely to be octahedrally rather than tetrahedrally coordinated in water [30][†]. This method of establishing solvation stereochemistries has been applied to the cations Ti^{3+}, V^{3+}, Cr^{3+}, and Fe^{3+}; Mn^{2+}, Co^{2+}, Ni^{2+}, and Cu^{2+} in aqueous solution [28].

This approach provided early evidence for octahedral environments for several transition metal cations in solution, but it must be admitted that the method leaves something to be desired. The general practice has been to compare the cation solution spectrum only with spectra of the cation in octahedral and tetrahedral surroundings, ignoring other coordination numbers and stereochemistries. In addition it is not always possible to distinguish between a fully solvated ion and a complex, for example $M(OH_2)_6^{n+}$ and $M(OH_2)_5L^{n+}$. These shortcomings have severely limited the usefulness of this method for estimating solvation numbers. Instead of helping in establishing new facts about the transition metal cations, it merely supports values obtainable in several other (and preferable) ways. Specifically its usefulness in lanthanide and actinide chemistry has proved very limited [28]. One modest success has been the application to uranium(IV). The spectrum of uranium(IV) in molten pyridinium hydrochloride is closely similar to that of uranium(IV) in solid $(Me_4N)_2UCl_6$, but the spectrum of uranium(IV) in aqueous perchloric acid differs markedly from these. The unexciting conclusion is that the primary hydration number of the uranium in $U^{4+}aq$ differs from six.

It is possible to estimate cation solvation numbers from the variation in the ultraviolet-visible spectra of solvated cations with solvent compositon in ranges of mixed solvents. Spectra of praseodymium salts in water, in aqueous ethanol, and in ethanol indicate a solvation number of six for the Pr^{3+} cation in these media [32].

---

† The only ultraviolet-visible spectrum reported for a square-planar tetra-aquo-cation is that of $Pd^{2+}aq$ [31].

Table 3.4 Ultraviolet–visible spectra of nickel(II) in tetrahedral, octahedral, and aqueous environments.

		ν/cm^{-1}				
Ni^{2+} in ZnO (tetrahedral)		13500	15100	16100	17400	[30]
Ni^{2+} in MgO (octahedral)			15000		24600	[30]
Ni^{2+} in water	8700		14500		25300	[5]

3.1.4 Solvent Spectra

Alternatively the spectra of solvated cations can be monitored by examining the spectra of the solvents to which the cations have been added. The great advantage of this approach is that the restriction to transition metal cations no longer applies. For example, it can be used in studying what happens when salts of the alkali and alkaline earth metals are added to aqueous solvent mixtures containing cyclopentanone or acetone [33]. The effects are small, even for salt concentrations up to about 8M, with the $n \rightarrow \pi^*$ transition of acetone only being shifted by up to about 10 nm. Series of spectra at different concentrations of a given salt exhibit an isosbestic point. This study made it possible to establish some sort of relative order of cation–solvent interactions, here

$$Li^+ \sim Ca^{2+} \sim Mg^{2+} > Na^+ > K^+ .$$

3.1.5 Luminescence

Very much less is known about fluorescence and phosphorescence of metal ions in solution than about their absorption properties. Several lanthanide cations exhibit radiative emission in aqueous solution; Gd^{3+}, Tb^{3+}, and Eu^{3+} emit strongly, whereas Sm^{3+} and Dy^{3+} emit only weakly. Marked enhancements are reported for analogous D_2O and methanol solutions [34]. A dependence of the time for luminescence quenching on temperature has been noted for Tb^{3+} and Eu^{3+} in solvents of low Gutmann donor number such as acetonitrile or acetone. This can probably be ascribed to the formation of ion-pairs or complexes in these media [35].

Amongst the non-lanthanide elements, luminescent properties have been described for several cations in aqueous solution, including Tl^+, Sn^{2+}, and Pb^{2+} [36]. However, while the luminescent properties of some cations, for example Tl^+ [36], in solution have been utilised in methods of analysis, there has been remarkably little investigation of the chemistry involved in emission ultraviolet-visible spectroscopy of solvento-cations.

Metal ions are also effective agents for quenching luminescence of other species. Rates of quenching of the luminescence of the UO_2^{2+}aq cation by a series of aquated metal cations correlate with the ionisation potentials of the respective metal ions [37].

3.2   INFRARED AND RAMAN SPECTROSCOPY

There are several ways in which vibrational spectroscopy can given information about the occurence and extent of ion-solvent, and of ion–ion, interactions in solution. The infrared and Raman spectra of a solution of an electrolyte may show bands which can neither be assigned to the individual ions present nor to the solvent. When the spectrum of the solution is compared with that of the pure solvent, significant frequency shifts arising from the perturbation of a solvent vibrational mode by an interacting ion may also be revealed. An empirical method for estimating solvation numbers has been developed which is based on the variation of such frequency shifts with system composition. One might remark in this introductory paragraph that the time-scale of vibrational spectroscopy is very different from that of n.m.r. spectroscopy. There is not the fast exchange versus slow exchange distinction for vibrational spectroscopy which is so marked a feature of n.m.r. spectroscopy (see Chapter 2).

3.2.1   Cation–solvent Vibrations

In the vibrational spectra of aqueous solutions of metal salts it is often possible to pick out new broad, low frequency, bands which can be assigned to cation-water vibrations [38]. Thus in aqueous solutions containing magnesium(II) salts, bands appear at 240, 315, and 360–365 cm^{-1} [39]. These bands can be assigned as ν_5, ν_2, and ν_1 by comparison with the reported values of $\nu_5 = 233$, 240, 250, $\nu_2 = 318$, and $\nu_1 = 374$ cm^{-1} for cation-water bands in the spectra of hexa-aqueomagnesium(II) in crystalline salts [40][†], and with spectra of MgO_6 octahedra in glasses [42]. The disappearance of bands when strong complexing agents are added to the solution can also be of help in identification and assignment. For example, the addition of chloride to aqueous solutions containing Zn^{2+} causes the band attributed to ν_1 ($Zn^{2+} - OH_2$) to disappear [43]. The symmetrical stretching frequency ν_1 depends markedly on the size, the charge, and the stereochemistry of the solvated metal cation, as is apparent from the selected values quoted in Table 3.5. On changing the solvent from ordinary water to D_2O the expected frequency shift occurs. For instance, for Mg^{2+} $\nu_1(H_2O) = 360$–365, whereas $\nu_1(D_2O) = 344$, while for Al^{3+} $\nu_1(H_2O) = 520$–526 whereas $\nu_1(D_2O) = 503$ cm^{-1} [38].

† In view of the fact that a recent treatment of the vibrational spectrum of $Ni(OH_2)_6^{2+}$ in its $SnCl_6^{2-}$ salt claims to be the first 'fully-proven assignment for a hexa-aquo-species $M(OH_2)_6^{n+}$' [41], one should exercise caution in assigning aquo-cation solution vibrational spectra by comparison with long-standing assignments of spectra of crystals containing $M(OH_2)_6^{n+}$ cations.

Table 3.5 Dependence of ν_1 (cation-solvent) (cm^{-1}) on the nature of the cation for water [38] and for liquid ammonia ([44] − perchlorate salts; [45] − tetrafluoroboronate, nitrate, or iodide salts).

	Water	Liquid ammonia		Water	Liquid ammonia
Li^+		241–249^a	Be^{2+}	530–543	485
Na^+		194	Mg^{2+}	360–365	328–330
			Ca^{2+}		265–266
			Sr^{2+}		243–250
			Ba^{2+}		215
Ag^+		260–263			

	Water	Liquid ammonia		Water	Liquid ammonia
Mn^{2+}	395		Al^{3+}	520–526	
Fe^{2+}	389		Ga^{3+}	475	477
Ni^{2+}	405		In^{3+}	400	440
Cu^{2+}	440				
Zn^{2+}	385–400	435–440			
Cd^{2+}		342			
Hg^{2+}		415			
Pb^{2+}		315			

a A recent study of lithium salts in liquid ammonia assigned the band at 245 cm^{-1} to N–Li–N deformation, and bands around 360 and 560 cm^{-1} to Li–N stretches [46].

Cation-solvent vibrations can similarly be identified for solutions of ionic compounds in non-aqueous solvents. Here, especially with solvents of low dielectric constant, one has to watch for complications from ion–ion interactions. With solutions of alkali metal salts in liquid ammonia, dimethyl sulphoxide, or pyridine for example, the positions of the new cation-solvent bands are independent of the anion present. But in 'worse' solvents such as tetrahydrofuran, the existence of cation–anion as well as cation–solvent interactions is indicated by the dependence of the frequencies of the vibrational bands associated with the latter on the nature of the anion present. This is illustrated in Table 3.6 [44, 47–49], as is the intermediate case of acetone. Here the onset of ion-pairing is marked by the variable frequencies for the lithium–acetone band in halide solutions in contrast with the almost constant value of 425 cm^{-1} in acetone solutions of the perchlorate, tetraphenylboronate, or thiocyanate salts of lithium. Pyridine provides another example of intermediate behaviour, where the importance of ion-pairing and complex formation has been shown to depend markedly on the nature of the cation [50].

Table 3.6 Dependence and independence of ν(cation–solvent) (cm^{-1}) on the nature of the anion present.

Solvent	Liquid ammonia (good) [44]	Dimethyl sulphoxide (good) [47]	Acetone[a] (poor) [48]	Tetrahydrofuran (bad) [49]
Li[Co(CO)$_4$]				413
LiBPh$_4$			424	412
LiClO$_4$	241	429	425	
LiNCS	240		425	
LiNO$_3$	242	429	420	407
LiI		429	423	373
LiBr		429	412	378
LiCl		429	409	387

a Specifically for ^{7}Li$^+$ salts.

The dependence of cation–solvent frequencies on the nature of the cation for solutions in liquid ammonia is shown in Table 3.5. The dependence of such frequencies on the nature of the solvent and of the cation (alkali metals only) is illustrated by the values in Table 3.7. Values quoted in this Table for 'poor' solvents are derived from observations reported for solutions of salts containing large, and therefore reluctant to pair, anions such as perchlorate or tetra-phenylboronate. The frequencies depend remarkably little on the nature of the solvent, but show the expected decrease in going from the small Li$^+$ to the large Cs$^+$. Force constants corresponding to these cation-solvent interactions have been estimated for Li$^+$, Na$^+$, and K$^+$ in tetrahydrofuran and in dimethyl sulphoxide. Considering only results from solutions in which ion-pairing is thought to be minimal, the respective force constants are 0.71–0.72 mdyn Å^{-1} for Li$^+$, 0.55–0.56 for Na$^+$, and 0.54 for K$^+$ in tetrahydrofuran [49]; for Li$^+$ 0.77 mdyn Å^{-1}, and for Na$^+$ 0.62 in dimethyl sulphoxide [49]. The trends in these force constants, indeed the trends in the cation–solvent frequencies of Table 3.7, do not simply parallel bulk dielectric effects. They need to be considered in terms of a more detailed model, considering both specific cation–solvent interactions and repulsive contributions for solvate units embedded in a continuum of bulk

Table 3.7 Dependence of cation–solvent vibrational frequences (cm^{-1}) on the nature of the cation and of the solvent.

	Liquid ammonia	Dimethyl sulphoxide	Acetone	Sulpholane	Tetrahydrofuran	Propylene carbonate	Pyridine
Li⁺	241[a] [44]	429 [47]	420–425 [48]		412–413 [49]	397 [55]	390[b] [50]
Na⁺	194[a] [44]	198–206 [47] 199–203 [49] 195 [51]	195–196 [48]	186 [53]	198 [49] 175 [54]	186 [55]	175–180 [50, 54]
K⁺		153–154 [47]	148 [48]	155 [53]	142 [49]	144 [55]	133 [54]
Rb⁺		123–129 [47]				115 [55]	
Cs⁺		109–118 [47, 52]				112 [55]	

a Earlier approximate estimates in ref. [46]; b doubtful, there is a pyridine band around 405 cm^{-1}.

solvent with bulk dielectric constant [49]. The specific cation–solvent interactions are not themselves simple. This is because they depend on the basicity of the solvent, including electron-withdrawing or -releasing effects of substituent groups, as well as on steric factors and the mass of the solvent molecule. These latter effects are illustrated by the results for the two series of related solvents, derived from pyridine and from dimethyl sulphoxide as parents, collected in Table 3.8. Here, as in the case of all the other solvents investigated, it is the

Table 3.8   Dependence of Li^+-solvent frequencies (cm^{-1}) on the nature of the solvent for solvent series derived from pyridine and from dimethyl sulphoxide.

Pyridine	390 [50]	Dimethyl sulphoxide	429 [47]
4-Methyl pyridine	387–391 [50]	Dipropyl sulphoxide	416–421 [52]
3,4-Dimethyl pyridine	383 [50]	Dibutyl sulphoxide	425–426 [52]
2,4-Dimethyl pyridine	362 [50]		
2-Chloro pyridine	355 [50]		

lithium cation whose cation–solvent frequencies show the largest variation (amongst alkali metal caions, that is) with solvent nature. While it is possible to state the various factors which affect cation–solvent frequencies, it does not yet seem possible to present a full and coherent picture of the factors which determine them.

Table 3.9   Solvent isotope effects on ν(cation–solvent) (cm^{-1}); from [48].

	Acetone	Acetone-d_6	DMSO	DMSO-d_6
$^7Li^+$	420–425	388–390	429	421
Na^+	192–196	186–192		

A more subtle modification of the solvent is isotopic substitution. One would expect the replacement of H by D to affect the cation–solvent vibrational frequency, since the mass difference should cause a drop in this frequency. There is indeed a marked difference in frequency between acetone and acetone-d_6 (Table 3.9), and between dimethyl sulphoxide and its hexadeuterated derivative (Table 3.9). Just as isotopic substitution in the solvent molecule affects cation–solvent frequencies, so does isotopic substitution of the cation. However only for ^6Li and ^7Li is the difference big enough to be detected easily (Table 3.10).

Table 3.10 Cation isotope effects on ν(cation–solvent) (cm^{-1}); from [48], [50].

	$^6\text{Li}^+$	$^7\text{Li}^+$
Acetone	434–436	420–425
4-Methylpyridine	416	387–391
2-Chloropyridine	373	355

3.2.2   Perturbation of Solvent Vibrations by Cations

Water. One would expect cations to have the biggest effect on ν_2, and anions on ν_1, because cations interact through the oxygen and anions via the hydrogen (Fig. 3.6). The effect of cations on ν_2 decreases thus

$$Ce^{4+} > Al^{3+} > Ba^{2+} > Mg^{2+} \sim Li^+ > Na^+ \sim K^+ \sim Rb^+ \sim Cs^+ .$$

This order is approximately consistent with the usual combination of cation charge and radius effects [38, 56]. Presumably interactions at the hydrogen end of the coordinated water molecule are also of some significance, as these primary-secondary hydration interactions are one aspect of the structure -making and -breaking effects of cations. However, to get an idea of the effects of cations on the structure of water it is better to monitor the intermolecular libration bands in the 450–900 cm^{-1} region. The frequencies of these bands do reflect the effects of added cations on solvent structure. Electrolytes, and thence ions, can be classified as structure-formers or structure-breakers on the basis of their effects on the halfwidths of infrared bands. Ions in the latter category decrease bandwidths, as does a rise in temperature [57].

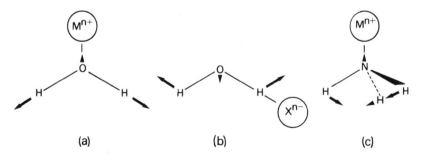

Fig. 3.6 The relation of modes of vibration of solvent molecules to their co-ordination to cations (a) ν_1 for water, (b) ν_2 for water, and (c) δ_s for ammonia.

Ammonia. Cation effects on the vibrational frequencies of ammonia are small, except for the symmetric bending vibration (Fig. 3.6(c) shows this). The frequency of this band, which is 1046 cm^{-1} in liquid ammonia, increases by up to about 300 cm^{-1} on coordination of the ammonia to a metal cation [44–46].

Acetonitrile. The coordination of a cation to acetonitrile affects both the C—C and C≡N frequencies, particularly the latter (Table 3.11) [58, 59]. Again, the relative shifts vary as expected with the charges and sizes of the cations, though the direction of the shift for the C≡N band is in the opposite direction from that expected by the authors [58]! In fact inspection of ν(C≡N) shifts for some mixed ligand complexes containing coordinated acetonitrile shows that the magnitude and even the direction of these shifts is strongly affected by the nature and oxidation state of the metal cation. For example, ν(C≡N) for Ru(NH$_3$)$_5$(MeCN)$^{3+}$ is 2286 cm^{-1}, but for Ru(NH$_3$)$_5$(MeCN)$^{2+}$ it is only 2239 cm^{-1}; ν(C≡N) in acetonitrile itself is 2254 cm^{-1}. Such variations can be understood in terms of a balance between σ and π bonding between the metal cation and the coordinated acetonitrile, and the consequent effect on the C≡N bond in the latter.

Table 3.11   The effects of cation coordination on the C—C and C≡N stretching frequencies (cm^{-1}) of acetonitrile.

	ν(C—C)	ν(C≡N)	
	[58]	[58]	[59]
Li$^+$	+14 to 15	+22 to 23	+21
Na$^+$	+7 to 8	+13 to 14	+10
K$^+$	+6	+10	
Rb$^+$	+5	+ 9	
Cs$^+$		+ 9	
Mg^{2+}	+ 20	+38	+36
Co^{2+}	+ 18	+35	

Acetone. Again, as for acetonitrile, coordination of a metal ion affects both C=O and skeletal C—C—C vibrational frequencies. The latter is shifted by +8, +16, and +24 cm^{-1} by Na$^+$, Li$^+$, and Mg^{2+} respectively [60].

3.2.3   Solvent Coordination Site

Vibrational spectroscopy has been used to deduce the site of attachment of a coordinated solvent molecule to a cation for several solvents. For example, it has been used to decide between O- and S- coordination for dimethyl sulphoxide, and between O- and N- coordination for amides. The argument is based upon the changes observed in appropriate regions of the infrared (Raman) spectrum of the solvent molecule on addition of various salts.

The carbonyl stretching band of dimethylformamide occurs at 1686 cm^{-1} both in neat dimethylformamide and in dimethylformamide+dioxan mixtures. When lithium perchlorate is added to such mixtures, a new band appears at 1670 cm^{-1} whose intensity increases as the Li$^+$:DMF ratio increases [61]. This suggests that the Li$^+$ cation is interacting with the oxygen of the dimethylformamide. Such a conclusion is consistent with parallel n.m.r. evidence (Chapter 2.4.5), with the expected charge distribution within the dimethylformamide molecule (Fig. 3.7), and with a ^{13}C n.m.r. demonstration that protonation of this molecule takes place predominantly at the oxygen atom [62]. Similarly there is infrared spectroscopic evidence for dimethyl sulphoxide coordinating, at least to 'hard' cations, through oxygen [47, 63]. On the basis of appropriate infrared spectra, it has been postulated that dimethylacetamide solvates metal cations via its oxygen atom [64, 65]. This arrangement is preferable on steric grounds to bonding through the nitrogen.

3.2.4 Estimation of Solvation Numbers

In a few cases it has proved possible to estimate from solvent vibrational bands a solvation number for a cation in mixtures consisting of a strongly and a weakly coordinating solvent. This is a purely empirical exercise that bears a strong resemblance to that described in Chapter 2.4.4 for estimating solvation numbers for cations in similar solvent mixtures from n.m.r. shift dependences

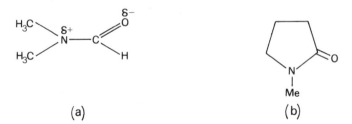

(a)   (b)

Fig. 3.7   (a) NN–Dimethylformamide; (b) 1–methyl–2–pyrrolidinone.

on cation:solvent ratios. The most fully described infrared example is that of Li$^+$ in 1-methyl-2-pyrrolidinone+dioxan mixtures. The heterocyclic ketone's (see diagram (b) of Fig. 3.7) carbonyl stretching frequency varies with the composition of the system as shown in Fig. 3.8. The marked break in the plot at a Li$^+$:1-methyl-2-pyrrolidinone mole ratio of just over 4 is suggested to reflect a solvation number of four for the majority of Li$^+$ cations in this system. Happily this conclusion is in good agreement with the mean solvation number of 4.3 determined for this system by the analogous n.m.r. shift method. Not only is the agreement satisfactory, but the actual value of about four seems chemically reasonable as well [66]. Similar methods lead to solvation numbers of about six for Li^+, Na^+, K^+, and Ag^+ with respect to propylene carbonate, in mixtures of this solvent with nitromethane [67], and of four for Li^+ with respect to acetone, where this solvent is again mixed with nitromethane [68].

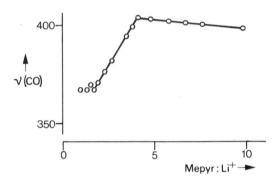

Fig. 3.8 The dependence of carbonyl stretching frequency ($\nu(CO)/cm^{-1}$) of 1–methyl–2–pyrrolidinone (Fig. 3.7(b) Mepyr) on the mole ratio Mepyr:Li^+ in solvent mixtures consisting of Mepyr and dioxan [66].

Solvation numbers may also be guessed directly from the relative intensities of the bulk and coordinated solvent bands. Once more the parallel with the n.m.r. spectroscopy peak area method is close, though the infrared technique is less well suited to precise estimation. From spectra of solutions of silver salts in acetonitrile, a solvation number of four for Ag^+ in this solvent has been obtained [69]. Similarly, solvation numbers for Li^+, Na^+, and Mg^{2+} of four, four, and six respectively in acetonitrile and in acetone have been estimated [59, 60]. In a rather different method, this time analogous to the ultra-violet method for guessing solvation numbers (Chapter 3.1.3), observed Raman spectra for solvated cations are compared with spectra for likely analogues of known structure. The Raman spectrum of Fe^{3+}aq is consistent with an octahedral arrangement of six water molecules around the metal ion [70], while that of Zn^{2+}aq is very similar to the spectra of solids known to contain $Zn(OH_2)_6^{2+}$ [43]. The Raman spectrum of UO_2^{2+}aq is consistent with the presence of six water molecules co-ordinated in the plane perpendicular to the $O-U-O$ axis [71].

3.2.5 Ion–ion Interactions

In Chapter 3.2.1 (Table 3.6) it was indicated that cation–solvent vibrational frequencies can be affected by ion-pair formation, particularly in poor solvents. While information on ion–ion interactions can be obtained in this way, a more fruitful approach has been the inspection of changes in the vibrational spectra of polyatomic anions consequent on ion-pair formation. The nitrate ion has proved particularly popular in this respect. This area of research is outside the scope of the present text; an authoritative account will be found in reference [38].

3.3   ELECTRON SPIN RESONANCE SPECTROSCOPY

The great usefulness of n.m.r. spectroscopy in investigating cation solvation is apparent in Chapter 2. However one significant disadvantage of n.m.r. spectroscopy is the need to use fairly strong solutions in most applications. To avoid complex formation and ion-pairing it is necessary to work with such salts as perchlorates and tetraphenylboronates, rather than with the usually more accessible, and often more soluble, salts such as nitrates and halides. Even if one does keep to non-complexing and non-ion-pairing anions, the relatively strong solutions are still well removed from the idealised model of ionic solvation depicted in Fig. 1.2, where each cation is within its own primary and secondary solvation shells and separated by bulk solvent from the next nearest solvated ions. The technique of e.s.r. spectroscopy is very much more sensitive than n.m.r. spectroscopy, and could in principle be used for very dilute solutions. At first sight it seems to promise a useful and sensitive probe for cation–solvent interactions, but closer inspection reveals a number of formidable obstacles.

The first and most obvious restriction is to paramagnetic ions. There are further restrictions on its usefulness, imposed by the need for long relaxation times for only slow relaxation results in signals sufficiently narrow for observation. The spin-lattice relaxation time T_1 is sensitive to the presence of low-lying electronic excited states. The larger the gap, the more difficult is relaxation, and so high-spin d^5 cations, for example Mn^{2+}, are very suitable. This is because there is no higher sextet state accessible above the ^6S ($^6A_1$) ground state for a symmetrical octahedral environment. Cations with a d^3 configuration, for instance V^{2+} or Cr^{3+}, also seem suitable. On the other hand many other cations, among them $V^{2+}(d^3)$, $Co^{2+}(d^7)$, and many lanthanide cations, relax too easily and give poor or unobservable e.s.r. signals. N.m.r. and e.s.r. spectroscopy are complementary from this point of view, for those cations with relaxation properties precluding the observation of satisfactory e.s.r. spectra are just the ones in whose presence serviceable or good n.m.r. spectra can be obtained. Thus Mn^{2+} can give excellent e.s.r. signals, but broadens solvent ^1H or ^{17}O resonances beyond the limits of detection, whereas Co^{2+} gives no useful e.s.r. signal, but allows n.m.r. observations of nuclei in adjacent solvent molecules (see Chapter 2).

The symmetry of the species being investigated is important in determining relaxation times and thus linewidths. Symmetry is not desirable in this context, and the usual octahedral hexa-solvento-cation arrangement is unsuited to monitoring by e.s.r. spectroscopy. Only in favourable cases such as $Mn^{3+}aq$ does one obtain a good e.s.r. spectrum (Fig. 3.9) with six lines, because ^{55}Mn has a nuclear spin of 5/2, and a hint of fine structure corresponding to the various M_s levels. However, the highly symmetrical environment, both in the geometric and in the electronic sense, results in both A and g being relatively insensitive to environmental change. The environmental change which does have a significant effect is that of going from an octahedral to a tetrahedral geometry. Then the ^{55}Mn nuclear hyperfine coupling constant A is reduced from 93–98 G for octahedral complexes to 75–79 G for tetrahedral complexes [72]. This change in geometry is a much bigger one than those we are interested in when considering solvated cations, however. From the relative values of $g_\perp$ and $g_\parallel$ it is possible to tell, at least in the solid state, whether the Jahn-Teller distortion of the $Cu(OH_2)_6^{2+}$ cation involves tetragonal elongation or compression.

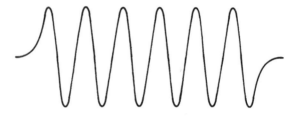

Fig. 3.9   The e.s.r. spectrum of $Mn(OH_2)_6^{2+}$.

The case of Fe^{3+}aq is complicated by hydrolysis to $FeOH^{2+}$aq as well as by the low natural abundance of ^{57}Fe. The spectrum of Cu^{2+}aq is not simple to interpret, though the $g_\perp$ and $g_\parallel$ values of 2.397 and 2.083 respectively have been used to estimate the energies ΔE_1 and ΔE_2 of Fig. 3.10, and thence the degree of covalent bonding involved [73]. Interpretation of the e.s.r. spectrum of Gd^{3+}aq is also complicated, this time by the uncertainity in the hydration number of Gd^{3+}.

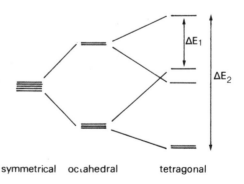

symmetrical   ocιahedral   tetragonal

Fig. 3.10   The splitting of the d levels in octahedral and in tetragonal (e.g. Cu^{2+}aq) ligand fields.

Very few spectra have been reported for cations in non-aqueous media. A study of Mn^{2+} in methanol yielded results for complexes $Mn(MeOH)_5X^+$. The spectrum of $Mn(MeCN)_6^{2+}$ was observed incidentally to an investigation of MnX_4^{2-} complexes. In $Mn(py)_6^{2+}$ the coordinated nitrogens lead to the collapse to a single broad line of the six-line spectrum of the type shown in Fig. 3.9. Linewidths were found to increase with increasing solvent viscosity for solutions of Mn^{2+} in a series of nine non-aqueous solvents. For aqueous methanol and aqueous acetonitrile plots of linewidths against mole fraction composition showed maxima, even after allowing for viscosity variation. These maxima are ascribed to mixed solvates [74].

As soon as the symmetry of the observed species is reduced, the chances of observing a good e.s.r. spectrum increase, but the relevance to the present book decreases. Hexa-solvento-titanium(III) species in water or in methanol give no observable signal at room temperature, but hydrolysed species such as $Ti(OH_2)_4(OH)_2^+$ [75] and complexes such as $Ti(OH_2)_4Cl_2^+$ and $Ti(MeOH)_4Cl_2^+$ [76] give satisfactory spectra. E.s.r. should therefore be of use in the study of mixed solvento-complexes of cations in mixed solvents. So far only Cu^{2+} in aqueous ammonia [77] and Mn^{2+} in aqueous methanol, acetonitrile, or dimethylformamide [78] appear to have been studied in this way.

Perhaps the most promising source of cation–solvent interaction information derives from mixed solvent–ligand complexes of the $VO(acac)_2$ and $Cu(acac)_2$ type. For the former complexes there are the difficulties that certain solvents may interact with the vanadyl–oxygen rather than, or as well as, with the vacant coordination site on the vanadium. Plots of A_{iso} and of g_{av} against solvent E_T values have helped in sorting out this problem. There is also the possibility of $VO(LL)_2$ complexes interacting with solvents to give species of the type shown in Fig. 3.11(a) rather than the expected six-coordinate product [79]. In the case of $Cu(acac)_2$, potentially hydrogen-bonding solvents are thought to interact in the manner shown in Fig. 3.11(b).

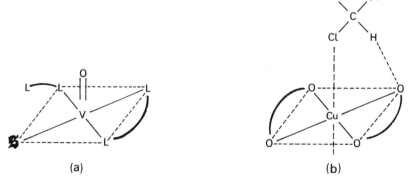

(a)   (b)

Fig. 3.11   (a) Suggested five-coordinate structure for solvated $VO(LL)_2$ complexes; (b) complex-solvent interaction for β-diketone complexes of copper(II) in potentially hydrogen-bonding solvents which can also interact with the copper.

Thus the technique of e.s.r. spectroscopy has so far proved of disappointingly little value in the probing of the nature and extent of cation-solvent bonding. The nearest area in which it has proved of significant value is that of ion-pairs between alkali metal and alkaline earth cations and organic anions. Here the predominant interest is always in the organic moiety, but the e.s.r. spectra do give some information about the whereabouts and kinetic characteristics of the cation in certain instances [80].

3.4 MÖSSBAUER SPECTROSCOPY

This spectroscopic technique has been very little used for the examination of electrolyte solutions. A study of europium(III) in aqueous chloride media was interpreted in terms of aquo-chloro-complexes [81]. Solvation of iron(II) in water and in some mixed solvents has been studied indirectly via Mössbauer spectra of rapidly frozen samples. Rapid freezing of aqueous solutions gave two phases, one of which had a composition implying a hydration number of 18 for Fe^{2+} [82]. The isomer shifts for solvento-iron(II) species derived from a range of organic solvents depend on solvent donicities. Iron(II) seems to be preferentially solvated by methanol in methanol-dimethylformamide mixture, while iron(II) in hexamethylphosphoramide-water mixtures prefers to exist as $Fe^{2+}aq$ and $Fe(hmpa)_n^{2+}$ rather than in the form of mixed solvates [83].

3.5 NEUTRON INELASTIC SCATTERING SPECTROSCOPY

Some preliminary work on electrolyte solutions suggests that this technique may yield information relevant to ion-solvent interactions and to the effects of ions on solvent structure. These initial experiments were carried out with aqueous solutions of lithium chloride and of potassium iodide [84].

FURTHER READING

Ultraviolet-visible spectroscopy

Hartmann, H. and Schläfer, H. L., Angew. Chem., 66, 768 (1954).
Taube, H., in Progress in Stereochemistry, Volume 3, ed. de la Mare, P.B.D., and Klyne, W., Butterworths (1962).
Jørgenson, C. K., Absorption Spectra and Chemical Bonding, Pergamon (1962).
Jørgenson, C. K., Adv. Chem. Phys., 5, 33 (1963).
König, E., Structure and Bonding, 9, 175 (1971).

Infrared and Raman spectroscopy

Irish, D. E., in Ionic Interactions, Volume 2, ed. Petrucci, S., Academic Press, (1971) Ch. 9.
Verrall, R. E. in Water: A Comprehensive Treatise – Volume 3: Aqueous Solutions of Simple Electrolytes, ed. Franks, F., Plenum Press (1973) Ch. 5.

Electron spin resonance spectroscopy

Lewis, W. B. and Morgan, L. O., in *Transition Metal Chemistry, Volume 4*, ed.
Carlin, R. L., Edward Arnold (1968).
Goodman, B. A. and Raynor, J. D., in *Advances in Inorganic Chemistry and
Radiochemistry, Volume 13*, ed. Emeléus, H. J. and Sharpe, A. G.,
Academic Press (1970).

REFERENCES

[1] Jørgenson, C. K., *Inorganic Complexes*, Academic Press (1963).
[2] E.g. Hill, H. A. O. and Day, P., *Physical Methods in Advanced Inorganic
Chemistry*, Wiley/Interscience (1968); Jørgenson, C. K., *Molec. Phys.*, 2,
96 (1959).
[3] Bagnall, K. W., *The Actinide Elements*, Elsevier (1972).
[4] Foos, J., Kertes, A. S., and Peleg, M., *J. Inorg. Nucl. Chem.*, 36, 837 (1974);
Chrysochoos, J., *J. Chem. Phys.*, 60, 1110 (1974).
[5] Figgis, B. N., *Introduction to Ligand Fields*, Wiley/Interscience (1966).
[6] Liehr, A. D. and Ballhausen, C. J., *Ann. Phys.*, 3, 304 (1958).
[7] Fackler, J. P. and Holah, D. G., *Inorg. Chem.*, 4, 954 (1965).
[8] Fackler, J. P. and Chawla, I. D., *Inorg. Chem.*, 3, 1130 (1964).
[9] Dingle, R., *Acta Chem. Scand.*, 20, 33 (1966).
[10] Sasaki, Y. and Sykes, A. G., *J. Chem. Soc. Dalton*, 1048 (1975).
[11] Gutmann, V., Beran, R., and Kerber, W., *Mh. Chem.*, 103, 764 (1972).
[12] Meek, D. W., Drago, R. S., and Piper, T. S., *Inorg. Chem.*, 1, 285 (1962).
[13] Gutmann, V. and Bohunovsky, O., *Mh. Chem.*, 99, 740 (1968).
[14] Schläfer, H. L. and Opitz, H. P., *Z. Elektrochem.*, 65, 372 (1961).
[15] Abu-Eittah, R. and Arafa, G., *J. Inorg. Nucl. Chem.*, 32, 2721 (1970).
[16] Jørgenson, C. K., *Adv. Chem. Phys.*, 5, 33 (1963).
[17] Waddington, T. C., *Non-aqueous Solvents*, Nelson (1969) p. 5.
[18] Drago, R. S., Meek, D. W., Joesten, M. D., and LaRoche, L., *Inorg. Chem.*,
2, 124 (1963).
[19] Gutmann, V. and Bardy, H., *Mh. Chem.*, 99, 763 (1968).
[20] Drago, R. S., Meek, D. W., Longhi, R., and Joesten, M. D., *Inorg. Chem.*, 2,
1056 (1963).
[21] Gutmann, V. and Melcher, G., *Mh. Chem.*, 103, 624 (1972).
[22] Chhonkar, N. S., *Z. Phys. Chem.*, 250, 290 (1972).
[23] Hartmann, H., Schläfer, H. L. and Hansen, K. H., *Z. Anorg. Allg. Chem.*,
284, 153 (1956).
[24] Hartmann, H. and Schläfer, H. L., *Z. Naturf.*, 6A, 754 (1951).
[25] Racah, C., *Phys. Rev.*, 61, 186 (1942); 62, 438 (1942); 63, 367 (1943).
[26] Kettle, S. F. A., *Coordination Compounds*, Nelson (1961) pp. 102, 107.
[27] Van Ingen Schenau, A. D., Groeneveld, W. L., and Reedijk, J., *Rec. Trav.
Chim.*, 91, 88 (1972).

[28] Taube, H. in *Progress in Stereochemistry, Volume 3*, ed. de la Mare, P. B. D. and Klyne, W., Butterworths (1962) p. 95; Taube H. *Electron Transfer Reactions of Complex Ions in Solution*, Academic Press (1970) Ch. 1; Lincoln, S. F., *Coord. Chem. Rev.*, **6**, 309 (1971).

[29] Schmitz-Dumont, O., Brokopf, H. and Burkhardt, K., *Z. Anorg. Allg. Chem.*, **295**, 7 (1958).

[30] Schmitz-Dumont, O., Gössling, H, and Brokopf, H., *Z. Anorg. Allg. Chem.*, **300**, 159 (1959).

[31] Rasmussen, L. and Jørgenson, C. K., *Acta Chem. Scand.*, **22**, 2313 (1968).

[32] Chernova, R. K., Sukhova, L. K. and Efimova, T. N., *Russ. J. Inorg. Chem.*, **19**, 677 (1974).

[33] Rao, C. N. R., Rao, K. G., and Reddy, N. V. R., *J. Amer. Chem. Soc.*, **97**, 2918 (1975).

[34] E.g. Becker, R. S., *Theory and Interpretation of Fluorescence and Phosphorescence*, Wiley/Interscience (1969) Ch. 13.

[35] Tachin, V. S., Ermolaev, V. L. and Bodunov, E. N., *Russ. J. Inorg. Chem.*, **20**, 189 (1975).

[36] Mayne, P. J. and Kirkbright, G. P., *J. Inorg. Nucl. Chem.*, **37**, 1527 (1975); Sill, C. W. and Peterson, H. E., *Analyt. Chem.*, **21**, 1266 (1949).

[37] Burrows, H. D., Formosinho, S. J., Miguel, M. da G. and Coelho, F. P., *J. Chem. Soc. Faraday I*, **72**, 163 (1976).

[38] Irish, D. E. in *Ionic Interactions, Volume 2*, ed. Petrucci, S., Academic Press (1971) Ch. 9.

[39] Da Silveira, A., *C. R. Hebd. Séanc. Acad. Sci., Paris*, **194**, 1336 (1932); Da Silveira, A., Marques, M. A., and Marques, N. M., *ibid.*, **252**, 3983 (1961); *Molec. Phys.*, **9**, 271 (1965).

[40] Mathieu, J. –P., *C. R. Hebd. Séanc. Acad. Sci., Paris*, **231**, 896 (1950); Hillaire, P., Abenoza, M. and Lafont, R., *ibid.*, **273B**, 225 (1971).

[41] Adams, D. M. and Lock, P. J., *J. Chem. Soc. A*, 2801 (1971); Adams, D. M. and Trumble, W. R., *Inorg. Chim. Acta*, **10**, 235 (1974).

[42] Rao, C. N. R., Bhujle, V. V., Goel, A., Bhat, U. R. and Paul, A., *J. Chem. Soc. Chem. Commun.*, 161 (1973).

[43] Irish, D. E., NcCarroll, B. and Young, T. F., *J. Chem. Phys.*, **39**, 3436 (1963).

[44] Plowman, K. R. and Lagowski, J. J., *J. Phys. Chem.*, **78**, 143 (1974).

[45] Gans. P. and Gill, H. B., *J. Chem. Soc. Dalton*, 779 (1976).

[46] Gardiner, D. J., Hester, R. E. and Grossman, W. E. L., *J. Chem. Phys.*, **59**, 175 (1973).

[47] Maxey, B. W. and Popov, A. I., *J. Amer. Chem. Soc.*, **89**, 2230 (1967).

[48] Ming Keong Wong, McKinney, W. J., and Popov, A. I., *J.Phys. Chem.*, **75**, 56 (1971).

[49] Edgell, W. F., Lyford, J., Wright, R., Risen, W., and Watts, A., *J. Amer. Chem. Soc.*, **92**, 2240 (1970).

[50] Handy, P. R. and Popov, A. I., *Spectrochim. Acta*, **28A**, 1545 (1972).

[51] Wuepper, J. L. and Popov, A. I., *J. Amer. Chem. Soc.*, **92**, 1493 (1970).

[52] Maxey, B. W. and Popov, A. I., *J. Amer. Chem. Soc.*, **91**, 20 (1969).

[53] Buxton, T. L. and Caruso, J. A., *J. Phys. Chem.*, **77**, 1882 (1973).

[54] French, M. J. and Wood, J. L., *J. Chem. Phys.*, **49**, 2358 (1968).

[55] Greenberg, M. S., Wied, D. M., and Popov, A. I., *Spectrochim. Acta*, **29A**, 1927 (1973).

[56] Nightingale, E. R. in *Chemical Physics of Ionic Solutions*, ed. Conway, B. E. and Barradas, R. G., Wiley/Interscience (1966) pp. 87–100; Holinski, R. and Brehler, B., *Z. Anorg. Allg. Chem.*, **406**, 69 (1974).

[57] Taniewska-Osinska, S. and Grochowski, R., *Russ. J. Phys. Chem.*, **48**, 1280 (1974).

[58] Coetzee, J. F. and Sharpe, W. R., *J. Solution Chem.*, **1**, 77 (1972).

[59] Perelygin, I. S. and Klimchuk, M. A., *Russ. J. Phys. Chem.*, **47**, 1138, 1402 (1973); **48**, 363, 1466 (1974).

[60] Perelygin, I. S. and Klimchuk, M. A., *Russ. J. Phys. Chem.*, **49**, 76 (1975).

[61] Lassigne, C. and Baine, P., *J, Phys. Chem.*, **75**, 3188 (1971).

[62] McClelland, R. A. and Reynolds, W. F., *J. Chem. Soc. Chem. Commun.*, 824 (1974).

[63] Maxey, B. W. and Popov, A. I., *J. Amer. Chem. Soc.*, **90**, 4470 (1968).

[64] Drago, R. S., Carlson, R. L., and Purcell, K. F., *Inorg. Chem.*, **4**, 15 (1965).

[65] Bull, W. E., Madan, S. K., and Willis, J. E., *Inorg. Chem.*, **2**, 303 (1963); Bello, J. and Bello, H. R., *Nature*, **190**, 440 (1961); **194**, 681 (1962).

[66] Wuepper, J. L. and Popov, A. I., *J. Amer. Chem. Soc.*, **91**, 4352 (1969).

[67] Yeager, H. L., Fedyk, J. D. and Parker, R. J., *J. Phys. Chem.*, **77**, 2407 (1973).

[68] Baum, R. G. and Popov, A. I., *J. Solution Chem.*, **4**, 441 (1975).

[69] Chang, T. -C. G. and Irish, D. E., *J. Solution Chem.*, **3**, 161 (1974).

[70] Sharma, S. K., *J. Inorg. Nucl. Chem.*, **35**, 3831 (1973).

[71] Sutton, J., *Nature*, **169**, 235 (1952).

[72] Chan, S. I., Fung, B. M. and Lütje, H., *J. Chem. Phys.*, **47**, 2121 (1967); Lynds, L., Crawford, J. E., Lynden-Bell, R. M., and Chan, S. I., *ibid.*, **57**, 5216 (1972).

[73] McLauchlan, K. A., *Magnetic Resonance*, Oxford University Press (1972) p. 41.

[74] Stockhausen, M., *Ber. Bunsenges. Phys. Chem.*, **77**, 338 (1973).

[75] Premović, P. I. and West, P. R., *Can. J. Chem.*, **52**, 2919 (1974).

[76] Goldberg, I. B. and Goeppinger, W. F., *Inorg. Chem.*, **11**, 3129 (1972).

[77] Tikhomirova, N. N., Zamaraev, K. I. and Berdnikov, V. M., *Russ. J. Struct. Chem.*, **4**, 407 (1963).

[78] Burlamacchi, L., Martini, G. and Romanelli, M., *J. Chem. Phys.*, **59**, 3008 (1973).

[79] Miller, G. A. and McClung, R. E. D., *Inorg. Chem.*, **12**, 2552 (1973).

[80] Basolo, F. and Pearson, R. G., *Mechanisms of Inorganic Reactions, 2nd Ed.*, Wiley/Interscience (1967) p. 489; Garst, J. F. in *Solute-Solvent Interactions* ed. Coetzee, J. F. and Ritchie, C. D., Marcel Dekker (1969) Ch. 8; Sorenson, S. P. and Bruning, W. H., *J. Amer. Chem. Soc.*, **95**, 2445 (1973); and refs. therein.

[81] Greenwood, N. N., Turner, G. E., and Vertes, A., *Inorg. & Nucl. Chem. Letters*, **7**, 389 (1971).

[82] Fröhlich, K. and Keszthelyi. L., *J. Chem. Phys.*, **58**, 4614 (1973).

[83] Vértes, A., Pálfalvy, M., Burger, K. and Molnár, B., *J. Inorg. Nucl. Chem.*, **35**, 691 (1973).

[84] White, J. W., *Ber. Bunsenges. Phys. Chem.*, **75**, 379 (1971).

Chapter **4**

METAL ION SOLVATION:
NON-SPECTROSCOPIC METHODS OF INVESTIGATION

4.1   INTRODUCTION

A variety of methods for examining ion solvation are described in this Chapter. First come the electrochemical methods based on the movement of solvated ions through the solution. Here transference numbers and conductivities can be measured with considerable precision. However a number of assumptions, some of a doubtful nature, have to be introduced before sizes of solvated ions and thence solvation numbers can be estimated. Several other approaches to ion solvation involve making deductions from such thermodynamic parameters as entropies, compressibilities, and heat capacities. Once again this entails making assumptions of considerable magnitude before solvation numbers can be derived. Indeed it usually seems safer to work in the opposite direction here, using knowledge of ion solvation derived from other sources to explain values and trends in thermodynamic parameters. The application of diffraction methods to the problem of ion solvation is limited, but does give one quantity of fundamental importance. This is the distance between an ion and its nearest neighbour solvent molecules.

The last section of this Chapter is devoted to miscellaneous methods outside the classes covered by Chapters 2 and 3 and the earlier sections of the present Chapter. Of these methods isotopic dilution provides an important direct route to solvation numbers for a few cations, but all the other methods dealt with here are more or less indirect and of limited value. In some cases there seems little justification for deriving solvation numbers from the experimental data. Most methods of estimating solvation numbers are mentioned in Chapters 2, 3 and 4, but those of limited value are dealt with only briefly. More details can be found elsewhere [1].

When solvation numbers estimated in the various sections of these three Chapters are compared, large disagreements for a given cation in a given solvent are often revealed. Sometimes this results from the unreliability of the assumptions used in the various methods of estimation, sometimes it is a reflection of the fact that different techniques actually measure different solvation numbers (see the various solvation regions depicted in Fig. 1.2). This latter question will be taken up in Chapter 5.

4.2   TRANSPORT PROPERTIES

4.2.1   Introduction

All the transport processes discussed in this section depend on the ease with which solvated ions can move through the solvent. All therefore depend on the effective size of the solvated ion, and are thus potential sources of information on the extent of solvation of cations.

Conductances, mobilities, and transference numbers. These three parameters are interrelated. It is a straightforward task to measure with high precision the conductivity of a solution of an electrolyte in aqueous, non-aqueous, or mixed aqueous media. This can be done even at low concentrations (c) of electrolyte, so that extrapolations to conductivities at infinite dilution (Λ^0) are short and can be made satisfactorily by using simple equations of the type

$$\Lambda = \Lambda^0 - A\sqrt{c}.$$

Because conductivities estimated for conditions of infinite dilution are used, this obviously rules out all worries about interference from ion-pairing.

The derived Λ^0 values for salts can be split into internally consistent sets of single ion values λ_+^0 and λ_-^0 (Kohlrausch's Law of the independent migration of ions, 1876). The problem here is to choose the most satisfactory procedure in the cation–anion split. For aqueous solutions the usual practice is to measure transference numbers, t_+ and t_-, by the Hittorf, moving boundary, or concentration cell with liquid junction methods [2]. Recently a quick and fairly good method for estimating transference numbers by paper electromigration has been described [3]. Once t_+ and t_- have been determined, then λ_+^0 and λ_-^0 follow from the expressions $\lambda_+^0 = t_+^0 \Lambda^0$ and $\lambda_-^0 = t_-^0 \Lambda^0$. An alternative approach is sometimes used for aqueous solutions and usually used for non-aqueous solutions. This is to split the conductivity for a solution of a salt consisting of a large cation and a large anion, for example tetraisoamylammonium tetraphenylboronate, into equal contributions from each ion, in other words to assume $t_+^0 = t_-^0 = 0.50$ (or $\lambda_+^0 = \lambda_-^0 = 0.50\Lambda^0$). Sometimes called the **Coplan-Fuoss split**, this method is arbitrary, but intuitively reasonable, and has been shown to be an entirely acceptable procedure in at least some situations [4].

Ionic mobilities (u) are directly related to ionic conductances ($u_\pm = \lambda_\pm/F$). Once the individual solvated ion conductances or mobilities have been estimated, than a qualitative discussion of the relative degrees of solvation of different cations can be undertaken (see below) [5].

To estimate solvation numbers for ions further calculations and assumptions are required. The usual approach is to use Stokes's Law, though because this applies to a particle moving in a hydrodynamic vacuum it is not strictly applicable to a solvated cation moving through a solvent consisting of discrete but connected molecules. Stokes's Law permits the calculation of an effective radius for the solvated cation from its mobility, u, and the solvent viscosity, η:

$$r = 1/(6\pi\eta u)$$

$$\text{or} \quad r = |z| F^2/(6\pi N_A \eta \lambda^0).$$

Thence the number of solvent molecules per ion can be calculated from estimated volumes of the ion itself and of the solvent molecules, allowing where possible for electrostriction effects on solvent molecular volumes and for packing of awkwardly shaped solvent molecules around the central ion. As discussed in Chapter 5, these methods estimate the average number of solvent molecules which move with a cation, rather than the number of solvent molecules bonded to the cation in a primary solvation sphere. Thus they may reflect primary and secondary solvation.

This outline of the obtaining and interrelation of conductances, mobilities, and transference numbers has been kept as brief as possible so as not to obscure the overall pattern. Several much more detailed and comprehensive treatments are available [1, 6]. In the next section we shall present a selection of results for each of these properties, and see what picture of ionic solvation emerges from each.

Viscosity and diffusion. These properties are not directly related to the electrochemical transport properties just discussed. They therefore provide separate and independent estimates of effective sizes of solvated ions, as reflected in their ease of movement through solvents.

4.2.2   Transference Numbers

These form the basis of one of the earlier methods for estimating solvation numbers. The original idea dates from 1900 (Nernst) and the first report of successful application from 1909 (Washburn). There are three ways of determining transference numbers. The original method involves adding an inert reference substance, generally a sugar such as raffinose or α-methyl-D-glucoside, to an electrolyte solution in a Hittorf cell. The movement of solvating solvent with the ions can be monitored by following the changing concentration of the sugar in the vicinity of the electrodes. Hence the cation and anion transference numbers t_+ and t_- can be determined. The complementary transport of solvent is characterised by its Washburn number, w_W. Alternative methods for estimating transference numbers include moving boundary experiments, preferable to the Hittorf method, and e.m.f. measurements [7].

Cation transference numbers reflect the relative extent of solvation. For the alkali metal cations in aqueous solution, transference numbers increase in the order

$$Li^+ < Na^+ < K^+ < Rb^+ \sim Cs^+ \; .$$

Values are $t(Li^+) = 0.34$, $t(Na^+) = 0.40$, $t(K^+) = 0.49$, $t(Rb^+) = 0.50$, and $t(Cs^+) = 0.50$ from the chlorides [8] and $t(Li^+) = 0.32$, $t(Na^+) = 0.38$, $t(K^+) = 0.48$, $t(Rb^+) = 0.51$, and $t(Cs^+) = 0.50$ from the iodides [9]. This trend is consistent with the lowest current bearing capacity for the slowest moving, because most heavily hydrated, $Li^+aq.$† Many more values of t_+ obtained from measurements on solutions of a variety of salts in aqueous solution, and some assorted results for non- and mixed aqueous solutions, can be found in reference [8]. This reference also cites several bibliographies of transference numbers.

Once the amount of water transported by the ions has been established, it only remains to apportion this between cations and anions in order to obtain ion solvation numbers. Some more or less arbitrary assumption is needed at this point. Several groups of workers have estimated cation hydration numbers on the assumption that the hydration number of Cl^- is 4. Thence hydration numbers for the alkali metal cations varying between 4 to 5 for Cs^+ and 13 to 14 for Li^+ were obtained [11]. The same assumption led to estimates of hydration numbers for the alkaline earth cations ranging from 11 for Ba^{2+} to 20 for Mg^{2+} [12]. The most recent, and most extensive, series of cation hydration numbers determined from transference numbers is reproduced in Table 4.1 [13]. It is important to avoid making measurements in strong electrolyte solutions. Apparent hydration numbers drop significantly in concentrated solution, for example from 9 to 5 for Li^+ when the medium is changed from M to 2.8M sulphuric acid [14].

Table 4.1 Cation hydration numbers estimated from transference numbers measured in 0.3N solutions. A hydration number of 5 has been assumed for the chloride ion [13].

Li^+	22	Mg^{2+}	36	Cu^{2+}	34
Na^+	13	Ca^{2+}	29		
K^+	7	Sr^{2+}	29	Zn^{2+}	44
Cs^+	6	Ba^{2+}	28	Cd^{2+}	39

The techniques used for measuring transference numbers in aqueous solution can also be used, with minor modifications, for measuring transference numbers in non-aqueous media. Hence cation solvation numbers may similarly be estimated in such solutions. A few examples are cited in Table 4.2 [15, 16].

† Yet the hydration of Li^+ cannot be all that extensive, for it is possible to achieve some $^6Li^+/^7Li^+$ separation by electromigration techniques [10].

A popular assumption for splitting measured values for salts into ionic components is to take a solvation number of zero for large tetraalkylammonium cations. This method for estimating transference numbers is restricted to 'good' solvents of high dielectric constant, as it cannot be used once ion-pairs or higher aggregates are present in significant concentration. For this reason, it is impossible to determine precise transference numbers, and hence solvation numbers, in acetic acid for example [17].

Table 4.2 Cation solvation numbers estimated from measured transference numbers in non-aqueous media.

Solvent:	N-Methyl acetamide	N-Methyl formamide	Dimethyl formamide
Assumed solvation number:	$R_4N^+ = 0$ [15]	$^nBu_4N^+ = 0$ [16]	$^nBu_4N^+ = 0$ [16]
Li^+	5.1	5.5	5.2
Na^+	3.5	2.6	3.3
K^+	3.3	1.0	2.9
Cs^+	2.6		
Mg^{2+}	10.3		
Ca^{2+}	8.6		
Sr^{2+}	8.6		
Ba^{2+}	9.0		

There have been a few measurements of transference numbers in mixed aqueous solvents. Such determinations have been made in order to probe preferential solvation [18] and solvent structure [7]. In aqueous methanol, transference numbers for alkali metal cations show a small, steady increase as the percentage of methanol rises from 0 to 40%, but the Washburn number for the water shows a maximum between 5 and 10% methanol [7]. Similarly the Washburn number of the water goes through a maximum in dioxan+water solutions. These effects are small for sodium chloride as solute, but marked when sodium iodide is the solute. Some dependence of Washburn number on solvent structure seems indicated [7]. Values of t_+ determined from solutions of potassium bromide in t-butyl alcohol+water mixtures and from solutions of potassium chloride in ethanol+water mixtures increase slightly then decrease as the mole fraction of the organic component (x_2) increases. The maxima occur at $x_2 \cong 0.065$ for t-butyl alcohol and at $x_2 \cong 0.11$ for ethanol. These mole fractions are in the regions where other extrema of physical properties have been reported and correspond to compositions of maximum structure formation in the respective mixed solvents [19].

4.2.3 Ionic Conductances

Aqueous solution. Limiting equivalent conductances, λ_+^0, for aquo-cations are listed in Table 4.3 [20–22]. These values are mutually consistent, based ultimately on a fixed 'best' value for the chloride ion $\lambda_-^0(Cl^-) = 76.35$ cm^2 ohm^{-1} mol^{-1}, and 'best' Λ^0 values from the literature [20]. The dependence of the λ_+^0 values on the effective size of the solvated cations rather than on their bare (crystal) radii is obvious, with the small but well solvated Li$^+$ cation having a smaller λ_+^0 value than the larger Cs$^+$. The pattern is similar for the alkaline earth series of cations. Comparisons between M$^+$, M^{2+}, and M^{3+} cations show that the effect on λ_+^0 of increasing the charge on the cation is more than balanced by the consequent heavier solvation, so that M^{2+}aq and M^{3+}aq cations are less effective current carriers than M$^+$aq. The effect of temperature on λ_+^0 is illustrated by the values in Appendix 6.2 of reference [2].

Table 4.3 **Limiting equivalent conductances (cm^2 ohm^{-1} mol^{-1}) of metal cations in aqueous solution, at 25°C. Values have been taken from reference [20] except where otherwise stated.**

Li$^+$	38.6	Be^{2+}	45	In^{3+}	(62.1) [a]	
Na$^+$	50.1	Mg^{2+}	53.1			
K$^+$	73.5	Ca^{2+}	59.5	La^{3+}	69.7 [b]	
Rb$^+$	77.8	Sr^{2+}	59.5	↓	↓	
Cs$^+$	77.3	Ba^{2+}	63.6	Yb^{3+}	65.6	
		Pb^{2+}	69.5			
Ag$^+$	61.9					
Tl$^+$	74.7					
		Co^{2+}	55			
		Cu^{2+}	53.6			
		Zn^{2+}	52.8			

a ref. [21]; *b* values for twelve cations in this series have been reported [22].

Non-aqueous solutions. Because the measurement of the conductance of a solution is straightforward, the results of a large number of such measurements have been reported. It is even possible to obtain results in 'bad' solvents, for example benzene, by the use of appropriate salts, for example tetraalkyl aluminates such as NaAlR$_4$ [23]. As for aqueous solutions, such conductance results are usually of high precision. Even if there are complications arising from ion-pairing or complex formation, extrapolation to zero concentration can generally be carried out with ease and confidence. The difficulties start when one tries to split measured Λ^0 values for salts into individual λ^0 values for the constituent ions. There are very few transference numbers known for ions in non-aqueous solvents, though they are known for a few cations in methanol, ethanol, and dimethylformamide, and have been guessed in n-propanol. So normally some other method of splitting Λ^0 values into λ^0_+ and λ^0_- has to be used. The most popular is to assume $t_+(R_4N^+) = t_-(BPh_4^-) = 0.50$ (see similar assumptions in Chapter 7.1.2).

A selection of cation λ^0_+ values for a range of non-aqueous solvents is given in Table 4.4 [24-44]. The dependence of λ^0_+ on temperature has been determined for alkali metal cations in N-methylacetamide and in dimethoxyethane [45]. Observations show the expected increase in λ^0_+ as the temperature is increased. Conductivities are affected by such solvent properties as viscosity and solvent structure as well as by the degree of solvation of the ions. Thus the very low λ^0_+ values for cations in sulphuric acid may be attributed largely to that solvent's high viscosity, with the current almost entirely carried via HSO_4^- jumps [44].

Trends of values in Table 4.4 suggest that in general λ^0_+ values for a series of cations of the same charge increase steadily as the ionic radii increase. There are only a few exceptions to this generalisation in Table 4.4. The trend is explained in terms of decreasing solvation with increasing cation radius, here as in aqueous systems (see above). Obviously such a trend cannot continue indefinitely, as eventually the increasing size of bare cations will start to have an effect, leading to a trend of decreasing λ^0_+. In fact metal cation radii are never big enough for the turnover point to be reached, but if tetraalkylammonium cations are included in a series then the full plot of the dependence of λ^0_+ on cation radius does show an initial increase followed by an eventual decrease. This is illustrated for M^+ cations in propylene carbonate [36, 46] in Fig. 4.1. A similar picture applies in several other solvents, for instance liquid sulphur dioxide [25].

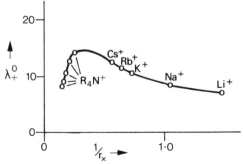

Fig. 4.1 The variation of ionic conductance, λ^0_+, on the reciprocal of the cation crystal radius, $1/r_X$, in propylene carbonate [36].

Table 4.4 Values of λ_+^0 ($\text{cm}^2\,\text{ohm}^{-1}\,\text{mol}^{-1}$) for cations in non-aqueous solvents, at 25°C unless otherwise stated[a].

Solvent	Li^+	Na^+	K^+	Rb^+	Cs^+	Ag^+	Tl^+	Mg^{2+}	Ca^{2+}	Sr^{2+}	Ba^{2+}	Zn^{2+}	Pb^{2+}	Ref
Liquid hydrogen cyanide		135.5	132.4											[24]
Liquid sulphur dioxide	79.9	119	179											[25]
Acetone	70.8	77.49	80.6					70.2	83.6		85.0			[26]
Acetonitrile	72.8	76.9	83.4		97.6			94.8				94.8		[26]
Water	38.6	50.1	73.5	77.8	77.3	61.9	74.7	53.1	59.5	59.5	63.6	52.8	69.5	e
Methanol[c]	39.6	45.2	52.4			50.2								[27]
		45.57	52.62	56.79	61.31									[28]
		45.7	53.8		62.3									[26]
Dimethylformamide	39.6	29.9	30.8	32.4	34.5	35.2	38.6	57.6	60.0	59.0		59.6		[27]
Pyridine	25.0[d]	26.8	32.0			34.3								[30]
Dimethylacetamide	24.9	25.6	25.2											[27]
Formic acid	20.30	25.80	25.30	26.98	27.81	27.19								[31]
Ethanol	19.36	20.97	23.99	24.95	26.47									[32]
Nitrobenzene	16.3	20.30	23.51											[28]
Dimethyl sulphoxide		17.8	13.9											[33]
n-Propanol[c]		13.1		12.91	13.69			9.40						[27]
	8.5	10.42	12.20											[34]
		10.32	12.45											[26]
Formamide	8.89	10.1	12.7	12.8	13.5		15.8							[35]
Propylene carbonate	6.62	9.45	11.17	11.90	12.66									[36]
N-Methylacetamide[c]	6.6	8.20	8.50	9.05	9.65	8.52								[37]
Hexamethylphosphoramide	9.95	8.2	8.4					9.8	10.2	10.4	10.1			[38]f
	6.8													[39]
														[40]
Sulpholane	3.63	5.9	6.1	5.97	6.52	5.15								[41]
	4.33	5.29	5.57											[42]
Ethylene glycol	2.11	3.61	4.05	4.16	4.27	4.4					0.39		2.66	[43]
		3.11	4.62											
Sulphuric acid	0.36	0.45	0.88				5.31							[44]

a Details of the $\Lambda^0 \rightarrow \lambda_+^0 + \lambda_-^0$ split will be found in the references cited; b solvents are arranged in descending order of λ^0 (Na^+); c duplicate and triplicate sets of results have been included to give an idea of consistency; d 23.62 [29]; e from table 4.3; f at 30° C.
3.37

Even when it is impossible to split Λ^0 values into λ_+^0 and λ_-^0 values, as when t_+ and t_- are unknown and measurements on R_4N^+ salts are impossible, something about relative solvation of cations can be deduced from trends in Λ^0 values for series of salts with a common anion. Thus Λ^0 values for alkali metal salts in acetic acid (Table 4.5) indicate the usual trend in solvation of the Li^+, Na^+, and K^+ cations [47].

Table 4.5 **Values of conductance,** $\Lambda^0 (cm^2\ ohm^{-1}\ mol^{-1})$, **for solutions of alkali metal salts in acetic acid, at** $30°C$ **[47].**

	Λ^0 (bromides)	Λ^0 (formates)	Λ^0 (acetates)
Li^+	29	23	13
Na^+	37	31	20
K^+	41	35	24

The Walden products of conductance and viscosity, $\lambda_+^0 \eta$ and $\lambda_-^0 \eta$, are sometimes preferred to simple conductances in discussions of effects of ions on solvent structure. Values of Walden products and their variation with temperature indicate whether ions are nett structure-formers or structure-breakers. This approach has been used in mixed aqueous [19] and in non-aqueous [48] media.

4.2.4 Solvation Numbers

Ionic mobilities are calculated from conductances and transference numbers (Chapter 4.2.1). From them are estimated Stokes radii for solvated ions, and thence solvation numbers.

Stokes radii. Some illustrative values of the apparent radii of solvated cations, derived as described above, are listed in Table 4.6. Duplicate series of independent estimates normally agree well, except for the case of Li^+ in hexamethylphosphoramide. However the Stokes radii for these solvated cations are considerably greater than the crystallographic cation radii, no matter which of the alternative sets of the latter (Chapter 1.1.1) be chosen for the comparison. This indicates that a solvation shell does indeed move with a cation when that cation moves through a solvent. The results in Table 4.6 give an idea of the relative volumes of solvation shells around cations, but before solvation numbers can be calculated from these results some correction is needed. For while the Stokes radii in Table 4.6 are all intuitively reasonable numbers, estimated Stokes radii of bromide, iodide, and perchlorate in dimethylformamide come out as negative numbers. The Stokes radius of Tl^+ in water is positive, but it is less than the crystal radius of this cation. Such anomalies are the consequence of the approximations involved in applying the Stokes-Einstein equation in the present situation. Some allowance must therefore be made.

Table 4.6 Stokes radii (Å) of solvated metal cations.

Ion	r_{xtal}	Water [27]	Acetonitrile [26]	Dimethyl sulphoxide [50]	Methanol [26]	Methanol [28]	Ethanol [28]	n–Propanol [26]	n–Propanol [34]	N-Methyl acetamide [37]	Dimethyl acetamide [31]	Dimethyl formamide [27]	Dimethyl formamide [50]	Acetone [26]	Hexamethyl phosphoramide [40]	Hexamethyl phosphoramide [41]	Propylene carbonate [50]
Li$^+$	0.60	2.4	2.97	3.18	3.79					4.02	4.39	4.1	4.11^a	3.72	3.61	7.26	3.37
Na$^+$	0.95	1.8	3.08	3.03	3.28	3.30	3.75	4.07	4.02	3.30	3.46	3.4	3.44	3.49	4.16	4.97	3.10
K$^+$	1.33	1.3	2.84	2.91	2.79	2.86	3.24	3.37	3.43	3.20	3.53	3.3	3.34	3.36	4.04	4.72	2.75
Rb$^+$	1.48			2.59		2.65	3.05		3.24	3.01	3.31	3.2	3.18			4.41	2.31
Cs$^+$	1.69		2.43	2.46	2.41	2.46	2.87		3.06	2.81	3.21	3.0	2.98			4.04	2.20
Ag$^+$	1.26				3.0^b					3.16	3.28	2.9					
Tl$^+$	1.40	1.2										2.7				5.11	
Mg^{2+}	0.65		5.00		5.21									7.71			
Ca^{2+}	0.99				5.00									6.47			
Sr^{2+}	1.13				5.08												
Ba^{2+}	1.35													6.37			
Zn^{2+}	0.74		5.00		5.03												

a Another estimate of the Stokes radius here is 4.37Å [29]; b ref. [27].

Two methods are in common use, due to Stokes and to Nightingale. Stokes's earlier method is simply based on the assumption that large tetraalkylammonium cations are unsolvated (for which there is some support from n.m.r. and infrared spectroscopy [51]), that their Stokes radii in solution are thus equal to their crystal radii, and that the Stokes radii of the solvated cations can then be adjusted to fit in with these assumptions [2]. This procedure works satisfactorily for large cations, but not for those with crystal radii less than about 2.5Å, so it is not satisfactory for the majority of metal cations.

Nightingale's method starts with the similar assumption that all tetraalkylammonium cations except Me_4N^+ are unsolvated. He then constructed a calibration curve based on plotting apparent Stokes radii against known crystal radii. Such a curve can be extrapolated to relatively small cation radii [52]. Some corrected Stokes radii are quoted in Table 4.7. Comparison of these values with the respective uncorrected radii in Table 4.6 shows that the corrected radii are significantly bigger. These methods of correcting Stokes radii are rather empirical; it is hoped that a less empirical method can be established [50] by considering solvent structure and solvent cavities.

Table 4.7 Corrected Stokes radii (Å) for solvated metal cations; S = Stokes's correction, N = Nightingale's correction (cf. text).

	Methanol [26]	N-Methylacetamide [37]		Dimethylacetamide [31]		Hexamethyl-phosphoramide [41]	
		S	N	S	N	S	N
Li^+	4.73	5.45	5.42	5.36	5.34	6.31	6.34
Na^+	4.43	4.83	4.80	4.58	4.59	5.06	5.07
K^+	4.12	4.72	4.70	4.66	4.65	4.94	4.94
Rb^+		4.57	4.56	4.47	4.47	4.77	4.76
Cs^+	3.89	4.41	4.42	4.40	4.39	4.57	4.57
Ag^+		4.69	4.67	4.46	4.45	5.14	5.17
Mg^{2+}	5.62						
Ca^{2+}	5.50						
Sr^{2+}	5.54						
Ba^{2+}	5.52						

Solvation numbers. These can be estimated by simple geometry and arithmetic from the Stokes radii just discussed. The difference between the volume of the solvated and 'naked' cation is assigned to the solvating solvent molecules. Consideration needs to be given to packing and electrostriction in estimating the

Table 4.8 Solvation numbers of cations estimated from ionic transport experiments via Stokes radii.

Solvent	Alkali metal cations						Other cations [a]
	Li^+	Na^+	K^+	Rb^+	Cs^+		
Water	7.0	5.0	5			[2]	Ln^{3+} 13–14 [53]
	14	9	7			[54]	Mg^{2+} 12 [26]
	21	10				[55]	
Methanol	7	6	5	4–5	4	[34]	Mg^{2+} 15 [26]
	9–11	5–6	4			[56, 57]	
						[58]	
Ethanol	6	5	4	3–4	3	[34]	
		4–5				[56, 57]	
n–Propanol	4–5	4	3	3	2–3	[34]	Mg^{2+} 14 [26]
Acetone	4–5	4–5	4			[56, 57]	Mg^{2+} 16 [26]
Acetonitrile	9	5–6	3–4			[56, 57]	Mg^{2+} 14 [26]
Furfural		5	4			[56]	
Pyridine		4	2–3			[56]	
Formamide	5.4	4.0	2.5	2.3	1.9	[35]	Ag^+ 2.3 [37]
	5.3	3.6	3.6	3.4	3.0	[37]	
Methyl ethyl ketone	4	3	2			[57]	
Dimethyl sulphoxide	5.5	3.4	3.2		2.5	[59]	
Sulpholane	1.4	2.0	1.5	1.4	1.3	[60]	
Dimethylformamide	3.2					[29]	
Dimethylacetamide	4.1	2.6	2.7	2.3	2.2	[31]	Ag^+ 2.3 [31]
Hexamethylphosphoramide	3.4	1.8	1.6	1.3	1.2	[41]	

solvation number. A selection of solvation numbers determined this way is given in Table 4.8; early values are from simple, while post-1960 values are from corrected, Stokes radii.

An entirely different method for the estimation of hydration numbers from conductivity measurements has been proposed and developed by Gusev [61]. The dependence of conductance on concentration in solutions containing a metal salt and an acid is usually of the form depicted in Fig. 4.2, with a marked

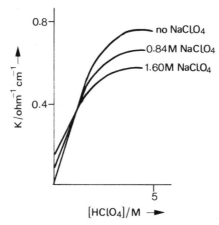

Fig. 4.2   The dependence of conductance, κ, on the concentration of perchloric acid in $NaClO_4$-$HClO_4$ solutions [61].

discontinuity as shown. It is proposed that at this point of discontinuity all the free, i.e. non-solvating, water has been used up. Beyond this point there are no

Table 4.9   Cation hydration numbers estimated from the Gusev conductivity method [61].

Li^+	2–3	Be^{2+}	4	Al^{3+}	11	Th^{4+}	20
Na^+	2.5–3.5	Mg^{2+}	8	In^{3+}	9		
Rb^+	4.8	Ca^{2+}	8				
Cs^+	5.5	Sr^{2+}	8	Cr^{3+}	12		
		Ba^{2+}	8				
Ag^+	3			La^{3+}	12		
		Zn^{2+}	6	Nd^{3+}	12		
		Cd^{2+}	6	Dy^{3+}	10–11		
		Hg^{2+}	6	Er^{3+}	11		
		Ni^{2+}	8			UO_2^{2+}	7

free water molecules left for transporting protons through the solution. The water:cation mole ratio at the discontinuity gives an indication of the hydration number of the cation. Results obtained by this method are summarised in Table 4.9; the hydration number of the proton is between 9 and 10. Hydration numbers are, where they can be compared, larger than those from n.m.r. peak area estimates, this conductivity method thus appears to give information about primary plus secondary solvation.

4.2.5 Viscosities

The dissolution of an ionic compound in a solvent affects its viscosity. Two factors may prove important, ion–ion and ion–solvent interactions. The latter may affect the viscosity both through the tying-down of some of the solvent in ion solvation shells, and through the more long-range effects of ions on the structural properties of the solvent. Nearly half a century ago it was demonstrated that the viscosity of aqueous solutions of electrolytes (η) could be related to that of pure water (η_0) by the expression now commonly referred to as the **Jones and Dole equation** [62] :

$$\eta = \eta_0 (1 + A\sqrt{c} + Bc).$$

Here c is the electrolyte concentration, and A and B are constants characteristic of the electrolyte. The same equation proves to be a good approximation for solutions in some non-aqueous solvents, though an extra Dc^2 term is needed in some systems, for instance in N-methylformamide [63]. The constant A in the above equation has been assigned, a trifle empirically, to ion–ion interactions, and the constant B to ion-solvent interactions. In practise the $A\sqrt{c}$ term is dominant at low solute concentrations; the Bc term predominates at higher solute concentrations. This is illustrated, for the typical case of potassium chloride solutions [64], in Fig. 4.3.

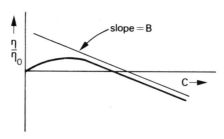

Fig. 4.3 The dependence of viscosity, η, on concentration, c, for aqueous solutions of potassium chloride [64].

It is the ion–solvent constant B which is of prime concern in this book. It is possible to subdivide B into components, such as intrinsic, void, and electrostriction contributions [65], but this amount of detail is beyond our present requirements. Having obtained B coefficients for salts, there arises the customary problem of separating these into self-consistent single ion values. Several

Table 4.10 Single ion viscosity B coefficients $(dm^3 mol^{-1})$ for cations in aqueous solution $(25^{\circ}C)$.

Assumption [a]	KCl [67]	Et_4N^+ [65]
Li^+	+0.147	+0.124
Na^+	+0.086	+0.066
K^+	−0.007	−0.026
Rb^+	−0.029	−0.053
Cs^+	−0.045	−0.064
Ag^+	+0.091	

a See text.

methods have been used, initially based on assumptions concerning lithium iodate. Early methods have also been based on potassium chloride, while more recently assumptions regarding such large ions as those of the tetraalkylammonium type have been employed:

(i) Lithium iodate [66]. Here the split takes place before viscosities are even considered. The lithium cation is said to have a mobility only slightly (about 3%) greater than the iodate anion. By assuming similar volumes for hydrated Li^+ and IO_3^- and applying Stokes's Law, it follows that the B coefficient for Li^+ should be approximately 10 (that is 3^2) per cent greater than that for IO_3^-. Thence the measured $B(LiIO_3)$ of 0.283 gives estimated values of $B(Li^+) = 0.147$ and $B(IO_3^-) = 0.136$.

(ii) Potassium chloride [67, 68]. The small B coefficient of potassium chloride, −0.014, is thought intuitively to arise from small equal effects of K^+ and Cl^-. This gives $B(K^+) = B(Cl^-) = -0.007$, whence $B(Li^+) = 0.147$ in excellent agreement with the lithium iodate value. This method has been used for aqueous solutions as well as for solutions in such non-aqueous solvents as methanol [69]. The assumptions $B(Rb^+) = B(Cl^-)$ or $B(Cs^+) = B(Cl^-)$ have been claimed to be preferable to $B(K^+) = B(Cl^-)$ [52].

(iii) Alkylammonium salts. It has been assumed, for example, that Et_4N^+ is unsolvated [51] and obeys the Einstein equation well in water [65], that $B(^nBu_4N^+) = B(Ph_4B^-)$ in acetonitritrile [70], and that $B(R_4N^+)$, $B(R_4B^-)$ are proportional to the volumes of the ions as they are effectively not solvated [71].

Table 4.11 Single ion viscosity B coefficients[a] ($dm^3 mol^{-1}$) for cations in non-aqueous solvents (25°C).

	Acetonitrile [69]	Methanol	Dimethyl sulphoxide	Propylene carbonate	N-Methyl formamide [71]	Dimethyl formamide	N-Methyl propionamide	Formamide [35]	[72]
Li^+		0.30^b			0.07		0.41	0.34	0.49
Na^+	0.44^c	0.27	0.35		0.09			0.36	0.48
K^+		0.23^d		0.60	0.11	-0.15	0.53	0.18	0.29
Rb^+								0.18	
Cs^+		0.03			0.15			0.21	

a R_4N^+, R_4B^- assumptions used; b or 0.43 [73]; c assumes $B(Me_4N^+) = 0.25$; d $B(K^+) = 0.382$, assuming $B(K^+) = B(Cl^-)$ [69].

Values of single ion B coefficients for cations in water are listed in Table 4.10 [65, 67] and for cations in non-aqueous solvents in Table 4.11 [71-73]. In many investigations the experimenters have been (understandably) reluctant to use one of the arbitrary single ion separation methods outlined above. In such cases it is a simple matter to find the *differences* between the B coefficients of various cations, as long as the anion remains the same. Table 4.12 contains some examples of such differences (ΔB). Table 4.13 contains further ΔB values, this time for mixed aqueous solvents. There is very little information on differences between B coefficients in water and in D_2O; it is known that the B coefficients for potassium iodide and for caesium iodide are both very slightly more negative in D_2O than in H_2O [75].

Table 4.12 Differences, ΔB, between single ion viscosity B coefficients (dm^3 mol^{-1}) for pairs of cations in solution ($25°C$).

		$\Delta B(Na^+-K^+)$	$\Delta B(K^+-Cs^+)$
N-Methylformamide	[63]	-0.016	
Methanol	[72]	0.003	
Water	[72]	0.093	
Sulphuric acid	[74]	0.175	
Formamide	[72]	0.186	
Ethylene glycol	[75]		0.113
Propane-1,3-diol	[75]	0.362	0.169
Propane-1,2-diol	[75]	0.406	0.175
Glycerol	[75]	0.542	0.223

Discussions of single ion B coefficients and ΔB values are all predominantly in terms of the effects of the cations on solvent structure. The relative values for a single ion in different solvents, both single and mixed, and for series of ions in a given solvent, can be rationalised to a reasonable extent in these terms. Thus B is normally negative for structure breakers, positive for structure makers, in water. In methanol all B values are positive, as there is less structure to break and thus it is difficult to decrease the viscosity of the solvent. It is necessary to consider the 'fault zone', region C of Fig. 1.2, in such considerations [72]. But it is widely realised that such a simple picture as this is not entirely satisfactory. Other specific factors need to be considered, such as the relative sizes of the solvent molecules [63] and the nature of solvent intermolecular interactions, for instance the chain association in methanol or in N-methylformamide versus the three dimensional structure of water, diols, or formamide [72]. The temperature dependences of viscosity B values and their relation to solvation and solvent structural changes have been considered [77].

Table 4.13   Differences, ΔB, between single ion viscosity B coefficients (dm^3 mol^{-1}) for pairs of cations in mixed aqueous solvents [76].

	$\Delta B(Li^+-Na^+)$	$\Delta B(Na^+-K^+)$	$\Delta B(K^+-Rb^+)$	$\Delta B(Rb^+-Cs^+)$	$\Delta B(Na^+-\frac{1}{2}Mg^{2+})$
Water	0.064	0.08, 0.093	0.023	0.011	-0.10
Methanol					
10%	0.074				
20%	0.082	0.090	0.033	0.013	
40%	0.139	0.114	0.021	0.011	
Dioxan					
10%		0.07			-0.13
20%		0.07			-0.22
30%		0.12			-0.34

Both single ion viscosity B values and ionic conductances are determined by the ease of movement of solvated ions through the solvent; it is reassuring that there exists a reasonable good (though not linear) correlation between these two quantities.

Very little effort seems to gone put into deriving solvation numbers of cations from their viscosity B coefficients. Indeed even comprehensive reviews of solvation numbers make no reference to such estimates. The chief value of viscosity B coefficients lies in their indication of qualitative solvation trends and of the effects of ions on solvent structural properties.

4.2.6   Diffusion

This appears to be one of the least satisfactory ways for obtaining hydration numbers for cations [1,2]. Diffusion coefficients can be determined by a variety of experimental methods, including the time dependence of conductivities, the use of porous diaphragms, and radioactive tracer techniques. Lack of confidence in this line of approach stems from observing, for example, the variation of the hydration number of Li^+ between 5 and 62 depending on the method used. Ranges of reported values for other cations are somewhat smaller, but still far too wide for any confidence in the results. As usual, intercomparison of results for various cations determined by the same method gives reasonable trends. For example, from diffusion through porous diaphragms, hydration numbers of 62, 44.5, and 29.3 have been derived for Li^+, Na^+, and K^+ respectively [78].

4.3   THERMOCHEMICAL APPROACHES

Thermochemical consequences of the solvation of cations form the content of Chapter 7, so at this stage one has to be careful not to prejudge or prejudice what will be said there, nor to generate any circular arguments or derivations. Moreover the complicated chemistry set out in Chapter 7 makes one wary of using thermochemistry as a route to the determination of cation solvation numbers. The present trend seems to recognise this, and to use established facts about ionic solvation in interpreting and rationalising thermodynamic parameters for ions in solution. This healthy tendency can be illustrated by recent investigations and discussions of specific heats for ions in solution [79]. Nonetheless, the use of thermochemical parameters as a source of solvation numbers was, and in some circles still is, a frequently attempted exercise [80]. We must therefore report here the results of some of these estimations.

4.3.1   Entropies

Entropies of cation solvation have been the most commonly used thermochemical route to solvation numbers. The effects of the cation on the surrounding solvent molecules are directly, though not necessarily simply, reflected in the appropriate single ion entropy of solvation. The various types of 'entropy of solvation' and the procedures involved in deriving single ion values are detailed in Chapter 7.

The transfer of a solvent molecule from bulk solvent to the solvation shell of a cation results in a considerable loss of entropy, a loss similar to that consequent upon freezing the solvent. This can be seen by comparing values of $T.\Delta S^{\ominus}$ for the reaction

$$MX \cdot nH_2O \ (c) \ \longrightarrow \ MX \ (c) \ + \ nH_2O \ (l) \tag{4.1}$$

with that for

$$nH_2O \ (c) \ \longrightarrow \ nH_2O \ (l) \ . \tag{4.2}$$

Values of $T.\Delta S^{\ominus}$ at 25°C for equation (4.1) are (in kcal per mole of water of crystallisation) for $Na_2SO_4 . 10H_2O$ 1.8, for $ZnSO_4 . 6H_2O$ 2.2, for $Mn_2S_2O_6 . 6H_2O$ 1.9, and for $MgCl_2 . 6H_2O$ 1.7. These values are close to the value of 1.8 kcal mol^{-1} for equation (4.2) at 25°C. The extension of this sort of idea from solid hydrates to cations in solution permits estimates of hydration numbers to be made, with results as given in Table 4.14. There is some correlation between the extent of solvation suggested by entropies of solvation and that indicated by viscosity B coefficients and by ionic mobilities [67].

Table 4.14   Cation hydration numbers estimated from entropies [80].

Li^+	5	Ag^+	4	Mg^{2+}	13	Sn^{2+}	9	Al^{3+}	21
Na^+	4			Ca^{2+}	10	Pb^{2+}	8		
K^+	3	Tl^+	3	Ba^{2+}	8				
Rb^+	3					Fe^{2+}	12		
				Zn^{2+}	12	Cu^{2+}	12		
				Cd^{2+}	11				

4.3.2   Compressibilities

These form another popular thermochemical approach to the estimation of hydration numbers, since they can be measured fairly readily by ultrasonic techniques. The use of compressibilities to estimate solvation numbers, for example via Passynski's equation [81], is based on the lower compressibility of solvating solvent molecules. These are already significantly compressed, compared with bulk solvent molecules, by electrostriction. It is often assumed that the compressibility of primary solvation solvent molecules is zero. Cation hydration numbers derived by this route are listed in Table 4.15 [1, 82-84].

Table 4.15   Cation hydration numbers estimated from adiabatic compressibility measurements (25°C).

Assumption	$K^+ = Cl^- = 3.2$ [82]	$NO_3^- = 2$ [82]	$I^- = 0$ [1, 83]	$NR_4^+ = 0$ [84]
Li^+	2.7	3.6	5-6	2
Na^+	3.9	4.8	6-7	3
K^+		4.1	6-7	2
Cs^+				1
Ag^+	3.1	4.0		
Be^{2+}			8	11
Mg^{2+}	7.0	7.9	16	8
Ca^{2+}				8
Ba^{2+}			16	8
Zn^{2+}				6
Cd^{2+}				5
Cu^{2+}				5
Al^{3+}			31	
Ce^{3+}				13

An alternative route uses ionic vibration potentials [85], leading to the hydration numbers shown in Table 4.16. This is arguably a good route to single ion values, because the ionic vibration potentials give the sum of the cation and anion contributions, whereas the compressibility measurements give the differences between the cation and anion contributions. This method for estimating ion solvation numbers has been criticised for the simplicity of the model used [86], and subsequently defended [87].

Table 4.16 Cation hydration numbers ^{a} derived from adiabatic compressibilities and ionic vibration potentials [85].

Li^+	4.0	Mg^{2+}	10.0	La^{3+}	15.0	Th^{4+}	22.0
Na^+	4.5	Ca^{2+}	9.0	Ce^{3+}	18.0		
K^+	3.5	Sr^{2+}	9.0				
Rb^+	3.0	Ba^{2+}	11.0				
Cs^+	2.5						

a Hydration numbers of the halide anions vary between 4.0 for F^- and 1.5 for I^-.

Solvation numbers, or at least their trends, derived from compressibility measurements generally bear some resemblance to those determined by other methods. Unexpectedly large values sometimes reported for large tetraalkyl-ammonium cations mirror the effect of these ions on the structure of the surrounding water. The balance between ion solvation and the effects of ions on solvent structure has been fully discussed [88], and the variation of hydration number with temperature has been probed by isothermal compressibility measurements. This experimental technique gives answers similar to the ultrasonic technique (adiabatic compressibility). Thus, at $25°C$ and assuming that the compressibilities of the solvated Na^+ and Cl^- ions are proportional to their (unsolvated) radii, hydration numbers of 3.7 for Na^+ and 7.7 for Mg^{2+} were obtained [89].

Adiabatic compressibilities have also been used to estimate solvation numbers for a few cations in a couple of non-aqueous solvents [82]. For Li^+, Na^+, and K^+ solvation numbers of 3.3, 3.9, and 3.6 are reported for methanol solution, while in ethanol solvation numbers of 1.4 and 0.9 were reported for Li^+ and Na^+. All of these estimates relate to an assumed solvation number of 2 for the planar nitrate ion.

4.3.3 Volumes

Volume is perhaps the thermochemical parameter most readily visualised. Molar volumes for solvated species can be derived from density measurements. As in the cases of entropies and compressibilities, the experimental result reflects all ion–solvent interactions. These include primary and secondary solvation and the effects of ions on the solvent structure of the water [90] or non-aqueous solvent, for instance N-methylacetamide [91]. When splitting experimentally determined values for whole salts into single ion components, it is often arbitrarily assumed that the partial molar volume for the hydrated proton $\overline{V}(H^+)$ is zero. Estimates of the absolute value of $\overline{V}(H^+)$ range from -0.2 down to -7.6 cm^3 mol^{-1} [92]. Values of $\overline{V}(M^+)$, for alkali metal cations in methanol, ethanol, and dimethylformamide as well as in water, are listed in Table 4.17. All the results in this Table come from one series of experiments and are so comparable, though they may not approximate too closely to the absolute values. Transference numbers were used in obtaining these single ion estimates from the experimental results. Table 4.17 shows how $\overline{V}(M^+)$ varies with the nature of the cation and the nature of the solvent, the latter variation largely determined by steric and packing considerations [93]. The shape of the plot of $\overline{V}(M^{3+})$ for the lanthanide(III) cations in water is odd, showing an inflection in the Nd^{3+} to Tb^{3+} region. This has been used as evidence supporting the hypothesis of a change in the hydration number of lanthanide(III) cations in this region of the rare earth series (see Chapter 5.3.8 for a full consideration of hydration numbers of Ln^{3+} cations) [94].

Table 4.17 Cation partial molal volumes (cm^3 mol^{-1}) in water and in non-aqueous solvents [93].

	Water	Methanol	Ethanol	Dimethylformamide
Li^+	-6.6	-17.9	-19.2	-26.4
Na^+	-6.9	-17.1	-9.8	-16.3
K^+	$+3.3$	-7.3	-0.4	-9.0
Rb^+	$+8.4$	-1.7	$+5.5$	-5.0
Cs^+	$+15.6$	$+4.8$	$+13.3$	$+1.5$

Although a number of $\overline{V}(M^{n+})$ values have been estimated for cations in both aqueous and non-aqueous solutions, there has been very little effort to estimate solvation numbers this way. One author has deduced solvation numbers of 4, 3, and 2 for Li^+, Na^+, and K^+ respectively in methanol, assuming zero solvation for large R_4N^+ cations [95]. A hydration number of 6 has been derived for In^{3+} [96].

4.3.4 Other Approaches

Other thermochemical routes to deductions about cation–solvent interactions are listed in Table 4.18 [97-101].

Table 4.18 Other thermochemical parameters used in the interpretation of cation solvation.

Parameter	Solvent		
Redox potential	water	effect of added salt on potential	[97]
Free energy	water	calculated and observed values compared for cations of known hydration number, then hydration numbers of unknown cases estimated by calculation	[98]
Activity	water glycol	related to Debye-Hückel	[99, 100]
Specific heat	methanol dimethylformamide	effects of cations on solvent structure probed	[101]

4.4   DIFFRACTION METHODS

Diffraction methods are uniquely valuable for determining that most fundamental property of crystalline solids — where the atoms are. Their success there depends on the orderly arrangement of atoms within a crystal, an orderly arrangement conspicuously lacking in solutions of electrolytes. Hence diffraction methods are enormously less powerful tools for probing the location of atoms in solution, but they are of some limited value nonetheless.

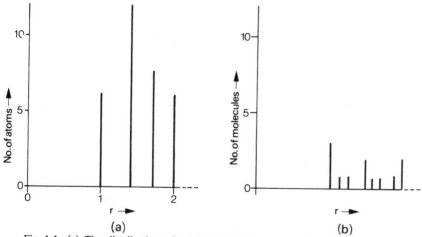

Fig. 4.4 (a) The distribution of neighbouring atoms in a solid with the sodium chloride structure; (b) a distribution of solvent molecules about an ion in solution.

The differences and similarities between the two situations of ions in a crystal and of ions in solution can be illustrated by the radial distribution functions for the two cases. For example, in the sodium chloride lattice one given sodium ion is surrounded by six chloride ions at a distance r, then twelve sodium ions at a distance $\sqrt{2}\,r$, next eight chlorides at $\sqrt{3}\,r$, and so on, as in Fig. 4.4(a). This distribution diagram is **identical** (apart from the small consequences of crystal imperfections) for all the sodium ions. Turning to a solution of sodium chloride in water then, and taking a very dilute solution so that only ion-solvent and not ion–ion interactions need be considered, a comparable diagram to represent the distribution of solvent molecules about one given sodium ion at a given instant of time might appear as shown in Fig. 4.4(b). In contrast to the sodium ion in the sodium chloride crystal, this diagram will be different for different sodium ions. Further, it will be continuously changing as water molecules move between the various environments. Summed over all sodium ions, the time-averaged radial distribution curve will look something like that shown in Fig. 4.5(a). Over the range (A) to (B) the radial distribution function $g(r)$ is zero, corresponding to the undesirability of interpenetration of the sodium ions and its solvating molecules. The maximum at (C) corresponds to the number of nearest, 'touching', water molecules. Eventually $g(r)$ tends to one, (D), when the distance of the solvent molecules from the sodium ion is too great for there to be any ordering or correlation with respect to that ion. It is the size of the maximum at (C) and the value of r at this point that gives information about primary solvation, and the shape of $g(r)$ beyond (C) that gives information about secondary solvation and ion effects on solvent structure.

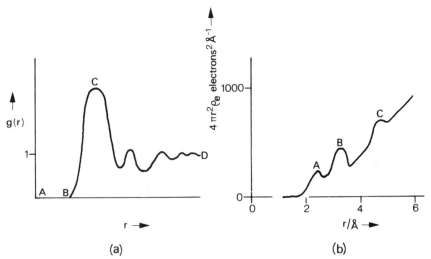

Fig. 4.5 (a) A general radial distribution function, $g(r)$, for solvent molecules around an ion in solution; (b) the radial distribution function, $4\pi r^2 \rho_e$, for water molecules about Ca^{2+} in aqueous solution [105].

4.4.1 X-Ray Diffraction

Early experiments suggested a hydration number of four, though with considerable uncertainty, for the Li^+ and K^+ cations [102]. Later experiments with better resolution suggested a hydration number of six for Er^{3+} [103]. More recently hydration numbers of 6 for Co^{2+} [104] and of between 6 and 8 for Mg^{2+}, Ca^{2+}, Sr^{2+} and Ba^{2+} [105] have been determined, and the tetrahedral arrangement of four water molecules around Li^+ confirmed. When Mg^{2+} and Ca^{2+} in water were studied, the effect of these cations on the secondary solvation region was apparent from the X-ray results. In the distribution curve of Fig. 4.5(b), peak A can be assigned to Ca^{2+}-water interaction, peak B to chloride-water and water-water (nearest neighbour) interactions, and peak C to second nearest neighbour interactions. It is this last peak, present in the curves for Mg^{2+} and Ca^{2+} but not those for Sr^{2+} and Ba^{2+}, which indicates ordering of the solvent by the added cations.

In all of the above cases, observations were made on concentrated solutions of chloride or bromide salts, where ion–ion interactions may confuse the interpretation. Radial distribution curves for uranium(IV) perchlorate solutions unequivocally indicate eight-coordination around the U^{4+} cation, and the absence of perchlorate from the primary solvation shell of the U^{4+} [106]. Radial distribution curves for solutions of the perchlorates of cadmium(II) [107] and of copper(II) [108] indicate hydration numbers of 6 for both cations, in regular octahedral and tetragonal arrangements respectively.

Cation–oxygen distances for several hydrated cations in aqueous solution have been estimated from radial distribution curves.†. The cobalt–oxygen distance determined from cobalt(II) chloride solution in water is 2.1Å; it has the same value in cobalt(II) bromide solution. The cobalt(II)–oxygen distance in a solution of cobalt(II) bromide in methanol is 2.0Å [104]. The cadmium–oxygen distance in aqueous solutions of cadmium(II) perchlorate is 2.31Å [107], a value which compares well with that of 2.29Å estimated from aqueous cadmium nitrate solution [109]. From the radial distribution curve for copper(II) perchlorate solution, the copper to water–oxygen distances are 1.94 and 2.43Å [108]. Caesium–oxygen distances have been estimated as 3.02Å from caesium iodide, 3.22Å from caesium bromide [110], and 3.13Å from caesium fluoride [111] solutions in water. The last-mentioned result comes from a particularly detailed X-ray study of aqueous solutions of caesium fluoride. The solution structure around the Cs^+ cations was likened to β-tridymite [112], one of the many forms of silica containing linked SiO_4 tetrahedra, with water and fluoride anions in lattice positions and Cs^+ cations in interstitial sites [111].

---

† The most recently published metal-oxygen distances for transition metal (II) cations in aqueous solution are: Mn^{2+} 2.20Å; Fe^{2+} 2.12Å; Co^{2+} 2.08Å; Ni^{2+} 2.04Å; Cu^{2+} 1.94Å (equatorial) and 2.38Å (axial); Zn^{2+} 2.08Å (Ohtaki, H., Yamaguchi, T., and Maeda, M., *Bull. Chem. Soc. Japan,* 49, 701 (1976)).

There are inconsistencies in the present picture for lanthanide cations. As mentioned above, Brady found a hydration number of six for Er^{3+} in erbium(III) chloride solution [103]. A coordination number of eight has also been suggested for the Er^{3+}, with two chlorides on opposite edges of the octahedron defined by the six coordinated water molecules [113]. The crystal structure of $GdCl_3.6H_2O$, which contains $Gd(OH_2)_6Cl_2^+$ units of this type [114], was cited as supporting evidence for this proposed structure for Er^{3+} in aqueous erbium(III) chloride solution. But a more recent X-ray investigation of aqueous lanthanum(III) chloride solution indicated eight nearest oxygens for each La^{3+}, at a distance of 2.48Å, with no chloride nearer than 4.7Å. It is not possible to say whether the eight oxygens are arranged in a cube, a square antiprism, or some other geometrical arrangement [115]††.

X-ray methods have also proved fruitful in probing the structures of solutions containing hydrolysable or polymerisable hydrated cations, for example $Hg^{2+}aq$ [116], $U^{4+}aq$, and $Th^{4+}aq$ [106, 117], and those containing complexes with mixed ligand plus solvent primary coordination shells. Examples in the latter category include Zn^{2+} or Cu^{2+} plus chloride in water [118].

There are times when the X-ray method is unsuccessful. Thus for iron(III) chloride solutions in water, no iron–nearest oxygen peak could be detected, although an iron-chloride peak was clearly defined [119].

4.4.2 Neutron Diffraction

The ease of detecting atoms by X-ray diffraction decreases regularly as atomic number decreases, since X-rays are scattered by electrons rather than by nuclei. It is extremely difficult to locate hydrogen atoms by X-ray methods. The ease of detection of atoms by neutron diffraction shows no such regularity. It is fortunate that both protons and deuterons cause much neutron scattering and can thus be located by neutron diffraction. In fact the incoherent nature of the scattering and the non-elastic recoil caused by protons means that deuteron scattering is easier to study [120].

To obtain experimental results sufficient for the extraction of useful information about the nature of an electrolyte solution, it is necessary to work with an electrolyte containing an element available in isotopic forms having widely different scattering characteristics. Nickel and copper are good elements to work with, iron and calcium possible. On the other hand the isotopes of potassium have similar scattering characteristics, while such isotopes as ^6Li are unsuitable because they absorb neutrons rather than scatter them. So one limitation on the usefulness of neutron diffraction studies of ions in solution is the availability, both in a chemical and a financial sense, of suitable isotopes. Other disadvan-

†† It has recently been shown that Gd^{3+} in aqueous solution has eight nearest neighbour oxygens, at a distance of 2.37Å. Again it is not possible to determine the coordination geometry from these X-ray diffraction observations (Steele, M. L. and Wertz, D. L., *J. Amer. Chem. Soc.*, **98**, 4424 (1976)).

tages include the need to use concentrated solutions, and the very small number of suitable neutron sources in existence.

Ion-solvent interactions in aqueous and many non-aqueous media are amenable to study via solvent and solute nuclei [120]. Because both oxygen and hydrogen (deuterium) cause significant neutron scattering, the radial distribution for the environment of the cation shows two peaks corresponding to water molecules in the primary solvation shell of a cation (contrast the analogous radial distribution curve for X-ray diffraction, Fig. 4.5). Cation-oxygen and cation-deuterium distances can thus be determined for the nearest D_2O molecules. If these distances can be determined with sufficient accuracy, then neutron diffraction experiments should be able to distinguish between pyramidal and planar geometries for $M-OH_2$ ($M-OD_2$) units in solution (see Chapter 2.2.2). It is possible to estimate hydration numbers from peak areas; the radial distribution function for concentrated solutions of nickel(II) chloride indicates a hydration number of just over five for the Ni^{2+} cation. The use of long wavelength neutrons ($\lambda \sim 20\text{Å}$) permits investigation of long range order in solvents and in solutions. Strong solutions of, for instance, nickel(II) chloride (6.38 mol dm^{-3}, that is to say saturated) [121] and sodium chloride produce the most interesting results here. These solutions have been found to have a remarkably ordered quasi-lattice structure, more ordered than an assembly of ion-pairs and more ordered than a molten salt, though of course less regular than the crystalline state.

In those areas, for example liquid metals, where it has been possible to compare results from X-ray and from neutron diffraction experiments, there has been a gratifying measure of agreement between the results from the two techniques. Both these diffraction methods have started to provide valuable fundamental information on ions in solution, and should give much further insight into the nature of electrolyte solutions.

4.5   OTHER METHODS

A variety of methods for probing the extent and nature of cation-solvent interaction which do not fit into the categories discussed so far remain. While many of these methods are of minor importance, that based on isotopic dilution experiments is of considerable significance. Though applicable in only a few situations, this method did provide an early way of estimating hydration numbers with some degree of confidence. Using this technique, a hydration number of six for Cr^{3+} was demonstrated a decade before the first estimates of solvation numbers were made by n.m.r. spectroscopy.

4.5.1   Isotopic Dilution

This method is based on distinguishing between bulk and firmly coordinated water by adding some labelled water to a solution and then determining the dilution of the label. The measured dilution will be determined by the volume of 'bulk' solvent, excluding any water firmly bonded to a substitution-inert ion. Obviously this method is restricted to those rare cations at which solvent exchange is very slow, of sufficient lethargy to permit the necessarily slow sampling and analysis methods. The use of low temperatures (for methanol) and of a modified method based on a flow system have permitted the extension of the method to a few cations of moderate solvent exchange rates. Some idea of the range of systems which have been studied, and of the limitations encountered, emerges from Table 4.19 [122–126].

Table 4.19 Isotopic dilution as a method for determining cation solvation numbers.

Cation	Solvent		
Cr^{3+}	water	hydration number = 6	[122]
Al^{3+}	water	hydration number = 6 (flow method)	[123]
Cd^{2+} Li^+	water	water exchange too fast	[124]
Cr^{3+}	liq. ammonia	Cr^{3+} salts insufficiently soluble	[125]
Fe^{3+} Co^{2+} Ni^{2+}	methanol	solvation numbers = 6 for each	[126]
Pr^{3+}	methanol	solvent exchange too fast	[126]

4.5.2 Dielectric Constants

This method starts with the handicap that measurements have to be made in concentrated solutions where ion–ion interactions are likely to be significant. Then it is assumed that water molecules in the primary solvation shell do not contribute to the observed dielectric constant, and that those in the secondary solvation shell make one-quarter of their usual (bulk solvent) contribution. Thereafter it is, as ever, necessary to make some assumption in order to separate the total salt result into single ion components. The hydration numbers in Table

Table 4.20 Cation hydration numbers from dielectric constant measurements on solutions of chlorides [127].

Li^+	6	Mg^{2+}	15.5	La^{3+}	23
Na^+	4	Ba^{2+}	14		
K^+	4				
Rb^+	3.5				

4.20 are based on the assumption that the chemically most reasonable split of the observed 11 waters per sodium chloride molecule is 8 for the Na^+ and 3 for the Cl^- [127].

4.5.3 Refractivity

This property is related to polarisability. Changes of refractivity on dissolving salts in water provided early evidence (1927) [128] for significant ion-water interactions, or at least suggested that the properties of water adjacent to ions were different from those of 'free' water. This early work also indicated that the solvation shell around a cation could be penetrated by anions to give ion-pairs and, at the limit when no solvent supervenes, complexes. Desultory studies over the following decades have led to estimates of refractometry indexes for a few alkali metal and alkaline earth cations [129]. A recent report of a refractivity study of aqueous solutions of alkali and alkaline earth metal cations claimed that refractometry can be used to assess solution stucture [130].

4.5.4 Distribution

In principle the effects of salts on the solubilities of non-electrolytes in water should give some idea of the hydration numbers of the constituent ions, in that the added salt will 'salt-out' the non-electrolyte by mopping up some of the water as M^{n+}aq and X^{n-}aq. Some apparently successful applications have been described [131], but often these results do not match well with those from other methods. Moreover it is extraordinarily difficult to accommodate the opposite effect of 'salting-in' in this approach. Solvent structure as well as simple ion solvation must be considered [132].

A more complicated extension involves monitoring the effect of salts on the distribution of a compound between water and a non-miscible organic layer. A suitable system would seem to be the distribution of acetic acid between water and amyl alcohol [133]. The same strictures apply here as to the simpler method outlined in the previous paragraph. Indeed they are underlined by the apparently negative hydration numbers obtain for some potassium salts, though such a result may be the consequence of the choice of 6 waters per sucrose as reference value. It must, however, be conceded that at least hydration numbers in the right order are obtained for the series of alkali metal cations.

4.5.5 Crystal Stoichiometries

This approach [134] is one of desperation in cases, particularly of little-studied solvents, where none of the methods described so far in this Chapter or in Chapters 2 and 3 can be used. Here one simply and crudely assumes that the solvation number of a cation in solution in a given solvent is equal to the number of molecules of solvent of crystallisation in appropriate salts crystallised from this solvent. Thus the preparation of $Ni(dmso)_6(NiCl_4)$ and of $Co(dmso)_6(ClO_4)_2$ from dimethyl sulphoxide solutions is not inconsistent with a solvation number of 6 for Ni^{2+} and for Co^{2+}. This method even sometimes works in aqueous systems [135]. A series of hexahydrates $MCl_2.6H_2O$ can be crystallised from aqueous solution, though the parallel series $MSO_4.7H_2O$ shows how certain anions can upset the picture. From liquid ammonia solution one can prepare $Cr(NH_3)_6(NO_3)_3$; the highest reported ammoniates of alkali metal and alkaline earth cations (Table 4.21 [136]) form a reasonable pattern which could well reflect cation ammoniation in liquid ammonia. Similar remarks can be made with respect to solvation of alkali metal cations by formic, acetic, and trichloroacetic acids [137].

Table 4.21   Highest ammoniates formed by alkali metal and alkaline earth bromides [136].

Li^+	5	Ca^{2+}	8
Na^+	5	Sr^{2+}	8
K^+	4	Ba^{2+}	8
Rb^+	3		

Aluminium perchlorate crystallises with 6, 6, 6, and 4 molecules respectively of trimethyl phosphate, dimethyl methylphosphonate, dimethyl phosphite, and hexamethylphosphoramide. Interestingly, n.m.r. studies of analogous solutions in nitromethane indicate solvation numbers of 6, 6, 6, and 4 for the Al^{3+}, so here agreement is perfect [138]. Crystal stoichiometries indicate that the solvation number of Li^+ with respect to hexamethylphosphoramide is also 4, of Na^+ 1 to 2 [139], and of several M^{2+} and M^{3+} cations up to 6 [140]. Many M^{2+} salts, for example of the series Mg^{2+} to Sr^{2+}, Mn^{2+} to Ni^{2+}, and Zn^{2+}, have six nitromethanes of crystallisation [141]. Crystal stoichiometries of appropriate salts of the MoF_6^- and WF_6^- anions crystallised from acetonitrile indicate an intuitively plausible order of decreasing solvation number [142]:

$$Tl^{3+}(6) > Ag^{2+}(4) \;=\; Cu^{2+}(4) > Ag^+(2) > Tl^+(0).$$

Many solvents tend to form solvates of lower coordination number than seems likely for the appropriate cations in solution. Thus while aluminium(III) salts can crystallise with six molecules of phosphorus oxychloride of crystallisation, most other salts seem limited to two molecules of phosphorus oxychloride of crystallisation. Again zinc(II) salts with two molecules of dinitrogen tetroxide of crystallisation are known, but salts of most other cations, for instance Fe^{2+}, Cu^{2+}, or UO_2^{2+}, have only one molecule of dinitrogen tetroxide of crystallisation. The crystal stoichiometry method does not look promising for hydrazine, in that the number of molecules of hydrazine of crystallisation seems dependent on the nature of the anion, as in, e.g., manganese(II) salts [143].

This method has a surprisingly large number of ardent propagandists, considering its serious drawbacks.

REFERENCES

[1] E.g. Hinton, J. F. and Amis, E. S., *Chem. Rev.*, **71**, 627 (1971).

[2] Robinson. R. A. and Stokes, R. H., *Electrolyte Solutions, 2nd. Ed.*, *revised* Butterworths (1968) Ch. 5.

[3] Lahiri, M. M., Seal, B. K. and Mukherjee, S. K., *Bull. Chem. Soc. Japan*, **46**, 1408 (1973).

[4] E.g. Springer, C. H., Coetzee, J. F. and Kay, R. L., *J. Phys. Chem.*, **73**, 471 (1969).

[5] See Appendix 6.1 of reference [2].

[6] Bockris, J. O'M., *Q. Rev. Chem. Soc.*, **3**, 173 (1949); Conway, B. E. and Bockris, J. O'M., in *Modern Aspects of Electrochemistry, Volume 1*, ed. Bockris, J. O'M. and Conway, B. E., Butterworths (1954) p. 47; Desnoyers, J. E. and Jolicoeur, C. in *Modern Aspects of Electrochemistry, Vol. 5*, ed. Bockris, J. O'M. and Conway, B. E., Butterworths (1969) p. 1; Price, E. in *The Chemistry of Non-aqueous Solvents, Vol. 1*, ed. Lagowski, J. J., Academic Press (1966) p. 67.

[7] Feakins, D. and Lorimer, J. P., *J. Chem. Soc. Faraday I*, **70**, 1888 (1974).

[8] Spiro, M. in *Techniques of Chemistry, Vol. I: Physical Methods of Chemistry, Part IIA: Electrochemical Methods*, ed. Weissberger, A. and Rossiter, B. W., Wiley/Interscience (1971) Ch. 4.

[9] Ray, B. R., Beeson, D. M., and Crandall, H. F., *J. Amer. Chem. Soc.*, **80**, 1029 (1958).

[10] Imai, M., *J. Inorg. Nucl. Chem.*, **37**, 123 (1975).

[11] McInnes, D. A., *The Principles of Electrochemistry*, Reinhold (1939) Ch. 4; Washburn, E. W. and Millard, E. B., *J. Amer. Chem. Soc.*, **37**, 694 (1915); Baborovsky, J., *Z. Phys. Chem.*, **168A**, 135 (1934); Baborovsky, J, and Kondela, G., *Chemické Listy*, **32**, 5 (1938).

[12] Baborovsky, J. and Viktorin, O., *Colln. Czech. Chem. Commun. Engl. Edn.*, **5**, 518 (1933).

[13] Rutgers, A. T. and Hendrikx, Y., *Trans. Faraday Soc.*, **58**, 2184 (1962).

[14] Troshin, V. P., *Elektrokhimiya*, **2**, 232 (1966).

[15] Gopal, R. and Bhatnager, O. N., *J. Phys. Chem.*, **69**, 2382 (1965).

[16] Ryabikova, V. M., Krumgal'z, B. S., and Mishchenko, K. P., *Russ. J. Phys. Chem.*, **45**, 1451 (1971).

[17] Lanning, W. C. and Davidson, A. W., *J. Amer. Chem. Soc.*, **61**, 147 (1939).

[18] Viehweger, U. and Emons, H. -H., *Z. Anorg. Allg. Chem.*, **383**, 183

(1971).
[19] Broadwater, T. L. and Kay, R. L., *J. Phys. Chem.*, **74**, 3802 (1970).
[20] See Chapter 6 and Appendix 6.1 of reference [2].
[21] Campbell, A. N., *Can. J. Chem.*, **51**, 3006 (1973); **54**, 3732 (1976).
[22] Spedding, F. H., Porter, P. E., and Wright, J. M., *J. Amer. Chem. Soc.*, **74**, 2055 (1952).
[23] Imhof, J., Westmoreland, T. D., and Day, M. C., *J. Solution Chem.*, **3**, 83 (1974).
[24] Coates, J. E. and Taylor, E. G., *J. Chem. Soc.*, 1245 (1936).
[25] Takesawa, S., Kondo, Y., and Tokura, N., *J. Phys.Chem.*, **77**, 2133 (1973).
[26] Zipp, A. P., *J. Phys. Chem.*, **78**, 556 (1974).
[27] Prue, J. E. and Sherrington, P. J., *Trans. Faraday Soc.*, **57**, 1795 (1961).
[28] Beronius, P., Wikander, G. and Nilsson, A. -M., *Z. Phys. Chem. Frankf. Ausg.*, **70**, 52 (1970).
[29] Paul, R. C., Singla, J. P., and Narula, S. P., *J. Phys. Chem.*, **73**, 741 (1969).
[30] Burgess, D. S. and Kraus, C. A., *J. Amer. Chem. Soc.*, **70**, 706 (1948).
[31] Paul, R. C., Banait, J. S., and Narula, S. P., *Aust. J. Chem.*, **28**, 321 (1975).
[32] Wehman, T. C. and Popov, A. I., *J. Phys. Chem.*, **72**, 4031 (1968).
[33] Witschonke, C. R. and Kraus, C. A., *J. Amer. Chem. Soc.*, **69**, 2472 (1947).
[34] Wikander, G. and Isacsson, U., *Z. Phys. Chem. Frankf. Ausg.*, **81**, 57 (1972).
[35] Notley, J. M. and Spiro, M., *J. Phys. Chem.*, **70**, 1502 (1966).
[36] Jansen, M. L. and Yeager, H. L., *J. Phys. Chem.*, **77**, 3089 (1973).
[37] Paul, R. C., Banait, J. S., Singla, J. P. and Narula, S. P., *Z. Phys. Chem., Frankf. Ausg.*, **88**, 90 (1974).
[38] Dawson, L. R., Wilhoit, E. D., Holmes, R. R., and Sears, P. G., *J. Amer. Chem. Soc.*, **79**, 3004 (1957).
[39] Gal, J. -Y., Moliton-Bouchetout, C. and Yvernault, T., *C. R. Hebd. Séanc, Acad. Sci., Paris*, **275C**, 253 (1972).
[40] Bruno, P., della Monica, M., and Righetti, E., *J. Phys. Chem.*, **77**, 1258 (1973).
[41] Paul, R. C., Banait, J. S., and Narula, S. P., *Z. Phys. Chem. Frankf. Ausg.*, **94**, 199 (1975).
[42] Della Monica, M., Lamanna, U., and Senatore, L., *J. Phys. Chem.*, **72**, 2124 (1968); della Monica, M., Lamanna, U., and Jannelli, L., *Gazz. Chem. Ital.*, **97**, 367 (1967).
[43] Santos, M. C. and Spiro, M., *J. Phys. Chem.*, **76**, 712 (1972).
[44] Sidebottom, D. P. and Spiro, M., *J. Phys. Chem.*, **79**, 943 (1975).
[45] Krumgal'z, B. S., *Russ. J. Phys. Chem.*, **47**, 528 (1973).
[46] Jansen, M. L. and Yeager, H. L., *J. Phys. Chem.*, **78**, 1380 (1974).
[47] Jones, M. M. and Griswold, E., *J. Amer. Chem. Soc.*, **76**, 3247 (1954).
[48] Singh, R. D. and Husain, M. M., *Z. Phys. Chem. Frankf. Ausg.*, **94**, 193 (1975).

[49] Greenwood, N. N., *Ionic Crystals, Lattice Defects, and Non-stoichiometry* Butterworths (1968) pp. 40–41.

[50] Matsuura, N., Umemoto, K., and Takeda, Y., *Bull. Chem. Soc. Japan*, **48**, 2253 (1975).

[51] E.g. Butler, R. N. and Symons, M. C. R., *Chem. Commun.*, 71 (1969); *Trans. Faraday Soc.*, **65**, 2559 (1969); Krumgal'z, B. S., Ryabikova, V. M., Akopyan, S. Kh., and Borisova, V. I., *Russ. J. Phys. Chem.*, **48**, 1538 (1974).

[52] Nightingale, E. R., *J. Phys. Chem.*, **63**, 1381 (1959).

[53] Choppin, G. R. and Graffeo, A. J., *Inorg. Chem.*, **4**, 1254 (1965).

[54] Baborovsky, J. and Velisek, J., *Chemické Listy*, **21**, 227 (1927).

[55] Devyatykh, G. G., Kuznetsova, Z. B., and Agafonova, A. L., *Tr. Khim., Khim. Tekhnol.*, **1**, 75 (1958).

[56] Ulich, H., *Trans. Faraday Soc.*, **23**, 388 (1927).

[57] Walden, P. and Birr, E. J., *Z. Phys. Chem.*, **153A**, 1 (1931).

[58] Skabichevskii, P. A., *Russ. J. Phys. Chem.*, **49**, 100 (1975).

[59] Gopal, R. and Jha, J. S., *J. Phys. Chem.*, **78**, 2405 (1974).

[60] della Monica, M. and Lamanna, U., *J. Phys. Chem.*, **72**, 4329 (1968).

[61] Gusev, N. I., *Russ. J. Phys. Chem.*, **45**, 1268, 1455, 1575 (1971); **46**, 1034, 1657 (1972); **47**, 52, 184, 687, 1309 (1973).

[62] Jones, G. and Dole, M., *J. Amer. Chem. Soc.*, **51**, 2950 (1929).

[63] Feakins, D. and Lawrence, K. G., *J. Chem. Soc. A.*, 212 (1966).

[64] Joy, W. E. and Wolfenden, J. H., *Proc. R. Soc.*, **134A**, 424 (1932).

[65] Stokes, R. H. and Mills, R., *Viscosity of Electrolytes and Related Properties*, Pergamon (1965); Desnoyers, J. E. and Perron, G., *J. Solution Chem.*, **1**, 199 (1972); Mandal, P. K., Seal, B. K., and Basu, A. S., *Z. Phys. Chem. Frankf. Ausg.*, **87**, 295 (1973).

[66] Cox, W. M. and Wolfenden, J. H., *Proc. R. Soc.*, **145A**, 475 (1934).

[67] Gurney, R. W., *Ionic Processes in Solution*, McGraw-Hill (1953).

[68] Kaminsky, M., *Z. Phys. Chem. Frankf. Ausg.*, **8**, 173 (1956); *Disc. Faraday Soc.*, **24**, 171 (1957).

[69] Criss, C. M. and Mastroaninni, M. J., *J. Phys. Chem.*, **75**, 2532 (1971).

[70] Tuan, D. F-T. and Fuoss, R. M., *J. Phys. Chem.*, **67**, 1343 (1963).

[71] Krumgal'z, B. S., *Russ. J. Phys. Chem.*, **47**, 956 (1973).

[72] McDowall, J. M. and Vincent, C. A., *J. Chem. Soc. Faraday I*, **70**, 1862 (1974).

[73] Feakins, D., Freemantle, D. J., and Lawrence, K. G., *J. Chem. Soc. Faraday I*, **70**, 795 (1974).

[74] Spiro, M., *J. Chem. Soc. Faraday I*, **71**, 988 (1975).

[75] Bare, J. P. and Skinner, J. F., *J. Phys. Chem.*, **76**, 434 (1972).

[76] Feakins, D., Smith, B. C., and Thakur, L., *J. Chem. Soc. A.*, 714 (1966).

[77] Krumgal'z, B. S., *Russ. J. Phys. Chem.*, **48**, 1163 (1974).

[78] Baborovsky, J., Viktorin, O, and Wagner, A., *Colln. Czech. Chem. Com-*

mun. Engl. Edn., **4**, 200 (1932).

[79] Babakulov, N. and Latysheva, V. A., *Russ. J. Phys. Chem.*, **48**, 587 (1974); Vasilev, V. A., Karapet'yants, M. Kh., Sanaev, E. S., and Novikov, S. N., *Russ. J. Phys. Chem.*, **48**, 1398 (1974).

[80] Conway, B. E. and Bockris, J. O'M. in *Modern Aspects of Electrochemistry*, *Vol. 1*, ed. Bockris, J. O'M. and Conway, B. E., Butterworths (1954) p. 65.

[81] Passynski, A., *Acta Phys.-chim. URSS*, **8**, 385 (1938).

[82] Allam, D. S. and Lee, W. H., *J. Chem. Soc. A*, 5, 426 (1966).

[83] Corey, V. B., *Phys. Rev.*, **64**, 350 (1943).

[84] Padova, J., *Bull. Res. Coun. Israel*, **10A**, 63 (1961).

[85] Bockris, J. O'M. and Saluja, P. P, S., *J. Phys. Chem.*, **76**, 2140, 2298 (1972); 77, 1598 (1973).

[86] Desnoyers, J. E., *J. Phys. Chem.*, **77**, 567 (1973): Yeager, E. and Zana, R. *J. Phys. Chem.*, **79**, 1228 (1975).

[87] Bockris, J. O'M. and Saluja, P. P. S., *J. Phys. Chem.*, **79**, 1230 (1975).

[88] Breitschwerdt, K. G. and Kistenmacher, H., *J. Chem. Phys.*, **56**, 4800 (1972).

[89] Millero, F. J., Ward, G. K., Lepple, F. K., and Hoff, E. V., *J. Phys. Chem.*, **78**, 1636 (1974).

[90] Choi, Y. S. and Bonner, O. D., *Z. Phys. Chem. Frankf. Ausg.*, **87**, 188 (1973).

[91] Gopal, R., Agarwal, D. K., and Kumar, R., *Bull. Chem. Soc. Japan*, **46**, 1973 (1973); Gopal, R. and Singh, K., *Z. Phys. Chem. Frankf. Ausg.*, **91**, 98 (1974).

[92] Millero, F. J., *Chem. Rev.*, **71**, 147 (1971); Hirata, F. and Arakawa, K., *Bull. Chem. Soc. Japan*, **46**, 3367 (1973).

[93] Kawaizumi, F. and Zana, R., *J. Phys. Chem.*, **78**, 627, 1099 (1974).

[94] Spedding, F. H., Cullen, P. F., and Habenschuss, A., *J. Phys. Chem.*, **78**, 1106 (1974).

[95] Padova, J., *J. Chem. Phys.*, **56**, 1606 (1972).

[96] Celeda, J. and Tuck, D. G., *J. Inorg. Nucl. Chem.*, **36**, 373 (1974).

[97] Malysko, J. and Duda, L., *Mh. Chem.*, **106**, 633 (1975).

[98] Morf, W. E. and Simon, W., *Helv. Chim. Acta*, **54**, 794 (1971).

[99] See chapter 9 (p. 60) of reference [2].

[100] Sen, U., *J. Chem. Soc. Faraday I*, **69**, 2006 (1973).

[101] Chang, S. and Criss, C. M., *J. Solution Chem.*, **2**, 457 (1973).

[102] Brady, G. W., *J. Chem. Phys.*, **28**, 464 (1958).

[103] Brady, G. W., *J. Chem., Phys.*, **33**, 1079 (1961).

[104] Wertz, D. L. and Kruh, R. F., *J. Chem. Phys.*, **50**, 4313 (1969); *Inorg. Chem.*, **9**, 595 (1970).

[105] Albright, J. N., *J. Chem. Phys.*, **56**, 3783 (1972); Licheri, G., Piccaluga, G., and Pinna, G., *ibid.*, **63**, 4412 (1975).

[106] Pocev, S. and Johansson, G., *Acta Chem. Scand.*, **27**, 2146 (1973).

[107] Ohtaki, H., Maeda, M., and Ito, S., *Bull. Chem. Soc. Japan*, **47**, 2217 (1974).

[108] Ohtaki, H. and Maeda, M., *Bull. Chem. Soc. Japan*, **47**, 2197 (1974).

[109] Bol, W., Gerrits, G. J. A., and van Panthaleon van Eck, C. L., *J. Appl. Crystallogr.*, **3**, 486 (1970).

[110] Lawrence, R. M. and Kruh, R. F., *J. Chem. Phys.*, **47**, 4758 (1967).

[111] Bertagnolli, H., Weidner, J. -U., and Zimmermann, H. W., *Ber. Bunsenges. Phys. Chem.*, **78**, 2 (1974).

[112] Wells, A. F., *Structural Inorganic Chemistry, 4th Ed*, Oxford University Press (1975) p. 805.

[113] Morgan, L. O., *J. Chem. Phys.*, **38**, 2788 (1963).

[114] Marezio, M., Plettinger, H. A., and Zachariasen, W. H., *Acta Cryst.*, **14**, 234 (1961).

[115] Smith, L. S. and Wertz, D. L., *J. Amer. Chem. Soc.*, **97**, 2365 (1975).

[116] Johansson, G., *Acta Chem. Scand.*, **25**, 2787, 2799 (1971).

[117] Pocev, S., *Acta Chem. Scand.*, **28A**, 932 (1974).

[118] Bell, J. R., Tyvoll, J. L., and Wertz, D. L., *J. Amer. Chem. Soc.*, **95**, 1456 (1973); Wertz, D. L. and Bell, J. R., *J. Inorg. Nucl. Chem.*, **35**, 137 (1973).

[119] Brady, G. W., *J. Chem. Phys.*, **29**, 1371 (1958).

[120] Bacon, G. E., *Neutron Diffraction, 2nd Ed.*, Clarendon Press (1962, reprinted 1967) Ch. 8; *Research, London*, **7**, 257 (1954); March, N. H. and Tosi, M. P., *Phys. Lett. A.*, **50**, 224 (1974); Enderby, J. E., *Proc. R. Soc.*, **345A**, 107 (1975).

[121] Neilson, G. W., Howe, R. A., and Enderby, J. E., *Chem. Phys. Lett.*, **33**, 284 (1975).

[122] Hunt, J. P. and Taube, H., *J. Chem. Phys.*, **19**, 602 (1951); Taube, H., *J. Phys. Chem.*, **58**, 523 (1954).

[123] Baldwin, H. H and Taube, H., *J. Chem. Phys.*, **33**, 206 (1960).

[124] Swinehart, J. H., Rogers, T. E., and Taube, H., *J. Chem., Phys.*, **38**, 398 (1963).

[125] Swaddle, T. W., Coleman, L. F., and Hunt, J. P., *Inorg. Chem.*, **2**, 950 (1963).

[126] Rogers, T. E., Swinehart, J. H., and Taube, H., *J. Phys. Chem.*, **69**, 134 (1965).

[127] Harned, H. S. and Owen, B. B., *The Physical Chemistry of Electrolytic Solutions, 3rd. Edn.*, Reinhold (1958).

[128] Fajans, K., *Trans. Faraday Soc.*, **23**, 357 (1927).

[129] Batsanov, S. S., *Refractometry and Chemical Structure*, van Nostrand, (1966) pp. 155-161.

[130] Tereshkevich, M. O., Kuprik, A. V., Kuratova, T. S., and Ivashina, G. A., *Russ. J. Phys. Chem.*, **48**, 1476 (1974).

[131] Herington, E. F. G. and Kynaston, W., *J. Chem. Soc.*, 3137 (1952);

Eucken, A. and Hertzberg, G., *Z. Phys. Chem.*, **195**, 1 (1950).

[132] Long, F. A. and McDevit, W. F., *Chem. Rev.*, **51**, 119 (1952).

[133] Sugden, J. N., *J. Chem. Soc.*, 174 (1926).

[134] Hunt, J. P., *Metal Ions in Aqueous Solution*, Benjamin (1963) pp. 24–25;
Waddington, T. C., *Non-aqueous Solvents*, Nelson (1969) Ch. 1; Lincoln,
S. F., *Coord. Chem. Rev.*, **6**, 309 (1971).

[135] Emons, H. -H., Berger, C., and Ponsold, B., *Z. Phys. Chem.*, **256**, 430
(1975).

[136] Lagowski, J. J. and Moczygemba, G. A. in *The Chemistry of Non-aqueous
Solvents, Vol. 2*, ed. Lagowski, J. J., Academic Press (1967) Ch. 7.

[137] Ivanova, E. F., Nguen Suan Net, Kruglyak, A. I., Chan Ngok Hai, and
Mamund, K. M., *Russ. J. Phys. Chem.*, **48**, 1202 (1974).

[138] Delpuech, J. J., Khaddar, M. R., Peguy, A., and Rubini, P., *J. Chem. Soc.
Chem. Commun.*, 154 (1974).

[139] Leuhrs, D. C. and Kohut, J. P., *J. Inorg. Nucl. Chem.*, **36**, 1459 (1974).

[140] de Bolster, M. W. G. and Groeneveld, W. L., *Rec. Trav. Chim.*, **90**, 477
(1971).

[141] Norbury, A. H. Personal communication.

[142] Prescott, A., Sharp, D. W. A., and Winfield, J. M., *J. Chem. Soc. Dalton*,
934, 936 (1975).

[143] Galvic, P., Bole, A. and Slivnik, J., *J. Inorg. Nucl. Chem.*, **35**, 3979
(1973).

Chapter 5

SOLVATION NUMBERS

5.1   INTRODUCTION

The determination of solvation numbers by a variety of physical techniques has been a recurring theme throughout Chapters 2 to 4. In those Chapters several sets of results have been detailed to illustrate the values obtained from individual approaches. Solvation numbers of anything from 700 down to 0, and even occasionally negative values, have been reported at various stages for metal cations in solution. It is the aim of the first part of this Chapter to compare and correlate results obtained from different techniques. Then an attempt is made to build up an overall picture of the dependence of solvation numbers on the nature of the cation, especially its size, charge, and position in the Periodic Table. Finally there is brief mention of secondary solvation and of ion association in solution.

5.2   COMPARISON OF METHODS AND RESULTS

There are often considerable, even dramatic, differences between solvation numbers for a given cation when determined by different physical techniques [1]. At times there are marked differences even between various investigators' results using the same technique. Such differences are illustrated in Table 5.1, where hydration numbers for the Na^+, Mg^{2+}, and Al^{3+} cations are collected together. The range of values for each of these ions is well outside anyone's range of experimental uncertainity. This must indicate either that gross inconsistencies exist in assumptions or interpretation, or that different techniques are in fact measuring different quantities.

The Mg^{2+} and Al^{3+} cations were chosen for inclusion in Table 5.1 since there is definite evidence from n.m.r. spectroscopy (peak areas, Chapter 2.2.1) for a primary hydration or coordination number of six for both these cations. The n.m.r. peak area approach, X-ray diffraction experiments (Chapter 4.4.1), ultraviolet-visible spectroscopy (Chapter 3.1.1), and isotopic dilution techniques (Chapter 4.5.1) are all methods giving information specifically about the first coordination, or primary solvation, shell around a cation (region (A) in Fig. 1.2 on page 20). It is obvious from simple geometry that the number of solvent

Table 5.1   Dependence of hydration numbers of Na^+, Mg^{2+}, and Al^{3+} on the method of estimation. Values have been taken from the appropriate sections of Chapters 2 and 4.

	N.m.r. peak areas	Transference numbers	Entropies	Compressibilities	Mobilities
Na^+		7-13	4	4-7	2-10
Mg^{2+}	6	12-14	13	7-16	10-13
Al^{3+}	6		21	31	

molecules which can bond directly to a cation is limited. In the majority of cases this number is six, but in the case of particularly small cations, for example Be^{2+} and probably Li^+, the maximum coordination number is only four with respect to most solvents. In the case of particularly large solvent molecules a reduction in coordination number is likely. Thus Al^{3+} has only four molecules of hexamethylphosphoramide in its primary coordination sphere [2]. The larger cations, including those of the lanthanide and actinide elements, may have coordination numbers greater than six in solution, possibly in the range eight to ten (*v.i*).

The sphere of influence of the cation will not cut off abruptly at the periphery of the first coordination, or primary solvation, shell. More distant solvent molecules, the secondary solvent shell (region (B) of Fig. 1.2), also interact with the cation. Many physical techniques detect both primary and secondary solvation of ions in solution. When an ion moves through a solution of its own volition (diffusion) or under the influence of an external force (for instance transference), it carries with it both its primary solvation shell and at least some secondary solvent molecules. Thus measurements of the movement of ions will indicate a larger solvation number than that corresponding just to the primary solvation shell. This is apparent from the examples cited in Table 5.1. It may be that the average number of solvent molecules moving with a cation depends on the experimental conditions used in the measurement method, and thus that apparent solvation numbers for a given ion in a given solvent may vary somewhat with the technique used.

Primary coordination numbers, as measured from n.m.r. spectra, do not usually vary with temperature or ionic strength. The hydration number of Ni^{2+} has been shown, by ^{17}O n.m.r. spectroscopy, to be six throughout the temperature range -30 to $+200°C$ [3]. Likewise the primary hydration numbers of Fe^{2+} and of Co^{2+} are six over extensive temperature ranges [3, 4]. The only cation whose primary coordination number may vary with temperature is Co^{2+}, where evidence has been cited for the coexistence of tetrahedral and octahedral species at elevated temperatures [5] (Chapter 5.3.5) . Solvation numbers determined from transport properties of ions are sensitive to temperature and to ionic strength, because these affect the secondary solvation shell. Examples of lower hydration numbers at increased temperature, derived from transport property measurements, can be found in reference [1]. The small but real effect of ionic strength on hydration numbers, as determined from colligative properties, is illustrated in Table 5.2 [6]. Larger hydration numbers in the more dilute solutions where there is less competition for the solvating water molecules seem eminently reasonable.

Table 5.2 Variation of estimated hydration numbers with salt concentration (from reference [6]).

	Mg^{2+}	Ca^{2+}	Sr^{2+}
$[MCl_2]$ = 0.5M	24.1	23.0	22.7
$[MCl_2]$ = 0.25M	27.6	26.6	26.3

The foregoing discussion has implied a fairly rigid distinction between primary solvating molecules coordinated to the cation and the secondary solvation shell. In some situations the distinction may be difficult to make. It has often been suggested that the primary hydration shell of lanthanide(III) cations may contain water molecules at various distances from the cation. There may, for instance, be six near water molecules at the corners of an octahedron or a trigonal prism, and two or three slightly more distant water molecules just beyond faces or edges of the octahedron or trigonal prism. These latter molecules may be only slightly nearer the cation than the next nearest water molecules in the secondary solvation shell. The recent X-ray diffraction demonstration of eight oxygen atoms equidistant from the cation in La^{3+}aq [7] may disfavour the 'six plus two or three' model for lanthanide cations in solution, but it does not necessarily rule it out in all circumstances, nor for all cations.

The model of an ion in solution given in Fig. 1.2 is dynamic, not static. Solvent molecules move between the various regions depicted at rates characteristic of the ion, the solvent, and the conditions. If the physical technique used to examine the system has a relatively long time-scale, then it will not be able to differentiate between solvent molecules in different environments. N.m.r. spectroscopy, where the frequency is of the order of 10^4 sec^{-1}, can only distinguish between cation primary and bulk solvent when the half-life for residence on the cation of a solvent molecule is significantly greater than 10^{-4} sec. This sort of time indeterminacy may introduce some confusion into the estimation and interpretation of solvation numbers as determined by certain techniques. There is a more fundamental indeterminacy when the residence time of a solvent molecule in the primary solvation shell is so short that it is comparable with the reorientation time characteristic of the bulk solvent.

The effects of different assumptions on estimated values of hydration numbers are illustrated in Tabel 5.3. Although a change in assumption may have a marked effect on individual hydration numbers, the trend within a series of cations is not affected. *Differences* between cation solvation numbers are, for given cation charge, independent of the assumptions made. The dependence of solvation number on assumptions made must always be borne in mind when considering solvation numbers estimated by any of the numerous methods where the actual experimental results refer to cation plus anion rather than to an individual single ion.

Table 5.3 Effect of reference assumptions on the estimation of hydration numbers (HN).

	Cs^+	Mg^{2+}	Ca^{2+}	Sr^{2+}	Ba^{2+}
Transference numbers					
HN of $Cl^- = K^+ = 6$	5	34	27	27	26
HN of $Cl^- = 5$ [8]	6	36	29	29	28
HN of $Cl^- = 0$	11	46	39	39	38
Thermochemical[a] [1]					
HN of $Cl^- = 0$		13.7	12.0	10.7	7.7
HN of $I^- = 0$		19.0	17.0	15.5	15.0

a From activity coefficient measurements.

Although there is a range of values for the hydration number of each cation, depending on methods and assumptions used, trends of values for series of cations are normally similar regardless of the method of investigation. This can be seen from the entries in Tables 5.1, 5.3, and 5.4, and in the remaining sections of this Chapter. The values in Table 4.15 of the previous Chapter, obtained from one series of ultrasonic determinations of compressibilities, also give a good idea of trends in hydration numbers. For cations of a given charge, hydration numbers tend to decrease as cation size increases, though this pattern is by no means universal. For cations of similar size, the hydration number increases greatly as cation charge increases. The patterns for non-aqueous solvents are similar.

5.3   SOLVATION NUMBERS

5.3.1   *sp*-Elements: Group I

These exist as solvated M^+ cations in a wide range of solvents, and are probably the most fully studied group of cations with respect to solvation number investigation. As solvent exchange is always very fast there is no chance of determining primary coordination numbers by the n.m.r. peak area or isotopic dilution methods. However X-ray diffraction indicates a coordination number of four for Li^+ in aqueous solution (Chapter 4.4.1).

A selection of reported hydration numbers for these alkali metal cations, determined by a variety of methods [1, 8-13], is shown in Table 5.4. Differences between different series of estimates are sometimes large, for reasons discussed in the previous section. Nonetheless a general pattern of hydration numbers decreasing steadily from Li^+ to Cs^+ is apparent, though the conductivity and compressibility results do not conform to this trend.

Table 5.4 Hydration numbers for the Group IA cations.

	Li^+	Na^+	K^+	Rb^+	Cs^+	
Transference numbers	22	13	7		6	[8]
	13-14	7-9	4-6		4	[9, 10]
Mobilities	3.5-7	2-4				[11]
	7-21	5-10	5-7			a
Conductivities	2-3	2.5-3.5		4.8	5.5	b
Diffusion	5	3	1	1	1	[1]
Compressibilities	2.7	3.9	3.2			[12]
Entropies	5	4	3	3		[11]
N.m.r.	$3.4-5.0^c$	3.0-4.6	1.0-4.6	1.6-4.0	1.0-3.9	[13]

a From Stokes radii, Chapter 4.2.4; b Gusev's method, Chapter 4.2.4. c Pulsed magnetic resonance techniques have recently indicated a hydration number of 4.9 for Li^+ [Sutter, E. J., Updegrove, D. M., and Harman, J. F., *Chem. Phys. Lett.*, **36**, 49 (1975)].

A selection of solvation numbers in non-aqueous solvents is given in Table 5.5 [1, 14-26]. Again there is a decrease in solvation number on descending the Periodic Table from Li^+ to Cs^+ in the majority of cases. There are some surprisingly low solvation numbers, for example for ethanol (compressibilities) and for sulpholane (conductances). These low numbers cannot simply be dismissed as the unfortunate consequences of unwise assumptions in splitting experimental values for salts into single ion components, since the overall solvation numbers for salts are themselves small. For example, the sum of the ion solvation numbers for $LiNO_3$ and $NaNO_3$ in ethanol are only 3.4 and 2.9 respectively [12].

Most methods of estimating solvation numbers give the sort of trend already mentioned for hydration numbers and sketched as curve (A) of Fig. 5.1. Other trends have been suggested in a few cases. The opposite trend, of solvation num-

Table 5.5 Solvation numbers for alkali metal cations in non-aqueous solvents.

	E_T/kcal mol^{-1}	Li$^+$	Na$^+$	K$^+$	Rb$^+$	Cs$^+$	Method[a]	Ref.
Formic acid	57.9	4	6	6	6	8	N	[14]
Formamide	56.6	5.4	4.0	2.5	2.3	1.9		[15]
Methanol	55.5	7–8[b]	4–6	2–5	4–5	4		[16],c
		3.3	3.9	3.6			C	[12]
N-Methylformamide	54.1	5.5	2.6	1.0		2.6		[17]
N-Methylacetamide	52.0	5.1	3.5	3.3		2.6		[1]
Ethanol	51.9	6	3–5	2–4	3–4	3		[16],c
		1.4	0.9				C	[12]
n-Propanol	50.7		4	3				c
Propylene carbonate	46.6	6	6	6	3	2–3	I	[18]
Acetonitrile	46.0	9	6	3				[1]
		6–9	4–6	2–4				[16]
		4	4				IA	[19]
Dimethyl sulphoxide	45.0	2	6				N	[20]
Sulpholane	44.0	1.4	2.0	1.5	1.4	1.3		[1]
Dimethylformamide	43.8	4					NN	[20]
		5.2	3.3	2.9				[17]
Acetone	42.2	4–5	3–5	3–4			I	[16]
		4						[21]
		4	4				IA	[19]
Methyl ethyl ketone	41.3	4	3–4	2–3				[16]
Pyridine	40.2		4	2–3				[16]
Phenol			5–6	4				[16]
Furfuraldehyde			5	4				[16]
1-Methyl-2-pyrrolidinone		4					NN	[20]
		4					I	[22]
Sulphuric acid		2	3	2	1	0	D	[23]
		3.4	3.0	2.1			O	[24]
Hexamethylphosphoramide			1.8	1.6	1.3	1.2		[25]
Liquid ammonia		3.4	4				d	[26]

a From transport properties (assumptions discussed in Chapter 4) except where otherwise indicated: C = compressibilities (assuming solvation number of 2 for NO$_3^-$), D = densities (assuming solvation number of 3 for Na$^+$), I = infrared spectra in mixed solvents (Chapter 3.2.4), IA = infrared peak areas (Chapter 3.2.4), N = n.m.r. shifts (Chapter 3.2.4), NN = n.m.r. shifts (arbitrary assumption that solvation numbers are smallest integers which fit the observed shift bers increasing steadily from Li^+ to Cs^+, has been proposed for formic acid (Table 5.5) as for water in Gusev's conductivity studies (*v.s.*), but seems unlikely. A curve (curve (B) of Fig. 5.1) peaking at Na^+ or perhaps K^+, as deduced from some ultrasonic and n.m.r. observations, seems more acceptable. Such a curve seems convincing for large solvent molecules such as sulpholane (Table 5.5) where steric restrictions might well limit the number of solvent molecules in close proximity to the small Li^+ cation.

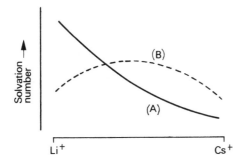

Fig. 5.1   Solvation number trends for the alkali metal cations.

It is impossible to detect an overall pattern in the variation of solvation numbers with solvent nature. Presumably solvation numbers for a given cation are determined mainly by the size and shape of the solvent molecules, and by their donor strengths. Yet plots of solvation numbers for a given cation against parameters representing the solvating powers of the solvents, for example Reichardt's spectroscopic E_T values or Gutmann's thermochemical solvent donor numbers (Chapter 1.3), rarely give satisfactory correlations.

From time to time attempts are made to estimate solvation numbers of these cations from theoretical calculations rather than from experiment. Calculations of free energies of hydration of the cations have been interpreted as favouring a hydration number of six for each of the alkali metal cations. However, these calculations have subsequently been strongly criticised [27]. Calculations using potential energy surface methods suggest that for Li^+ a solvation number of four, with tetrahedral geometry, is energetically favoured over such alternative geometries and solvation numbers as are represented by octahedral, trigonal bipyramidal, or trigonal coordination [28].

5.3.2   *sp*-Elements: Group II

The Be^{2+} cation is noteworthy in that, as shown by n.m.r. spectroscopy, it has a coordination number of four in water and in several non-aqueous solvents [29]. Solvent exchange at Mg^{2+} is also slow enough, at sufficiently low temperatures, for estimation of its primary solvation or coordination number as six in several solvents, including water, liquid ammonia, methanol, and ethanol [29]. There has been some argument as to whether n.m.r. spectra indicate a solvation number of five or six for Mg^{2+} in acetone [30]. This cation also has a coordination number of six in acetonitrile, from infrared spectra peak areas [19] (Chapter 3.2.4). As already reported in Chapter 5.2, the hydration number of Mg^{2+} determined from transport experiments is considerably greater than six, since several water molecules in the secondary hydration shell move with the cation. Similarly, solvation numbers of Mg^{2+} in several non-aqueous solvents also come out as much larger than six when determined from transport properties. Values determined via Stokes radii (Chapter 4.2.4) are about fifteen for several solvents (Table 4.8).

By the time we reach the alkaline earth cations Ca^{2+}, Sr^{2+}, and Ba^{2+} rates of solvent exchange are too fast for n.m.r. peak area determination of primary solvation numbers to be possible. However for these cations, and also for Be^{2+} and Mg^{2+}, there are many estimates of hydration numbers by non-spectroscopic techniques. Some of these hydration numbers are given in Table 5.6. With one exception, the expected trend of decreasing hydration number with increasing cation size is found for each method of estimation. X-ray diffraction experiments have indicated hydration numbers between 6 and 8 for this series of cations [32]. There is very little information on solvation numbers of these cations in non-aqueous media. Measurements of apparent molal volumes have led to the proposal that the solvation numbers of Ca^{2+}, Sr^{2+}, and Ba^{2+} in sulphuric acid are 8, 8, and 5 respectively [23].

The Zn^{2+} cation appears to be six-coordinate in water, from n.m.r. [33] and from Raman [34] evidence, and six in methanol [35]. However there is some evidence that Zn^{2+} is only five-coordinated in some mixed aquo-ligand complexes [36]. Both Zn^{2+} and Cd^{2+} have estimated hydration numbers between 10 and 12 from transport properties and from ionic entropies [1, 11] but hydration numbers of 6 are indicated by Gusev's conductivity method (Chapter 4.2.4) for these cations and for Hg^{2+}. Yet smaller hydration numbers (Zn^{2+} 3.9, Cd^{2+} 4.6,

Table 5.6   Hydration numbers for Group IIA cations.

	Be^{2+}	Mg^{2+}	Ca^{2+}	Sr^{2+}	Ba^{2+}	
N.m.r. peak areas	4	6				[29]
Transference numbers		12–14	8–12	8–10	3–5	[9]
Mobilities		10.5–13	7.5–10.5		5–9	[11]
Conductivities (Gusev)	4	8	8	8	8	a
Entropies		13	10		8	[11]
Diffusion	10	9	9	9	8	[1]
Activity coefficients		5.1	4.3	3.7	3.0	[1]
^{1}H n.m.r. linewidths		3.8	4.3	5.0	5.7	[31]

a See Chapter 4.2.4.

Hg^{2+} 4.9) were obtained by the ^1H n.m.r. linewidth method [31], but it should be remembered that this method gives a hydration number of only 3.8 for Mg^{2+}. Further details of estimates of hydration numbers of Zn^{2+} are included in Table 5.7, where this cation is included for comparison with adjacent transition metal(II) cations.

5.3.3 *sp*-Elements: Group III

Solvent exchange at Al^{3+} and Ga^{3+} is generally slow enough, particularly at low temperatures, for the primary solvation numbers of these cations to have been measured in a range of solvents by the n.m.r. peak area method. A primary solvation number of six has been established for Al^{3+} in water, methanol, ethanol, acetonitrile, dimethylformamide, dimethyl sulphoxide, and liquid ammonia, using variously ^1H, ^{14}N, and ^{17}O n.m.r. spectroscopy [29]. A primary solvation number of six has also been demonstrated for Ga^{3+} in water, acetonitrile, and dimethylformamide [29] and in methanol [37]. Again a primary solvation number of six has been found for In^{3+} in aqueous acetone, where only the water solvates the cation of course [38]. Density measurements were also consistent with $In(OH_2)_6^{3+}$ [39]. Hexamethylphosphoramide is an unusually bulky solvent, so much so that the primary solvation number of Al^{3+} is only four in this solvent. This was demonstrated by ^{27}Al n.m.r. spectroscopy, using hexamethylphosphoramide+nitromethane solvent mixtures rather than neat hexamethylphosphoramide [2].

Studies by classical physical chemistry techniques to establish effective overall hydration numbers have been rare for members of this series of cations. Such results as are available indicate the expected heavy secondary solvation due to the high charge on these cations. Thus hydration numbers for Al^{3+} of 11.9, 13, 21, and 31 have been estimated from activity coefficient, diffusion, entropy, and compressibility studies respectively [1, 29].

Thallium(I) as a cation in solution bears a certain resemblance to the alkali metal cations of Group I. Its solvation properties have been much less studied than those of the Group I cations, and indeed less studied than those of the other 1+ cation so beloved of physical chemists, Ag^+. From entropy considerations a hydration number of 3 has been suggested for Tl^+. This value seems reasonable, being equal to that derived by this method for the similar size cations K^+ and Rb^+ [11] (ionic radii are Tl^+ 1.40Å, K^+ 1.33Å, Rb^+ 1.48Å [40]). Diffusion experiments lead to values of one for the hydration numbers of Tl^+, K^+, and Rb^+ [1].

5.3.4 *sp*-Elements: Groups IV to VII

The incidence of simple solvated cations in these Groups is, as mentioned in Chapter 1, infrequent. The two best approximations are Sn^{2+} and Pb^{2+} in aqueous solution. Their respective hydration numbers have been estimated as 9 and 8 from entropy considerations [11]; compare the estimated hydration number of 8 for Ba^{2+} according to the same method and assumptions. Ionic radii of 1.12Å for Sn^{2+} and of 1.20Å for Pb^{2+} are comparable with ionic radii of 1.13Å and 1.35Å for Sr^{2+} and Ba^{2+} respectively [40]. The ionic mobility of Pb^{2+} suggests a hydration number of between 4 and 7.5 (Ba^{2+} is again similar) [11]. ^{1}H n.m.r. line-broadening suggests 5.7 [31]. Solutions containing tin(IV) dissolved in aqueous acetone have been claimed, on the basis of their n.m.r. spectra, to contain hexahydrated Sn^{4+} cations [33].

Evidence for the existence of the simple Bi^{3+}aq cation in aqueous solution is, to put it generously, flimsy. Yet this cation has been used as the point of reference for estimating hydration numbers for several 3+ cations from the results of diffusion experiments [41]. By the time Groups VI and VII are reached, the only possible candidates for inclusion here are exotic species such as Po^{2+} whose hydration numbers have yet to be investigated.

5.3.5 Transition Metals: 2+ Cations

It is a normal and reasonable assumption that these cations have an octahedral primary solvation shell in solution in water and in polar non-aqueous solvents. Take the cases of the Fe^{2+}, Co^{2+}, and Ni^{2+} cations, where relaxation times permit the observation of suitable n.m.r. spectra. Here coordination numbers of six have been established by the peak area method for water (all three cations), for methanol and dimethyl sulphoxide (Co^{2+} and Ni^{2+}), and for acetontrile and dimethylformamide (Co^{2+}) [29]. Hexahydration of Co^{2+}, of Ni^{2+}, and of Mn^{2+} has also been demonstrated from ultraviolet-visible spectra (Chapter 3.1.3). A primary solvation shell containing six methanol molecules around Co^{2+} or Ni^{2+} is indicated by isotopic dilution experiments (Chapter 4.5.1). However the primary coordination sphere may not be a regular octahedron in the case of some cations. Cations with a d^4 or d^9 electronic configuration, specifically Cr^{2+} and Cu^{2+}, have tetragonally distorted primary solvation shells as a consequence of the Jahn-Teller effect. There have been suggestions of a primary coordination number of five for Cu^{2+} under special circumstances [42] (see the adjacent Zn^{2+} cation, Chapter 5.3.2).

The crystal field stabilisation of a tetrahedral set of ligands is always less than that for an octahedral set of the same ligands. However the difference is smallest for the d^7 configuration, Co^{2+} in the present context. Evidence has been presented, though not universally accepted, indicating significant amounts of tetrasolvento-species in equilibrium with hexasolvento-species for Co^{2+} in water and in dimethylacetamide, at least at elevated temperatures [5]. Equilibrium constants ($K = $ [tetrahedral]/[octahedral]) have been determined spectrophotometrically in the latter solvent. At 70°C $K = 0.19$, at 25°C $K = 0.028$, and by $-10°$C the amount of tetrahedral species has become immeasurably small. A plot of logarithms of K values against reciprocal temperature is not linear, so it is not possible to specify an enthalpy change for the tetrahedral ⇌ octahedral transformation.

If the solvent molecules are sufficiently bulky, then the primary coordination shell may contain less than six solvent molecules. An example of this is again provided by hexamethylphosphoramide. The solvation number for Co^{2+} in this solvent is only four, a situation analogous to that mentioned in Chapter 5.3.3 for Al^{3+}. The other type of situation in which a coordination number of four is favoured is that of such d^8 cations as favour square-planar geometry. Ni^{2+} does not form such solvento-ions, but Pd^{2+} is believed to have a square-planar primary hydration sphere in aqueous solutions of, for example, palladium(II) perchlorate.

Hydration numbers are tabulated in Table 5.7. The Zn^{2+} cation has also been included in this Table because it bears a close resemblance to its immediate predecessor Cu^{2+}. For a given method of investigation, hydration numbers for the first row transition metal 2+ cations vary little along the row. This is just the behaviour that one would expect, since the charge is constant and the cation radii cover only a small range, from 0.80Å for Mn^{2+} to 0.74Å for Zn^{2+} [40].

Table 5.7   Hydration numbers for 2+ cations of the transition metals.

	Mn^{2+}	Fe^{2+}	Co^{2+}	Ni^{2+}	Cu^{2+}	Zn^{2+}	
N.m.r. peak areas		6	6	6		6	a
Transference numbers[b]						10-12.5	[10]
Transference numbers[c]					34	44	[1]
Mobilities		10-12.5			10.5-12.5	10-12.5	[8]
Entropies		12			12	12	[1]
Redox potentials					12		[43]
Diffusion[d]	12	12	12	12	11	11	[1, 44]
Diffusion[e]				8.5	7.4	9	[1, 41]
Activity coefficients	11	12	13	13		20^f	[1]

a From Chapter 2.2.1; b assuming zero hydration for large organic cations; c assuming a hydration number of 5 for Cl^-; d assuming hydration numbers of 1 for both K^+ and Cl^-; e assuming zero hydration for Ba^{2+}; f a hydration number of 5.3 was estimated in another study, in which zero hydration of Cs^+ was assumed.

5.3.6   Transition Metals: 3+ Cations

Hydration numbers are listed in Table 5.8, including primary from n.m.r. spectroscopy, isotopic dilution, or electronic spectra comparison methods, and primary plus secondary from other methods [45-52]. The expected primary hydration number of six is found for these cations. The diffusion hydration numbers for Cr^{3+} and for Fe^{3+} may be compared with the value of 13 estimated

Table 5.8 Hydration numbers for 3+ cations of the transition metals.

	Sc^{3+}	Ti^{3+}	V^{3+}	Cr^{3+}	Fe^{3+}	Y^{3+}	Rh^{3+}
N.m.r. peak areas	4–5^a [33, 45]		6^{b,c} [47]	6^c [49]		2.5^a [33, 45]	
Isotopic dilution				6 [50]			6 [52]
Ultraviolet-visible		6 [46]	6 [48]	6 [29, 51]	6 [29, 51]		
Diffusion^a				16.5 [41]	10.5 [41]		

a Probably low due to the presence of ion-pairing nitrate; b ^{1}H n.m.r.; c ^{17}O n.m.r.; d assuming a hydration number of zero for Bi^{3+}

by this method for the analogous sp-element cation Al^{3+}. Primary solvation numbers of six have been demonstrated for several of these cations in non-aqueous media (Table 5.9 [53-56]). Estimation of the hydration number of Co^{3+} is rendered virtually impossible by this cation's propensity for oxidising water. In liquid ammonia solution Co^{3+} has the expected solvation number of six, as has been shown by n.m.r. spectroscopy [29].

Table 5.9　Solvation numbers for 3+ cations of the transition metals in non-aqueous solvents.

	Solvents	Solvation number	Methods[a]	
Sc^{3+}	DMSO, DMF, DMA	6	X	[53]
Ti^{3+}	MeOH, EtOH	6	U, X	[54]
V^{3+}	MeOH, EtOH, i-BuOH	6	U	[48]
Mn^{3+}	DMSO, DMF	6	U, X	[55]
Fe^{3+}	MeOH	6	I	[56]

[a] I = isotopic dilution, U = ultraviolet-visible spectroscopy, X = crystal stoichiometries.

5.3.7　Other Transition Metal Cations

Transition metal(IV) cations are rarely simple to deal with (see Chapter 1) and estimates of their solvation numbers seem rarely to have been made (see reference [57], where zirconium(IV) is discussed).

Transition metal(I) cations are also rare, with species such as Cu^+ being unstable in aqueous solution. This Cu^+ cation has been assumed to be four-co-ordinate in water [58], and recently an estimate of a hydration number of five has been made from redox potentials [43]. However there is one important and much studied cation in this category, and that is Ag^+. Solvent exchange at this cation is very fast, so there can be, sadly, no direct n.m.r. spectroscopic estimates of its primary solvation number. Estimates of the hydration number of Ag^+ range from 0.3 (arguable) [59] via, for example, 2 by diffusion [1] and 3.1 by ultrasonics [12] to 4 from entropy considerations [11]. The pH dependence of the solubility of silver hydroxide has been interpreted in terms of equilibrium between unhydrated Ag^+ and $Ag(OH_2)_2^+$ [60]. Where direct comparison is possible, the hydration number of Ag^+ is about the same as that for K^+ (ionic radii Ag^+ 1.26, K^+ 1.33Å [40]). Solvation numbers of Ag^+ and K^+ are also similar in methanol, ethanol, acetonitrile, and ethyl acetate (from mobilities [11]), in formamide and dimethylacetamide (Table 4.8, Chapter 4.3.4) and in sulphuric acid (from apparent molar volumes [18] and from osmotic co-efficients [24]). The solvation number of Ag^+ in acetonitrile is 4 according to peak areas from solution infrared spectra [61] (Chapter 3.2.4). In propylene carbonate it is 6 from infrared spectra in propylene carbonate+nitromethane solutions [18] (Chapter 3.2.4). Again these numbers are the same as those reported for Na^+ and K^+ in analogous situations.

5.3.8 Lanthanide and Actinide Cations

These cations exhibit a variety of coordination numbers and stereochemistries in the solid state [62]. Though there are many examples of octahedral six-coordination there are also a large number of examples of higher coordination numbers. Seven and eight coordination, in each case in a variety of geometrical arrangements, are particularly common. Nine, ten, and twelve coordination are known for elements from both series, while a compound containing an actinide element with a coordination number of fourteen, $U(BH_4)_4$, has been reported [63]. In view of this information from the solid state it would hardly be surprising to find a varied and complicated pattern of solvation numbers for lanthanide and actinide cations in solution. Also, especially for solvation numbers greater than six, there is the possibility of more than one type of primary solvating solvent molecule at slightly different distances from the cation.

Lanthanides. It is not possible to determine hydration numbers of lanthanide cations directly by the n.m.r. peak area method in aqueous solution. However such determinations can be made for some of these cations in aqueous acetone or aqueous Freon. In mixtures of suitable composition, n.m.r. spectra of such solutions can be obtained at temperatures as low as $-120°C$. At $-100°C$ a solution of lutecium perchlorate in aqueous acetone gives a two-peak spectrum (Chapter 2.2), whence a hydration number of six for Lu^{3+} has been derived from the ratio of the bulk to coordinated solvent peaks [64]. Similarly a hydration number of six for La^{3+} has been estimated from n.m.r. spectra run at temperatures of -110 to $-115°C$ [65]. Paramagnetic $M^{3+}aq$ ions are more difficult to study in this way, though it has proved possible to estimate hydration numbers of six and four respectively for Ce^{3+} and Yb^{3+} [64, 65]. In all these cases the possibility of significant ion-pairing or even complex formation, not impossible at such low temperatures in mixed solvents with considerably lower dielectric constants than water, should be borne in mind. True hydration numbers could be higher than those reported.

X-ray diffraction studies of solutions containing La^{3+} have indicated a hydration number of eight for this cation [7]. Proton relaxation times indicate a hydration number of eight or nine for Gd^{3+} [66]. Crystal stoichiometries long ago suggested that a hydration number of nine was possible for lanthanide(III) cations. This suggestion followed from the knowledge of the existence of $Ln(OH_2)_9^{3+}$ units in the crystal structures of such salts as $Nd(BrO_3)_3.9H_2O$ [67].

On the other hand, a Job's method investigation of the spectrum of Pr^{3+} in water, in ethanol, and in aqueous ethanol indicated a hydration number of six for this cation [68].

Thus there are indications of hydration numbers of six, eight, or perhaps nine, for lanthanide(III) cations. Controversy as to whether the hydration numbers of the lanthanide(III) cations vary along the series continues. The commonest suggestion here is that the aquo-cations are nine-coordinate for the earlier members of the series, and eight-coordinate for the later, smaller, members. The suggestion has also been made that $La^{3+}aq$ [69] and, perhaps [70], $Nd^{3+}aq$ [71] may exist as an equilibrium mixture of two forms, differing in geometry and perhaps also in coordination number. Molar volumes, activity coefficients, transference numbers, conductances [72], and thermochemical studies [73] have all provided evidence which can be interpreted in terms of a change in coordination number around the middle of the series of lanthanide(III) cations. In contrast, the near constancy of $\bar{C}_p^{\ominus}$ values along the series [74], the regular pattern of ^{17}O n.m.r. shifts [70], the trend of ionic entropies [75], and calculations based on a simple electrostatic model [76] all suggest a constant hydration number throughout. The present complicated situation, and correspondingly complicated explanation, are the subject of current discussion [77].

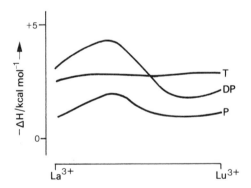

Fig. 5.2   The variation of the enthalpy of formation of complexes of the lanthanide(III) cations with atomic number; T = tropolonate, P = picolinate, DP = dipicolinate (Fig. 5.3) [78].

Irregularities in the trends of enthalpies and entropies of complex formation along the series (Fig. 5.2) are more satisfactorily explained by a change in coordination number of the cations in the complexes rather than in the aquo-ions themselves. Thus plots of enthalpies or entropies of complex formation with picolinate, dipicolinate, acetate, or glycolate against atomic number have kinks around the middle, but analogous plots for tropolonate or kojate (Fig. 5.3) do not [78]. A similar pattern has been described for oxodiacetate and iminodiacetate versus maleate [79]. The current view is that there is a change in co-ordination number along the lanthanide series when the ligand concerned is, for example, picolinate, acetate, or tartrate [80], but not for such ligands as tropolonate. A similar contrast may obtain even for such simple ligands as chloride versus nitrate [81]. Kinetic studies of Ln(edta)⁻aq complexes also suggest a change in coordination number [82]. Such a dependence of cation coordination number on the nature of the ligand recalls the behaviour of zinc(II) (*v.s.*).

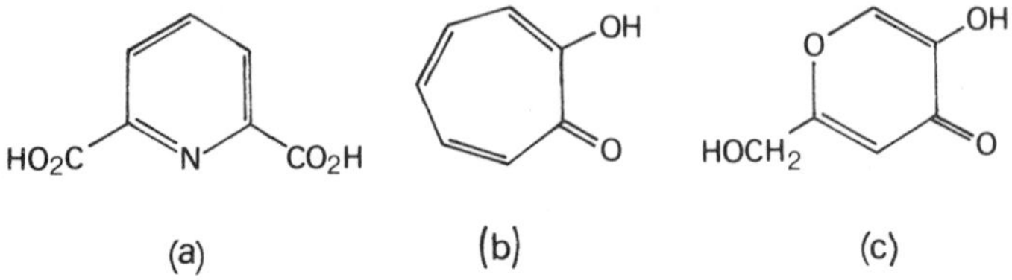

(a)   (b)   (c)

Fig. 5.3   (a) Dipicolinic acid; (b) tropolonic acid; (c) kojic acid.

One set of hydration numbers from transport properties suggests that the amount of water moving with the cation is independent of the nature of the lanthanide(III) cation; an overall hydration number of between 13 and 14 is indicated throughout (Table 4.8 of Chapter 4.2.4). A second series derived from transport properties suggests a small increase in overall hydration numbers, from 12.8 to 13.9, as the series is traversed [1]. Thus, particularly if the primary hydration number decreases along the series, secondary solvation may increase slightly towards the right hand end of this series of cations. Such a situation is not inconsistent with the decreasing radii of the cations themselves in this direction.

Actinides. The body of information relating to solvation numbers of actinide cations is much more slender. An n.m.r. peak area estimate of the hydration number of Th^{4+}, in an aqueous acetone solution of its perchlorate at $-100°C$, indicated a value of 9 [33]. An indirect n.m.r. linewidth method (Chapter 2.4.6) suggested a hydration number of 10 for this cation [31]. Still for Th^{4+}, Gusev's conductivity method (Chapter 4.2.4) gives a primary plus secondary hydration number of 20, and compressibility plus ionic vibration potential yield a value of 22 [13, 83].

A general review of the available evidence pertaining to hydration numbers of U^{4+} and Np^{4+} has suggested that two forms of each of these aquo-cations may exist, differing in geometry and possibly in coordination number [84] †. There may [77] or may not [76, 77] be a change in the primary hydration number of the actinide(III) cations along the series (see lanthanides above).

† A hydration number of eight has several times been claimed, on rather slender evidence, for Np^{4+} [Sullivan, J. C., Gordon, S., Cohen, D., Mulac, W., and Schmidt, K. H., *J. Phys. Chem.*, **80**, 1684 (1976) and references therein].

For the UO_2^{2+} cation, low temperature n.m.r. spectra indicate a hydration number of 4 [33] [††], Raman spectroscopy 6 [85]. These are considered to be primary hydration numbers. However, similar hydration numbers are obtained by methods in which the secondary hydration shell may also be making a contribution. Thus activity coefficient measurements suggest a hydration number of 7.4 (relative to an assumed hydration number of zero for Cs^+) [1]. Similarly, hydration numbers of between 2 and 6, and of 7, have been derived from ultrasonic [86] and conductivity (Chapter 4.2.4) methods respectively.

Solvation numbers of actinide cations in non-aqueous media have been still less studied. Both n.m.r. spectroscopy and crystal stoichiometries demonstrate a solvation number of 2 for the UO_2^{2+} cation in tributyl phosphate, but the presence of nitrate which may well be complexed to the cation means that confidence in this result is less than absolute [87].

5.4   SECONDARY SOLVATION AND ION ASSOCIATION

There is little information about secondary solvation around metal complexes. N.m.r. investigations of $Ni(en)_3^{2+}$ in water and in dimethyl sulphoxide [88], of $Cr(en)_3^{3+}$ in dimethyl sulphoxide [89], of $Ru(NH_3)_6^{3+}$ in water [90], of $Pt(NH_3)_2(butane-1,3-diamine)^{2+}$ in various solvents [91], and of the tris(4,7-dimethyl-1,10-phenanthroline)chromium(II) cation in methanol and water [92] have been reported. Ultrasonic studies of cobalt(III) [93, 94], of platinum(II) and palladium(II) [95], and of platinum(IV) [93] complexes have been undertaken in aqueous solution. Secondary hydration numbers of about 20 were estimated by one group of workers for $Co(NH_3)_6^{3+}$, $Co(NH_3)_5Cl^{2+}$, and $Pt(NH_3)_5Cl^{3+}$ [93], but lower values in the region of 2 to 5 by a second group for a range of cobalt(III) complexes including $Co(NH_3)_6^{3+}$ [94]. The usefulness of n.m.r. spectroscopy in monitoring the composition of the secondary solvation shell around complex ions, for example $Cr(NCS)_6^{3-}$ [96], $Cr(ox)_3^{3-}$ [97], and trans-$Cr(NH_3)_2(NCS)_4^-$ [98], in mixed aqueous solvents has also been demonstrated.

Analogous studies of secondary solvation around solvated metal cations, in single or in mixed solvents, have been rare and sometimes unsatisfactory [99]. Proton n.m.r. T_1 values for aqueous solutions containing Ni^{2+} are consistent with primary and secondary hydration shells with an average nickel-oxygen distance for secondary waters 1.6 times the nickel–oxygen distance for primary waters. This is just the ratio expected from simple geometry [100]. There is also some information on the primary and secondary solvation around Mn^{2+} and Cr^{3+} from proton T_1 measurements [101].

X-ray diffraction studies of aqueous solutions have given metal–water dis-

†† Recently it has been claimed that low temperature ^{1}H n.m.r. spectra of solutions of UO_2^{2+} in aqueous acetone indicate a hydration number of 6. [Shcherbakov, V. A., and Shcherbakova, L. L., Radiokhimiya, 18, 207 (1976)].

tances for the secondary hydration shell around the $PtCl_6^{2-}$ and $PtBr_6^{2-}$ anions [102], and some indication of secondary water molecules taking up positions just beyond the centres of the octahedral faces defined by the primary waters of solvation of some aquo-cations. A similar geometrical arrangement was proposed for dimethyl sulphoxide around Cr^{3+}. In dimethyl sulphoxide solutions containing Cr^{3+} there is a solvent exchange process which can be monitored by variable temperature n.m.r. spectroscopy. The rate constant at $25°C$ is 1.3×10^3 sec^{-1} and the activation enthalpy 6.9 kcal mol^{-1}. Such kinetic parameters cannot possibly refer to exchange of primary dimethyl sulphoxide of solvation in this case (see Chapter 11.3), so they have been assigned to exchange of dimethyl sulphoxide between the secondary solvation shell of the Cr^{3+} and bulk solvent [103]. A similar exchange process has been proposed for $Cr(en)_3^{3+}$ in dimethyl sulphoxide solution. Here a rate constant of 1.2×10^3 sec^{-1} and an activation enthalpy of 6.6 kcal mol^{-1}, both very similar to the Cr^{3+} in dimethyl sulphoxide system just mentioned, are reported [89].

In strong solutions of salts there is a possibility that an anion may enter the secondary solvation shell of a cation. This gives an ion-pair, a species exactly analogous to a cation in a mixed solvent having a secondary solvation shell containing molecules of both solvent components. As in the cases of secondary solvation discussed above, the greater part of the information on ion-pairing refers not to solvento-cations but to complexes of metal ions with non-solvent ligands.

The formation of ion-pairs [104], otherwise known as outer-sphere complexes, can be treated as a problem in electrostatics. Starting from the Maxwell-Boltzmann distribution of ions of a given charge around an ion of opposite charge, and assuming an arbitrary distance for the limit of significant ion association, the equilibrium constant K_{ip} for ion-pair formation can be derived as

$$K_{ip} = \frac{4N_A}{1000} \int \exp\left[-\frac{z_+ z_- e^2}{DkT}\right] r^2 \, dr.$$

In this expression N_A is Avogadro's number, z_+ and z_- are the charges on the ions, e is the charge of the electron, D is the dielectric constant of the medium, k is Boltzmann's constant, T is the temperature, and r the distance between the centres of the two ions [105]. Thus ion-pair formation is favoured by high charges on the ions, small radii, and low solvent dielectric constant. 1:1 electrolytes do not give significant ion-pairing in water, but if their radii are small may do so in non-aqueous media. This pattern is exemplified by most of the large number of K_{ip} values reported, especially for the $Co(NH_3)_6^{3+}$ and $Co(en)_3^{3+}$ cations [106]. However there are a significant number of exceptions to this general pattern, indicating the importance of other factors beyond simple electrostatics in at least some cases [105].

Ion-pairing constants can be estimated by a variety of experimental techniques, including ultraviolet-visible and n.m.r. spectroscopies, polarimetry, polarography, conductivity, and solubility measurements. Agreement between different methods is not always good. In all cases meaningful results can only be obtained when the inner-sphere complex persists long enough for the outer-sphere association equilibrium to be established and monitored. Thus the inner-sphere complex must be kinetically inert for the determination of K_{ip} values by the methods listed above. This requirement immediately reduces the list of qualifying complexes to those of such centres as Cr^{3+}, Co^{3+}, Rh^{3+}, and Pt^{4+}. Attempts to determine K_{ip} values for the association of labile aquo-cations with anions will in general result in the formation of inner- rather than outer-sphere complexes. As far as aquo-cations are concerned, there is one obvious choice for study, Cr^{3+}aq. For this aquo-cation, ion association has been qualitatively demonstrated for a few anions, for example PF_6^- and CrO_4^{2-}. Values of K_{ip} have been reported only for association with thiocyanate, where $K_{ip} = 7M^{-1}$ at $25°C$, and sulphate, where K_{ip} is, as expected, slightly larger at $40M^{-1}$, again at $25°C$ [106].

The association of labile solvento-cations with potential ligands is an important feature of mechanisms of complex formation. Occasionally K_{ip} (K_{os}) values can be estimated from kinetic experiments (ultrasonic or stopped-flow), but more usually they are estimated (Chapter 12.1). The existence of ion-pairing between labile solvento-cations and polyatomic anions can sometimes be inferred from the cation perturbation of anion vibration frequencies (Chapter 3.2.5). There is an extensive literature relating to ion-pairs between alkali metal cations and organic anions in aprotic solvents. Here the emphasis is very firmly on the properties, especially e.s.r. spectra, of the organic anions, so this topic is not covered here.

FURTHER READING

Amis, E. S. in *Techniques of Chemistry*, ed. Weissberger, A., *Volume VIII; Solutions and Solubilities* ed. Dack, M. R. J., Part I, Ch. 3.

Lincoln, S. F., *Coord. Chem. Rev.*, **6**, 309 (1971).

Hinton, J. F. and Amis, E. S., *Chem. Rev.*, **71**, 627 (1971).

Padova, J. I. in *Modern Aspects of Electrochemistry, Vol. 7*, ed. Conway, B. E., and Bockris, J.O'M., Butterworths (1972).

REFERENCES

[1] Hinton, J. F. and Amis, E. S., *Chem. Rev.*, **71**, 627 (1971).

[2] Delpuech. J. –J., Khaddar, M. R., Peguy, A. A. and Rubini, P. R., *J. Chem. Soc. Chem. Commun.*, 154 (1974); *J. Amer. Chem. Soc.*, **97**, 3373 (1975).

[3] Chmelnik, A. M. and Fiat, D., *J. Amer. Chem. Soc.*, **93**, 2875 (1971);

Neely, J. W. and Connick, R. E., *J. Amer. Chem. Soc.*, **94**, 3419 (1974).

[4] Matwiyoff, N. A. and Darley, P. E., *J. Phys. Chem.*, **72**, 2659 (1968).

[5] Drago, R. S. and Matwiyoff, N. A., *Acids and Bases*, Heath (1968) p. 118; Gutmann, V., Beran, R., and Kerber, W., *Mh. Chem.*, **103**, 764 (1972).

[6] Bourion, F. and Hun, O., *C. R. Hebd. Séanc. Acad. Sci., Paris*, **198**, 1921 (1934); Bourion, F. and Rouyer, E., *ibid.*, **198**, 1944 (1934); Rouyer, E., *ibid.*, **198**, 1156 (1934).

[7] Smith, L. S. and Wertz, D. L., *J. Amer. Chem. Soc.*, **97**, 2365 (1975).

[8] Rutgers, A. T. and Hendrikx, Y., *Trans. Faraday Soc.*, **58**, 2184 (1962).

[9] Glasstone, S., *Textbook of Physical Chemistry, 2nd ed.*, Macmillan (1956) p. 921.

[10] Remy, H., *Trans. Faraday Soc.*, **23**, 381 (1927).

[11] Conway, B. E. and Bockris, J. O'M. in *Modern Aspects of Electrochemistry, Vol. 1*, ed. Bockris, J. O'M., Butterworths (1954) Ch. 2.

[12] Allam, D. S. and Lee, W. H., *J. Chem. Soc. A*, 5, 426 (1966).

[13] Bockris, J. O'M. and Saluja, P. P. S., *J. Phys. Chem.*, **76**, 2298 (1972); and references therein.

[14] Rode , B. M., *Z. Anorg. Allg. Chem.*, **399**, 239 (1973).

[15] Motley, J. M. and Spiro, M., *J. Phys. Chem.*, **70**, 1502 (1966).

[16] Ulich, H., *Trans. Faraday Soc.*, **23**, 388 (1927).

[17] Ryabikova, V. M., Krumgal'z, B. S., and Mishchenko, K. P., *Russ. J. Phys. Chem.*, **45**, 1451 (1971).

[18] Yeager, H. L., Fedyk, J. D., and Parker, R. J., *J. Phys. Chem.*, **77**, 2407 (1973).

[19] Perelygin, I. S. and Klimchuk, M. A., *Russ. J. Phys. Chem.*, **47**, 1138, 1402 (1973).

[20] Lassigne, C. and Baine, P., *J. Phys. Chem.*, **75**, 3188 (1071); Maxey, B. W. and Popov, A. I., *J. Amer. Chem. Soc.*, **90**, 4471 (1968); Wuepper, J. L. and Popov, A. I., *J. Amer. Chem. Soc.*, **92**, 1493 (1970).

[21] Baum, R. G. and Popov, A. I., *J. Solution Chem.*, **4**, 441 (1975).

[22] Wuepper, J. L. and Popov, A. I., *J. Amer. Chem. Soc.*, **91**, 4352 (1969).

[23] Flowers, R. H., Gillespie, R. J., Robinson, E. A., and Solomons, C., *J. Chem. Soc.*, 4327 (1960); Gillespie, R. J. in *Chemical Physics of Ionic Solutions* ed. Conway, E. E. and Barradas, R. G., Wiley/Interscience (1966) p. 599.

[24] Lee, W. H. in *The Chemistry of Non-aqueous Solvents, Vol. 2*, ed. Lagowski, J. J., Academic Press (1967) Ch. 3, p. 114.

[25] Paul, R. C., Banait, J. S., and Narula, S. P., *Z. Phys. Chem. Frankf. Ausg.*, **94**, 199 (1975).

[26] Cotton, F. A. and Wilkinson, G., *Advanced Inorganic Chemistry, 3rd ed.*, Wiley/Interscience (1972) pp. 198–199.

[27] Morf, W. E. and Simon, W., *Helv. Chim. Acta.*, **54**, 794 (1971); Rosseinsky, D. R., *A. Rep. Chem. Soc.*, **68A**, 81 (1971).

[28] Kollman, P. A. and Kuntz, I. D., *J. Amer. Chem. Soc.*, **96**, 4766 (1974).

[29] Lincoln, S. F., *Coord. Chem. Rev.*, **6**, 309 (1971).

[30] Green, R. D. and Sheppard, N., *J. Chem. Soc. Faraday II*, **68**, 821 (1972); Toma, F., Villemin, M., and Thiéry, J. M., *J. Phys. Chem.*, **77**, 1294 (1973); Covington, A. D. and Covington, A. K., *J. Chem. Soc. Faraday I*, **73**, 831 (1975).

[31] Swift, T. J. and Sayre, W. G., *J. Chem. Phys.*, **44**, 3567 (1966).

[32] Albright, J. N., *J. Chem. Phys.*, **56**, 3783 (1972).

[33] Butler, R. N. and Symons, M. C. R., *Trans. Faraday Soc.*, **65**, 945 (1969); Fratiello, A., Lee, R. E., and Schuster, R. E., *Inorg. Chem.*, **9**, 391 (1970).

[34] Irish, D. E., McCarroll, B., and Young, T. F., *J. Chem. Phys.*, **39**, 3436 (1963).

[35] Al-Baldawi, S. A. and Gough, T. E., *Can. J. Chem.*, **47**, 1417 (1969).

[36] Cayley, G. R. and Hague, D. N., *Trans. Faraday Soc.*, **67**, 2896 (1971).

[37] Richardson, D. and Alger, T. D., *J. Phys. Chem.*, **79**, 1733 (1975).

[38] Fratiello, A., Davis, D. D., Peak, S., and Schuster, R. E., *Inorg. Chem.*, **10**, 1627 (1971).

[39] Celeda, J. and Tuck, D. G., *J. Inorg. Nucl. Chem.*, **36**, 373 (1974).

[40] Greenwood, N. N., *Ionic Crystals, Lattice Defects, and Non-Stoichiometry* Butterworths (1968) pp. 40–41.

[41] Spandau, H. and Spandau, C., *Z. Phys. Chem.*, **192**, 211 (1943).

[42] Nord, G. and Matthes, H., *Acta Chem. Scand.*, **28A**, 13 (1974).

[43] Malyszko, J. and Duda, L, *Mh. Chem.*, **106**, 633 (1975).

[44] Gapon, E. N., *Z. Anorg. Allg. Chem.*, **168**, 125 (1927); *J. Russ. Phys-Chem. Soc.*, **60**, 237 (1928).

[45] Fratiello, A., Lee, R. E., Nishida, V. M., and Schuster, R. E., *J. Chem. Phys.*, **50**, 3624 (1969).

[46] Hartmann, H. and Schläfer, H. L., *Z. Phys. Chem.*, **197**, 116 (1951).

[47] Chmelnik, A. M. and Fiat, D., *J. Magnetic Resonance*, **8**, 325 (1972).

[48] Hartmann, H. and Schläfer, H. L., *Z. Naturf.*, **6A**, 754 (1951).

[49] Alei, M., *Inorg. Chem.*, **3**, 44 (1964).

[50] Hunt, J. P. and Taube, H., *J. Chem. Phys.*, **18**, 757 (1950).

[51] Hartmann, H. and Schläfer, H. L., *Angew. Chem.*, **66**, 768 (1954).

[52] Plumb, W. and Harris, G. M., *Inorg. Chem.*, **3**, 542 (1964).

[53] Kůtek, F., *Colln. Czech. Chem. Commun. Engl. Edn.*, **32**, 3767 (1967); Kůtek, F. and Petrů, F., *ibid.*, **33**, 296 (1968).

[54] Pittel, B. and Schwarz, W. H. E., *Z. Anorg. Allg. Chem.*, **396**, 152 (1973).

[55] Prabhakaran, C. P. and Patel, C. C., *J. Inorg. Nucl. Chem.*, **30**, 867 (1968).

[56] Rogers, T. E., Swinehart, J. H., and Taube, H., *J. Phys. Chem.*, **69**, 134 (1965).

[57] Fratiello, A., Vidulich, G. A., and Mako, F., *Inorg. Chem.*, **12**, 470 (1973); Claude, R. and Vivien, D., *Bull. Soc. Chim. Fr.*, 65 (1974).

[58] Dockal, E. R., Everhart, E. T., and Gould, E. S., *J. Amer. Chem. Soc.*,

93, 5661 (1971).

[59] Akitt, J. W., *J. Chem. Soc. Dalton,* 175 (1974); Brown, R. D. and Symons, M. C. R., *ibid.,* 426 (1976).

[60] McGowan, J. C., *Rec. Trav. Chim.,* **85**, 777 (1966).

[61] Chang, T. -C. G. and Irish, D. E., *J. Solution Chem.,* **3**, 161 (1974).

[62] Muetterties, E. L. and Wright, C. M., *Q. Rev. Chem. Soc.,* **21**, 109 (1967).

[63] Bernstein, E. R., Hamilton, W. C., Keiderling, T. A., La Placa, S. J., Lippard, S. J. and Mayerle, J. J., *Inorg Chem.,* **11**, 3009 (1972).

[64] Fratiello, A., Kubo, V., and Vidulich, G. A., *Inorg. Chem.,* **12**, 2066 (1973).

[65] Fratiello, A., Kubo, V., Peak, S., Sanchez, B., and Schuster, R. E., *Inorg. Chem.,* **10**, 2552 (1971).

[66] Morgan, L. O., *J. Chem. Phys.,* **38**, 2788 (1963).

[67] Helmholz, L., *J. Amer. Chem. Soc.,* **61**, 1544 (1939).

[68] Chernova. R. K., Sukhova, L. K. and Efimova, T. N., *Russ. J. Inorg. Chem.,* **19**, 677 (1974).

[69] Nakamura, K. and Kawamura, K., *Bull. Chem. Soc. Japan,* **44**, 330 (1971).

[70] Reuben, J. and Fiat, D., *J. Chem. Phys.,* **51**, 4909, 4918 (1969).

[71] Karraker, D. G., *Inorg. Chem.,* **7**, 473 (1968).

[72] *The Structure of Electrolytic Solutions,* ed. Hamer, W. J., McGraw-Hill (1953) Ch. 22; Spedding, F. H. and Rard, J. A., *J. Phys. Chem.,* **78**, 1435 (1974); Spedding, F. H., Cullen, P. F., and Habenschuss, A., *ibid.,* **78**, 1106 (1974).

[73] Stavely, L. A. K., Markham, D. R. and Jones. M. R., *J. Inorg. Nucl. Chem.,* **30**, 231 (1968).

[74] Grenthe, I., Hessler, G., and Ots, H., *Acta Chem. Scand.,* **27**, 2543 (1973).

[75] Morss, L. G., *J. Phys. Chem.,* **75**, 392 (1971).

[76] Goldman, S. and Morss, L. R., *Can. J. Chem.,* **53**, 2695 (1975).

[77] Sinha, S. P., *Helv. Chim. Acta,* **58**, 1978 (1975); Mioduski, T. and Siekerski, S., *J. Inorg. Nucl. Chem.,* **37**, 1647 (1975); Fidelis, I., *Inorg. Nucl. Chem. Lett.,* **12**, 475 (1976).

[78] Stampfi, R. and Choppin, G. R., *J. Coord. Chem.,* **1**, 173 (1971).

[79] Dellien, I. and Malmsten, L.-Å., *Acta Chem. Scand.,* **27**, 2877 (1973).

[80] Yun, S. S. and Bear, J. L., *J. Inorg. Nucl. Chem.,* **37**, 1757 (1975).

[81] Reidler, J. and Silber, H. B., *J. Chem. Soc. Chem. Commun.,* 354 (1973).

[82] Ryhl, T., *Acta Chem. Scand.,* **27**, 303 (1973).

[83] Bockris, J. O'M. and Saluja, P. P. S., *J. Phys. Chem.,* **76**, 2140, 2298 (1972); **77**, 1598 (1973).

[84] Rykov, A. G., Vasil'ev, V. Ya. and Blokhin, N. B., *Russ. J. Inorg. Chem.,* **16**, 1539 (1971).

[85] Sutton, J., *Nature,* **169**, 235 (1952).

[86] Ernst, S. and Jezowska-Trzebiatowska, B., *J. Phys. Chem.,* **79**, 2113

(1975); *Z. Phys. Chem.*, **256**, 330 (1975).

[87] Siddall, T. H. and Stewart, W. E., *Inorg. Nucl. Chem. Lett.*, **3**, 279 (1967).

[88] Cramer, R. E. and Harris, R. L., *Inorg. Chem.*, **12**, 2575 (1973).

[89] Watkins, C. L., Vigee, G. S., and Harris, M. E., *J. Phys. Chem.*, **77**, 855 (1973).

[90] Waysbort, D. and Navon, G., *J. Phys. Chem.*, **77**, 960 (1973).

[91] Appleton, T. G. and Hall, J. R., *Inorg. Chem.*, **10**, 1717 (1971).

[92] La Mar, G. N. and Van Hecke, G. R., *Chem. Commun.*, 274 (1971).

[93] Kolobov, N. P. and Marenina, K. N., *Russ. J. Phys. Chem.*, **46**, 566 (1972).

[94] Kawaizumi, F. and Miyahara, Y., *Bull. Chem. Soc. Japan*, **44**, 1979 (1971).

[95] Marenina, K. N., Mironov, V. E., and Knyazeva, N. N., *Russ. J. Phys. Chem.*, **48**, 1535 (1974).

[96] Behrendt, S., Langford, C. H., and Frankel, L. S., *J. Amer. Chem. Soc.*, **91**, 2236 (1969); Frankel, L. S., *J. Phys. Chem.*, **73**, 3897 (1969); **74**, 1645 (1970); Frankel, L. S., Langford, C. H., and Stengle, T. R. *ibid.*, **74**, 1376 (1970).

[97] Sastri, V. S. and Langford, C. H., *J. Phys. Chem.*, **74**, 3945 (1970).

[98] Sastri, V. S., Henwood, R. W., Behrendt, S., and Langford, C. H., *J. Amer. Chem. Soc.*, **94**, 753 (1972).

[99] E.g. Chibiskova, V. I. and Pavlov, N. N., *Russ. J. Inorg. Chem.*, **16**, 694 (1971).

[100] Strehlow, H. and Frahm, J., *Ber. Bunsenges. Phys. Chem.*, **79**, 57 (1975).

[101] Craighead, K. L., Jones, P. and Bryant, R. G., *J. Chem. Phys.*, **63**, 1586 (1975).

[102] Lawrence, R. M. and Kruh, R. F., *J. Chem. Phys.*, **47**, 4758 (1967); Maeda, M., Akaishi, T. and Ohtaki, H., *Bull. Chem. Soc., Japan*, **48**, 3193 (1975).

[103] Vigee, G. S. and Ng, P., *J. Inorg. Nucl. Chem.*, **33**, 2477 (1971).

[104] Davies, C. W., *Ion Association*, Butterworths (1962); Nancollas, G. H., *Interactions in Electrolyte Solutions*, Elsevier (1966).

[105] Beck, M. T., *Chemistry of Complex Equilibria*, Van Nostrand Reinhold (1970) Ch. 9; Basolo, F. and Pearson, R. G., *Mechanisms of Inorganic Reactions*, Wiley/Interscience (1958) Ch. 9; Fuoss, R. M. and Kraus, C. A., *J. Amer. Chem. Soc.*, **79**, 3304 (1957).

[106] Beck, M. T., *Coord. Chem. Rev.*, **3**, 91 (1968).

Chapter **6**

MIXED SOLVENTS AND SELECTIVE SOLVATION

6.1   INTRODUCTION

Cations in mixtures of coordinating solvents are likely to have primary solvation shells containing molecules of both (or all) components of the solvent mixture. However, if there is a big difference in solvating power, then one solvent only may be found in the primary solvation shell, as in the case of cations in aqueous acetone. Most of the detailed information about cation primary solvation in such circumstances derives from n.m.r. spectroscopic observations, as with cations in single solvents (Chapter 2). In special cases of cations where solvent exchange is very slow, compounds containing cations with mixed solvent primary solvation shells can be isolated.

6.2   IDENTIFICATION OF MIXED SOLVATES

6.2.1   Isolation

Chromium(III) is one of most favourable cases. Rates of solvent exchange at this d^3 cation are slow, and half-lives for solvent exchange are measured in hours (Chapter 11). It is thus possible to separate variously solvated species from solutions in equilibrium by ion-exchange, and then to precipitate these cations and characterise them. From solutions of chromium(III) perchlorate in series of ethanol+water solvent mixtures, mixed solvento-species $Cr(OH_2)_{6-n}(EtOH)_n^{3+}$ with n = 1, 2, or 3 have been isolated [1]. In aqueous ethanol [1] and in aqueous methanol [2] the Cr^{3+} favours water in its primary solvation shell and so complexes with a high alcohol:water ratio in the primary coordination sphere of the Cr^{3+} are difficult to prepare.

All the possible mixed solvates $Cr(OH_2)_{6-n}(dmso)_n^{3+}$ can be isolated from chromium(III) solutions in various dimethyl sulphoxide+water mixtures. Moreover, for the species with n = 2, 3, or 4 it proved possible to detect the geometrical isomers in each case [3]. A similar pattern has been described for chromium(III) in pyridine-N-oxide+water mixtures [4].

6.2.2   Identification by N.m.r. Spectroscopy

An early n.m.r. study of a cation in a solvent mixture was concerned with Mg^{2+} in aqueous methanol [5]. Solvent ^1H resonances were observed for co-ordinated methanol and for coordinated water molecules as well as for 'bulk' methanol and 'bulk' water at low temperatures, say $-75°C$ (Fig. 6.1). The mixed solvent aspect of this work was only incidental to the main object, detecting the primary solvation shell of this cation. However the n.m.r. spectra do show that both water and methanol can coordinate to Mg^{2+} in aqueous methanol. What these early spectra do not show is whether the solution contains mixed solvates $Mg(OH_2)_{6-n}(MeOH)_n^{2+}$ or merely a mixture of $Mg(OH_2)_6^{2+}$ and $Mg(MeOH)_6^{2+}$. Chemical intuition may favour the former, but experimental proof requires more detailed spectroscopic information than is available in this case.

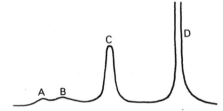

Fig. 6.1 The ^1H n.m.r. spectrum of a solution of composition 1 mole $Mg(ClO_4)_2$: 17.1 moles methanol:3.8 moles water at $-75°C$. Resonance A is assigned to hydroxyl-protons of solvating methanol, B to solvating water protons, C to hydroxyl-protons of bulk methanol, and D to methyl-protons of bulk methanol [5].

Mixed solvates have been identified in the case of aluminium(III) in acetonitrile+water mixtures [6]. Six different water-proton resonances were seen (Fig. 6.2), and the ratios of their areas were found to vary in a reasonable manner with variation in the composition of the system. The mono-aquo-solvate has the single narrow signal expected for a unique isomer containing only one type of proton. The di-, tri-, and tetra- aquo-species have broader and more complicated resonances, suggesting that some at least of the isomers contain more than one type of water molecule. Surprisingly, the penta-aquo-solvate exhibits only one sharp line, despite the expectation of two resonances, one from water *cis-* to the coordinated acetonitrile, the other from water *trans-* to acetonitrile. Again unexpectedly, the hexa-aquo-solvate has a broad resonance. This breadth of signal may be attributable to Al-H coupling as the line sharpens on double irradiation of the ^{27}Al nuclei. The less symmetrical mixed solvates will have greater field gradients at the aluminium and thus more efficient quadrupolar relaxation, so broadening due to Al-H coupling is less likely to be significant. The double resonance techniques used in this investigation provide unequivocal evidence that the signals really do arise from water molecules actually coordinated to the Al^{3+} [7].

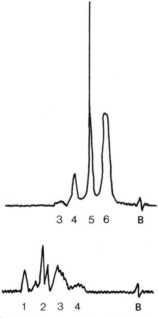

Fig. 6.2   Part of the [1]H n.m.r. spectra of solutions of aluminium(III) perchlorate in aqueous acetonitrile. The water:Al^{3+} ratio is 4.4 for the upper spectrum, 1.0 for the lower. The numbers under the peaks are the suggested numbers of coordinated water molecules per Al^{3+} cation. The peaks marked B, from traces of benzene present, provide a convenient reference point [6].

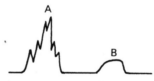

Fig. 6.3   Part of the [1]H n.m.r. spectrum of an acetonitrile + dimethyl sulphoxide mixture containing Al^{3+}. Peak A is assigned to dimethyl sulphoxide protons, with six lines corresponding to solvates containing from one to six molecules of coordinated dimethyl sulphoxide. Peak B is assigned to coordinated acetonitrile protons [8].

Proton n.m.r. spectra of dimethyl sulphoxide + acetonitrile mixtures containing Al^{3+} show separate signals for the dmso protons in each of the five mixed solvates (Fig. 6.3), though similar separation is not observed for the coordinated acetonitrile signal. However, with Al^{3+} in ethylene carbonate + acetonitrile mixtures, distinct resonances for each solvent in each mixed solvate have been observed. In the latter systems the ^{27}Al n.m.r. spectra also show evidence for a range of mixed solvates [8]. Solutions of Al^{3+}, Ga^{3+}, or In^{3+} in aqueous trimethyl phosphate give n.m.r. spectra indicating separate signals from each solvent in each mixed solvate (Fig. 6.4) [9].

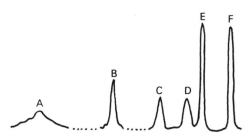

Fig. 6.4 Part of the ^{1}H n.m.r. spectrum of a solution of aluminium(III) perchlorate in a trimethyl phosphate+water+acetone solvent mixture at $-10°C$. The assignment of the peaks is A coordinated water, B bulk water, C and D coordinated TMP, E and F bulk TMP[9].

In the previous examples the central cation has been diamagnetic, with consequent small shifts between the resonances for different solvates. In principle a paramagnetic cation could be used, producing strong contact or pseudo-contact interactions with nuclei of the coordinated solvent molecules. This should give much larger shifts and, if line-broadening can be kept at a minimum, larger separations between signals. In practice the relaxation times for systems containing such paramagnetic cations as Mn^{2+} or Cu^{2+} are unfavourable and resultant n.m.r. spectra of solvent nuclei are very broad, to the point of being undetectable. Fortunately this is not the case for all paramagnetic cations. Serviceable solvent nuclei n.m.r. spectra can be obtained from systems containing Co^{2+} or certain lanthanide cations (Chapter 2.3.2). Particularly detailed information has been obtained from a proton n.m.r. spectroscopic study of Co^{2+} in aqueous methanol [10]. In solutions containing relatively small amounts of water, ^{1}H n.m.r. spectra indicate the presence of $Co(OH_2)(MeOH)_5^{2+}$ and $Co(OH_2)_2(MeOH)_4^{2+}$ as well as $Co(MeOH)_6^{2+}$ (Fig. 6.5). Two signals can be detected for the bis-aquo-species, corresponding to the *cis-* and *trans-* isomers; the precise assignment is tentative as these results were obtained at the limits of resolution of the instrument employed.

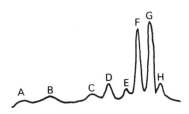

Fig. 6.5 Part of the ^{1}H n.m.r. spectrum of a 0.24M solution of cobalt(II) perchlorate in aqueous methanol containing 21g water per 1000g methanol. Peaks A, B, C are assigned to methanol hydroxyl-protons for the tetra-, hexa- and penta-methanol solvates respectively; E and H, D and G, and F to methanol methyl-protons for the tetra-, penta-, and hexa- methanol solvates respectively [10].

6.3 METHODS FOR STUDYING SELECTIVE SOLVATION

In the previous section we discussed cases where definite n.m.r. evidence indicates mixed solvates for appropriate cations in certain solvent mixtures. In the present section we examine more indirect ways of obtaining information about cation solvation in mixed solvents.

6.3.1 N.m.r. Spectroscopy

This technique provides the most direct evidence that mixed solvates exist for a few cations at which solvent exchange is sufficently slow relative to n.m.r. frequencies. It is also useful in fast solvent exchange situations, though here supplying less direct and more limited information. Such information can be derived either from chemical shifts or from linewidths. It is advantageous to monitor the cation nucleus where possible, though this is less easy to work with than the solvent nucleus. However, the former has fewer possible environments to consider.

Chemical shifts. Selective solvation of a cation by one component of a solvent mixture may be indicated if a plot of chemical shift against solvent composition displays marked curvature. Such a case is shown in Fig. 6.6 [11], which indicates a marked preference of the Na^+ cation for dimethyl sulphoxide over nitromethane. From similar plots of ^{23}Na chemical shifts for appropriate solvent mixtures, an order of relative solvation preferences for the Na^+ cation has been built up (Table 6.2).

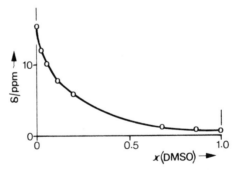

Fig. 6.6 Sodium-23 chemical shifts, δ, measured in dimethyl sulphoxide + nitromethane solvent mixtures, for various mole fractions x (DMSO) [11].

Linewidths. The dependence of ^{23}Na linewidths on solvent composition, examined for a range of systems similar to those cited in the previous paragraph, takes one of the two forms shown in Fig. 6.7(A) or (B). Dependence (A) is attributed to the formation of long-lived complexes between the Na^+ cation and dimethyl sulphoxide or dimethylformamide, whereas dependence (B) suggests that there are no persistent complexes [12]. This approach has also been successfully used for solutions containing paramagnetic cations. Observations of ^{31}P linewidths when a nickel(II) salt was dissolved in a range of tributyl phosphate+cosolvent mixtures have lead to Ni^{2+} solvation preferences being established (Table 6.2, *v.i.*) [13]. The Cu^{2+} cation has no ethanol in its primary solvation shell in ethanol+D_2O mixtures containing more than 55% by weight of D_2O [14].

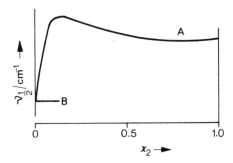

Fig. 6.7 The dependence of the linewidth of the sodium-23 resonance, $\nu_{1/2}$, on solvent composition in binary solvent mixtures containing acetonitrile as one component, mole fraction x_2 of the other component. Curve A applies to cosolvents dimethylformamide or dimethyl sulphoxide; curve B to cosolvents methanol or water [12].

6.3.2   E.s.r. Spectroscopy

This should prove a sensitive probe, especially for low concentrations of asymmetric species. However, e.s.r. studies of metal cations in mixed solvents are rare, and sometimes subject to strange assumptions. A report on the e.s.r. spectra of Mn^{2+} in methanol+water mixtures assumed that the composition of the primary solvation shell was the same as that of the bulk solvent [15].

6.3.3   Ultraviolet-visible Spectroscopy

The recorded spectra of Co^{2+}, introduced as its nitrate, in aqueous alcohols have been mathematically analysed into their component bands. Assuming six-coordination of the cation in all cases, absorption maxima, bandwidths, and molar extinction coefficients were computed for the various mixed solvates [16]. One must admit to some reservations over this analysis, because a number of unlikely mixed solvates containing surprisingly large amounts of t-butyl alcohol or n-heptanol are postulated as significant components.

6.3.4   Infrared and Raman Spectroscopy

The far infrared spectra of dimethyl sulphoxide+pyridine solutions containing Na^+ (Fig. 6.8) suggest preferential solvation of this cation by the dimethyl sulphoxide. The pattern of the results and the conclusion both parallel the ^{23}Na n.m.r. observations described above (Chapter 6.3.1) [11]. Preferential solvation by water of Li$^+$ in aqueous acetonitrile has been demonstrated in a similar manner [17]. The relative solvating powers of a series of non-aqueous solvents for Li$^+$ have been established by infrared studies of Li$^+$ salts in appropriate binary solvent mixtures (Table 6.2, *v.i.*) [18].

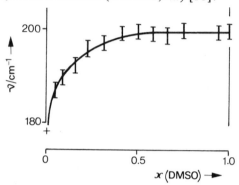

Fig. 6.8 Variation of the frequency, ν, of the Na$^+$ solvation band from infrared spectra of sodium tetraphenylboronate solutions in dimethyl sulphoxide + pyridine solvent mixtures [11].

6.3.5 Transport Properties

Selective solvation has been monitored by conductance and transference measurements [19]†. Conductance results reflect selective solvation through

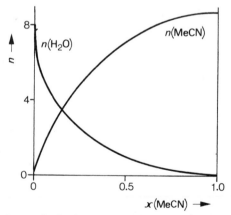

Fig. 6.9 Dependence of solvation numbers for the Ag$^+$ cation on solvent composition in aqueous acetonitrile; n is the number of coordinated solvent molecules for each solvent. These values have been estimated from transport experiments [20].

---

† The monitoring of both primary and secondary solvation of Cr^{3+} in methanol + water mixtures using cation exchange techniques has recently been described [Melsa, C. M. and Baltisberger, R. J., *Analyt. Chem.*, **48**, 435 (1976)] variations of deduced Stokes radii. Transference (Hittorf) results reflect it through variation in solvent composition in the electrode chambers, monitored by chemical analysis, density, or refractive index measurements. The results of a typical investigation of the Hittorf type are shown in Fig. 6.9. Here the total solvation number of the Ag^+ cation is fairly constant, between 7 and 9, over the whole solvent composition range. The curvature of the plots shows a marked preference on the part of Ag^+ for acetonitrile [20]. Similar experiments showed a similar result for Ca^{2+} in aqueous methanol. In this case the total solvation number of the Ca^{2+} was about 15 over the whole range of solvent mixtures, and there was preferential solvation by the water [21]. Likewise Ag^+ shows a marked preference for the water in aqueous ethanol [22].

6.4   BEHAVIOUR OF MIXED SOLVENTS

Table 6.1 indicates the relative solvating powers of water and a range of non-aqueous solvents for a number of cations [23–41]. This Table provides a generous selection of examples culled from the literature, but is far from being exhaustive. Relative solvating powers of non-aqueous solvents are indicated by the selected examples listed in Table 6.2 [42–45]. These Tables also give some idea of the range and variety of applications of the techniques for the study of selective solvation mentioned earlier in this Chapter.

It is not possible to provide a succinct summary covering all the information included in, let alone that omitted from, Tables 6.1 and 6.2. The relative solvating powers of two solvents for a given cation depend on a number of properties of both the cation and solvents. Such properties as sizes and shapes, charges, polarisabilities, basicities, and specific cation-solvent interactions may all be involved. Relative solvating powers correlate much better with properties such as solvent basicity than with, for example, dielectric constants or dipole moments. This topic of the relative affinity of two solvents for a given cation will be mentioned again in Chapter 7. There it will be approached from the thermodynamic angle of transfer parameters of cations from one solvent to another and from water to binary aqueous mixtures. This quantitiative thermodynamic approach, together with the alternative quantitative approach of equilibrium constants (next section), and the purely qualitative and descriptive approach of the present section, all represent different facets of the same problem.

All studies cited in Table 6.1 and 6.2 deal with binary solvent mixtures. There have also been a few studies of cations in ternary solvent mixtures. Fortunately for interpreters and readers, such studies [46] have been restricted to mixtures consisting of two solvating solvents plus one which can be safely ignored because it is non-solvating.

Table 6.1 Selective solvation of metal cations in binary aqueous mixtures: N = preferential solvation by the non-aqueous component, NP = no marked preferential solvation, otherwise preferential solvation by water[a].

Cosolvent:	D_2O	Hydrogen peroxide	Methanol	Dioxan	Dimethyl sulphoxide	Aceto-nitrile	Dimethyl-formamide
Li^+	NP ^{7}Li nmr [23]	^{7}Li nmr [24]	tf [25]	N itp [30]	^{7}Li nmr [32] NP act [33]	ir [17] ^{7}Li nmr [36]	
Na^+		ise [24] ^{23}Na nmr [24]	tf [25] ^{23}Na nmr [26] emf [26]	^{1}H nmr [31]	N ^{23}Na nmr [12]	^{23}Na nmr [36]	^{23}Na nmr [12]
Cs^+		N ^{133}Cs nmr [24]	tf [25] ^{133}Cs nmr [26] emf [26]	^{1}H nmr [31]	N ^{133}Cs nmr [32]		
Ag^+						N tf [20] N ^{1}H nmr [37]	
Tl^+			^{205}Tl nmr [27]				
Ca^{2+}			tf [21]	^{1}H nmr [31]			
Mn^{2+}			esr [28]			esr [28]	NP esr [28]
Co^{2+}			^{1}H nmr [10]	^{1}H nmr [31]		^{1}H nmr [38]	
Al^{3+}					^{1}H nmr [34]		
La^{3+}				^{1}H nmr [31]			
Nd^{3+}			uvv [29]				
Eu^{3+}						fl [39] ^{1}H nmr [40]	uvv [41]
Er^{3+}					^{1}H nmr [35]		
Th^{4+}				^{1}H nmr [31]			

a Abbreviations for methods used: *act* = activity coefficients of transfer; *emf* = emf measurements; *fl* = fluorescence; *ir* = infrared spectroscopy; *ise* = ion selective electrodes; *itp* = ionic transfer potentials; *tf* = transference numbers; *uvv* = ultraviolet-visible spectroscopy.

Table 6.2 Preferential solvation in mixtures of non-aqueous solvents.

Cation	Solvent order	Methods	
Li^+	$NH_3 > MeNH_2 > Me_2NH > Me_3N > MeCN > MeNO_2$	i.r.	[18]
	acetone $> MeNO_2$	i.r., Raman, ^7Li n.m.r	[42]
Na^+	$HMPA > DMSO > py > MeCN > MeNO_2$	^{23}Na n.m.r., i.r.	[11, 43]
	DMSO, DMF $> MeOH > MeCN$, acetone	^{23}Na n.m.r.	[12]
	THF $>$ cyclohexane	conductance, i.r., n.m.r.	[44]
Fe^{2+}	$MeOH > DMF$	Mössbauer	[45]
Co^{2+}	DMSO $>$ alcohols, amides $>$ THF $>$ acetone, MeCN, dioxan	^1H n.m.r.	[38]
Ni^{2+}	trioctylamine $>$ glycol $>$ water $>$ DMF $>$ MeOH $>$ TBP $>$ acetic acid $>$ acetone	^{31}P n.m.r.	[13]

6.5   QUANTITATIVE ASPECTS

The preference of a cation for one component of a binary mixture can often be expressed quantitatively in the form of one or more equilibrium constants involving two or more differently solvated species. The analysis of the experimental results may be a matter of considerable difficulty. For a cation which is always octahedrally coordinated there are, ignoring isomers, seven possible species to be considered if the solvating powers of the two components are comparable. Where pairs of solvents have markedly different solvating powers, the number of species observed may be considerably smaller and the analysis more manageable. If the different mixed solvates have readily distinguishable n.m.r. signals, then equilibrium constants of useful precision can be obtained. Table 6.3 gives references to a variety of determinations of equilibrium constants relating to mixed solvates in mixed solvents. This Table indicates both the type of system for which data are available, and the range of experimental techniques which have proved useful. The reader must assess for himself the likely precision of the published results in each individual case. There is one special case for which a large number of results of reasonable accuracy are available, that of aqueous ammonia solutions. Standard textbooks and compilations of stability constant data include many stability constants for the formation of (aquated) ammine complexes $M(NH_3)_m^{n+}$ aq for a large number of metal cations.

Data relating to binary mixtures of non-aqueous solvents are rare. There are some values for D_2O+glycerol [56], for mixtures of acetonitrile and another organic solvent [50, 51, 57], and for mixtures of methanol plus another organic solvent [58].

6.6   COMPLEX FORMATION

The mixed solvates discussed in earlier sections of this Chapter are merely special cases of complexes ML_xL_y' in which L is one solvent, L' another. Logically, our next step should be to consider complexes in which L is a 'normal' ligand and L' a solvent, especially water. Such a step would also conclude the sequence earlier established which began with non-interacting cations and anions in very dilute solution. These start to interact as the concentration increases, giving first various forms of ion-pairs and ultimately, when there is direct interaction between cation and anion, complexes. However, the subject of inorganic complexes in solution which we are now approaching so gingerly is extremely large. A reasonable treatment would be disproportionately lengthy in the present volume, especially considering that the subject is at the borders of our terms of reference. We shall therefore content ourselves with a reminder that this subject is adequately covered at an elementary level in several standard textbooks, well treated in more specialised treatises [59], and excellently documented [60].

Table 6.3 Examples of systems for which quantitative equilibrium data relating to selective solvation of metal cations in binary aqueous solvent mixtures are available.

Cosolvent	Cation	Methods	
Methanol	Mn^{2+}	e.s.r.	[28]
Ethylene glycol	Cr^{3+}	ultraviolet-visible	[47]
Dimethylformamide	Mn^{2+}	e.s.r.	[28]
Dimethyl sulphoxide	Li^+, Na^+, Ag^+	solubility, ^1H n.m.r., vapour pressure, calorimetry	[48, 49]
	Al^{3+}	^1H n.m.r.	[34]
Acetonitrile	Li^+, Na^+, K^+	^1H n.m.r., cation sensitive electrodes, conductance	[50–52]
	Li^+, Na^+, Ag^+	solubility, ^1H n.m.r., vapour pressure, calorimetry	[49]
	Mg^{2+}, Ca^{2+}, Sr^{2+}, Ba^{2+}	^1H n.m.r.	[50]
	Mn^{2+}	e.s.r.	[28]
Nitromethane	Cu^{2+}	voltammetry, spectrophotometry	[53]
Propylene carbonate	Li^+, Na^+, Ag^+	solubility, vapour pressure, calorimetry	[49]
Sulpholane	Li^+, Na^+, Ag^+	solubility, vapour pressure, calorimetry	[49]
Hexamethylphosphoramide	Be^{2+}	^1H n.m.r.	[54]
Trimethylphosphate	Al^{3+}	^1H n.m.r.	[54]
Benzene	Ag^+	^1H n.m.r.	[55]

FURTHER READING

Schneider, H. in *Solute-Solvent Interactions, Vol. 2,* ed. Coetzee, J. F. and Ritchie, C. D., Marcel Dekker (1976).

REFERENCES

[1] Kemp, D. W. and King, E. L., *J. Amer. Chem. Soc.,* **89,** 3433 (1967).
[2] Jayne, J. C. and King, E. L., *J. Amer. Chem. Soc.,* **86,** 3989 (1964).
[3] Scott, L. P., Weeks, T. J., Bracken, D. J., and King, E. L., *J. Amer. Chem. Soc.,* **91,** 5219 (1969).
[4] Weeks, T. J. and King, E. L., *J. Amer. Chem. Soc.,* **90,** 2545 (1968).
[5] Swinehart, J. H. and Taube, H., *J. Chem. Phys.,* **37,** 1579 (1962).
[6] Supran, L. D. and Sheppard, N., *Chem. Commun.,* 832 (1967).
[7] Akitt, J. W., *J. Chem. Soc. A,* 2347 (1971).
[8] Strehlow, H., Knoche, W., and Schneider, H., *Ber. Bunsenges. Phys. Chem.,* **77,** 760 (1973).
[9] Crea, J. and Lincoln, S. F., *Inorg. Chem.,* **11,** 1131 (1972).
[10] Luz, Z. and Meiboom, S., *J. Chem. Phys.,* **40,** 1058 (1964).
[11] Erlich, R. H., Greenberg, M. S., and Popov, A. I., *Spectrochim. Acta,* **29A,** 543 (1973).
[12] Green, R. D. and Martin, J. S., *Can. J. Chem.,* **50,** 3935 (1972).
[13] Vashman, A. A., *Russ. J. Inorg. Chem.,* **17,** 1064 (1972).
[14] Shamonin, Ya. Ya. and Yan, S. A., *Dokl. Akad. Nauk SSSR,* **152,** 677 (1963).
[15] Bard, J. R. and Wear, J. O., *Z. Naturf.,* **26B,** 1091 (1971).
[16] Antipova-Karataeva, I. I. and Rzhevskaya, N. N., *Russ. J. Inorg. Chem.,* **17,** 853 (1972).
[17] Baron, M. -H. and de Loze, C., *J. Chim. Phys.,* **68,** 1293 (1971).
[18] Regis, A. and Corset, J., *Can. J. Chem.,* **51,** 3577 (1973).
[19] Schneider, H. in *Solute-Solvent Interactions* ed. Coetzee, J. F. and Ritchie, C. D., Marcell Dekker (1969) Ch. 5.
[20] Strehlow, H. and Koepp, H. -M., *Z. Elektrochem.,* **62,** 373 (1958).
[21] Schneider, H. and Strehlow, H., *Z. Elektrochem.,* **66,** 309 (1962).
[22] Childs, W. V. and Amis, E. S., *J. Inorg. Nucl. Chem.,* **16,** 114 (1960).
[23] Loewenstein, A., Shporer, M., Lauterbur, P. C., and Ramirez, J. E., *Chem. Commun.,* 214 (1968).
[24] Covington, A. K., Newman, K. E., and Wood, M., *J. Chem. Soc. Chem. Commun.,* 1234 (1972); Covington, A. K., Lilley, T. H., Newman, K. E., and Porthouse, G. A., *J. Chem. Soc. Faraday I,* **69,** 963 (1973).
[25] Viehweger, U. and Emons, H. -H., *Z. Anorg. Allg. Chem.,* **383,** (1971).
[26] Covington, A. K., Newman, K. E., and Lilley, T. H., *J. Chem. Soc. Faraday I,* **69,** 973 (1973).
[27] Dechter, J. J. and Zink, J. I., *J. Amer. Chem. Soc.,* **97,** 2937 (1975).

[28] Burlamacchi, L., Martini, G., and Romanelli, M., *J. Chem. Phys.,* **59**, 3008 (1973).

[29] Sayre, E. V., Miller, D. G., and Freed, S., *J. Chem. Phys.,* **26**, 109 (1957).

[30] Rat, J. C., Villermaux, S. and Delpuech, J. J., *Bull. Soc. Chim. Fr.* 815 (1974).

[31] Fratiello, A. and Douglass, D. C., *J. Chem. Phys.,* **39**, 2017 (1963).

[32] Covington, A. K., Lantzke, I. R., and Thain, J. M., *J. Chem. Soc. Faraday I,* **70**, 1869 (1974).

[33] Courtot-Coupez, J. and Madec, C., *C. R. Hebd. Séanc. Acad. Sci., Paris,* **274C**, 1673 (1972).

[34] Olander, D. P., Marianelli, R. S., and Larson, R. C., *Analyt. Chem.,* **41**, 1097 (1969). Chi, H., Ng, C. -H. and Li, N. C., *J. Inorg. Nucl. Chem.,* **38**, 529 (1976).

[35] Fratiello, A., Kubo, V., and Vidulich, G. A., *Inorg. Chem.,* **12**, 2066 (1973).

[36] Bloor, E. G. and Kidd, R. G., *Can. J. Chem.,* **46**, 3425 (1968); Maciel, G. E., Hancock, J. K., Lafferty, L. F., Mueller, P. A., and Musker, W. K., *Inorg. Chem.,* **5**, 554 (1966).

[37] Schneider, H. and Strehlow, H., *Z. Phys. Chem. Frankf. Ausg.,* **49**, 44 (1966).

[38] Fratiello, A., Lee, R. E., Miller, D. P., and Nishida, V. M., *Molec. Phys.,* **13**, 349 (1967).

[39] Haas, Y. and Stein, G., *J. Phys. Chem.,* **75**, 3677 (1971).

[40] Haas, Y. and Navon, G., *J. Phys. Chem.,* **76**, 1449 (1972).

[41] Lugina, L. N., Davidenko, N. K. and Yatsimirskii, K. B., *Russ. J. Inorg. Chem.,* **18**, 1453 (1973).

[42] Wong, M. K., McKinney, W. J., and Popov, A. I., *J. Phys. Chem.,* **75**, 56 (1971); Baum, R. G. and Popov, A. I., *J. Solution Chem.,* **4**, 441 (1975).

[43] Greenberg, M. S. and Popov, A. I., *Spectrochim. Acta,* **31A**, 697 (1975).

[44] Olander, J. A. and Day, M. C., *J. Amer. Chem. Soc.,* **93**, 3584 (1971).

[45] Vértes, A., Pálfalvy, M., Burger, K. and Molnár, B., *J. Inorg. Nucl. Chem.,* **35**, 691 (1973).

[46] E.g. Fratiello, A., Vidulich, G. A., Cheng, C. and Kubo, V., *J. Solution Chem.,* **1**, 433 (1972).

[47] Klonis, H. B. and King, E. L., *Inorg. Chem.,* **11**, 2933 (1972).

[48] Benoit, R. L. and Buisson, C., *Inorg. Chim. Acta,* **7**, 256 (1973).

[49] Benoit, R. L. and Lam, S. Y., *J. Amer. Chem. Soc.,* **96**, 7385 (1974).

[50] Stockton, G. W. and Martin, J. S., *J. Amer. Chem. Soc.,* **94**, 6921 (1972).

[51] Izutsu, K., Nomura, T., Nakamura, T., Kazama, H., and Nakajima, S., *Bull. Chem. Soc. Japan,* **47**, 1657 (1974).

[52] Chantooni, M. K. and Kolthoff, I. M., *J. Amer. Chem. Soc.,* **89**, 1582 (1967).

[53] Larson, R. C. and Iwamoto, R. T., *Inorg. Chem.,* **1**, 316 (1962).

[54] Delpuech, J. J., Peguy, A., and Khaddar, M. R., *J. Magnetic Resonance*, **6**, 325 (1972).

[55] Andrews, L. J. and Keefer, R. M., *J. Amer. Chem. Soc.*, **71**, 3644 (1949); Foreman, M. I., Gorton, J., and Foster, R., *Trans. Faraday Soc.*, **66**, 2120 (1970).

[56] Kieboom, A. P. G., Spoormaker, T., Sinnema, A., van der Toorn, J. M., and van Bekkum, H., *Rec. Trav. Chim.*, **94**, 53 (1975).

[57] Nakamura, T., *Bull. Chem. Soc. Japan*, **48**, 1447 (1975).

[58] Kuntz, I. D. and Cheng, C. J., *J. Amer. Chem. Soc.*, **97**, 4852 (1975).

[59] Rossotti, F. J. C. and Rossotti, H. S., *The Determination of Stability Constants*, McGraw-Hill (1961); Beck, M. T., *Chemistry of Complex Equilibria*, Van Nostrand Reinhold (1970).

[60] Sillén, L. G. and Martell, A. E., *Stability Constants of Metal Ion Complexes*, Special Publications 17 and 25, The Chemical Society, London (1964, 1971).

Chapter 7

THERMOCHEMISTRY OF METAL ION SOLVATION

7.1 INTRODUCTION

In this Chapter and the next we shall be concerned with thermodynamic aspects of the chemistry of metal ions in solution. Two main features will be covered. In this Chapter the thermodynamics of interaction between cations and solvent molecules will be considered. The following Chapter will review the thermodynamics associated with the transfer of electrons to and from solvated cations.

The main thermodynamic parameters (X) discussed will be Gibbs free energies (G), enthalpies (H), and entropies (S). Less widely studied parameters such as volume (V), heat capacities $(C_p$ or $C_v)$, and compressibilities (K) will make brief appearances.

$$M^{n+}(g) + \text{solvent} \longrightarrow M^{n+}(\text{solvated}) \text{ in solution}, \Delta X_{\text{solvation}}$$

$$M(\text{standard state}) + \text{solvent} \xrightarrow{-ne^-} M^{n+}(\text{solvated}) \text{ in solution}, \Delta X_f^{\ominus}$$

$$M^{n+}(\text{solvated}) \xrightarrow{+me^-} M^{(n-m)+}(\text{solvated}), \Delta X_{\text{redox}} \ .$$

The first process will be of prime concern in this Chapter, the third in Chapter 8. Because the second process includes redox properties as well as solvation properties, $\Delta X_f^{\ominus}$ values will be presented towards the end of Chapter 8.

Values of $\Delta X(\text{solvation})$ and of $\Delta X_f^{\ominus}(M^{n+}$, solvated) for single ions are not directly measurable, though various attempts have been made to estimate them, as detailed below. Their relation to measurable quantities is outlined diagrammatically in Fig. 7.1. The question of separating sums of cation plus anion solvation parameters, $\Sigma \Delta X$ (ion solvation) in Fig. 7.1, into single ion values is discussed in Chapter 7.1.2. We have $\Delta X(\text{solution})$ values which are often easy to obtain but difficult to interpret, and $\Delta X(\text{ion solvation})$ values which are difficult to obtain with accuracy but are of considerable chemical interest and of direct relevance to this book. $\Delta X(\text{ion solvation})$ will therefore be dealt with first and at length, but $\Delta X(\text{solution})$ will be considered only briefly. This is because occasionally solubility trends alone are available to give information relating to cation (and anion) solvation in some little-studied solvents.

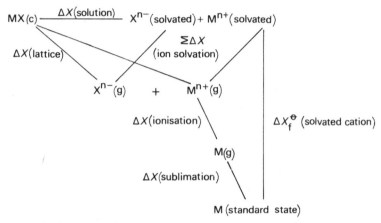

Fig. 7.1   The interconnection of thermochemical parameters relevant to cation solvation.

7.1.1   Theoretical Estimation of Cation Solvation Parameters

The ΔX(cation solvation) values which will be presented and discussed in subsequent sections of this Chapter are all based on experimental measurements. There have been many attempts to find satisfactory ways of estimating and calculating these quantities from appropriate models. These have usually been markedly unsuccessful, though recent calculations from complicated models have given values which agree tolerably well with generally accepted experimental ones. Here we shall indicate the simple 'theoretical' approach to estimating ΔX(cation solvation) values [1, 2], list its drawbacks and use it no further.

The change in the Gibbs free energy on transferring a gaseous ion of charge z and radius r into a medium whose dielectric constant is D is $-\dfrac{N_A z^2 e^2}{8\pi\epsilon_0 r}\left(1 - \dfrac{1}{D}\right)$. This expression is due to Born [3]. The entropy change can now be obtained by differentiation with respect to temperature:

$$\Delta S = \frac{N_A z^2 e^2}{8\pi\epsilon_0 Dr}\left(\frac{\partial \ln D}{\partial T}\right)_P.$$

Thence one can derive the corresponding enthalpy of transfer of the ion from the gas phase to the medium of dielectric constant D by simple arithmetic from the free energy and entropy terms. In practice these values agree much more closely with those obtained from experiment if the ion radius r (from crystal measurements) is replaced by an effective radius r_{eff} equalling $r + 0.85(Å)$ [4]. Various modifications to these simple Born calculations have been suggested [5, 6], particularly to make some allowance for the oversimplified picture of a charge in a dielectric continuum for the post-transfer state. Despite the crudity of this type of calculation, it is often still used to give a rough idea of relative values, especially for non-aqueous systems.

The application of molecular orbital calculations to this area of chemistry is very much in its infancy. Hydration energies have been computed for some simple species such as $Li(OH_2)_n^+$ and $Na(OH_2)_n^+$ where $n = 1$ to 6 [7], and for $Li(NH_3)^+$ [8]. Progression from here to a consideration of cations fully furnished with their primary and secondary solvation shells is a daunting prospect. Nonetheless, calculations have recently been made for Li^+, Na^+, and K^+ with between 2 and 10 molecules of water per cation. These have suggested primary hydration numbers of 4, 5-6, and 5-7 respectively for these cations, with the remaining water molecules relegated to a secondary hydration shell [9].

7.1.2 Practical Estimation of Cation Solvation Parameters

Sums of cation plus anion pairs of Gibbs free energies of solvation, $\Sigma \Delta G$(ion solvation) of Fig. 7.1, can be obtained directly from solubility or e.m.f. measurements, if necessary with appropriate estimation of activity coefficients. $\Sigma \Delta H$(ion solvation) values can be arrived at using one of two routes, depending on the solubility of the salt in question. If sufficiently soluble, then one can measure ΔH(solution) by calorimetry, a direct, accurate, and eminently satisfactory method. If the salt is only sparingly soluble, then indirect estimation of ΔH(solution) via use of the van't Hoff isochore may be technically easier. From measurements of the solubility of the salt at various temperatures, the variation of its solubility product K_s with temperature can be determined:

$$\Delta G^{\ominus} = -RT\ln K_s$$

then

$$\frac{\partial \ln K_s}{\partial (1/T)} = -\frac{\Delta H^{\ominus}}{R}.$$

As the mean activity coefficient $\gamma_{\pm}$ for such sparingly soluble salts is close to 1, $\partial \ln \gamma_{\pm}/\partial(1/T)$ will be close to zero, and this estimate of ΔH(solution) will be acceptable. For more soluble salts $\gamma_{\pm} < 1$, $\partial \ln \gamma_{\pm}/\partial(1/T)$ will have a significant influence on the apparent value for ΔH(solution), and $\gamma_{\pm}$ at various temperatures must be determined [10]. Whether ΔH(solution) is measured calorimetrically or by the van't Hoff method, the lattice enthalpy of the salt must be calculated [11] in order to arrive at $\Sigma \Delta H$(ion solvation). Entropy values are derived from experimental enthalpies and free energies. A vast amount of information on solubilities and enthalpies of solution of salts in water is available from numerous sources, and a useful compilation of analogous information for electrolytes in non- and mixed aqueous media exists [12].

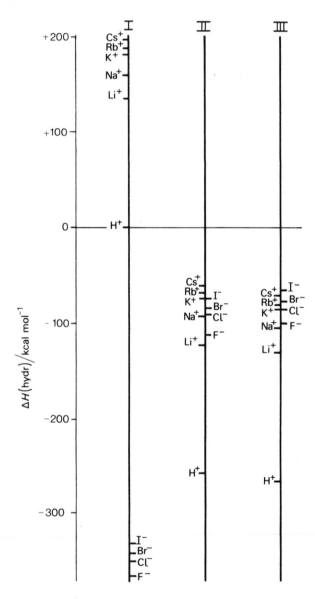

Fig. 7.2 The dependence of single ion hydration enthalpies on the choice of zero: series I refers to ΔH(hydration) for $H^+ = 0$; series II to an estimate of -260.7 kcal mol^{-1} for ΔH(hydration) of H^+; and series III to equal values of ΔH(hydration) for K^+ and Cl^-.

Once $\Sigma \Delta X$(ion solvation) has been calculated, then this has to be separated into individual ionic components. Examination of the constancy of $\Delta X(M_a^{n+}A^{n-}) - \Delta X(M_b^{n+}A^{n-})$, where M_a and M_b are two metals and A represents one of a series of anions, shows that it is reasonable to undertake such a separation. This separation can be achieved in a number of ways, which can broadly be divided into two types:

(i) Most early efforts concentrated on guessing a good value of ΔX(solvation) for one or two ions. Organisationally simple but entirely arbitrary assignments could be made, for example that ΔH(hydration) of the proton was zero. Rough ion-size methods were also used, such as assuming that ΔX(solvation) of K^+ and Cl^- were equal. Reasonably sophisticated estimates of ΔX(solvation) for a specific ion were made as well, generally for H^+ [13].

(ii) Recently the fashion has swung another way [14, 15], particularly for non-aqueous systems. Now large organic ions of the R_4N^+ and R_4B^- type are used as reference ions and it is assumed that ΔX(solvation) for one cation and one anion of this type are both small and equal. The most common assumption is that ΔX(solvation) of Ph_4As^+ equals ΔX(solvation) of Ph_4B^- [15]. A similar ferrocene-ferrocinium split, though once popular for obtaining certain transfer solvation functions for ions, is currently out of favour [14].

Specific examples of these methods of assignment will be described at appropriate points in the sections which follow. It is worth bearing in mind that the choice of reference value is not important if one is only comparing cations. $\Delta X(M_a^{n+}) - \Delta X(M_b^{n+})$ will be the same no matter what single ion split is used (so long as charge is conserved in the split). Figure 7.2 shows this. It also shows how unreal the choice of ΔH(hydration) of H^+ as zero is and how much nearer the truth it is to assume equal ΔH(hydration) for K^+ and for Cl^-.

7.2   AQUEOUS SOLUTIONS

7.2.1   Cation Hydration Enthalpies

Single ion hydration enthalpies for cations seem to be best obtained by way of estimating the hydration enthalpy of the proton [13]. Halliwell and Nyburg [16] offer the best method for this, by making use of *all* the alkali metal halide values together in their efforts to estimate realistic ion hydration enthalpies. They started by dividing ΔH(ion solvation) into single ion contributions, here called ΔH(conv), using a conventional assignment (Born, and zero for H^+). Then, and this is the kernel of the method, the **differences** between cation values and anion values were used to obtain a good estimate for the absolute single ion hydration enthalpy for the proton ΔH^+(abs):

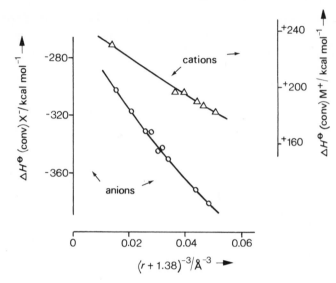

Fig. 7.3 The variation of ΔH(conv) of hydration of cations and of anions with effective ionic radius (r is the crystal radius) [16].

$$\Delta H^{\ominus}(\text{conv})M^{+}(g) - \Delta H^{\ominus}(\text{conv})X^{-}(g) \;=\; \Delta H^{\ominus}(\text{abs})M^{+}(g) - \Delta H^{\ominus}(\text{abs})X^{-}(g)$$
$$- 2\Delta H^{\ominus}(\text{abs})H^{+}(g).$$

Either of the terms on the left-hand side can be plotted against some function of the ionic radius to get smooth curves, for instance as shown in Fig. 7.3. Here 1.38Å is chosen from a large number of crystallographic estimates for the radius

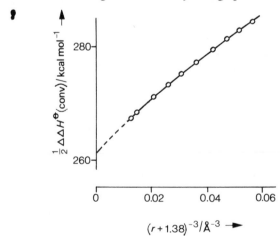

Fig 7.4 Plot of the differences between ΔH(conv) of hydration of cations and anions against the reciprocal of the cube of the effective ionic radius [16].

of water of crystallisation as the effective radius of solvating water (cf. Born). Now the difference between the two curves can be plotted against the same function of ionic radius, Fig. 7.4. Extrapolation of this curve to infinite ionic radius then gives (twice) the value for the single ion hydration enthalpy for the proton. The value obtained by Halliwell and Nyburg was -260.7 ± 2.5 kcal mol^{-1} from salts containing a wide range of 1+ cations and 1− anions. Later treatment of results for salts containing 2+ cations and 2− anions confirmed this estimate [17].

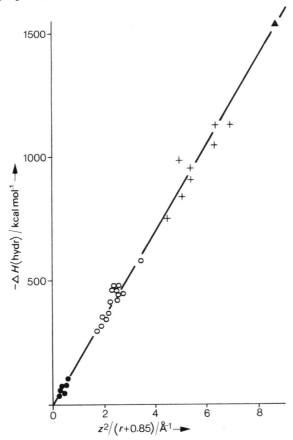

Fig. 7.5 Dependence of single ion hydration enthalpies, ΔH(hydr), on the cation charge:radius ratio; • 1+ cations, ○ 2+ cations, + 3+ cations, ▲ Ce^{4+} [13, 18].

Single ion hydration enthalpies for metal cations are collected in Table 7.1 [13, 18–21]. The entries here and in similar Tables later in this Chapter are selective, usually consisting of well-established values from standard reference sources with a sprinkling of recent values. The values in Table 7.1 have been adjusted to correspond with a reference value of $-260.7\,\text{kcal mol}^{-1}$ for the enthalpy of hydration of the proton[†]. The single ion hydration enthalpies for all cations are, of course, negative. The importance of simple electrostatics, the basis on which the scale of single ion values is established, is illustrated in Fig. 7.5 [18]. However, close inspection of the tabulated cation hydration enthalpies reveals that there are significant deviations from the simple electrostatic correlation of this Figure, especially for some transition metal cations and for a few 'soft' cations.

Transition metal cations. Fig. 7.6 shows how cation hydration enthalpies for the dipositive cations of the metals of the first transition series depend on atomic number. The dependence of the cation hydration enthalpies on the number of d electrons is clear on this scale, with the deviations from the 'electrostatic' line paralleling the crystal field stabilisation energies of the respective cations [22] (Table 7.2).

† A more recently recommended value of $\Delta H^{\bullet}(\text{hydr}) = -270.1\,\text{kcal mol}^{-1}$ for H^+ [Niedermeijer-Denessen, H. J. M. and de Ligny, C. L., *J. Electroanal. Chem. Interfac. Electrochem.*, **57**, 265 (1974)] suffers from the disadvantage of being based on the subsequently criticised 'ferrocene-ferrocinium' assumption [14].

Table 7.1 Single ion hydration enthalpies (kcal mol⁻¹) for metal cations; from Reference

H^+ -260.7

Li^+ -123.0	Be^{2+} -594.4	
Na^+ -96.9	Mg^{2+} -459.4	
K^+ -76.7	Ca^{2+} -380.6	Zn^{2+} -488.6
Rb^+ -70.8	Sr^{2+} -345.3	Cd^{2+} -431.6
Cs^+ -62.9	Ba^{2+} -311.6	
	Ra^{2+} -301.0	

	Cr^{2+} -442.1	Mn^{2+} -441.0
	Cr^{3+} -1052^c	

$Ce^{3+}-805.1^d$ $Pr^{3+}-815.3$ $Nd^{3+}-822.2$ $Pm^{3+}-830.8$ $Sm^{3+}-839.6$ $Eu^{3+}-847.4$ $Gd^{3+}-$
$Ce^{4+}-1550.8^c$

Pu^{3+} -822.3^e

a Values for the lanthanide(III) cations are from ref. [19]; *b* from ref. [20]; *c* from ref. [13]; *d* ref. [

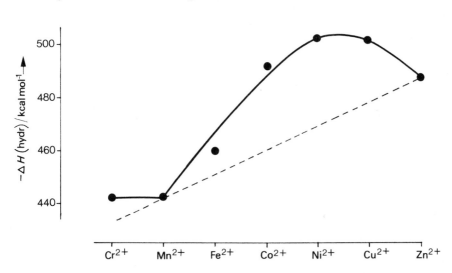

Fig. 7.6 The variation of cation hydration enthalpies along the first transition metal series for 2+ cations. The dashed line represents the trend expected from a simple electrostatic model in the absence of crystal field effects.

ɔt where otherwise stated. [a]

−1113.7			
−946.5	Ga^{3+} −1119.7		
−865.2	In^{3+} −982.0		
−784.6	Tl^{3+} −1000.0		
		Sn^{2+} −371.5 [b]	
	Tl^+ −77.9	Pb^{2+} −353.7	

			Cu^+ −142.0[c]
−458.9	Co^{2+} −491.0	Ni^{2+} −503.3	Cu^{2+} −502.0
−1046			
			Ag^+ −113.6[c]

−861.3 Dy^{3+}−868.8 Ho^{3+}−876.1 Er^{3+}−881.8 Tm^{3+}−887.9 Yb^{3+}−893.2 Lu^{3+}−898.3

No^{2+} −355[f]

s −849.1; e from ref. [19]; f from ref. [21].

Table 7.2 Variation of cation hydration enthalpies, ΔH(hydr), with atomic number for divalent cations of first row transition elements.

Atomic number	Cation		CFSE	ΔH(hydr) /kcal mol^{-1}
24	Cr^{2+}	d^4	6Dq	-442.1
25	Mn^{2+}	d^5	0	-441.0
26	Fe^{2+}	d^6	4Dq	-458.9
27	Co^{2+}	d^7	8Dq	-491.0
28	Ni^{2+}	d^8	12Dq	-503.3
29	Cu^{2+}	d^9	6Dq	-502.0
30	Zn^{2+}	d^{10}	0	-488.6

Hard and Soft Cations. The electrostatic picture of hydration and the consequent close correlation of cation hydration enthalpies with their size and charge is more likely to work well for **hard**, or Chatt class a, cations such as those of the alkali and alkaline earth metals, than for the **soft**, or Chatt class b, cations. The latter class includes such cations as Ag$^+$, Hg^{2+}, and Tl^{3+}, and here polarisation effects and covalent interactions may well make an appreciable contribution to the experimentally determined cation hydration enthalpy. The selected values given in Table 7.3 confirm this view [23]. Here cations of equal charge and approximately equal size are compared in **hard–soft** pairs. In each case there is a significantly larger hydration enthalpy for the **soft** cation, with differences ranging up to 80 kcal mol^{-1}.

Table 7.3 Cation hydration enthalpies (ΔH(hydr)) for 'hard' and 'soft' cations [23].

Cation^a	Radius/Å	ΔH(hydr) /kcal mol^{-1}	Cation^a	Radius/Å	ΔH(hydr) /kcal mol^{-1}
K$^+$(H)	1.33	-73	Ca^{2+}(H)	0.99	-362
Ag$^+$(S)	1.26	-112	Cd^{2+}(S)	0.97	-428
Mg^{2+}(H)	0.65	-437	Sr^{2+}(H)	1.13	-327
Cu^{2+}(S)	0.69	-499	Hg^{2+}(S)	1.10	-443

a H = hard, S= soft.

Water orientation. The actual geometry of the cation hydrate may affect its hydration enthalpy (see Chapter 2.2.2 and reference [13]). This may be the factor which leads to the odd trend of reported cation hydration enthalpies for the Group III sequence of cations Al^{3+} to Tl^{3+}.

7.2.2 Cation Hydration Free Energies

Gibbs free energies of solution of salts can be obtained from e.m.f. or solubility (and activity) data. Then free energies of solvation of single ions can be derived using some more or less arbitrary cation-anion division. The situation is analogous to ion hydration enthalpies. However there is an alternative route available to single ion Gibbs free energies of solvation. This involves consideration of surface potentials [24], a method which has proved more useful in respect of non-aqueous and mixed aqueous media.

Values for the Gibbs free energies of hydration of cations are listed in Table 7.4 [13, 20, 25]. These values are relative to an estimated free energy of hydration of the proton of -260.5 kcal mol^{-1}. Free energies of hydration of cations are dominated by enthalpies rather than entropies (see below).

7.2.3 Cation Hydration Entropies

Enthalpies and Gibbs free energies are readily obtained experimentally, entropies less so. The numbers presented in this section are derived from enthalpy and free energy measurements. There are three common ways of presenting data concerning hydration entropies of ions:

(i) $M^{n+}(g) +$ solvent $\longrightarrow M^{n+}$(solvated) in solvent, $\Delta S^{\ominus}$(solv). This is the analogue of the enthalpies and free energies discussed in the two preceding sections.

(ii) Many authors prefer to discuss partial molar entropies, $\overline{S}^{\ominus}$, of the solvated cation in its standard state. Thanks to the third law of thermodynamics, $\overline{S}^{\ominus}$ values for elements can be established on a common scale. There is no corresponding inter-element link for enthalpies or free energies.

(iii) Two other, related, processes have discussed. In one an ion is transfered from its crystal lattice to solution:

$$M^{n+} [\text{in } MX(c)] + \text{solvent} \longrightarrow M^{n+}(\text{solvated}) \text{ in solution}, \Delta S_s.$$

In the other an ion is transferred from a hypothetical ideal liquid state of unit molality to a solution of unit molality:

$$M^{n+}(\text{ideal liquid}) \longrightarrow M^{n+}(\text{solution of unit molality}), \Delta S_{tr}.$$

These entropy changes are linked:

$$\Delta S_{tr} = \Delta S_s - \Delta S_{addn},$$

where the term ΔS_{addn} includes the entropy changes corresponding to the solid to liquid phase change and subsequent dilution to unit molality [26].

Table 7.4 Single ion Gibbs free energies of hydration (kcal mol^{-1}) for metal cations, at 25°C; from references [13] and [20].

H^+ −260.5

Li^+ −122.1	Be^{2+} −583.6				
Na^+ −98.2	Mg^{2+} −455.5	Al^{3+} −1103.3			
K^+ −80.6	Ca^{2+} −380.8		Zn^{2+} −484.6	Ga^{3+} −1106.0	
Rb^+ −75.5	Sr^{2+} −345.9		Cd^{2+} −430.5	In^{3+} −973.2	
Cs^+ −67.8	Ba^{2+} −315.1		Hg^{2+} (−436.3)	Tl^{3+} −978.5	
	Ra^{2+} −307				Sn^{2+} −372.7
					Pb^{2+} −357.8
					Bi^{3+} [a]

Tl^+ −82.0

Ti^{2+} −427	V^{2+} −442	Cr^{2+} −446.1	Mn^{2+} −437.8	Fe^{2+} −456.4	Co^{2+} −479.5	Ni^{2+} −494.2	Cu^{2+} −498.7	
		Cr^{3+} −1039.5		Fe^{3+} −1035.5				Pd^{2+} −474

[a] A value for Bi^{3+}, relative to a different zero, is given in ref. [25].

Entropies of cation hydration, $\Delta S^{\ominus}$(hydr). Single ion values for a range of cations are listed in Table 7.5 [26-28]. To derive them the standard molal entropy for H^+aq is assigned as zero. Because $S^{\ominus}(H^+aq)$ is thought [26, 29] to be close to zero, the values in Table 7.5 may well be close to absolute values. In all cases the process of hydrating a gaseous cation is accompanied by a decrease in entropy, which presumably arises from the solvent's loss of freedom. 'Real' standard entropies of some ions in water have been derived from temperature coefficients of electrode/solution potential differences (see below). From these $\Delta S^{\ominus}(M^+aq)$ values of -24, -17, and -8 cal deg^{-1} mol^{-1} have been calculated for Li^+aq, Na^+aq, and K^+aq respectively [30]. Interestingly, the sum of the entropies of hydration of K^+ and Cl^- is less negative than that for two argon atoms, the uncharged isoelectronic equivalent. This result is at first sight surprising, and it shows that the structure-broken region between the hydrated cation and bulk water is important in determining the thermodynamic parameters of at least some ions in aqueous solution.

Table 7.5 Single ion entropies of hydration (cal deg^{-1} mol^{-1}) for metal cations, at $25°C$ [26-28].

H^+ -26^a						
Li^+ -28.4						
Na^+ -20.9	Mg^{2+} -64		Al^{3+} -111			
K^+ -12.4	Ca^{2+} -50	$Zn^{2+}$$-64$		Ga^{3+} -122		
Rb^+ -9.6	Sr^{2+} -49	$Cd^{2+}$$-55$		In^{3+} -102		
Cs^+ -8.8	Ba^{2+} -38	$Hg^{2+}$$-43$	La^{3+} -88	Tl^{3+} -150		
				Tl^+ -12	Pb^{2+} -37	
	Mn^{2+} -58	Fe^{2+} -65	Co^{2+} -76	Ni^{2+} -77	Cu^{2+} -62	
	$Fe^{3+}$$-110$					
						Ag^+ -22

a Noyes [20] estimates -18.9 cal deg^{-1} mol^{-1} from plots of ionic entropies against reciprocal radii, and making an allowance for entropies of hydration of analogous uncharged species.

Standard partial molal entropies of hydrated cations. Table 7.6 lists $\bar{S}^{\ominus}(M^{n+}aq)$ values [31-46]. These are all relative to zero for H^+aq. As mentioned above, $\bar{S}^{\ominus}(H^+aq)$ is probably close to zero, with most estimates in the region of -3 to -5 cal $deg^{-1}mol^{-1}$ [26, 29], and so the Table 7.6 values are close to the absolute values.[†] Where entropies of hydrated cations are not available from the

† 'Real' standard ionic entropies derived from the temperature variation of potential differences between electrodes and their surrounding solutions are $S^{\ominus}(H^+aq) = +5$, $S^{\ominus}(Li^+aq) = +8$, $S^{\ominus}(Na^+aq) = +19$, and $S^{\ominus}(K^+aq) = +29$ cal deg^{-1} mol^{-1} [30].

appropriate experimental data, they may be estimated by empirical formulae due to Latimer [47] and to Powell and Latimer [48]:

$$\bar{S}^{\ominus}(M^{n+}aq) = \tfrac{3}{2} R \ln m - \frac{270z}{(r+x)^2} + 37 .$$

Here m is the atomic weight of the cation, x is a constant equal to 2.0Å for cations, r is the crystal radius in Å, and z is the charge on the cation. This formula gives $S^{\ominus}$ in units of cal deg^{-1} mol^{-1}.

It has long been recognised that $\Delta S^{\ominus}(\text{hydr})$ and $\bar{S}^{\ominus}(M^{n+}aq)$ values are

---

Table 7.6   Standard partial molal entropies (cal deg^{-1} mol^{-1}) of hydrated cations; from ref

H$^+$ 0		
Li$^+$ 2.7^b	Be^{2+} -31.0^b	
Na$^+$ 14.0^b	Mg^{2+} -33.0^b	
K$^+$ 24.1^b	Ca^{2+} -12.7^b	Zn^{2+} -26.2^b -26.9^d, -26.1^e
Rb$^+$ 28.8^b	Sr^{2+} -7.8^b	Cd^{2+} -18.9^c, -17.4^d
Cs$^+$ 31.8^b	Ba^{2+} 2.3^b	Hg^{2+} -8.66^f
	Ra^{2+} 13^b	
		Hg$_2^{2+}$ 15.72^f

	Mn^{2+} -17.6^b	Fe^{2+} -25.6^d, -32.9^b
		Fe^{3+} -67^d, -75.5^b

					Eu^{2+} -10^o, $+1^r$
Ce^{3+} 49.1	Pr^{3+} -49.9	Nd^{3+} -49.4	Pm^{3+} (-51)	Sm^{3+} -52.4	Eu^{3+} -52.7 Gd^{3+} -53
Ce^{4+} -72^r					
Th^{3+} -40.9	Pa^{3+} -41.3	U^{3+} -42.1	Np^{3+} -43.3	Pu^{3+} -45^q	Am^{3+} -48.7 Cm^{3+} $(-4$
Th^{4+} -101^r		U^{4+} -81^p	Np^{4+} -78^p	Pu^{4+} -91^q	
			NpO$_2^+$ -6^h	PuO$_2^+$ -13^q	
		U$_2^{2+}$ -20^p	NpO$_2^{2+}$ -20^h	PuO$_2^{2+}$ -26^q	

---

a Lanthanide(III) values from ref. [32] (values in Natn. Bur. Stand. Technical Note 270/7 (1973) are, cept for Gd^{3+} at -49.2, similar) and actinide(III) values from ref. [33]; *b* ref. [34]; *c* selected best val [35]; *d* ref. [36]; *e* ref. [37]; *f* ref. [38]; *g* ref. [39]; *h* ref. [33]; *i* values in aqueous hydrochloric ac *j* ref. [40]; *k* ref. [41]; *l* ref. [42]; *m* ref. [43]; *n* values ranging from -44 to $+10$ have been proposed affected by the size and charge of the cation. Increasing charge or decreasing radius lead to a greater immobilisation of water molecules around the cation, in other words to a greater entropy loss. In 1930 Ulich proposed an 'iceberg' model for hydration around a cation [49]. Entropies of cation hydration were then related to the hydration number of the cation and the entropy of freezing of water. Frank and Evans developed this approach in their classic paper of 1945 [50]. The estimation of further values of $\bar{S}^{\ominus}(M^{n+}aq)$ during the following few years permitted Powell to undertake an extensive correlation of hydration entropies with cation charge z and radius r [51]. Although the Born equation indicates that solvation entropies should depend on z^2, in fact they depend on z. Similarly solvation entropies correlate with reciprocals of effective radii, r_{eff},

---

:e [31] unless otherwise stated. ^{a}

$^{3+}$ -76.9^b			
$^{3+}$ -61^b	Ga^{3+} -79^b		
$^{3+}$ $-63.4, -60^b$	In^{3+} -63^b	Sn^{4+} -28^i	
$^{3+}$ -52.2^g			
$^{3+}$ -43.3^h			
		Sn^{2+} -4^i	
	Tl^+ 30^b	Pb^{2+} 4.8^i	Bi^{3+} -45.4^k

- -

		Cu^+ 9.7^b	
$^{2+}$ $-25.2^l, -26.6^d, -27$ Ni^{2+} $-30.8^b, -31.6^d$		Cu^{2+} $-22.2^m, -23.6^d$	
$^{3+}$ -73^b			
		Ag^+ 17.5^b	
	Pd^{2+} 7^n	Ag^{2+} $(-21)^b$	

- -

$^{3+}-55.2$ $Dy^{3+}-55.3$ $Ho^{3+}-54.4$ $Er^{3+}-58.1$ $Tm^{3+}(-58)$ $Yb^{3+}-56.8$ $Lu^{3+}-63.4$

---

ich the most recent estimate [44] has been quoted despite strictures expressed concerning the experi-
ntal method used [45]; o ref. [46]; p based on $S^{\ominus}(Gd^{3+}aq) = -43$ [33]; q estimates, to be treated with
tion [33]; r Morss, L. R. and McCue, M. C., *J. Chem. Engng. Data,* 21, 337 (1976); r Natn. Bur. Stand.
:hnical Note 270/7 (1973).

which are considerably larger (by about 1.3Å) than the crystal radii of the cations. Powell plotted cation partial molal hydration entropies against z/r_{eff} for a large number cations of charge +1 to +4 to achieve a remarkably good straight line [51]. Since then the theoretical treatment of ion solvation entropies has become increasingly sophisticated. Calculations have been performed on increasingly more complicated (that is to say realistic) models incorporating both primary and secondary solvation as well as effects of ions on solvent structure [52].

The $\bar{S}^{\ominus}(M^{n+}aq)$ values for the lanthanide(III) cations (Table 7.6) show a less smooth variation with atomic number than might be expected. The deviations from a smoothly graduated dependence are small, but are probably real because all the values have been derived in the same manner. The internal electronic entropies are between 0 and 6 cal deg^{-1} mol^{-1} and vary irregularly with atomic number, and if the $\bar{S}^{\ominus}(M^{n+}aq)$ values of Table 7.6 are corrected for these a much smoother dependence of hydration entropies on atomic number (or $1/r$) emerges [39]. Comparative discussion of hydration entropies of actinide cations is made difficult by the uncertainties in many of the reported values [33].

For series of related cations, entropies and enthalpies of hydration show an approximate correlation. This is just what one would expect if the Born model, or some approximation to it, has any validity.

Table 7.7 Values of $T\Delta S_s$, kcal mol^{-1}, for the transfer of cations from crystal lattice to solution [26].

Li$^+$ -1.4	Be^{2+} -18.4				
Na$^+$ $+1.0$	Mg^{2+} -11.5		Al^{3+} -27.1		
K$^+$ $+3.2$	Ca^{2+} -7.6	Zn^{2+} -11.7	Sc^{3+} -20.7		
Rb$^+$ $+4.1$	Sr^{2+} -7.2	Cd^{2+} -9.4	Y^{3+} -23.2		Sn^{2+} -6.4
Cs$^+$ $+4.3$	Ba^{2+} -4.2				Pb^{2+} -4.4
				Tl$^+$ $+4.0$	

	Mn^{2+} -9.9	Fe^{2+} -12.4	Co^{2+} -12.5	Ni^{2+} -13.8	Cu^{2+} -10.9
Cr^{3+} -25.6		Fe^{3+} -25.0			
					Ag$^+$ $+0.5$

Lattice-to-solution hydration entropies. Values for $T\Delta S_s$ when a cation is transferred from its crystal lattice into solution are collected in Table 7.7 [26]. These values have been estimated on two assumptions, that the ΔS_s values for the Ph$_4$As$^+$ cation and the Ph$_4$B$^-$ anion are equal, and also that the entropy of a given ion is independent of the lattice within which it finds itself.[†] The as-

---

† Working from this assumption one eventually finds that $\bar{S}^{\ominus}(H^+aq)$ is -4 cal deg^{-1} mol^{-1}, which is in good agreement with the values cited earlier.

sumption that ΔS_s values for K^+ and for Cl^- are equal gives a similar set of values. The difference between the use of these two assumptions gives rise to discrepancies of only 0.3 and 1.9 cal deg^{-1} mol^{-1} for Li^+ and Mg^{2+} respectively [53]. As expected, $T\Delta S_s$ values again depend on cation charge and radius. It has been claimed that a plot of $T\Delta S_s$ against z^2/r is a good straight line, except for 1+ cations so large that their size may interfere with the structure of the surrounding water. In fact such a plot approximates to linearity only for 1+ and 2+ cations as Fig. 7.7(a) shows. A better linear plot is obtained for $T\Delta S_s$ against z/r (Fig. 7.7(b) illustrates this); compare Powell's similar plot for cation partial molal hydration enthalpies [51] above. Here only Be^{2+} and Al^{3+}, the cations with the largest z/r, are significantly removed from the correlation line.

These ΔS_s values do not allow for the entropy difference between the gas and unit molality solution states discussed in (iii) above. The quantity derived there, ΔS_{tr}, is to be preferred because it monitors only ion-solvent factors, uncomplicated by phase change and dilution. ΔS_{tr} reflects only ion-solvent interactions and any effects of the ion on solvent–solvent interactions. Values of ΔS_{tr} are listed in Table 7.8 [26].

Table 7.8 Values of ΔS_{tr}, cal deg^{-1} mol^{-1}, for metal cations from a hypothetical ideal state of unit molality to water (unit molality) [26].[a]

Li^+ −15.7	Be^{2+} −72.7					
Na^+ −7.6	Mg^{2+} −49.6		Al^{3+} −101.9			
K^+ −0.3	Ca^{2+} −36.5	Zn^{2+} −50.3	Sc^{3+} −80.5			
Rb^+ +2.7	Sr^{2+} −35.2	Cd^{2+} −42.5	Y^{3+} −88.8			
Cs^+ +3.4	Ba^{2+} −25.1					Sn^{2+} −32.5
					Tl^+ +2.4	Pb^{2+} −25.8
	Mn^{2+} −40.9	Fe^{2+} −52.6	Co^{2+} −52.9	Ni^{2+} −57.3		Cu^{2+} −47.6
Cr^{3+} −96.9		Fe^{3+} −94.8				
						Ag^+ −9.3

a Similar idealised parameters are discussed in references [28] and [50].

For M^+aq cations, ΔS_{tr} values lie between −15.7 and +3.4 cal deg^{-1} mol^{-1}. This range suggests that ΔS_{tr} represents a balance between the high ordering (entropy decrease) of the secondary solvation shell and surrounding bulk solvent. This series of ΔS_{tr} values for M^+aq cations correlates with the relative structure -making and -breaking properties of these cations as established by other, independent, approaches. The higher charge on M^{2+} and M^{3+} cations produces a nett orientation, and so negative value of ΔS_{tr}, in all cases.

7.2.4 Cation Hydration Volumes

Partial molar volume changes on hydration can be obtained from solution

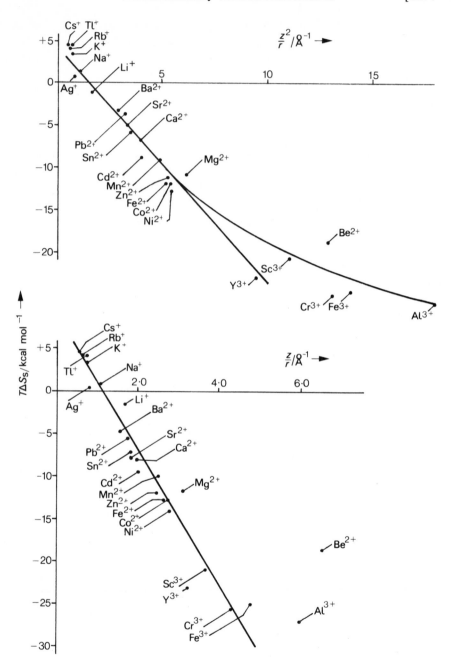

Fig. 7.7 Plots of $T\Delta S_S$ values for M^{n+}aq against z^2/r and against z/r.

Table 7.9 Partial molal hydration volumes ($cm^3\,mol^{-1}$) for metal cations; relative to zero for H^+aq; at 25°C; from reference [56].

H^+ 0						
Li^+ −0.88	Be^{2+} −12.0					
Na^+ −1.21	Mg^{2+} −21.17	Al^{3+} −42.2				
K^+ 9.02	Ca^{2+} −17.85		Zn^{2+} −21.6			
Rb^+ 14.07	Sr^{2+} −18.16		Cd^{2+} −20.0			
Cs^+ 21.34	Ba^{2+} −12.47	La^{3+} −39.10	Hg^{2+} −19.3	Tl^+ 10.6	Pb^{2+} −15.5	
	Mn^{2+} −17.7	Fe^{2+} −24.7	Co^{2+} −24.0	Ni^{2+} −24.0	Cu^{2+} −27.76	
Cr^{3+} −39.5		Fe^{3+} −43.7				Ag^+ −0.7

Pr^{3+} −42.53 Nd^{3+} −43.31 Sm^{3+} −42.33 Gd^{3+} −40.41 Tb^{3+} −40.24 Dy^{3+} −40.83 Ho^{3+} −41.76 Er^{3+} −42.86 Yb^{3+} −44.22
Th^{4+} −53.5

density measurements. As ever, the experimental $\bar{V}^{\ominus}$ values for whole salts have to be subjected to an arbitrary split so as to obtain single ion values. As ever, one of the methods used is to take $\bar{V}^{\ominus}(H^+aq)$ as zero. Early values of cation hydration volumes based on this assumption can be found in references [54] and [55]; a modern set of values is given in Table 7.9 [56]. The absolute value of $\bar{V}^{\ominus}(H^+aq)$ has been estimated as -1.49 cm^{3}mol^{-1} [54]. Absolute values for $\bar{V}^{\ominus}(M^{n+}aq)$ based on this surprisingly low value for the hydrated proton are given in Table 7.10 [54]. This Table also contains some $\bar{V}^{\ominus}(M^{n+}aq)$ values calculated for cations with hydration numbers which have been established by n.m.r. spectroscopy [57]. In the few cases where direct comparison is possible, there is some measure of agreement between these two essentially independent sets of absolute single ion values.

Table 7.10 Absolute values of cation hydration volumes, $\bar{V}^{\ominus}(M^{n+}aq)$, cm^3 mol^{-1} (based on $\bar{V}^{\ominus}(H^+aq) = -1.49$ cm^{3}mol^{-1} [54] or n.m.r. spectroscopy [57]).

[54]		[54]	[57]	[54]		[57]
Li^+	-2.4	Be^{2+}	-18.4	Al^{3+}		-57.3
Na^+	-3.0	Mg^{2+} -23.9	-29.4	La^{3+}	-42.8	-51.0
K^+	$+7.3$	Ca^{2+} -20.7				
Rb^+	$+12.3$	Sr^{2+} -21.2		Cr^{3+}		-53.0
Cs^+	$+19.7$	Ba^{2+} -15.3		Fe^{3+}		-57.2
Ag^+	-2.5	Cd^{2+} -16				

The single ion $\bar{V}^{\ominus}(M^{n+}aq)$ values of Tables 7.9 and 7.10 conform to a qualitative pattern which is determined by cation charge and radius. They range from large positive values for large unicharged cations such as Cs^+ to large negative values for the tripositive cations. Quantitative correlation with cation charge and size has been the subject of much debate over the past decades, together with explanations of the observed trends. Only now does a reasonably consistent picture seem to have emerged, in which all the effect of ions on the surrounding solvent are taken into account. Early authors discussed single ion $\bar{V}^{\ominus}(M^{n+}aq)$ values in terms of the volumes of the bare (gaseous) cation and of the electrostricted solvent surrounding it in solution, $\bar{V}^{\ominus} = V^0 - V^e$. This worked tolerably well for the alkali metal cations, which were often the only ones considered. Later Glueckauf showed that the electrostriction effect had to be calculated in terms of an effective radius considerably larger than the crystal radius of the cation. Such a state of affairs also obtains with cation hydration enthalpies, free energies, and entropies, and there is no more reason to cavil at this element of empirical adjustment here than above.

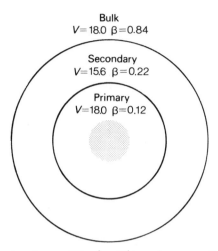

Fig. 7.8 Estimated volumes $(V/cm^3 mol^{-1})$ and compressibility coefficients $(\beta/cm^3 mol^{-1} kbar^{-1})$ for water molecules in the primary and secondary solvation shells around a transition metal 3+ cation [58].

Finally Akitt has discussed this topic in terms of three contributory factors. The first is the size of the 'bare' cation. The second is the change in volume undergone by the primary solvent molecules in their transfer from bulk solvent to the primary solvation shell. The third factor relates to electrostriction and the secondary solvation shell (see Fig. 7.8). Akitt's discussion centres on cations Be^{2+}, Mg^{2+}, and Al^{3+}, where n.m.r. spectroscopy has firmly established the composition of their primary hydration shells. The second factor is the one that proves most useful in explaining oddities in $\bar{V}^{\ominus}(M^{n+}aq)$ trends. The unexpectedly small negative value for Be^{2+} can be attributed to its low primary hydration number of four. Cation hydration volumes for the alkali metal cations and the 3+ cations can be accommodated in the same charge-radius correlation if the cation radii taken are the crystal radii and the $M(OH_2)_6^{3+}$ radii respectively. The value for Ag^+ does not come within such a correlation, but once again the unusual coordination about this cation may be invoked in explanation [57].

7.2.5   Cation Hydration Compressibilities and Heat Capacities

These quantities are considered together because they are both one stage further removed from direct experimental measurement than enthalpies, free energies, or volumes. Compressibilities are derivatives of volumes, $K = -\partial V/\partial P|_T$, and heat capacities (at constant pressure) are derivatives of enthalpies, $C_p = \partial H/\partial T|_P$. The available results are few in number, and less precise and more difficult to treat. This is unfortunate because compressibilities are closely linked to electrostriction (Fig. 7.8) and so directly relevant to the solvation shells around ions. Compressibilities (adiabatic) can be derived from ultrasonic experiments, and can be satisfactorily split into internally consistent sets of single ion values. Table 7.11 [54] lists such values for hydrated cations, based on the usual convention that $\bar{K}^{\ominus}(H^+aq)$ is zero.

Table 7.11 Cation hydration compressibilities, $\bar{K}^{\ominus}(M^{n+}aq)$, relative to $\bar{K}^{\ominus}(H^+aq) = 0$ [54, 55].

$10^4 \bar{K}^{\ominus}(M^{n+}aq) / cm^3 mol^{-1} bar^{-1}$			
Li^+ -34	Be^{2+} -23	Zn^{2+} -70	Ce^{3+} -152
Na^+ -42	Mg^{2+} -83	Cd^{2+} -57	
K^+ -37	Ca^{2+} -71	Cu^{2+} -62	
Cs^+ -27	Ba^{2+} -99		

One early attempt to estimate absolute hydrated ion compressibilities [54] was based on an uncomfortably large number of assumptions, as the author fully recognised. Absolute values for $\bar{K}^{\ominus}(M^{n+}aq)$ can be obtained from Table 7.11 values via $\bar{K}^{\ominus}(H^+aq) = +31 \times 10^{-4} cm^3 mol^{-1} bar^{-1}$, but these agree only badly with later estimates. More recently single ion values have been derived via detailed models of hydration around Cl^- [59] and around large organic cations such as pyridinium and the R_4N^+ group [60]. The former model led to an absolute value of $-17 \times 10^{-4} cm^3 mol^{-1} bar^{-1}$ for Cl^-aq; the latter to a value of -13×10^{-4}. Absolute single ion values for hydrated cations based on the former reference value are listed in Table 7.12. These values can be rationalised in terms of the structure -making and -breaking effects of the various cations [59]. The effects of Ba^{2+} and La^{3+} are somewhat below expectation, based on the M^+ values. 'Dielectric saturation' [61] can be called on to explain this minor anomaly.

Table 7.12 Absolute values of cation hydration compressibilities, $\bar{K}^{\ominus}_{abs}(M^{n+}aq)$, based on the reference value of $\bar{K}^{\ominus}_{abs}(Cl^-aq) = -17 \times 10^{-4} cm^3 mol^{-1} bar^{-1}$ [59].

$10^4 \bar{K}^{\ominus}_{abs}(M^{n+}aq) / cm^3 mol^{-1} bar^{-1}$	
Na^+ -34	Ba^{2+} -79
K^+ -27	
Rb^+ -24	La^{3+} -120
Cs^+ -20	

Several publications have given estimates of partial molal heat capacities for cations in water [31, 34, 54, 62, 63], and fortunately there is a reasonable measure of agreement between different authors. For example, $\bar{C}_p^{\ominus}(Na^+aq)$

values range only from +5 to +11 cal deg^{-1} mol^{-1} (relative to $\bar{C}_p^{\ominus}(\text{H}^+\text{aq})=0$). Much wider discrepancies exist between estimates for the absolute ionic partial molal heat capacities. Consequently the 'absolute' value of $\bar{C}_p^{\ominus}(\text{H}^+\text{aq}) = 28$ cal deg^{-1}mol^{-1} should be regarded with some reservations, although it is the reference point for the most comprehensive list of cation $\bar{C}_p^{\ominus}$ values (Table 7.13). Relative values are consistent with normally accepted effects of cations on water structure in their vicinity.

Table 7.13 **Cation partial molal heat capacities (cal deg^{-1}mol^{-1}) in aqueous solution at 25°C, relative to $\bar{C}_p^{\ominus}(\text{H}^+\text{aq}) = 28$ [63].**

H$^+$ 28				
Li$^+$ 41	Be^{2+} 76			
Na$^+$ 37	Mg^{2+} 62		Al^{3+} 89	
K$^+$ 31	Ca^{2+} 54		Sc^{3+} 79	
Rb$^+$ 30	Sr^{2+} 52	Cd^{2+} 54		
Cs$^+$ 26	Ba^{2+} 44			
	Ra^{2+} 40			
				Tl$^+$ 28 Pb^{2+} 44
	Mn^{2+} 57			Cu^{2+} 59
	Cr^{3+} 88		Fe^{3+} 86	
Zr^{4+} 98				Ag$^+$ 35
	Gd^{3+} 72			
	U^{4+} 93			

a Natn. Bur. Stand. Technical Note 270/6 (1973) contains $\bar{C}_p^{\ominus}$ values for several lanthanide-(III) cations, but these values are expressed relative to a different reference point ($\bar{C}_p^{\ominus}(\text{H}^+\text{aq}) = 0$).

7.3   SOLUTIONS IN D$_2$O

Once one moves away from ordinary water, whether to D$_2$O or to non-aqueous solvents, discussion of thermodynamic parameters for ions in solutions are framed in one of two ways. The first considers single ion values for $\Delta X(\text{M}^{n+}\text{solv})$ in a manner similar to that employed above for $\Delta X(\text{M}^{n+}\text{aq})$ values. The second way, with due deference to the central role played by ordinary water in chemistry, discusses values of ΔX for the transfer of ions to the other solvent. The reader is warned that the author will employ either mode of discussion in the course of this and subsequent sections, depending on the form of the data available.

Although the simple Born treatment of ion solvation predicts that both the enthalpy and the entropy ($\partial D/\partial T$ is the same for H$_2$O and D$_2$O) of transfer of cations from H_2O to D_2O will be zero, this is not the case. Interpretation of thermodynamic parameters of ion transfer from H_2O to D_2O is generally conducted in terms of solvent structure. In general, experimental results are consistent with the view that D_2O has greater structuredness, at least in the 25°C region where measurements are usually made. Added ions are thought to have increased structure -making or -breaking effects in D_2O. Such explanations sometimes deserve the epithet 'glib' [64].

7.3.1   Enthalpies

The enthalpy of transfer of lithium fluoride from H_2O to D_2O is exothermic. Enthalpies of transfer of other alkali halides from H_2O to D_2O are endothermic. These results are in keeping with the generally accepted qualitative notion that while small ions such as Li^+ and F^- are nett structure-formers, larger ions are structure-breakers. Table 7.14 [64] lists single ion enthalpies of transfer indicating the relative effects of various metal cations in modifying structure. Most of these values have been obtained by direct calorimetric measurements of enthalpies of solution for appropriate salts in H_2O and in D_2O. However, the values for Ag^+ and Cu^{2+} are derived from the temperature dependence of solubility products of silver bromate and copper(II) iodate. The separation of the experimentally determined enthalpies into the single ion values given in Table 7.14 was based on the arbitrary (though see Section 7.3.2, footnote) assumption that the enthalpy of transfer of Li^+ from H_2O to D_2O is zero.

Table 7.14   Single ion enthalpies of transfer ($\Delta H_{tr}^{\ominus}$) of cations from H_2O to D_2O, at 25°C. Values are relative to $\Delta H_{tr}^{\ominus}(Li^+) = 0$ [64].[a]

$\Delta H_{tr}^{\ominus}$ / kcal mol^{-1}			
Li^+	0	Mg^{2+}	0.050
Na^+	0.160	Ca^{2+}	0.190
K^+	0.215	Ba^{2+}	0.600
Rb^+	0.255		
Cs^+	0.285	Cu^{2+}	0.190
Ag^+	0.090		

a Another, similar, set of values can be found in Ref. [63].

7.3.2   Free Energies

Some disagreement exists between the values for the free energies of transfer of alkali metal cations from H_2O to D_2O as derived from e.m.f. measurements [64] and from n.m.r. spectroscopy [65] respectively. Both sets of results are shown in Table 7.15, and both are calculated on the assumption that the Gibbs free energy of transfer of Li^+ from H_2O to D_2O is zero.[†] It is impossible to decide between these rival sets of values. Although the n.m.r. results both indicate a smooth trend and are the more recent, smoothness and novelty alone are insufficient guarantors of veracity. It is interesting that the older results are associated with regular trends in their enthalpies and entropies of transfer, in contrast to the more commonly encountered situation where irregularly varying enthalpy and entropy series compensate to give regularly varying free energies.

Table 7.15 Single ion Gibbs free energies of transfer ($\Delta G_{tr}^{\ominus}$) of cations from H_2O to D_2O, at 25°C, from e.m.f. [64] and n.m.r. [65] measurements. Values are relative to $\Delta G_{tr}^{\ominus}(Li^+) = 0$. [a]

	$\Delta G_{tr}^{\ominus}$ / kcal mol^{-1}	
	E.m.f. [64]	N.m.r. [65]
Li^+	0	0
Na^+	0.095	0.013
K^+	0.107	0.042
Rb^+	0.103	0.050
Cs^+	0.048	0.064

a A somewhat more extensive set of values (different zero) can be found in Ref. [63].

7.3.3 Entropies

Entropies of transfer reported for the alkali metal cations are given in Table 7.16.

Table 7.16 Single ion entropies of transfer ($\Delta S_{tr}^{\ominus}$) of cations from H_2O to D_2O, based on the assumption $\Delta S_{tr}^{\ominus}(Li^+) = 0$ [64].

	$\Delta S_{tr}^{\ominus}$ / cal deg^{-1} mol^{-1}
Li^+	0
Na^+	0.22
K^+	0.36
Rb^+	0.51
Cs^+	0.80

† This assumption derives from the report that the ^7Li chemical shift for Li^+ is the same in H_2O as in D_2O [66]. But such a scale of values gives a value for chloride ion markedly different from that derived from consideration of the ionisation of HCl, DCl in H_2O, D_2O [67].

7.3.4   Other Functions

Ionic volume and adiabatic compressibility transfer functions have been determined for the alkali metal cations and for Ba^{2+} (Table 7.17) [68]. Again both sets of results can be interpreted on the solvent structure model.

Table 7.17   Ionic volume transfer functions, ΔV_{tr}, and adiabatic compressibility transfer functions, K_{tr}, for cations from H_2O to D_2O, at $25°C$ [68].

	$\Delta V_{tr} / cm^3 mol^{-1}$	$10^4 K_{tr} / cm^3 mol^{-1} bar^{-1}$
Na^+	-0.1	-6.9
K^+	$+0.3$	-5.3
Rb^+	$+0.4$	-4.9
Cs^+	$+0.5$	-4.2
Ba^{2+}	$+0.6$	-12.0

7.4   NON-AQUEOUS SOLUTIONS

Enthalpies, free energies, and entropies of solvation of cations in non-aqueous solvents will be considered in this section. Both the relatively meagre amount of data and differences in single-ion-splitting assumptions between different authors generally preclude full and detailed discussion. Nonetheless, certain general trends emerge. Hopefully the current activity in this field will lead to a comprehensive supply of experimental results for a wide range of electrolytes, together with a uniformity of assumptions in deriving single ion values. Only then can an overall picture be developed.

For each of the functions H, G, and S, the available data will be presented and discussed in two parts. First estimates of (absolute) cation solvation parameters will be considered, and then estimates of parameters of transfer of cations from water to various non-aqueous solvents. Whereas the former quantities are of fundamental importance, the latter provide useful comparisons with the much studied and best (though still only partially) understood aqueous systems.

7.4.1   Cation Solvation Enthalpies

Enthalpies of solution, and hence cation solvation enthalpies, have been determined for a range of non-aqueous solvents, including liquid ammonia, amides, alcohols, and even liquid sodium.

Liquid ammonia. Self-consistent sets of single ion enthalpies of ammoniation, $\Delta H_{amm}(M^{n+})$ can be obtained from calorimetric determinations of enthalpies of solution for a series of chlorides, bromides, iodides, and nitrates. Such single ion values for alkali metal and alkaline earth cations can be accommodated by the modified Born equation [3, 4] if effective cation radii 0.61Å bigger than the crystal radii are used (see the situation in aqueous system, Chapter 7.1.1). As the modified Born equation seemed to correlate these single ion values tolerably well, it was used to estimate an absolute single ion enthalpy of ammoniation for the Na^+ cation. ΔH_{amm} values for other cations may then be calculated from the differences between the experimentally determined enthalpies of solution of salts MX and NaX in liquid ammonia. The resulting values are reported in Table 7.18. They agree reasonably well with single ion values based on the assumption that $\Delta H_{amm}(Rb^+) = \Delta H_{amm}(Br^-)$. They agree less well with values calculated from a detailed model of ammoniated cations in liquid ammonia [71]. The values in Table 7.18 are considerably less negative than early estimates [72] for enthalpies of ammoniation of cations, and a few kcal mol^{-1} less negative than recent estimates [73] for the alkali metal cations.

Table 7.18   Cation solvation enthalpies (kcal mol^{-1}) in liquid ammonia, estimated for $-33°C$ [69].[a]

Li^+	-133	Ca^{2+}	-403
Na^+	-105	Sr^{2+}	-370
K^+	-84	Ba^{2+}	-336
Rb^+	-76		
Cs^+	-70	Hg^{2+}	-490
		Pb^{2+}	-390
Ag^+	-138		

a   Earlier sets of values for a range of M^+ and M^{2+} cations can be found in ref. [70].

The $\Delta H_{amm}(M^{n+})$ values listed in Table 7.18 vary as expected with cation charge and size, a conclusion implicit in the method of estimation described above. Although ΔH_{amm} values for the alkali metal and alkaline earth cations fit the simple electrostatic correlation well, values for such cations as Ag^+, Hg^{2+}, and Pb^{2+} do not. They have ΔH_{amm} values more negative than would be consistent with the electrostatic correlation, suggesting significant covalent interaction within these ammoniates. The Hg^{2+} value is just over 100 kcal mol^{-1} more negative than forecast by the electrostatic model [69]. Enthalpies of ammoniation of cations are more negative than enthalpies of hydration [71, 72].

Amides. Cation solvation enthalpies for the alkali metal cations have been estimated for the amide solvents formamide, N-methylformamide, and NN-dimethylformamide. Sums of cation plus anion solvation enthalpies were obtained from measured enthalpies of solution [74, 75] and calculated lattice enthalpies in the normal way. Several related methods have then been used to

Table 7.19 Cation solvation enthalpies (kcal mol^{-1}) in non-aqueous amide solvents.[a]

	Formamide				N-methyl formamide		Dimethylformamide			Water		
	[74]	[73]	[77]	[78]	[74]	[73]	[74]	[78]	[73]	[74]	[78]	[73]
Li$^+$	−123.6	−128.5	−122.9		−126.3	−135.0	−129.2		−153.0	−121.2		−115.0
Na$^+$	−99.8	−105.2	−99.6	−103.8	−100.0	−109.5	−101.8	−107.8	−127.2	−94.6	−99.9	−89.4
K$^+$	−80.5	−85.3	−79.7	−83.7	−81.4	−90.8	−83.0	−89.1	−108.5	−75.8	−79.7	−69.1
Rb$^+$	−74.5	−79.6	−74.0		−74.5	−84.6	−77.3		−102.5	−69.2		−63.1
Cs$^+$	−67.0	−71.4	−65.8		−67.8	−76.2	−69.7		−93.9	−62.0		−54.7
Ag$^+$				−122.0				−125.8			−116.6	
Ba^{2+}				−327.0				−337.7			−317.4	
Zn^{2+}				−500.2				−509.6			−494.5	
Cd^{2+}				−444.1				−452.7			−437.6	

[a] Recently reported estimates for cation solvation enthalpies in dimethylacetamide are very close to the cation hydration enthalpies listed in the right hand column of this Table, see Paul, R. C., Banait, J. S., and Narula, J. P., J. Electroanal. Chem. Interfac. Electrochem., 66, 111 (1975).

separate the results into single ion contributions. One approach achieved the separation by using the modified Born equation [4], varying δ until the best plot of ion solvation enthalpies against reciprocal effective ionic radius $r + \delta$ was obtained [74]. In each case $\delta = 0.75\text{Å}$ gave the best fit; this result may be compared with $\delta = 0.85\text{Å}$ for water and $\delta = 0.61\text{Å}$ for ammonia. A similar approach based on extrapolation of anionic solvation enthalpies against reciprocal radius [76], together with one based on the difference method of Halliwell and Nyburg (see proton hydration enthalpy, Chapter 7.2.1) [77], have also been used for formamide. The results of all these efforts are collected together in Table 7.19, which also includes some similarly-derived hydration enthalpies for direct comparison. The similarity of the cation solvation enthalpies determined by any one method for a given cation in the three amide solvents is remarkable, but the agreement between methods is not so good. The small differences between these solvation enthalpies and the respective cation hydration enthalpies are ascribed to differences in hydrogen-bonding between the liquids.

Some wise comments on the Born model, its modification and its deficiencies, can be found in an article on enthalpies of solution of salts in these three amides [79]. Unfortunately this paper deals only with sums of cation plus anion solvation enthalpies, and does not separate them into single ion components.

Alcohols. Information here is sparse (Table 7.20) [78, 80]. As with the amides, cation solvation entropies depend very little on the nature of the solvent and are close to cation hydration enthalpies.

Table 7.20   Cation solvation enthalpies (kcal mol^{-1}) in methanol, ethanol, and, for comparison, water.

	Methanol	Ethanol [80][a]	Water	Methanol [78][b]	Water
Li$^+$	-127	-127	-125		
Na$^+$	-100	-99	-98	-104.8	-99.9
K$^+$		-79	-78	-84.1	-79.7
Ag$^+$				-121.6	-116.6
Ba^{2+}				-331.5	-317.4
Zn^{2+}				-505.5	-494.5
Cd^{2+}				-447.2	-437.6

a Assume $\Delta H(\text{solv})\text{M}^+/\Delta H(\text{solv})\text{X}^-$ is the same in different solvents and equals 1 for caesium iodide; b assume $\Delta H(\text{hydr})\text{H}^+ = -263.7$ kcal mol^{-1} and $\Delta H_{tr}(\text{Ph}_4\text{As}^+) = \Delta H_{tr}(\text{Ph}_4\text{B}^-)$ for transfer from water to methanol.

Dimethyl Sulphoxide and Acetonitrile. Again cation solvation enthalpies (Table

7.21) show little dependence on the nature of the solvent. Intersolvent differences are easier to see when results are presented in the form of transfer functions (see below). The results for the 2+ cations in Table 7.21 give some idea of the differences between 'hard' (Ba^{2+}), 'soft' (Cd^{2+}) and intermediate (Zn^{2+}) cation behaviour.

Table 7.21   Cation solvation enthalpies (kcal mol^{-1}) in dimethyl sulphoxide, acetonitrile, and, for comparison, water [78].

	Dimethylsulphoxide	Acetonitrile	Water
Na^+	−106.5	−103.0	−99.9
K^+	−88.0	−85.1	−79.7
Ag^+	−129.7	−129.2	−116.6
Ba^{2+}	−336.0	−319.3	−317.4
Zn^{2+}	−509.6	−489.7	−494.5
Cd^{2+}	−454.6	−435.7	−437.6

Liquid Sodium. The cation solvation enthalpy for Na^+ in liquid sodium has been estimated as −90 kcal mol^{-1}. This is slightly lower than the cation hydration enthalpy of Na^+, which is −97 kcal mol^{-1} using an analogous method of derivation [81].

General. Two general points emerge. The first is the similarity of solvation enthalpies for a given cation over the range of solvents covered here. The second is that even the present rather insecurely based trends of solvation enthalpies with solvent nature seem more likely to bear some relation to Gutmann donor numbers than to such empirical parameters as Y, Z, or E_T (Chapter 1.3).

7.4.2   Enthalpies of Transfer

These are based on direct calorimetric measurements either of enthalpies of solution or of precipitation of salts in the respective solvents. Such salts as trifluoromethylsulphonates are preferable because they minimise ion-pairing, especially for salts of 2+ (or 3+) cations. The split of salt values into single ion values is achieved by one of two methods. An enthalpy of transfer of zero for the Me_4N^+ cation can be assumed, but it is better to assume that the (small) enthalpies of transfer of the Ph_4As^+ cation and of the Ph_4B^- anion are equal. Results are collected into Tables 7.22 and 7.23 [6, 78, 82, 83], but there are insufficient experimental data for satisfactorily bringing all the results to a common reference. A further set of enthalpies of transfer of the alkali metal cations from water to N-methylformamide exist, where calculations are based on a model involving six-coordinated cations [86]. These values are about

1 kcal mol^{-1} less negative than those listed in Table 7.22.

Table 7.22 Single ion enthalpies of transfer (ΔH_{tr}/kcal mol^{-1}) of cations from water to non-aqueous solvents, based on $\Delta H_{tr}(Me_4N^+) = 0$. From Reference [82] except where otherwise stated, and at 25°C.

	Methanol	Ethanol	N-Methyl formamide	Dimethyl formamide	Dimethyl sulphoxide	Acetonitrile	Propylene carbonate[a]
Li$^+$	−5.59	−5.5	−4.06	−3.57	−2.65		+4.62
Na$^+$	−5.20	−5.8	−4.87	−3.75	−2.96	+0.21	+1.45
K$^+$	−4.80	−5.7	−6.21	−5.30	−4.68	−1.84	−1.35
Rb$^+$	−3.99		−5.88	−4.89	−4.35	−1.88	−1.98
Cs$^+$	−3.56		−5.40	−4.36	−4.05		−2.51

a From ref. [83].

The variation of ΔH_{tr} down the series of alkali metal cations is often not smooth. This may indicate that the electrostatic and the steric or packing interactions between cation and solvent are in competition. Particularly in the case of the bulkier solvent molecules, steric restrictions will exist on the number of solvent molecules which can be packed around the small Li$^+$ cation. In the cases of Ba^{2+}, Zn^{2+}, and Cd^{2+}, the relative ΔH_{tr} values for transfer from water to various solvents give some idea of the relevance of the hard and soft acids and bases concept to cation-solvent interactions.

Almost all the ΔH_{tr} values are negative. This reflects the advantage of re-forming the particularly strong three-dimensional structure of the water after the cation is transferred out of the aqueous solution. Differences between ΔH_{tr} values for cation transfer to various solvents stem from several factors, including the degree of structuredness of the non-aqueous solvent. Hydrogen-bonded alcohols may be contrasted with non-hydroxylic non-hydrogen-bonding solvents such as acetonitrile. The relative solvating abilities of the solvents are also important and in this connection N-methylpyrrolidone may be contrasted to oxygen bonding solvents such as the alcohols. In addition, there are occasional special features such as the specific interaction between acetonitrile and the Ag$^+$ cation.

Finally, the case of transfer of alkali metal cations from dimethylformamide and from methanol into propylene carbonate can be cited as a rare example of the determination of enthalpies of transfer of cations between non-aqueous solvents [87].

7.4.3 Cation Solvation Free Energies

These are obtainable from solubility or e.m.f. data. Single ion values can be derived in two different ways. One way establishes a single value or pair of

Table 7.23 Single ion enthalpies of transfer (ΔH_{tr}/kcal mol^{-1}) of cations from water to non-aqueous solvents, based on $\Delta H_{tr}(Ph_4As^+) = \Delta H_{tr}(Ph_4B^-)$. Values for M$^+$ from reference [6]; for M^{2+} from reference [78]; for hexamethylphosphoramide from reference [84].

	Methanol	Formamide	Dimethyl formamide	Dimethyl sulphoxide	Acetonitrile	Propylene carbonate^a	Sulpholane	N-Methyl pyrrolidone	Hexamethyl phosphoramide
Li$^+$	−5.3	−1.3	−7.7	−6.3		+0.9		−5.1	
Na$^+$	−4.9	−3.9	−7.9	−6.6	−3.1	−1.6	−3.6	−9.4	
K$^+$	−4.4	−4.0	−9.4	−8.3	−5.4	−5.0	−6.0	−10.5	−13.8
Rb$^+$	−3.7	−4.1	−9.0	−8.0	−5.5	−5.6	−6.4		
Cs$^+$	−3.3	−4.1	−8.8	−7.7		−6.2	−5.9		
Ag$^+$	−5.0	−5.4	−9.2	−13.1	−12.6	−3.0	−3.2		
Ba^{2+}	−14.1	−9.6	−20.4	−18.8	−2.0				−34.6
Zn^{2+}	−10.9	−5.7	−15.0	−14.9	+4.8				−23.3
Cd^{2+}	−9.7	−6.6	−15.1	−16.9	+2.0				−27.7

a A similar set of values for propylene carbonate can be found in reference [85].

values by arbitrary but chemically reasonable means, and this is the approach used in preceding sections of this Chapter. The second way deals with the transfer of one ion from the gas phase into the solvent [88]. This avoids the usual arbitrary split of salt value into single ion contributions, introducing in its stead the need to estimate surface potentials, which is by no means an easy task [24]. Results obtained using the first method, applied to e.m.f. measurements, are given in Table 7.24 [89, 90]. Results obtained by the second method are in good agreement when reasonable values for surface potentials are used [24]. The variation of values with the nature of the cation shows the expected dependence on cation charge and size. There is, as with ΔH_{tr} values, remarkably little variation with the nature of the non-aqueous solvent.

7.4.4 Free Energies of Transfer

ΔG_{tr}, Gibbs free energies of transfer of cations from water to other solvents, give a better indication of relative solvating abilities than do free energies of solvation, ΔG_{solv} (see previous section). The latter are relatively insensitive to solvent because they are dominated by gas phase free energies, but the overshadowing effect of these is absent from ΔG_{tr} values. However, ΔG_{tr} values are sensitive to small variations in the e.m.f. measurements from which they are normally obtained. At present the accuracy of ΔG_{tr} values derived in this way is somewhat limited by the precision with which the required e.m.f.s can be measured [91]. Values of ΔG_{tr} calculated from solubility data are subject to errors and uncertainties caused by any ion-pairing or ionic strength effects. As usual, one is dependent on the assumptions involved in making the single ion split. $\Delta G_{tr}(Cl^-)$ from water into methanol ranges between 3.0 and 8.6 kcal mol^{-1} depending on these assumptions.

The dependence of $\Delta G_{tr}(M^{n+})$ on both the nature of the cation and of the solvent is illustrated in Tables 7.25 [82, 92] and 7.26 [6]. The values in Table 7.25 are derived from solubility measurements and are based on the assumption that $\Delta G_{tr}(Me_4N^+) = 0$ from water to all other solvents. The values in Table 7.26 are derived from e.m.f. measurements and are based on the slightly preferable assumption that $\Delta G_{tr}(Ph_4As^+) = \Delta G_{tr}(Ph_4B^-)$. Table 7.27 values [93] are derived in an analogous manner to those in Table 7.26. $\Delta G_{tr}(M^{n+})$ values cannot be explained satisfactorily by the simple Born model for the solvation of ions.

The results in Tables 7.25 and 7.26 show that, as one would expect, there is a greater range of ΔG_{tr} values (in other words a greater difference between cation-solvent interactions) for Li^+ than for Cs^+ in a given series of solvents. The comparison between $\Delta G_{tr}(K^+)$ and $\Delta G_{tr}(Ag^+)$ values, set out unconventionally in Table 7.27 in terms of transfer from acetonitrile rather than from water, is illustrated in Fig. 7.9, where transfer values from water have been plotted. There is only one dramatic deviation from the approximate correlation of values. This is for the solvent dimethylthioformamide, and may be ascribed to the especially

Table 7.24 Cation solvation Gibbs free energies (kcal mol^{-1}), at 25°C [89] [a].

	Methanol	Ethanol	Butanol	Amyl alcohol	Acetone	Acetonitrile	Formic acid	Liquid ammonia	Hydrazine
H$^+$	−252.5	−251.5	−251.5	−252.0	−252.0	−249.0	−243.0	−278.0	−274.5
Li$^+$	−115.0	−113.0	−114.0	−115.5	−115.0	−114.5	−113.5	−121.0	−113.5
Na$^+$	−92.0	−89.0	−88.5	−85.0	−80.0	−90.5	−91.5	−96.5	−92.0
K$^+$	−75.0	−72.5	−69.5	−69.5	−68.0	−75.0	−73.0	−77.0	−75.0
Rb$^+$	−69.0	−66.5	−60.0	−65.0	−62.0	−70.5	−70.0	−71.0	−70.0
Cs$^+$	−59.5	−58.5	−57.0	−57.0	−57.0	−61.5	−61.0	−63.0	−61.5
Ag$^+$	−106.0	−106.0	−106.0	−106.0	−108.0	−115.0	−111.0	−131.5	−128.5
Ca^{2+}	−366	−360	−362	−360	−362	−350.0	−358.0	−355.5	−363.0
Zn^{2+}	−479.0	−472.0	−472	−470	−468	−472.0	−471.0	−530.0	−508.5
Cd^{2+}	−415.0	−413.5	−416	−412	−414	−412.0	−413.0	−457.0	−446.5

[a] Values from reference [89]; these values agree well with an earlier and less extensive tabulation [90] where comparison is possible.

Table 7.25 Single ion Gibbs free energies of transfer (ΔG_{tr}/kcal mol^{-1}, on the molar scale) of cations from water to non-aqueous solutions, based on $\Delta G_{tr}(Me_4N^+) = 0$ except where otherwise stated. From reference [82].

	Methanol	Ethanol	N-Methyl formamide	Dimethyl formamide	Acetone	Acetonitrile	Dimethyl sulphoxide
Li+	-0.89	-0.3	-2.9			+6.9	-3.0
Na+	+0.30	+0.4	-1.3	-0.5	+2.3	+4.0	-2.0
K+	+0.65	+1.0	-1.4	-0.5	+1.1	+2.4	-1.6
Rb+	+0.75	+1.0	-1.2	-0.5	+0.9	+1.9	-1.3
Cs+	+0.75	+1.0	-1.1	-0.5	+0.7	+1.7	-1.7

Methanol values for Li+, Na+, K+ from ref [92][a]: -1.31[a], -0.47[a], +0.50[a].

a From ref. [92].

Table 7.26 Single ion Gibbs free energies of transfer (ΔG_{tr}/kcal mol^{-1}, on the molar scale) of cations from water to non-aqueous solvents, based on $\Delta G_{tr}(Ph_4As^+) = \Delta G_{tr}(Ph_4B^-)$ [6].

	Methanol	Formamide	Dimethyl formamide	Dimethyl sulphoxide	Acetonitrile	Propylene carbonate	Sulpholane	N-Methyl pyrrolidone	Hexamethyl phosphoramide[a]
Li+	+0.9	-2.3	-2.3	-3.5	+7.1	+5.7	-0.7	-3.9	
Na+	+2.0	-1.9	-2.5	-3.3	+3.3	+3.6	-1.0	-3.3	-13.8
K+	+2.4	-1.5	-2.3	-2.9	+1.9	+1.4	-2.1	-2.4	
Rb+	+2.4	-1.3	-2.4	-2.6	+1.6	-0.7	-2.4		
Cs+	+2.3	-1.8	-2.2	-3.0	+1.2	-2.9			
Ag+	+1.8	-3.7	-4.1	-8.0	-5.2	+3.8	-0.9	-7.2	
Tl+	+1.0		-2.7	-4.9	+2.2	+2.5			

a From ref. [84]

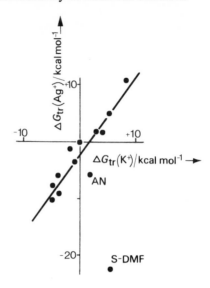

Fig. 7.9   Correlation of Gibbs free energies of transfer, ΔG_{tr}, from water to non-aqueous solvents for K^+ and for Ag^+ [72]; the point marked S–DMF is for thiodimethylformamide, that marked AN for acetonitrile.

Table 7.27   Comparison between the Gibbs free energies of transfer $(\Delta G_{tr}/\text{kcal mol}^{-1}$ on the molar scale) of the K^+ and Ag^+ cations from acetonitrile to other solvents, at $25°C$ [93].

	$\Delta G_{tr}(K^+)$	$\Delta G_{tr}(Ag^+)$
2,2,2-Trifluoroethanol	6.2	15.9
Dimethylthioformamide	3.1	−19.8
Nitromethane	2.3	9.8
Ethanol	1.3	5.6
Methanol	0.6	6.1
Acetonitrile	0	0
Propylene carbonate	−1.0	9.6
Sulpholane ^{a}	−1.6	6.7
Formamide	−3.0	1.6
Water	−3.3	4.3
Dimethylformamide	−4.8	1.0
N-Methyl-2-pyrrolidone	−4.8	−0.6
Dimethylacetamide	−5.5	−0.7
Dimethyl sulphoxide	−6.0	−3.5
Hexamethylphosphoramide	−7.3	−5.3

a At $30°C$.

favourable interaction between the Ag^+ cation and the sulphur atom of the solvent molecules. Acetonitrile, which may interact in different ways with Ag^+ and with K^+, is on the edge of the correlation band covering the remaining (O-bonded) solvents. Specific interaction between Ag^+ and acetonitrile or dimethyl sulphoxide is also reflected in another related investigation of cation transfer functions [94]. The order of $\Delta G_{tr}(M^+)$ values for transfer of cations from NN-dimethylformamide to its sulphur-bonding analogue NN-dimethylthio-formamide fall in the following order (numbers in brackets are ΔG_{tr} in kcal mol^{-1}):

$$Li^+(+15.4) > Na^+(+12.0) > K^+(+8.9) > Cs^+(+5.6) > Tl^+(-1.0) > Ag^+(-20.8).$$

This is just the order that would be expected from the 'hard' to 'soft' trend from left to right for the cations cited, though this bald statement must represent an oversimplified view of the situation [95].

Table 7.28 lists some solvent activity coefficients for cations, a related parameter which can give information on the strength of cation-solvent interaction. The solvent activity coefficients for the alkali metal cations show that hexamethylphosphoramide, dimethylformamide, and dimethyl sulphoxide are very good solvators of these cations. Formamide, acetonitrile, propylene carbonate, and water are rather less good, and methanol slightly less good again. This order can be related to the basicity of the solvents, including the degree of delocalisation of the negative end of their dipoles onto the atoms actually interacting with the cation. The order also bears some resemblance to that of the Gutmann donor numbers (Chapter 1.3) for these solvents. The rather different order of solvent activity coefficients for the Ag^+ cation highlights the importance of specific favourable interactions between this cation and certain solvents, for instance π-electron donors such as acetonitrile and dimethyl sulphoxide [96].

Table 7.28 Logarithms of solvent activity coefficients for cations, with respect to methanol, at $25°C$ [96]^a.

	Water	Acetonitrile	Propylene carbonate	Formamide	Dimethyl formamide	Dimethyl sulphoxide	Hexamethyl phosphoramide
Na^+		+1.4			−3.9	−3.6	−6.3
K^+	−1.5	−0.8	−1.7	−2.8	−3.7	−4.5	−4.8
Cs^+	−1.1	−1.3		−2.1	−3.3	−4.3	
Ag^+	−0.8	−6.3	−1.0	−3.7	−5.1	−8.2	−10

a These single ion values have been derived using a tetraethylammonium picrate assumption which gives results close to those obtained by using the usual $Ph_4As^+ \equiv BPh_4^-$ assumption [97].

7.4.5 Cation Solvation Entropies

Standard partial molal entropies, $\bar{S}^{\ominus}$, of a range of 1:1 electrolytes have been determined for salts in several non-aqueous solvents. These $\bar{S}^{\ominus}$ values have been split into 'absolute' single ion values [98] by means of guessing $\bar{S}^{\ominus}(H^+solv)$ in each solvent. $\bar{S}^{\ominus}(MX\ solv)$ was split for several assumed values of $\bar{S}^{\ominus}(H^+solv)$, both $\bar{S}^{\ominus}(M^+solv)$ and $\bar{S}^{\ominus}(X^-solv)$ plotted against the corresponding values in water, and the single ion $\bar{S}^{\ominus}$ values selected which gave the closest fit to a single straight line for both cations and anions. The $\bar{S}^{\ominus}(H^+aq)$ value producing the best fit is -13.0 cal deg^{-1} mol^{-1} (mole fraction standard state). This gives $\bar{S}^{\ominus}(H^+aq)$ $= -5$ cal deg^{-1} mol^{-1} (hypothetical molal standard state) which is in close agreement with $\bar{S}^{\ominus}(H^+aq)$ determined by other methods both similar and otherwise (see above). The $\bar{S}^{\ominus}(M^{n+}solv)$ values determined by the present method can therefore be considered as reasonable approximations to absolute values. They are listed in Table 7.29; the standard state here is mole fraction rather than unit molality, hence the differences between the $\bar{S}^{\ominus}(M^{n+}aq)$ values included in this Table and those tabulated earlier.

Table 7.29 Standard ionic entropies of solvation of cations in non-aqueous solvents. Values are in cal deg^{-1} mol^{-1} at 25°C; mole fraction standard state [98].

	Water	Liquid ammonia	Methanol	Ethanol	Formamide	N-Methyl formamide	Dimethyl formamide
Li$^+$	-9.6	-28.1	-17.7	-22.5	-5.3	-15.6	-23.3
Na$^+$	$+1.4$	-21.0	-10.5	-15.6	-4.2	-2.8	-14.2
K$^+$	$+11.5$	-14.7	-2.7	-8.0	$+4.6$	$+2.0$	-9.1
Rb$^+$	$+16.7$	-7.1	$+3.0$		$+8.2$		
Cs$^+$	$+18.8$	-7.1	$+5.2$		$+10.4$	$+8.2$	$+0.7$

For a given solvent, $\bar{S}^{\ominus}(M^+solv)$ values show the same trend to more positive values in the direction Li$^+$ to Cs$^+$ as do $\bar{S}^{\ominus}(M^+aq)$. This corresponds to less strong orientation and movement-restriction of solvent molecules around the larger cations. For each cation, solvents are generally in the order [98, 99]:

acetone < liquid ammonia < acetonitrile < dimethylformamide ~ ethanol
< methanol < N-methylformamide < formamide < water.

This order correlates well with neither dielectric constants, dipole moments, nor solvent basicities, but it does correlate with current notions concerning the degree of ordering of structure in the respective solvents. It is interesting that single ion entropies of solvation are linearly related to the respective single ion entropies of hydration:

$$\bar{S}^{\ominus}(M^+solv) = a + b\bar{S}^{\ominus}(M^+aq).$$

The absolute partial molar entropy of the solvated proton in liquid ammonia has been estimated as -25 cal deg^{-1} mol^{-1} at $25°C$, while the value of $\bar{S}^{\ominus}(H^+)$ in water is close to zero (Chapter 7.2.3). Thus absolute $\bar{S}^{\ominus}(M^{n+}$, liq. NH$_3$) values are considerably more negative than $\bar{S}^{\ominus}(M^{n+}aq)$ values [70].

7.4.6 Entropies of Transfer

As in the case of cation hydration entropies (Chapter 7.2.3), a choice of methods exists for obtaining and presenting data relating to entropies of transfer of single ions from one solvent to another. One of the more straightforward ways is to calculate ΔS_{tr} from ΔH_{tr} and ΔG_{tr}, using in all cases values relative to zero for a given ion. Values for ΔS_{tr} obtained this way using the Me$_4$N$^+$ cation as reference ion are listed in Table 7.30. This approach gives an acceptable means for comparing cation values in a given solvent, but is less satisfactory for comparisons of values for a given ion in a range of solvents. An allowance then has to be made for the differing solvation of the reference ion, such as Me$_4$N$^+$, in the various solvents. For example, ΔS_{tr} values for all cations are more negative when the transfer is from water to methanol than when it is to acetonitrile (Table 7.30). This indicates the different effects of the Me$_4$N$^+$ cation on the solvent structure in water and in methanol or acetonitrile as well as the effects of the alkali metal cations [82, 100]. Indeed a rather different set of ΔS_{tr} values has been derived using the assumption that $\Delta S_{tr}(Ph_4As^+) = \Delta S_{tr}(Ph_4B^-)$ [6].

A more complicated treatment of the data is required if the effects on ΔS_{tr} of a reference ion are to be eliminated. One method [98] involves the estimation of $\bar{S}^{\ominus}(H^+)$ in water and in the various non-aqueous solvents (see Chapter 7.2.3 and 7.4.5). When these have been established $\Delta S_{tr}(M^{n+})$ values can be derived from them [82, 100]. Such 'absolute' ΔS_{tr} values for metal cations are listed in Table 7.31 [82, 100, 101]. The two types of ΔS_{tr} values

Table 7.30 Standard ionic entropies of transfer of cations from water to non-aqueous solvents, relative to $\Delta S_{tr}(Me_4N^+) = 0$ [82, 100].

	Methanol	Ethanol	N-Methyl formamide	Dimethyl formamide	Dimethyl sulphoxide	Aceto nitrile	Acetone
Li$^+$	-15.7	-17	-10	-9.0	$+1$		
Na$^+$	-18.4	-21	-12	-10.9	-3	-12.7	-20
K$^+$	-18.3	-22	-16	-16.1	-10	-14.1	-18
Rb$^+$	-15.9		-16	-14.9	-10	-12.5	
Cs$^+$	-14.5		-14	-12.9	-8		

shown in Tables 7.30 and 7.31 show the expected greater solvent variation for Li^+ than for Cs^+. However there is no smooth trend from Li^+ to Cs^+ for a given solvent, but usually a maximum at K^+ or Rb^+. The variation in $\Delta S_{tr}(M^+)$ with solvent can be qualitatively rationalised in terms of the effect of the cation on the structure of the solvents. For example, the effects are greater when transfer is to a non-hydrogen bonded solvent such as acetone. Often values of $\Delta S_{tr}(M^{n+})$ from water to non-aqueous solvents reflect the importance of the 'fault zone' (Fig. 1.2, p.20) surrounding cations in water.

Table 7.31 'Absolute' single ion entropies of transfer of cations from water to non-aqueous solvents, in cal $deg^{-1} mol^{-1}$ at $25°C$ and from reference [100] unless otherwise stated.

	Methanol		Ethanol		N-Methyl formamide	Dimethyl formamide	Dimethyl sulphoxide	Aceto nitrile	Acetone
Li^+	-10.7	-10^a	-14		-4	-15.0	-5		
Na^+	-13.4	-17^a	-18	-21^a	-6	-16.9	-9	-18.7	-28
K^+	-13.3	-18^a	-19	-21^a	-10	-22.1	-16	-20.1	-26
Rb^+	-10.9	-15^a			-10	-20.9	-16	-18.5	
Cs^+	-9.5	-13^a			-8	-18.9	-14		

a From ref. [101]

Another way of discussing entropies of transfer of cations between solvents estimates such values for real solvents from hypothetical ideal non-interacting solvents of identical molecular weight [6]. Such transfer values are listed in Table 7.32; the assumption that entropies of transfer of Ph_4As^+ and of Ph_4B^- are equal is used in their derivation. These values give a clear picture of the relative effects of various cations on interactions in the respective solvents, and the ordering properties of the cations are reflected in them, untrammelled by solvent–solvent interactions. For every solvent there is a clear and consistent trend from Li^+ down to Cs^+, with the value for Li^+ always the most negative. Inter-solvent comparisons need to be considered in the light of Fig. 1.2, with both structure-making (regions A, B) and structure-breaking (region C) taken into account. The positive values for the transfer of Rb^+ and Cs^+ into real water show how important the structure-breaking contribution must be in these cases at least [6].

7.4.7 Volumes of Transfer

Partial molal volumes of transfer of the alkali metal cations Li^+, Na^+, and K^+ from water to a variety of dipolar aprotic solvents have been determined (Table 7.33). Important factors here, both of which may determine the solvation volume, are the strength of the cation–solvent interaction (and the consequent

Table 7.32 Single ion entropies of transfer (cal deg^{-1} mol^{-1}) of cations from ideal to real solvents (cf text) [6].

	Water	Methanol	Formamide	Dimethyl formamide	Dimethyl sulphoxide	Aceto nitrile	Propylene carbonate	Sulpholane	N-Methyl pyrrolidinone
Li$^+$	−15	−35	−10	−30	−21		−28		
Na$^+$	−7	−29	−12	−22	−15	−26	−21	−13	−22
K$^+$	0	−22	−6	−21	−15	−22	−18	−13	−21
Rb$^+$	+3	−16	−4	−16	−12	−18	−10	−7	
Cs$^+$	+4	−12	−2	−15	−9		−4	−4	
Ag$^+$	−9	−31	−13	−23	−23	−32	−29	−13	

electrostriction of solvating solvent) and the bulk of the solvent molecules [102].

Table 7.33 Partial molal volumes of transfer $(cm^3 mol^{-1})$ of metal cations from water to dipolar aprotic solvents, at 25°C. It is assumed that the partial molal volumes of transfer of the Ph_4As^+ and BPh_4^- ions are equal in each case [102].

	Li^+	Na^+	K^+
Acetonitrile	−12	−10	−11
Methanol	−11	−10	−10
Dimethylformamide	3	5	6
Formamide	4	5	6
Propylene carbonate		6	10
Dimethyl sulphoxide	8	14	12
Hexamethylphosphoramide	10	19	21

7.5   MIXED AQUEOUS SOLUTIONS

A fair number of enthalpies, free energies, and entropies of transfer of salts from water to binary aqueous mixtures exist. Some of these have, using more or less reasonable assumptions, been used to derive single ion transfer parameters. In general the results are rationalised in terms of the variation of solvent structure with composition and the reflection of this variation in cation solvation, as so often in discussions alluded to elsewhere in this Chapter. These rationalisations are often prompted when an extremum in the plot of ΔX_{tr} against mole fraction occurs at a mole fraction for which extrema in other physical properties, spectroscopic or thermodynamic, of the binary mixture have been observed.

7.5.1   Enthalpies of Transfer

Enthalpies of transfer of Na^+ from water to aqueous dimethyl sulphoxide [103] and of Ni^{2+} from water to aqueous dioxan [104] both show maxima at low mole fractions of organic cosolvent (Table 7.34). Addition of dimethyl sulphoxide or of dioxan to water is thought to increase the water structure at first, then to decrease it. The maxima in ΔH_{tr} here may be associated with solvent mixtures of maximum structuredness.

Table 7.34 Single ion enthalpies of transfer $(\Delta H_{tr}/\text{kcal mol}^{-1})$ of cations from water to mixed aqueous solvents (mole fraction x_2 organic component).[a]

Cation	Cosolvent	x_2	ΔH_{tr}	
Na^+	dimethyl	0.042	+0.4	[103]
	sulphoxide	0.097	+0.9	
		0.204	−0.5	
		0.276	−2.0	
		0.321	−2.5	
		0.386	−3.7	
		0.503	−5.3	
		1	−7.2	
Ni^{2+}	dioxan	0.020	+1.7	[104]
		0.042	+0.6	
		0.064	−0.4	
		0.120	−2.7	
		0.169	−4.6	

[a] Assume $\Delta H_{tr}(Ph_4P^+) = \Delta H_{tr}(Ph_4B^-)$.

7.5.2 Free Energies of Transfer

There are several sets of values for Gibbs free energies of transfer of cations from water to aqueous mixtures, derived from experimental data on e.m.f.s and solubilities by one of three routes. Feakins has reported $\Delta G_{tr}(M^{n+})$ values to aqueous methanol [105] and to aqueous dioxan [106] based on estimates for $\Delta G_{tr}(H^+)$. Wells has derived single ion ΔG_{tr} values in a similar manner, but using a more complicated model for estimating $\Delta G_{tr}(H^+)$. In this approach the transfer of H^+ in the form of tetrahedral $(H_3O)(H_2O)_4^+$ from water to solvent mixtures is calculated by a Born method, and then the replacement of solvating water by solvating methanol considered. Once $\Delta G_{tr}(H^+)$ has been thus calculated, values for $\Delta G_{tr}(X^-)$ are derived from established $\Delta G_{tr}(HX)$ values, and then $\Delta G_{tr}(M^{n+})$ from $\Delta G_{tr}(Cl^-)$ and $\Delta G_{tr}(MCl_n)$ [107]. Results are available for the transfer of alkali metal cations, Ag^+, Sr^{2+}, Ba^{2+}, Zn^{2+}, and Cd^{2+} to aqueous methanol [107], for K^+, Rb^+, Cs^+, and Ag^+ to aqueous acetone [108], and for Li^+, Na^+, and K^+ to aqueous glycol [109]. A third set of values for the transfer of alkali metal cations to aqueous methanol is available, derived by Case and Parson's surface potential method (Chapter 7.2.2) [24, 110].

A new method of determining free energies of transfer of metal cations based on n.m.r. spectra of their nuclei offers a promising alternative to the usual e.m.f. methods [111].

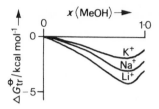

Fig. 7.10 Standard Gibbs free energies of transfer, $\Delta G_{tr}^{\ominus}$, of alkali metal cations from water to aqueous methanol, mole fraction x(MeOH) methanol [112].

It is too early to undertake a detailed discussion of these $\Delta G_{tr}(M^{n+})$ values. There are significant differences between the results derived by different methods, and some fairly gross assumptions are involved at times. Nevertheless a few general features are apparent. The pattern for transfer to aqueous methanol is illustrated in Fig. 7.10 [112]. This Figure shows extrema in methanol-rich mixtures, with Li^+ most sensitive to solvent environment variation. Most sources seem to indicate that the cation trend for transfer to aqueous methanol is regular going from Li^+ down to Cs^+. However it has been suggested that plots of $\Delta G_{tr}(M^+)$ against cation reciprocal radius have a maximum at K^+ or Rb^+ for transfer to some mixtures, for example to aqueous methanol [113] or to aqueous dimethyl sulphoxide [114]. There is an extremum at about 30% dimethyl sulphoxide in the plot of $\Delta G_{tr}(Na^+)$ against solvent composition [103, 115].

When more data become available in this area, it should prove possible to relate these to the thermodynamic parameters characterising mixed aqueous solvents. One would expect the overall pattern to reflect the differences between 'typically aqueous' (TA) and 'typically non-aqueous positive or negative' (TNAP, TNAN) mixtures (Chapter 1.1.2). There seems to be a tendency for cations to be stabilised, anions destablised, on transfer from water to TA mixtures.

Free energies of transfer of cations have found at least one practical application, in connection with selective solvation of Ag^+ in aqueous nitrile solvent mixtures and the importance of this in electrorefining [116].

7.5.3 Entropies of Solvation

Ion solvation entropies reflect solvent structure both in water and in non- and mixed aqueous media. One might expect therefore that cation solvation entropies would reflect the variation in solvent-solvent interactions with solvent composition in mixed aqueous solvents (see Chapter 1.3). Ion solvation entropies for the alkali metal cations in aqueous methanol [117] are listed in Table 7.35. These numbers are based on both a $K^+ \equiv Cl^-$ split, which is close to the earlier assumption of $\bar{S}^{\ominus}(H^+) = 0$, and on an ideal ionic gas standard state, which is not

Table 7.35 Standard partial molal ionic entropies of solvation of alkali metal cations in aqueous methanol. Values are for 25°C, and refer to a hypothetical ideal ionic gas standard state [117].

Weight %methanol:	0	10	20	43.1	68.3	100
Li^+	−42.3	−39.9	−40.1	−40.4	−43.6	−49.6
Na^+	−34.9	−34.3	−34.3	−35.7	−37.4	−46.1
K^+	−26.1	−26.3	−26.7	−28.6	−33.2	−40.7

close to earlier assumptions (contrast Chapter 7.2.3, 7.4.5, and 7.4.6). The dependence of $\bar{S}^{\ominus}(M^+\text{solv})$ on solvent composition is illustrated in Fig. 7.11. This shows that the maximum deviation from a linear dependence of $\bar{S}^{\ominus}(M^+\text{solv})$ on mole fraction occurs at methanol mole fractions in the region of 0.3 to 0.4. From such evidence as excess entropies of mixing of the components, maximum structure is thought to exist in this region. Effects are, as expected, greatest for Li^+. Presumably the deviations in $\bar{S}^{\ominus}(M^+\text{solv})$ against mole fraction plots would be greater in aqueous ethanol and dramatic in aqueous t-butyl alcohol.

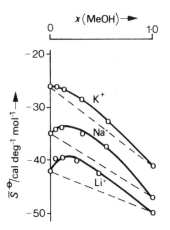

Fig. 7.11 Partial molal entropies of alkali metal cations in aqueous methanol, mole fraction x(MeOH) methanol [117].

7.6   SOLUBILITIES IN NON-AQUEOUS SOLVENTS

The amount of information on solubilities and on enthalpies and entropies of solution of inorganic salts in non-aqueous solvents is not large. Attempts to separate such quantities into single ion values are infrequent. Some of the information available in this area has been presented in earlier Sections on ΔX_{tr} parameters, where solvation differences for cations in aqueous and in non-aqueous environments have been stated in quantitative terms. The present Section will deal with the topic of cations in non-aqueous solvents in a much more qualitative way. We will present a selection from the available data on solubilities of salts and see how far any trends apparent can be rationalised in terms of lattice free energies [118] as opposed to ion solvation free energies. Where there are results for a series of salts with a common ion, some idea of the effect of cation variation can be obtained. Enthalpy and entropy data will also be mentioned, though these are sparse.

Solubilities of alkali metal halides in a selection of non-aqueous solvents are collected together in Table 7.36 [119-123]. The series of chlorides illustrates the variation with cation nature, the series of sodium salts gives the comparative variation with cation nature. Two general patterns emerge, with solubilities NaF < NaCl < NaBr < NaI, and sodium chloride or potassium chloride proving to be the least soluble of each set of chlorides. Lithium chloride is often extraordinarily more soluble than the other chlorides. All these trends are consistent with a fine balance of lattice free energies and ion solvation free energies, with the favourable solvation of the very small Li^+ tipping the balance in favour of solubility of lithium salts.

The solubilities of 2:1 and 1:2 salts in liquid ammonia are very much lower than those of the 1:1 halides. For example, barium chloride, sodium sulphate, and potassium sulphate all have negligible solubilities. This indicates that lattice free energies are dominant here. The liquid ammonia versus water comparison of solubilities is consistent with water being a better solvator of cations. Liquid ammonia, a less effective solvator, can only make ion solvation free energies win over lattice free energies in the most favourable case of 1:1 salts. This simple discussion works for straightforward salts such as those of the alkali metals and

Table 7.36 Solubilities of some alkali metal halides in non-aqueous solvents, ex-

	Water	Liquid ammonia	Liquid sulphur dioxide[a]	Methanol		Ethanol	n-Heptanol	Formic acid
	[119]	[120]	[121]	[120]	[12]	[12]	[12]	[120]
LiCl	55	3.02	0.012	41.0	21.0			27.5
NaCl	36	3.02		1.4	1.40			5.2
KCl	36	0.04	0.041	0.53	0.53			19.2
RbCl	94		0.33	1.34				56.9
CsCl	192			3.01	3.62	0.39	0.011	130.5
NaF	4	0.35	0.029	0.03	0.023			
NaCl	36	3.02		1.4	1.40			5.2
NaBr	46	138	0.014	17.4	16.1			19.4
NaI	184	162	15	83.0	62.5			61.8

[a] At $0°C$; [b] potassium salts, at $80°C$.

alkaline earths. However it is less satisfactory for both ammonium salts, where there is the possibility of cation-solvent hydrogen bonding, and for transition metal salts, where there may be stronger cation-solvent interactions [123]. Hydrazine proves a reasonable solvent for perchlorates and nitrates of the alkali and alkaline earth metals [125]. Liquid nitrogen dioxide, on the other hand, is a very bad solvent in which no simple metal salts dissolve to an appreciable extent.

Solubilities of fluorides in liquid hydrogen fluoride [126] show a similar pattern to that described for liquid ammonia. Liquid hydrogen chloride is a poorer solvent, with solvation energies less able to dominate over lattice energies and only some salts of very large cations soluble to any marked extent. Liquid sulphur dioxide resembles liquid ammonia and liquid hydrogen fluoride as a solvent for inorganic salts, with 1:1 electrolytes often soluble but 1:2 electrolytes reluctant to dissolve [121].

The solubilities of inorganic salts have been studied in several carboxylic acids and their derivatives [127]. Chlorides MCl are sparingly soluble in acetic acid, but some chlorides of divalent metals, including $ZnCl_2$ and $FeCl_3$, are fairly soluble. Formic acid and monochloroacetic acids [128], with dielectric constants higher than that of acetic acid, are slightly better solvents for inorganic salts, but propionic acid is worse. Solubility patterns in acyl chlorides are similar to those in the analogous carboxylic acids. For example, chlorides of the alkali and alkaline earth metals are sparingly soluble in acetyl chloride and in benzoyl chloride, but $ZnCl_2$, $HgCl_2$, $SnCl_2$, and especially $AlCl_3$ and $FeCl_3$, are readily soluble.

All these solubility trends and comparisons can be rationalised more or less

pressed as grams of salt per 100 grams of solvent, at 25°C.

Sulpholane [122]	Acetonitrile [120]	Acetone [120]	Formamide [12]	Acetamide[b] [123]	N-Methyl formamide [12]	Dimethyl formamide [12]
1.5	0.14	0.83	28.2		23.0	11 or 28
0.005	0.0003	0.000042	9.4		3.2	0.04
0.004	0.0024	0.000091	6.2	2.45	2.1	0.017 to 0.05
0.010	0.0036	0.00022				
0.019	0.0084	0.00041				0.0052
	0.003	0.0000025				0.0002
0.005	0.0003	0.000042	9.4	2.45^b	3.2	0.04
	0.040	0.011	35.6	10.3^b	28	3.2 or 10.3
	24.9	28.0	57 or 85	32.3^b	61	3.7 or 6.4

Table 7.37   Standard enthalpies of solution (kcal mol^{-1}) of inorganic chlorides and iodides in non-aqueous solvents at 25°C; from reference [12] unless otherwise stated.

	Water	Methanol	Ethanol	Formamide	N-Methyl formamide	Dimethyl formamide	Dimethyl sulphoxide	Propylene carbonate	Hexamethyl phosphoramide	Formic acid	Ethylene diamine	Liquid ammonia[b,c]
LiCl	−8.9	−12.0	−12.4	−8.9	−10.0	−11.6[a]	−10.9		−12.3			
NaCl	0.9	−2.3		−2.1	−1.2					−0.2	−4.9	−7.8
KCl	3.1	−1.2		0.8	0.3					−0.3	−2.9	−2.8
RbCl	4.0	1.3		0.7						−0.9	−4.1	−3.6
CsCl	4.3	2.1		0.8	0.9					−1.3	−4.4	−4.9
LiI	−15.1	−16.9	−20.7	−18.3	−21.1	−19 or −26[a]	−24.2	−15.1	−24.2		−28.8	
NaI	−1.9	−7.4	−6.8	−7.4	−8.3	−12.8	−11.5	−5.0			−15.7	
KI	4.9	−0.2	−0.4	−1.5	−3.2	−7.9	−6.3				−5.9	
RbI				0.2	−1.6	−6.6						
CsI	7.9	4.2		2.0	0.7	−4.3	−2.8					

a Ref. [129] gives −11.4 for LiCl and −26.1 for LiI; *b* ref. [130]; *c* a more detailed, but older, compilation can be found on pp. 231–233 of ref. [71].

successfully in several terms, for instance, in terms of the size and charge of the cation and anion, of all of the numerous and diverse properties of solvents which have been invoked earlier in this Chapter, and, where relevant or necessary, of specific cation-solvent interactions.

Enthalpies of solution. There are a few scattered data on enthalpies of solution of inorganic salts in non-aqueous solvents. Two sequences of salts with a common anion each illustrate differences between the characteristics of the various alkali metal cations (Table 7.37) [129, 130].

Entropies of solution. There is little published information here too. Standard partial molal entropies of several alkali metal halides in a range of non-aqueous solvents (alcohols, amides, dimethyl sulphoxide, propylene carbonate, and formic acid) are listed in Reference [12]. Entropies of solution of alkali metal chlorides in liquid ammonia and in several amine and ether solvents are reported in Reference [130].

FURTHER READING

Rosseinsky, D. R., *Chem. Rev.*, **65**, 467 (1965); *A. Rep. Chem. Soc.*, **68A**, 81 (1971).

Friedman, H. L. and Krishnan, C. V. in *Water: A Comprehensive Treatise – Vol. 3: Aqueous Solutions of Simple Electrolytes*, ed. Franks, F., Plenum Press (1973) Ch. 1.

REFERENCES

[1] Phillips, C. S. G. and Williams, R. J. P., *Inorganic Chemistry, Vol. 1*, Clarendon Press (1965) p. 161.

[2] Dasent, W. E., *Inorganic Energetics*, Penguin Books (1970) p. 141.

[3] Born, M., *Z. Phys.*, **1**, 45 (1920).

[4] Latimer, W. M., Pitzer, K. S., and Slansky, C. M., *J. Chem. Phys.*, **7**, 108 (1939).

[5] E.g. Millen, W. A. and Watts, D. W., *J. Amer. Chem. Soc.*, **89**, 6051 (196 (1967).

[6] Cox, B. G., Hedwig, G. R., Parker, A. J., and Watts, D. W., *Aust. J. Chem.*, **27**, 477 (1974).

[7] Saluja, P. P. S. and Scheraga, H. A., *J. Phys. Chem.*, **77**, 2736 (1973); Bauge, K. and Støgård, Å., *Acta Chem. Scand.*, **27**, 2683 (1973); and references therein.

[8] Støgård, Å., *Acta Chem. Scand.*, **27**, 2669 (1973).

[9] Kistenmacher, H., Popkie, H., and Clementi, E., *J. Chem. Phys.*, **61**, 799 (1974).

[10] Williamson, A. T., *Trans. Faraday Soc.*, **40**, 421 (1944).

[11] E.g. Waddington, T. C. in *Advances in Inorganic Chemistry and Radio-chemistry, Vol. 1,* ed. Emeléus, H. J. and Sharpe, A. G., Academic Press (1959); Conway, B. E. and Bockris, J. O'M. in *Modern Aspects of Electrochemistry, Vol. 1* ed. Bockris, J.O'M. and Conway, B. E., Butterworths (1954) pp. 47–48.

[12] Criss, C. M. in *Physical Chemistry of Organic Solvent Systems,* ed. Covington, A. K and Dickinson, T., Plenum (1973) Ch. 2.

[13] Rosseinsky, D. R., *Chem. Rev.,* **65**, 467 (1965).

[14] Alfenaar, M., *J. Phys. Chem.,* **79**, 2200 (1975).

[15] Grunwald, E., Baughman, G. and Kohnstam, G., *J. Amer. Chem. Soc.,* **82**, 5801 (1960).

[16] Halliwell, H. F. and Nyberg, S. C., *Trans. Faraday Soc.,* **59**, 1126 (1963); Morris, D. F. C., *Structure and Bonding,* **4**, 63 (1968).

[17] Lister, M. W., Nyburg, S. C., and Poyntz, R. B., *J. Chem. Soc. Faraday I,* **70**, 685 (1974).

[18] See, e.g., p. 160 of reference [1].

[19] Morss, L. R., *J. Phys. Chem.,* **75**, 392 (1971).

[20] Noyes, R. M., *J. Amer. Chem. Soc.,* **84**, 513 (1962).

[21] Silva, R. J., McDowell, W. J., Keller, O. L., and Tarrant, J. R., *Inorg. Chem.,* **13**, 2233 (1974).

[22] Figgis, B. N., *Introduction to Ligand Fields,* Interscience/Wiley (1966) pp. 227-228.

[23] Gutmann, V., *Allg. Prakt. Chem.,* **21**, 116 (1970); *Fortschr. Chem. Forsch.* **27**, 59 (1972).

[24] Case, B. and Parsons, R., *Trans. Faraday Soc.,* **63**, 1224 (1967).

[25] Barton, A. F. M. and Wright, G. A. in *Proceedings of the First Australian Conference on Electrochemistry,* ed. Friend, J. A. and Gutmann, F., Pergamon (1965) p. 124.

[26] Cox, B. G., and Parker, A. J., *J. Amer. Chem. Soc.,* **95**, 6879 (1973).

[27] Hunt, J. P., *Metal Ions in Aqueous Solution,* Benjamin (1963) p. 16.

[28] Robinson, R. A. and Stokes, R. H., *Electrolyte Solutions, 2nd Edn. revised,* Butterworths (1968) p. 70.

[29] E.g. Noyes, R. M., *J. Chem. Educ.,* **40**, 2 (1963).

[30] Schiffrin, D. J., *Trans. Faraday Soc.,* **66**, 2464 (1970).

[31] Natn. Bur. Stand. Circular 500 (1952).

[32] Johnson, D. A., *J. Chem. Soc. Dalton* 1671 (1974).

[33] Fuger, J. in *MTP International Review of Science, Inorganic Chemistry, Series 1, Vol. 7,* ed. Bagnall, K. W., Butterworths, London, and University Park Press, Baltimore (1972) Ch. 5.

[34] Natn. Bur. Stand. Technical Notes 270/3 (1968), 270/4 (1969), 270/5 (1971) and 270/6 (1971); *J. Chem. Thermodynamics,* **8**, 603 (1976).

[35] Wulff, C. A., *J. Chem. Engng. Data,* **12**, 82 (1967).

[36] Larson, J. W., Cerutti, P., Garber, H. K., and Hepler, L. G., *J. Phys.*

Chem., **72**, 2902 (1968).

[37] Berg, R. L. and Vanderzee, C. E., *J. Chem. Thermodynamics*, **7**, 229. (1975).

[38] Vanderzee, C. E. and Swanson, J. A., *J. Chem. Thermodynamics*, **6**, 827 (1974); Hepler, L. G. and Olofsson, G., *Chem. Rev.*, **75**, 585 (1975).

[39] Hinchley, R. J. and Cobble, J. W., *Inorg. Chem.*, **9**, 917 (1970).

[40] Vasil'ev, V. P., Kozlovskii, E. V., and Shitova, V. V., *Russ. J. Phys. Chem.*, **45**, 109 (1971).

[41] Vasil'ev, V. P. and Ikonnikov, A. A., *Russ. J. Phys. Chem.*, **45**, 162 (1971).

[42] Vasil'ev, V. P., Raskova, O. G., Belonogova, A. K., and Vasil'eva, V. N., *Russ. J. Inorg. Chem.*, **19**, 1331 (1974).

[43] Vasil'ev, V. P. and Kunin, B. T., *Russ. J. Inorg. Chem.*, **17**, 1129 (1972).

[44] Izatt, R. M., Eatough, D. J., Morgan, C. E., and Christensen, J. J., *J. Chem. Soc. A*, 2514 (1970).

[45] Rosseinsky, D. R., *A. Rep. Chem. Soc.*, **68A**, 81 (1971).

[46] Morss, L. R. and Haug, H. O., *J. Chem. Thermodynamics*, **5**, 513 (1973).

[47] Latimer, W. M., *Oxidation Potentials, 2nd Edn.*, Prentice-Hall, New York (1952) pp. 365–366.

[48] Powell, R. E. and Latimer, W. M., *J. Chem. Phys.*, **19**, 1139 (1951).

[49] Ulich, H., *Z. Elektrochem.*, **36**, 497 (1930).

[50] Frank, H. S. and Evans, M. W., *J. Chem. Phys.*, **13**, 507 (1945).

[51] Powell, R. E., *J. Phys. Chem.*, **58**, 528 (1954).

[52] Bockris, J. O'M. and Saluja, P. P. S., *J. Phys. Chem.*, **76**, 2298 (1972).

[53] See p. 259 of reference [1].

[54] Noyes, R. M., *J. Amer. Chem. Soc.*, **86**, 971 (1964).

[55] Owen, B. B. and Brinkley, S. R., *Chem. Rev.*, **29**, 461 (1941).

[56] Millero, F. J., *Chem. Rev.*, **71**, 147 (1971).

[57] Akitt, J. W., *J. Chem. Soc. A*, 2347 (1971).

[58] Stranks, D. R., *Pure appl. Chem., 15th Int. Conf. Coord. Chem.*, **38**, 303 (1974).

[59] Mathieson, J. G. and Conway, B. E., *J. Solution Chem.*, **3**, 455 (1974).

[60] Laliberté, L. H. and Conway, B. E., *J. Phys. Chem.*, **74**, 4116 (1970).

[61] Glueckaurf, E., *Trans. Faraday Soc.*, **64**, 2423 (1968).

[62] De Ligny, C. L., Alfenaar, M., and van der Veen, N. G., *Rec. Trav. Chim.*, **87**, 585 (1968); Fortier, J. -L., Leduc, P. -A., and Desnoyers, J. E., *J. Solution Chem.*, **3**, 323 (1974); Criss, C. M. and Cobble, J. W., *J. Amer. Chem. Soc.*, **86**, 5390 (1964).

[63] Friedman, H. L. and Krishnan, C. V. in *Water: A Comprehensive Treatise – Vol. 3: Aqueous Solutions of Simple Electrolytes*, ed. Franks, F., Plenum Press (1973) Ch. 1.

[64] Arnett, E. M. and McKelvey, D. R. in *Solute-Solvent Interactions*, ed. Coetzee, J. F. and Ritchie, C. D., Marcel Dekker (1969) Ch. 6.

[65] Voice, P. J., *J. Chem. Soc. Faraday I*, **70**, 498 (1974).

[66] Loewenstein, A., Shporer, M., Lauterbur, P. C. and Ramirez, J. E., *Chem. Commun.*, 214 (1968).

[67] Salomaa, P., *Acta Chem. Scand.*, **25**, 365 (1971).

[68] Mathieson, J. G. and Conway, B. E., *J. Chem. Soc. Faraday I*, **70**, 752 (1974).

[69] Senozan, N. M., *J. Inorg. Nucl. Chem.*, **35**, 727 (1973).

[70] Jander, J., *Anorganische und allgemeine Chemie in flüssigen Ammoniak*, ed. Jander, G., Spandau, H., and Addison, C. C., Vieweg, Braunschweig (1966).

[71] Coulter, L. V., *J. Phys. Chem.*, **57**, 553 (1953); Jolly, W. L., *Progr. inorg. Chem.*, **1**, 235 (1959).

[72] Jolly, W. L., *Chem. Rev.*, **50**, 351 (1952).

[73] Somsen, G. and Weeda, L., *J. Electroanal. Chem. Interfac. Electrochem.*, **29**, 375 (1971).

[74] Gill, D. S., Singla, J. P., Paul, R. C., and Narula, S. P., *J. Chem. Soc., Dalton*, 522 (1972).

[75] Somsen, G. and Coops, J., *Rec. Trav. Chim.*, **84**, 985 (1965).

[76] Somsen, G., *Rec. Trav. Chim.*, **85**, 517 (1966).

[77] Somsen, G., *Rec. Trav. Chim.*, **85**, 526 (1966).

[78] Hedwig, G. R. and Parker, A. J., *J. Amer. Chem. Soc.*, **96**, 6589 (1974).

[79] Finch, A., Gardner, P. J., and Steadman, C. J., *J. Phys. Chem.*, **71**, 2996 (1967).

[80] Mishchenko, K. P., *Acta Phys.-chim. URSS*, **3**, 693 (1935).

[81] Addison, C. C., *Chem. Brit.*, **10**, 331 (1974).

[82] Abraham, M. H., *J. Chem. Soc. Faraday I*, **69**, 1375 (1973).

[83] Wu, Y. -C. and Friedman, H. L., *J. Phys. Chem.*, **70**, 2020 (1966).

[84] Hedwig, G. R., Owensby, D. A., and Parker, A. J., *J. Amer. Chem. Soc.*, **97**, 3888 (1975).

[85] Krishnan, C. V. and Friedman, H. L., *J. Phys. Chem.*, **73**, 3934 (1969).

[86] de Ligny, C. L., Denessen, H. J. M., and Alfenaar, M., *Rec. Trav. Chim.*, **90**, 1265 (1971).

[87] Krishnan, C. V. and Friedman, H. L., *J. Phys. Chem.*, **75**, 3606 (1971).

[88] Randles, J. E. B., *Trans. Faraday Soc.*, **52**, 1573 (1956).

[89] Constantinescu, E., *Rev. Roumaine, Chim.*, **17**, 1819 (1972).

[90] Izmailov, N. A., *Russ. J. Phys. Chem.*, **34**, 1142 (1960).

[91] Bennetto, H. P. and Willmott, A. R., *Q. Rev. Chem. Soc.*, **25**, 501 (1972).

[92] Padova, J., *J. Chem. Phys.*, **56**, 1606 (1972).

[93] Owensby, D. A., Parker, A. J., and Diggle, J. W., *J. Amer. Chem. Soc.*, **96**, 2682 (1974).

[94] Matsuura, N. and Umemoto, K., *Bull. Chem. Soc. Japan*, **47**, 1334 (1974).

[95] Alexander, R., Owensby, D . A., Parker, A. J., and Waghorne, W. E., *Aust. J. Chem.*, **27**, 933 (1974).

[96] Parker, A. J., *Chem. Rev.*, **69**, 1 (1969).

[97] Alexander, R. and Parker, A. J., *J. Amer. Chem. Soc.,* **90**, 3313 (1968).

[98] Criss, C. M., Held, R. P., and Luksha, E., *J. Phys. Chem.,* **72**, 2970 (1968).

[99] Criss, C. M., *J. Phys. Chem.,* **78**, 1000 (1974).

[100] Abraham, M. H., *J. Chem. Soc. Chem. Commun.,* 888 (1972).

[101] Bax, D., de Ligny, C. L., and Alfenaar, M., *Rec. Trav. Chim.,* **91**, 452 (1972), Bax, D., de Ligny, C. L., and Remijnse, A. G., *ibid.,* **91**, 965 (1972).

[102] Dack, M. R. J., Bird, K. J., and Parker, A. J., *Aust. J. Chem.,* **28**, 955 (1975).

[103] Fuchs, R. and Hagan, C. P., *J. Phys. Chem.,* **77**, 1797 (1973).

[104] Underdown, D. R., Yun, S. S., and Bear, J. L., *J. Inorg. Nucl. Chem.,* **36**, 2043 (1974).

[105] Feakins, D., Lawrence, K. G., and Tomkins, R. P. T., *J. Chem. Soc. A,* 753 (1967), Andrews, A. L., Bennetto, H. P., Feakins, D., Lawrence, K. G., and Tomkins, R. P. T., *ibid.,* 1486 (1968).

[106] Bennetto, H. P., Feakins, D., and Lawrence, K. G., *J. Chem. Soc. A.,* 1493 (1968); Feakins, D., Hickey, B. E., Lorimer, J. P., and Voice, P. J., *J. Chem. Soc. Faraday I,* **71**, 780 (1975).

[107] Wells, C. F., *J. Chem. Soc. Faraday I,* **69**, 984 (1973).

[108] Wells, C. F., *J. Chem. Soc. Faraday I,* **70**, 694 (1974).

[109] Wells, C. F., *J. Chem. Soc. Faraday I,* **71**, 1868 (1975).

[110] Parsons, R. and Rubin, B. T., *J. Chem. Soc. Faraday I,* **70**, 1636 (1974).

[111] Covington, A. K., Lilley, T. H., Newman, K. E., and Porthouse, G. A., *J. Chem. Soc. Faraday I,* **69**, 963 (1973).

[112] Franks, F. and Ives, D. J. G., *Q. Rev. Chem. Soc.,* **20**, 1 (1966).

[113] Feakins, D. and Voice, P. J., *J. Chem. Soc. Faraday I,* **68**, 1390 (1972).

[114] Das, A. K. and Kundu, K. K., *J. Chem. Soc. Faraday I,* **70**, 1452 (1974).

[115] Fuchs, R., Plumlee, D. S. and Rodewald, R. F., *Thermochim. Acta,* **2**, 515 (1971).

[116] Parker, A. J., Diggle, J. W. and Avraamides, J., *Aust. J. Chem.,* **27**, 721 (1974).

[117] Franks, F. and Reid, D. S., *J. Phys. Chem.,* **73**, 3152 (1969).

[118] Johnson, D. A., *J. Chem. Educ.,* **45**, 236 (1968).

[119] Seidell, A., *Solubilities of Inorganic and Metal Organic Compounds, Vol. 1,* Van Nostrand (1940).

[120] Price, E. in *The Chemistry of Non-aqueous Solvents, Vol. 1,* ed. Lagowski, J. J., Academic Press (1966) Ch. 2, p. 70.

[121] Waddington, T. C., *Non-aqueous Solvents,* Nelson (1969) Ch. 3.

[122] Starkovich, J. A. and Janghorbani, M., *J. Inorg. Nucl. Chem.,* **34**, 789 (1972).

[123] Wallace, R. A., *Inorg. Chem.,* **11**, 414 (1972).

[124] Gutmann, V., *Coordination Chemistry in Non-aqueous Solutions,* Springer-Verlag (1968).

[125] Sakk, Zh. G. and Rasolovskii, V. Ya., *Russ. J. Inorg. Chem.*, **17**, 927 (1972).

[126] E.g. Ikrami, D. D., Dzhuraev, Sh., and Nikolaev, N. S., *Russ. J. Inorg. Chem.*, **17**, 591 (1972); Jache, A. W. and Cady, G. W., *J. Phys. Chem.*, **56**, 1106 (1952).

[127] *The Chemistry of Non-aqueous Solvents, Vol. 3*, ed. Lagowski, J. J., Academic Press (1970) Chs. 3, 5, and 6.

[128] Malhotra, K. C. and Sud, R. G., *J. Inorg. Nucl. Chem.*, **36**, 3767 (1974).

[129] Tsai, Y. A. and Criss, C. M., *J. Chem. Engng. Data*, **18**, 51 (1973).

[130] Strong, J. and Tuttle, T. R., *J. Phys. Chem.*, **77**, 533 (1973).

Chapter **8**

REDOX POTENTIALS

8.1   GENERAL

8.1.1   Introduction

Oxidation-reduction (**redox**) potentials provide a useful means of both expressing and systematising quantitative measures of the oxidising and reducing powers of various species in solution, including solvated cations. The e.m.f. of a cell under standard conditions, $E^{\ominus}$, is related to the Gibbs free energy change, $\Delta G^{\ominus}$, for the cell reaction:

$$\Delta G^{\ominus} = -nFE^{\ominus} .$$

Here n is the number of electrons transferred in the redox process and F is the Faraday, the conversion factor from electrical to thermochemical units. Thus the Gibbs free energy change for a redox reaction can be obtained directly from a measured e.m.f. if the reaction in question can be conducted in a suitable cell. This is often the case, for example when a metal is displaced from a solution of one of its salts by a more electropositive metal:

$$\text{e.g. } Cu^{2+} + Zn^{0} = Cu^{0} + Zn^{2+} .$$

Throughout this Chapter M^{0} stands for the pure metal, not for a solution or amalgram of metal(O); this is the conventionally designated standard state.

Such reactions can be split into two components, corresponding to the reactions occuring at the two electrodes of the cell:

$$Cu^{2+} + 2e^{-} = Cu^{0}$$
$$Zn^{0} - 2e^{-} = Zn^{2+} .$$

Redox potentials for reactions, for whole cells, should thus be separable into components. As always, some arbitrarily chosen assumption is necessary if this separation is to be achieved. Invariably, the potential for the half-reaction or electrode reaction

$$H^+ + e^- = \tfrac{1}{2}H_2$$

is taken as zero.[†] Though this choice of reference point is always the same, there are unhappily two opposite sign conventions for relating half-cell potentials to this zero. We shall use the one more commonly encountered in current text-books, though it is not the convention adopted in the book which for decades has served as the source of much redox potential data [2]. In the convention we shall use, when relating an electrode reaction to the $H^+/\tfrac{1}{2}H_2$ electrode the cell reaction is written with the hydrogen gas on the left hand side:

$$M^{(n+m)+} + \tfrac{m}{2}H_2 = M^{n+} + mH^+$$

or
$$M^{n+} + \tfrac{n}{2}H_2 = M^0 + nH^+.$$

To give actual examples:

$$Cr^{3+} + \tfrac{1}{2}H_2 = Cr^{2+} + H^+$$

$$Mg^{2+} + H_2 = Mg^0 + 2H^+.$$

The half-cell reactions are therefore written:

$$Cr^{3+} + e^- = Cr^{2+}$$

$$Mg^{2+} + 2e^- = Mg^0.$$

In tables these are often abbreviated to Cr^{3+}/Cr^{2+} and Mg^{2+}/Mg^0 respectively. With this convention, strongly oxidising couples have large positive redox potentials, strongly reducing couples large negative potentials.

Once the assumption of zero potential for the $H^+/\tfrac{1}{2}H_2$ electrode has been made, potentials for other electrodes can be obtained from suitable reactions and cells incorporating the hydrogen electrode as one half-cell. In practice it is usually easier to use some other reference electrode, for example the saturated calomel electrode, which has been accurately calibrated against the standard hydrogen electrode. From such cells and their e.m.f.s, standard electrode potentials can be obtained for a large number of M^{n+}/M^0 and $M^{(n+m)+}/M^{m+}$ couples. Where chemical or technical complications preclude direct measurements of electrode potentials, these can generally be estimated from Gibbs free energies of reactions determined in other ways [3, 4].

---

† It is possible to estimate absolute values for standard electrode potentials from work functions, for example for the process $e^-(g) \rightarrow e^-(Pt)$, though these are not accurately known. The best estimate for the absolute redox potential for the $\tfrac{1}{2}H_2/H^+$ couple is $-0.9V$, and from this absolute standard electrode potentials are $0.9V$ less positive (more negative) than the normally quoted potentials relative to the $\tfrac{1}{2}H_2/H^+$ couple as zero. However these absolute potentials may be up to as much as $0.5V$ in error [1].

The discussion so far has centred on potentials and free energies, an emphasis which reflects the dominance of ΔG in this area of chemistry. However, enthalpy and entropy changes for cell reactions can also be determined from the temperature variation of e.m.f.s.

8.1.2 Usefulness and Limitations

Standard electrode potentials are most useful in two main areas, in the provision of a quantitative scale of oxidising and reducing powers, and in the convenient storage of much information concerning redox properties of species in solution. It is possible to construct a large number of chemical reactions by taking pairs of half-cell reactions. For instance, a recent textbook claims that its table of thirty-three standard redox potentials stores information for 528 reactions [4]. The direction of each reaction and, where meaningful, the position of equilibrium, can both easily be determined from the potentials of the two half-cell components, as illustrated in the following examples.

Iron(II)/(III) and uranium(III)/(IV). Here

$$Fe^{3+} + e^- = Fe^{2+} \qquad E^\ominus = +0.77V \qquad \Delta G^\ominus = -0.77F,$$

$$U^{4+} + e^- = U^{3+} \qquad E^\ominus = -0.61V \qquad \Delta G^\ominus = +0.61F.$$

Subtracting,

$$Fe^{3+} + U^{3+} = Fe^{2+} + U^{4+} \text{ has } \Delta G^\ominus = -1.38F = -31.8 \text{ kcal mol}^{-1}.$$

Therefore uranium(III) will reduce iron(III) in acidic aqueous solution, a reaction essentially complete in view of the large negative free energy change in this direction.

Chromium(II)/(III) and europium(II)/(III). Here the standard electrode potentials for the two half cells are much more evenly matched:

$$Cr^{3+} + e^- = Cr^{2+} \qquad E^\ominus = -0.41V \qquad \Delta G^\ominus = +0.41F,$$

$$Eu^{3+} + e^- = Eu^{2+} \qquad E^\ominus = -0.43V \qquad \Delta G^\ominus = +0.43F.$$

Subtracting,

$$Cr^{3+} + Eu^{2+} = Cr^{2+} + Eu^{3+} \text{ has } \Delta G^\ominus = -0.02F = -0.5 \text{ kcal mol}^{-1}.$$

The equilibrium constant for this reaction can be calculated in the normal manner from the expression $\Delta G^\ominus = -RT\ln K$. From this it can be estimated that the reaction will only have proceeded a little beyond half-way at equilibrium; starting with equimolar solutions, about 60% of the chromium(III) will be reduced by the europium(II) when equilibrium has been attained.

Calculations of the above type are valuable both in rationalising and in forecasting redox chemistry in aqueous media. In principle they could be useful in several areas of applied chemistry. In practice their use in connection with electrochemical extraction processes is severely circumscribed by the big difference in medium and by such effects as electrode polarisation. They are rather more useful in connection with electrorefining of metals [5], though here due account has to be taken of their modification by any complexing agents present.

Many of the limitations in the use of standard electrode potentials are implicit in their name and definition. They apply to the standard conditions defined, that is 25°C, an aqueous solution, and an acid concentration corresponding to unit activity of H^+ ions.[†] Unit activity of H^+ ions requires the presence of an equivalent concentration of the requisite anion, and so forecasting must be limited to acid solutions containing non-complexing anions. This point is relatively unimportant for 1+ and 2+ cations, but is of major importance by the time 4+ cations are reached (see Ce^{4+} below.) There is also the usual complication that predictions from standard electrode potentials refer to the final equilibrium state of the system, giving no indication whatsoever of the rate of approach to equilibrium. As every reader is probably aware, aqueous solutions of permanganates or of peroxodisulphates are thermodynamically unstable with respect to oxidation of the water by these anions, although the rate of decomposition of such solutions is very slow.

8.1.3 Metal Ions in Solution

Two types of redox reaction are particularly relevant to the present book:

$$M^{n+} + n\,e^- = M^0$$

$$M^{(m+n)+} + n\,e^- = M^{m+}.$$

Any discussion of observed values for standard electrode potentials must remember that both of these half-cell reactions can be broken down into three components, as shown in Figs. 8.1(a) and 8.1(b). Therefore trends in redox potentials, even for a series of closely related cations, may not be simple, as will be illustrated later. Redox potentials for $M^{(n+m)+}/M^{m+}$ couples are slightly more central to our theme, because two of their three components refer to solvated cations. Redox potentials M^{n+}/M^0 are determined by both the ease of sublimation of the metal and by its ionisation potential(s), as well as by solvation of the M^{n+} cation.

---

† There are, of course, redox potentials cited for other temperatures, other pHs, and other solvents, but the only large body of results in existence is for the standard conditions named here.

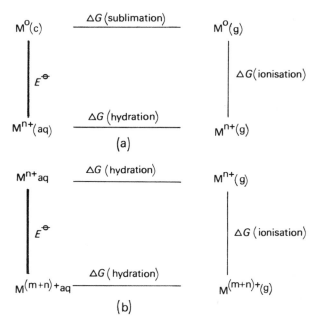

Fig. 8.1 Components of redox potentials (a) for M^{n+}/M^0 couples, and (b) for $M^{(m+n)+}/M^{n+}$ couples.

The majority of results refer to aqueous solution. Although it is in principle straightforward to obtain analogous sets of redox potential data pertaining to non-aqueous solutions experimentally, it can be beset with difficulties (see Section 3 of this Chapter). In practice a set of well established redox potentials for a range of inorganic species exists only in the case of liquid ammonia. Complications arise as soon as redox properties in different solvents are compared. At this point the problem of junction potentials is encountered, to which there is no unequivocally satisfactory answer. The same problem also presents itself when results obtained in aqueous mixtures are related to those in water.

8.2   VARIATION WITH CATION

8.2.1 M^{n+}/M^0 Redox Potentials

Table 8.1 lists values for standard redox potentials for M^{n+}/M^0 couples [6–15]. Most have been obtained from precise electrochemical measurements, although some have been obtained by indirect means. A few are of theoretical rather than practical interest, referring to fanciful species such as Sn^{4+}aq or Au^{3+}aq. It is interesting to note that sodium is a slightly less powerful reductant than the alkaline earth metals, while lanthanum is nearly as good as sodium and better than magnesium, and americium is comparable with the lanthanides.

sp-**Elements.** The values for alkali metals and their cations illustrate the conflicting contributions of the components of the M^{n+}/M^0 potentials as shown in Fig. 8.1(a). The irregular trend of the redox potentials down this group can be ascribed to competition between the ionisation potential contribution, most favourable for $Cs(g) \rightarrow Cs^+(g)$, and the hydration contribution, most favourable for $Li^+(g) \rightarrow Li^+aq$. The trends for these two properties do not compensate entirely, leaving sodium in the extreme position as least strongly reducing (Table 8.2 shows the most important contributory factors). The trends of M^{2+}/M^0 and M^{3+}/M^0 potentials down Group IIA(Be $\rightarrow$ Ra) and Group IIIA (Al $\rightarrow$ La) are regular in both cases, but Group IIB and IIIB trends are opposite to these.

Returning to M^+/M^0 redox potentials, there is a striking contrast between the very negative values for the alkali metals, -2.71 to -3.05V, and the values

Table 8.1   Standard redox potentials (V) for M^{n+}/M^0 couples. Values are from referer are approximate estimates.

H^+ 0			
Li^+ -3.045	Be^{2+} -1.85		
Na^+ -2.714	Mg^{2+} -2.37		
K^+ -2.925	Ca^{2+} -2.87		Zn^{2+} -0.75
Rb^+ -2.925	Sr^{2+} -2.89		Cd^{2+} -0.40
Cs^+ -2.923	Ba^{2+} -2.90		Hg^{2+} $+0.85$
	Ra^{2+} -2.92		
			Hg_2^{2+} $+0.79$
Ti^{2+} -1.63	V^{2+} (-1.18)	Cr^{2+} -0.91^b	Mn^{2+} -1.1
	V^{3+} $(-0.87)^b$	Cr^{3+} -0.74	Mn^{3+} -0.2
Zr^{4+} -1.53	Nb^{3+} (-1.1)	Mo^{3+} (-0.2)	
Hf^{4+} -1.70			

$Ce^{2+}(-2.4)^j$		$Sm^{2+}-2.65^b$	$Eu^{2+}(-2.78)^b$	$Gd^{2+}(-1.5)$
$Ce^{3+}-2.32$ $Pr^{3+}-2.34$ $Nd^{3+}-2.32$	$Pm^{3+}-2.31$	$Sm^{3+}-2.28$	$Eu^{3+}-1.98$	$Gd^{3+}-2.27$
$Ce^{4+}-1.32$				
		$U^{3+}-1.80$	$Np^{3+}-1.86$	$Pu^{3+}-2.07$
$Th^{4+}-1.90$	$Pa^{4+}-1.46^K$	$U^{4+}-1.51^b$	$Np^{4+}-1.33^b$	$Pu^{4+}-1.80$

a Ref. [8]; *b* calculated by the present author from pairs of published E° values obtained from referenc cited here and in the following $M^{(m+n)+}/M^{n+}$ Tables; *c* this is the ref. [6] value, ref. [7] gives -2.37

for thallium, -0.34V, and especially for silver, $+0.80$V. Sublimation, ionisation, and hydration are all important in determining these differences (Table 8.2).

Table 8.2 Principal enthalpic factors contributing to the M^{n+}/M^0 redox potentials for 1+ cations of the sp-elements.

	Li	Na	K	Rb	Cs	Tl	Ag
$\Delta H^{\ominus}$/kcal mol^{-1}							
$M(s) \rightarrow M(g)$	38	26	22	20	19	43	68
$M(g) \rightarrow M^{+}(g)$	126	120	102	98	91	142	176
$M^{+}(g) \rightarrow M^{+}aq$	-123	-97	-77	-71	-63	-78	-114
$cf.\ E^{\ominus}$/V	-3.04	-2.71	-2.92	-2.92	-2.92	-0.34	$+0.80$

] (lanthanides(III) from reference [7]) except where otherwise indicated. Values in brackets

$^{3+}$ -1.66				
$^{3+}$ -2.08	Ga^{3+} -0.53			
$^{3+}$ -2.37	In^{3+} -0.342	Sn^{4+} $+0.01^{e}$		
$^{3+}$ -2.52^{c}	Tl^{3+} $+0.73$			
$^{3+}$ -2.18^{d}		Sn^{2+} -0.136		
	Tl^{+} -0.3363	Pb^{2+} -0.126		Bi^{3+} $+0.16^{f}$

				Cu^{+} $+0.521$
$^{2+}$ -0.473^{a}	Co^{2+} -0.287^{a}	Ni^{2+} -0.228^{a}		Cu^{2+} $+0.345^{a}$
$^{3+}$ -0.06^{b}	Co^{3+} $+0.41^{b}$			
				Ag^{+} $+0.7991$
	Rh^{3+} $(+0.8)$	Pd^{2+} $+0.915^{g}$		Ag^{2+} $+1.39^{b}$
		Pt^{2+} $(+1.2)^{h}$		Au^{+} $+1.67^{i}$
				Au^{3+} $+1.50$

$^{2+}(-1.5)^{j}$					$Yb^{2+}-2.77^{b}$	
$^{3+}-2.27$	$Dy^{3+}-2.32$	$Ho^{3+}-2.37$	$Er^{3+}-2.33$	$Tb^{3+}-2.30$	$Yb^{3+}-2.23$	$Lu^{3+}-2.25$
$^{3+}-2.06^{d}$	$Cm^{3+}-2.07^{d}$	$Bk^{3+}-2.03^{d}$	$Cf^{3+}-2.01^{d}$	$Es^{3+}-1.99^{d}$	$Fm^{3+}-1.97^{d}$	

ref. [9]; e ref. [10]; f estimated from a Bi|BiOCl in dil. HCl cell [11]; g ref. [12]; h tentative estimate fered in ref. [13]; i ref. [14]; j estimated from polarographic measurements [9]; k in M HCl [15].

Transition metals. There is a general trend to more positive potentials here as the first row of elements is traversed from left to right. Insufficient data exist to determine whether there is a similar trend for the second and third rows. The first row trend is the same for M^{2+}/M^0 and for M^{3+}/M^0 couples. Thus for the former series the trend is consistent, though not smooth, from $-1.63V$ for Ti^{2+}/Ti^0 across to $+0.337V$ for Cu^{2+}/Cu^0. A similar trend runs from Sc^{3+}/Sc^0 at $-2.08V$ across to Co^{3+}/Co^0 at $+0.41V$. The irregularities in redox properties of transition metal cations will be discussed in the Section on $M^{(m+n)+}/M^{n+}$, as they are more pronounced in this sequence.

In the case of the lanthanide elements, the redox potentials for the M^{3+}/M^0 couples are all very similar, except for a small dip at Eu^{3+}/Eu^0 and a still smaller one at Yb^{3+}/Yb^0. The range for actinide M^{3+}/M^0 couples is greater, even for the limited range of elements where these are well established. In contrast to the analogous couples for the first transition series, going from the left towards the right of the Periodic Table the actinide couples have increasingly negative values.

8.2.2 $M^{(m+n)+}/M^{n+}$ Redox Potentials

A variety of these, mainly for $m = 1$ but also some for $m = 2$, have either been published or can readily be extracted from published data.

sp-**Elements.** Here stable oxidation states are normally 2 apart. Because 1+ and 3+ (or 2+ and 4+) oxidation states rarely both give stable aquo-cations, there is little to report in this area. The Sn^{4+}/Sn^{2+} couple has a standard electrode potential of $+0.15V$ [6], and the Tl^{3+}/Tl^+ couple $+1.25$ [6] or $+1.26$ [16] V. $Tl^{2+}aq$ has been widely discussed, both its existence as a transient species and its redox potentials relative to Tl^+aq and to $Tl^{3+}aq$. The most recent estimates for the Tl^{3+}/Tl^{2+} and Tl^{2+}/Tl^+ couples are $+0.33$ and $+2.22V$ respectively [17]. Mercury(II) is a mild oxidant, with Hg^{2+}/Hg_2^{2+} at $+0.92V$ [6].

Table 8.3 M^{3+}/M^{2+} redox potentials (V) for transition elements in aqueous solution [6]. [a]

Sc	Ti	V	Cr	Mn	Fe	Co	Ni
(-2.6^b)	ca -0.37 ca -1.1^a	-0.26	-0.41^c	$+1.51$	$+0.77^d$	$+1.82$	$(+4.2)$
					Ru $+0.22^e$		

a Values in brackets are estimates; *b* ref. [18]; *c* the most recent determination gave $-0.429V$ (in chloride medium) [19]; *d* the latest discussion quotes $+0.738V$ [16]; *e* ref. [13] gives $+0.25V$.

Transition metals. Standard electrode potentials are available for a series of M^{3+}/M^{2+} couples for first row transition elements (see Table 8.3 and Fig. 8.2). Here $E^{\ominus}$ values parallel the third ionisation potentials, in other words $M^{2+}(g) \rightarrow M^{3+}(g)$. It is from the 'kink' in the third ionisation potential sequence that the 'kink' in the M^{3+}/M^{2+} trend between manganese and iron derives. Interestingly, the irregularity is smaller in the redox potential sequence than in the third ionisation potentials; apparently some compensation is built into the redox system. Extra covalent interaction in the M^{3+}aq cations has been suggested. The

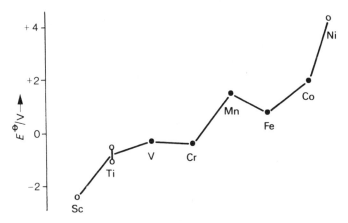

Fig. 8.2 The variation of the M^{3+}/M^{2+} redox potentials along the first row transition elements; • = reasonably accurate value, ○ = rough estimate only for $E^{\oplus}$.

great differences between the redox chemistries of copper and silver are reflected in the values of their M^{2+}/M^{+} standard electrode potentials, $+0.15$ and $+1.98$V respectively.

Table 8.4 M^{3+}/M^{2+} redox potentials (V) for lanthanide elements.

Sm	Eu	Yb
−1.55 [4]	−0.43 [4, 6]	−1.15 [4]
	−0.35 [20]	
	−0.379 [21]	

Lanthanides. Experimentally derived values for the M^{3+}/M^{2+} couples for lanthanides with a tolerably stable 2+ state are listed in Table 8.4. A great deal of effort has been devoted to obtaining estimates of the M^{3+}/M^{2+} standard electrode potentials for the complete series of lanthanide elements. One method

using charge-transfer spectroscopic data [22] gave results which were implausible. A subsequent approach used pulse radiolysis and kinetic studies and
assumed an outer-sphere mechanism (Chapter 13) for the reaction of M^{2+}
with nitrite, giving results which were more plausible but none-too-precise.
Seven lanthanide elements with unstable M^{2+} ions were found to have redox
potentials between -1.6 and -2.0V, while all estimates were ± 0.2V [23]. The
most recent estimates are derived from thermochemical calculations on an ionic
model [7] and spectroscopic studies [24]. They are listed in Table 8.5.

Gross irregularities in the sequence of M^{3+}/M^{2+} potentials occur at europium, ytterbium, and samarium. They can be explained by the desire of these
2+ cations to attain (Eu, Yb) or approach (Sm) the favourable f^7 and f^{14}
(half-filled and filled shell) configurations. The reader is referred elsewhere for
discussion of the finer details of this sequence of values [7, 18]. At present such
discussion is still handicapped by the lack of precise estimates for some
elements.

Table 8.5 **Estimated values for M^{3+}/M^{2+} redox potentials (V) for the whole
series of lanthanide elements. See the text and references for details of the
thermochemical calculations and spectroscopic approaches used.**

Element	Thermochem. [7]	Spectr. [24]	Element	Thermochem. [7]	Spectr. [24]
La	-3.8	-3.1	Gd	-3.6	-3.9
Ce	-3.5	-3.2	Tb	-3.5	-3.7
Pr	-3.0	-2.7	Dy	-2.6	-2.6
Nd	-2.8	-2.6	Ho	-2.9	-2.9
Pm	-2.5	-2.6	Er	-3.0	-3.1
Sm	-1.5	-1.6	Tm	-2.1	-2.3
Eu	-0.35	-0.35	Yb	-1.1	-1.1

Cerium is the only lanthanide element for which an experimentally determined M^{4+}/M^{3+} standard electrode potential is available. Only this element has a 4+
oxidation state which persists for a long time in aqueous solution. The important
feature here is the way that the redox potential varies with the nature of the acid
used to maintain standard conditions. In perchloric acid of unit activity of
hydrogen ions the redox potential is 1.70V, in nitric acid 1.61V, in sulphuric
acid 1.44V, and in hydrochloric acid 1.28V [25]. This variation can be ascribed
to the strong tendency for Ce^{4+}aq to form complexes, and perhaps to the onset
of hydrolysis (Chapter 9.2.1). Only the perchloric acid redox potential has a
chance of referring to the simple Ce^{4+}aq/Ce^{3+}aq couple. Redox potentials for the
Pr^{4+}/Pr^{3+} and Tb^{4+}/Tb^{3+} couples have been guessed as about $+2.9$V in each case
[4].

Actinides. The 3+, 4+, 5+, and 6+ oxidation states are all known in aqueous solution for uranium and the subsequent few elements. The first two exist as simple aquo-cations $M^{3+}aq$ and $M^{4+}aq$, the last two as oxocations MO_2^+aq and $MO_2^{2+}aq$. Full redox potential information is available, and is summarised in Table 8.6 [26]. The trend of redox potentials $M^{(n+1)+}/M^{n+}$ for a given element is not always as expected. Both for uranium and for plutonium the 6+ state is less strongly oxidising than the 5+ state, which is abnormal. The chemical consequences of such an irregularity are detailed in the Section on disproportion-action following (see p. 248). The differences between the redox potentials connecting the various oxidation states of plutonium are so small that all the states, 3+ to 6+ inclusive, can coexist in significant quantities.[†] There is the expected increase from left to right across the actinide series until berkelium is reached. Here, a drop in the M^{4+}/M^{3+} redox potential matches the drop in the fourth ionisation potential at berkelium. This behaviour parallels that in the first row transition elements and in the lanthanides, with a drop in potential as the half-filled shell configuration is passed.

The 2+ oxidation state is unknown at the beginning of the series, and is strongly reducing in the middle (Table 8.6). Towards the end of the series the 2+ oxidation state becomes more stable in solution. In the case of nobelium, the penultimate element in the actinide series, the 2+ state is the most stable. The redox pattern for the higher actinides resembles that for the analogous lanthanides, though with the stability of the 2+ state (filled shell, f^{14} configuration) even more marked for nobelium than for its lanthanide analogue ytterbium.

8.3   NON-AQUEOUS SOLVENTS

It is a fairly straightforward matter to set up a scale of redox potentials relative to the standard hydrogen electrode in a non-aqueous solvent. However, it is difficult to relate such a scale to the scale of standard electrode potentials in aqueous solution. So while obtaining a quantitative picture of the relative reducing and oxidising powers of cations in a given non-aqueous solvent is easy, acquiring an idea of the relative reducing or oxidising powers of a given cation in a range of solvents is not. Much of the experimental data has been collected and reviewed by Strehlow [28], who has been active in this field for some time. Methods of calculating redox potentials in non-aqueous solvents have been devised, and tested to a reasonable degree of satisfaction [29, 30]. While it is preferable to measure and discuss standard electrode potentials, assessing redox properties of cations in non-aqueous media by polarography is sometimes easier. At the end of the present Section some polarographic results will be presented, and where possible compared with electrode potentials measured in the normal way.

† A relatively straightforward method for calculating the plutonium(III)/(IV)/(V)/(VI) distribution under various conditions has recently been described in Silver, G. L., *J. Radioanal. Chem.*, 23, 195 (1974).

Table 8.6 Redox potentials (V) for the actinide elements. [a]

	Pa	U	Np	Pu	Am	Cm	Bk	Cf	Es	Fm
M^{3+}/M^{2+} [9][b]	−4.7				−2.3	−4.4	−2.8	−1.6	−1.3	−1.0
M^{4+}/M^{3+} [26]		−0.63	+0.16	+0.98	(+2.18)[c]	d	(+1.6)			
MO_2^+/M^{4+} [26]	(−0.1)	+0.58	+0.74	+1.17	+1.04					
MO_2^{2+}/MO_2^+ [26]		+0.06	+1.14	+0.91	+1.6					

a Values in brackets are estimates; b polarographically derived; c reported values range from +2.0V to +2.9V (Am^{4+}aq is not a stable species), with a recent estimate of +2.50V extrapolated from measurements in strong phosphoric acid (in which Am^{4+} is complexed and stable) [27]; d large and positive.

8.3.1 Inorganic Solvents

Liquid ammonia has probably been the solvent best studied in this respect. The primary $H^+/\frac{1}{2}H_2$ half-cell can be set up satisfactorily in liquid ammonia [31]. Several more convenient secondary electrodes have been developed and calibrated [28], for example silver/silver chloride [32]. Standard electrode potentials for a range of couples in liquid ammonia are compared with those in aqueous solution in Table 8.7. As free energies of sublimation of metals and of ionisation will be the same, regardless of solvent, differences between M^{n+}/M^0 redox potentials in liquid ammonia and in water can be attributed to solvation differences of the M^{n+} cations. The difference between the alkali metal couples and those for 'soft' metals such as copper and zinc is greater in water than in liquid ammonia. The overall order of redox potential values is similar in the two

Table 8.7 M^{n+}/M^0 redox potentials (V) in water, liquid ammonia, and hydrazine, at $25°C$ and relative to $H^+/\frac{1}{2}H_2 = 0$ in each case.

	Water [6]	Liquid ammonia [4, 33]	Hydrazine [28]
Li^+/Li^0	−3.05	−2.34	−2.20
Na^+/Na^0	−2.71	−1.89	−1.83
K^+/K^0	−2.93	−2.04	−2.02
Rb^+/Rb^0	−2.93	−2.06	−2.01
Cs^+/Cs^0	−2.92	−2.08	
Cu^+/Cu^0	+0.52	+0.36	+0.22
Ag^+/Ag^0	+0.80	+0.76	
Hg_2^{2+}/Hg^0	+0.79		+0.77
Mg^{2+}/Mg^0	−2.37	−1.74	
Ca^{2+}/Ca^0	−2.87	−2.17	−1.91
Sr^{2+}/Sr^0	−2.89	−2.3	
Ba^{2+}/Ba^0	−2.90	−2.2	
Zn^{2+}/Zn^0	−0.76	−0.54	−0.41
Cd^{2+}/Cd^0	−0.40	−0.20[a]	
Hg^{2+}/Hg^0	+0.86	+0.67	
Cu^{2+}/Cu^0	+0.34	+0.40	
Pb^{2+}/Pb^0	−0.13	+0.28	+0.35

a Ref. [28].

solvents, but because it is not exactly the same there will be some differences of chemical behaviour to be observed when dealing with some pairs of half reactions of like redox potentials. This is apparent when considering the disproportionation behaviour of some cations, as will be discussed in the disproportionation Section below.

The pattern of redox potentials reported for half-cells in hydrazine is, predictably, similar to that in liquid ammonia (Table 8.7). Some e.m.f.s of fringe relevance has been reported for a number of species in liquid hydrogen fluoride solution [34].

8.3.2 Organic Solvents

Alcohols. A number of redox potentials determined for inorganic couples in methanol and in ethanol exist. These values, derived from a number of sources, are collected together in Table 8.8; the original references can be tracked down through reference [28]. No experimental values are available for higher alcohols, but values have been calculated for them. The method of calculation used is one that gives reasonable agreement, generally to within about 0.2V, with experimentally determined values for aqueous, methanolic, and ethanolic media. Table 8.8 includes a comparison between calculated and observed values in ethanol, to illustrate this measure of agreement.

Table 8.8 M^{n+}/M^0 **redox potentials (V) in alcohols, at 25°C and relative to** $H^+/\frac{1}{2}H_2 = 0$ **in each case.**

	Experimental		Calculated		
	Methanol [28]	Ethanol [28]	Ethanol [29]	Butyl alcohol [29]	Amyl alcohol [29]
Li^+/Li^0	−3.10	−3.04	−3.14	−3.21	−3.23
Na^+/Na^0	−2.73	−2.66	−2.87	−2.85	−2.68
K^+/K^0	−2.92		−3.13	−2.98	−2.98
Rb^+/Rb^0	−2.91		−3.14	−2.85	−3.05
Cs^+/Cs^0			−3.07	−3.00	−2.98
Tl^+/Tl^0	−0.38	−0.34			
Cu^+/Cu^0	+0.49				
Ag^+/Ag^0	+0.76		+0.50	+0.50	+0.52
Ca^{2+}/Ca^0			−2.72	−2.77	−2.70
Zn^{2+}/Zn^0	−0.74		−0.89	−0.89	−0.83
Cd^{2+}/Cd^0	−0.43		−0.40	−0.45	−0.33

Redox potential values are similar to those determined in aqueous solution. This is hardly surprising in view of the similarity of thermochemical parameters for solvation of M^{n+} cations in water and in alcohols (see Chapter 7).

Other organic solvents. Available values are listed in Table 8.9. These numbers are all derived from experimental observations, except for those in acetone, which have been calculated by the same method as used for the ethanol and higher alcohol values in Table 8.8 [29]. Again the overall trend in each solvent is similar to that in water, but with sufficient minor divergencies to give differences in solution redox chemistry when some pairs of cations with comparable potentials are considered.

Table 8.9 M^{n+}/M^0 **redox potentials (V) in organic solvents, at 25°C and relative to** $H^+/\frac{1}{2}H_2 = 0$ **in each case.**

	Acetonitrile [28]	Formic acid [28]	Formamide [28]	Acetone[a] [29]
Li^+/Li^0	-3.23	-3.48		-3.21
Na^+/Na^0	-2.87	-3.42		-2.46
K^+/K^0	-3.16	-3.36	-2.87	-2.91
Rb^+/Rb^0	-3.17	-3.45	-2.86	-2.92
Cs^+/Cs^0	-3.16	-3.44		-2.98
Tl^+/Tl^0			-0.34	
Cu^+/Cu^0	-0.38[b,c]			
Ag^+/Ag^0	$+0.23$	$+0.17$		$+0.43$
Hg_2^{2+}/Hg^0		$+0.18$		
Ca^{2+}/Ca^0	-2.75	-3.20		-2.74
Zn^{2+}/Zn^0	-0.74	-1.05	-0.76	-0.79
Cd^{2+}/Cd^0	-0.47	-0.75	-0.41	-0.38
Cu^{2+}/Cu^0	-0.28[b]	-0.14	$+0.28$	

a Calculated values; b Cu^{2+}/Cu^+ has recently been measured [35]; c see ref. [36] for a recent redermination.

Redox potentials in Table 8.7 to 8.9 are all relative to the hydrogen electrode, which is the usual practice. However, there is an argument in favour of selecting some other reference electrode for the purposes of intersolvent comparisons. These would be easier if a reference couple less likely to be solvent sensitive were involved. Cs^+/Cs^0 would be preferable to $H^+/\frac{1}{2}H_2$ for such comparisons.

8.3.3 Polarography

An alternative approach to the assessment of redox characteristics in solution is provided by the technique of polarography, which has proved particularly popular for non-aqueous media. It is often easier to obtain experimental results this way than by the use of electrochemical cells. Because of non-reversibility, overvoltages at amalgam surfaces, and the presence of the necessary supporting electrolyte, standard electrode potentials are not obtained from polarography. In non-aqueous media, complications arising from ion association and complex formation are also much more likely. Reference points for polarographic scales of redox characteristics are also perforce different from the $H^+/\tfrac{1}{2}H_2$ zero used in establishing the scale of standard electrode potentials. It is customary to use the potential corresponding to the reduction of a large and therefore lightly solvated cation as zero in polarography. For solvento-metal cations the choice of zero often falls on Rb^+/Rb^0, or on a couple incorporating a large organic cation. The $Cr(biphenyl)_2^+/Cr(biphenyl)_2^0$ couple is the zero for several of the series of results tabulated in this Section.

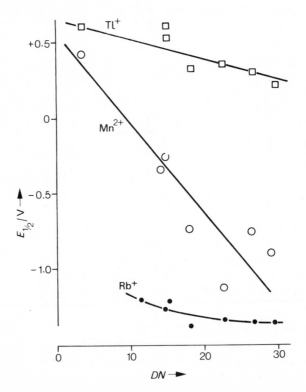

Fig. 8.3 The dependence of polarographic half-wave potential, $E_{1/2}$, on Gutmann solvent donor numbers, DN.

Table 8.10 Polarographic half-wave potentials, $E_{1/2}$/V, for the reduction of cations in alcohols, acetonitrile, sulpholane, and propylene carbonate, relative to Rb^+/Rb^0, at $25°C$.

	Water [38]	Methanol [38]	Ethanol [38]	Acetonitrile [37]	Sulpholane [38]	Propylene carbonate [39]
Li^+	−0.20	−0.27	−0.27	0.00	0.00	−0.02
Na^+	0.01	0.01	0.03	0.10	0.11	0.01
K^+	−0.01	−0.01	0.00	−0.01	0.01	0.13
Rb^+	0	0	0	0	0	0
Cs^+	0.04	0.04	0.00		0.01	
Ag^+				2.27	2.48	
Tl^+					1.67	
Mg^{2+}						0.25
Ba^{2+}	0.21	0.17	0.18	0.32	0.25	0.30
Zn^{2+}					1.54	
Mn^{2+}					0.97	
La^{3+}					0.44	

Table 8.11 Polarographic half-wave potentials, $E_{1/2}$/V, for reduction of a range of cations in acetonitrile, relative to the saturated calomel electrode, at $25°C$ [37][a].

Li^+ −1.95	Be^{2+} −1.6			
Na^+ −1.85	Mg^{2+} −1.84			Al^{3+} −1.42
K^+ −1.94	Ca^{2+} −1.82	Zn^{2+} −0.59		
Rb^+ −1.95	Sr^{2+} −1.76	Cd^{2+} −0.23		
	Ba^{2+} −1.63			
		Hg_2^{2+} +0.5[b]		Tl^+ −0.24
				Cu^+ −0.28[b]
				Ag^+ −0.32
Cr^{2+} −1.12	Mn^{2+} −1.12	Fe^{2+} −1.00	Co^{2+} −0.65	Ni^{2+} −0.33
Cr^{3+} −0.00				
	Eu^{2+} −1.67	Yb^{2+} −1.69		
Sm^{3+} −1.62	Eu^{3+} +0.15	Yb^{3+} −0.57		

a Alternative values for some of these cations and values for $E_{1/2}$ in other nitrile solvents may be found in refs. [40] and [41]; *b* ref. [42].

Table 8.12 Polarographic half-wave potentials, $E_{1/2}$/V, for reduction of cations in a range of solvents, whose Gutmann donor numbers, DN, are quoted. All $E_{1/2}$ values are at 25°C, relative to Cr(biphenyl)$_2^+$/Cr(biphenyl)$_2^0$ = 0 [41, 43].

Solvent	Dimethyl sulphoxide	Dimethyl formamide	Trimethyl phosphate	Water	Propylene carbonate	Acetonitrile	Benzonitrile	Nitromethane
DN	29.8	27	23	~18	15.1	14.9	11.9	2.7
Li$^+$	-1.37				-1.25	-1.20	-1.13	
Na$^+$	-1.40	-1.37	-1.37	-1.36	-1.10	-1.12	-1.05	
K$^+$	-1.37	-1.39	-1.36	-1.39	-1.08	-1.23	-1.16	
Rb$^+$		-1.37	-1.35	-1.38	-1.21	-1.25	-1.18	
Tl$^+$	+0.18	+0.25	+0.31	+0.28	+0.52	+0.46		+0.55
Ag^{+a}	+0.85	+0.95			+1.35	+0.90	+0.95	+1.45
Ba^{2+}	-1.36	-1.34	-1.33	-1.20	-0.91	-0.90		
Zn^{2+}	-0.37	-0.28	-0.12	-0.28	+0.22	+0.11	-0.91	+0.50
Cd^{2+}	+0.02	+0.12	+0.21	+0.12	+0.65	+0.47		+0.79
Mn^{2+}	-1.00	-0.86	-1.15	-0.74	-0.32	-0.35		-0.43
Co^{2+}	-0.71	-0.55	-0.69	-0.69	+0.03	+0.12		+0.50
Ni^{2+}	-0.35	-0.22	-0.20	-0.36	+0.28	+0.44		+0.57

a Ref. [42].

It has been demonstrated that polarographic half-wave potentials, $E_{1/2}$, parallel standard electrode potentials, $E^{\ominus}$, in solvents where comparable sets of data are available. Solvents include water, methanol, ethanol, and acetonitrile [37, 38]. In fact this demonstration is only moderately convincing, because the number and variety of M^{n+}/M^0 couples for which both $E_{1/2}$ and $E^{\ominus}$ values are known is limited. This limitation applies to all four solvents named.

Polarographic half-wave potentials for metal cations are collected together in Tables 8.10 to 8.12. Values for the alkali metals, for silver, and for barium in several organic solvents, including water-like solvents such as methanol and ethanol, and water itself, are compared in Table 8.10 [37, 38]. The most extensive series of results available for one solvent refer to acetonitrile, Table 8.11 [37]. Comparative data for a series of 1+ and 2+ cations in a range of solvents of varying Gutmann solvent donor numbers (Chapter 1.3) are reported in Table 8.12 [41–43]. There is some correlation between $E_{1/2}$ values and solvent donor numbers for a variety of cations, as Fig. 8.3 illustrates [44]. Such correlations are most successful when cases of specific cation-solvent inter-action, for example between Ag^+ and nitriles or sulphur-donor ligands, are ex-cluded. Half-wave potentials for the reduction of M^{3+} cations in acetonitrile solution also parallel those in aqueous solution, and parallel the $E^{\ominus}$ values for the respective M^{3+}/M^{2+} couples in aqueous solution (Table 8.13). Again the variation in $E_{1/2}$ with solvent for a given M^{3+}/M^{2+} couple can be correlated with Gutmann solvent donor numbers [41].

Table 8.13 Polarographic half-wave potentials, $E_{1/2}/V$, for M^{3+}/M^{2+} couples in acetonitrile and in water, relative to the saturated calomel electrode, at $25°C$ [37].

	$E_{1/2}$		$E^{\ominus}$ [a]
	Acetonitrile	Water	Water
Cr^{3+}/Cr^{2+}	0.00	−0.91	−0.41
Sm^{3+}/Sm^{2+}	−1.62	−1.80	−1.55
Eu^{3+}/Eu^{2+}	+0.15	−0.67	−0.38 to −0.43
Yb^{3+}/Yb^{2+}	−0.57	−1.17	−1.15

[a] Values from Tables 8.3 and 8.4.

8.4   SPONTANEOUS REDOX PROCESSES

8.4.1   Disproportionation

There are several well-established examples of aquo-cations which spon-taneously disproportionate to species of higher and lower oxidation states. One of the most studied and discussed cases is that of the Cu^+aq cation:

$$Cu^{2+}aq + e^- = Cu^+aq \quad E^\ominus = +0.153V \quad \therefore \Delta G^\ominus = -0.153F,$$

$$Cu^+aq + e^- = Cu^0 \quad\quad E^\ominus = +0.521V \quad \therefore \Delta G^\ominus = -0.521F.$$

Hence the Gibbs free energy change for the reaction,

$$2Cu^+aq = Cu^0 + Cu^{2+}aq,$$

is $-0.368F$ or -8.5 kcal mol^{-1}. Therefore the disproportionation of Cu^+aq has a negative Gibbs free energy change and is thermodynamically favourable. Interestingly, the analogous reaction for silver,

$$2Ag^+aq = Ag^0 + Ag^{2+}aq,$$

has a large positive standard free energy change in the left to right direction. This difference is ascribed in part to hydration differences of the various cations involved. Whereas the relatively high free energy of hydration of Cu^{2+} helps in the disproportionation of Cu^+, the lower free energy of hydration of the larger Ag^{2+} does not provide similar encouragement for Ag^+ to disproportionate [45]. It has been suggested that the Hg^+aq cation, which can be generated by pulse radiolysis, reverts to the more stable $Hg_2^{2+}aq$ cation by an analogous disproportionation previous to formation of the dimeric cation [46]:

$$2Hg^+aq \longrightarrow Hg^0 + Hg^{2+}aq,$$

$$Hg^0 + Hg^{2+}aq \longrightarrow Hg_2^{2+}aq.$$

The similarly transient species $Tl^{2+}aq$ is another cation eager to disproportionate [17]:

$$2Tl^{2+}aq \longrightarrow Tl^+aq + Tl^{3+}aq.$$

$Mn^{3+}aq$ is a rather more stable ion, but nonetheless disproportionates spontaneously into $2+$ and $4+$ oxidation state species.

Uranium(V), which exists in aqueous solution as UO_2^+aq, is unstable with respect to disproportionation:

$$2U(V) \longrightarrow U(IV) + U(IV).$$

Although this reaction is spontaneous, it is not by any means instantaneous. Uranium(V) persists long enough in aqueous solution for its chemistry to have been established. There has been a report on its spectrum [47]. The analogous reaction for plutonium has only a small negative Gibbs free energy change in the disproportionation direction. Small amounts of plutonium(V) coexist in aqueous solutions with plutonium(VI), plutonium(IV), and plutonium(III) (cf. Chapter 8.2.2). Neptunium(V) is stable with respect to disproportionation into neptunium(IV) and neptunium(VI).

On going from an aqueous to a non-aqueous environment, disproportionation behaviour may change dramtically. For instance, Cu^+ is stable to disproportionation in acetonitrile or in liquid ammonia. On the other hand mercury(I), which is just stable to disproportionation in water[†], is very unstable with respect to disproportionation in liquid ammonia (Fig. 8.5).

$$Hg^0 \longleftarrow Hg_2^{2+} \longleftarrow Hg^{2+}$$

water: $-0.789V$ $+0.92V$

liquid ammonia: $+1.5V$ $-0.2V$

Fig. 8.4 Redox potentials for mercury and its cations in water and in liquid ammonia.

8.4.2 Oxidation and Reduction of Solvent

The range of solvated cations whose chemistry can be studied in a given solvent is limited by the oxidation and reduction properties of the solvent. The range available for the existence of thermodynamically stable solvated cations can be defined by the redox potentials for the oxidation and reduction of the solvent in question. Fortunately the actual range of solvento-cations which can be studied is in general considerably wider than these thermodynamic considerations would indicate. There are two main reasons for this. First, the rates of oxidation or reduction of the solvent may be very slow. Associated with this usually is an **overvoltage** connected with the liberation of a gas from solution. It should also be remembered that $E^\ominus$ values apply only to standard state conditions, a difference in temperature or concentration may alter the stability picture markedly.

The thermodynamic limits for aqueous solution chemistry of cations are set by the following potentials:

$$O_2(g) + 4H^+aq + 4e^- = 2H_2O \qquad E^\ominus = +1.23V,$$

$$H^+aq + e^- = \tfrac{1}{2}H_2(g) \qquad E^\ominus = 0.$$

Despite the theoretical limits, aqueous solutions of oxidising cations such as $Ce^{4+}aq$ (Ce^{4+}/Ce^{3+} has $E^\ominus = 1.70V$ in perchloric acid) can be preserved for long periods. At the other extreme, the overvoltage required for liberation of hydrogen gas permits the existence of modestly reducing cations for indefinitely long periods. This latter statement applies to anaerobic conditions;

[†] The equilibrium constant for the reaction $Hg^{2+}aq = Hg^0 + Hg_2^{2+}aq$ is 0.011 [8].

the solubility of oxygen in water will affect the situation for aqueous solutions of reducing cations.

At first sight liquid ammonia looks an unpromising solvent, because the oxidising and reducing properties are only separated by 0.04V:

$$\tfrac{1}{2}H_2 + NH_3 = NH_4^+ + e^- \qquad\qquad E^\ominus = 0,$$

$$4NH_3 = \tfrac{1}{2}N_2 + 3NH_4^+ + 3e^- \qquad\qquad E^\ominus = 0.04V.$$

However a combination of kinetic inertia and overvoltage lead to a much wider practicable range than this. The dissolution of alkali metals in liquid ammonia to give, under appropriate conditions, solvated cations and solvated electrons, is a well known phenomenon. Liquid hydrogen sulphide is another solvent whose range would be expected to be considerably less than that for water on thermodynamic grounds. Not only the alkali metals but also most other metals (including even copper, silver, and mercury) liberate hydrogen from liquid hydrogen sulphide [48]. On the other hand, liquid hydrogen fluoride should have a very wide range for the stable existence of solvated cations as the $\tfrac{1}{2}F_2/F^-$ couple is so strongly oxidising [4, 49].

8.5   ENTHALPIES AND ENTROPIES FROM REDOX CELLS

In principle enthalpies and entropies for cell reactions can be obtained from measurements of the variation of cell e.m.f. with temperature:

$$\Delta G = -nFE,$$

$$\Delta H = -nF\,\frac{d(E/T)}{d(1/T)},$$

$$\Delta S = \frac{\Delta H - \Delta G}{T}.$$

The temperature variation properties of the standard hydrogen electrode have to be considered in deriving single ion values for cation hydration from such experiments. In practise $E^\ominus$ for this is taken as zero at all temperatures, so that enthalpies derived in this way include that for the $H^+/\tfrac{1}{2}H_2$ reaction. However no complication arises because the cation hydration enthalpies, especially $\Delta H_f^\ominus(M^{n+}\,aq)$, are conventionally quoted relative to the proton value.

This approach is often used to determine enthalpies and entropies of reactions which are more conveniently carried out as cell reactions rather than as 'real' reactions, for example the solid-solid reaction of lead iodide with silver chloride. However, it has rarely been used in the area of cation hydration thermochemistry. A recent application concerned the Pd^{2+}aq cation [12]. Measurements over a range of temperatures were made on the somewhat unorthodox cell,

$$
Pd \mid Pd(ClO_4)_2, HClO_4 \mid Pt \mid NaCl \mid \begin{array}{l} \text{saturated} \\ \text{calomel} \\ \text{electrode.} \end{array}
$$

From this values of $\Delta G_f^{\ominus}$, $\Delta H_f^{\ominus}$, and $\overline{S}^{\ominus}$ for Pd^{2+}aq were obtained as detailed above. Measurements of $E^{\ominus}$ for the Fe^{3+}/Fe^{2+} couple over the temperature range 50 to 350°C [50] indicate a value of -7.7 kcal mol^{-1} for the enthalpy change for the reaction

$$
Fe^{3+}aq + e^- = Fe^{2+}aq .
$$

8.6   ENTHALPIES AND FREE ENERGIES OF FORMATION OF AQUO-CATIONS

Most of the thermochemical parameters of solvation of metal cations were considered in Chapter 7. The exceptions were the enthalpies and Gibbs free energies of formation of solvated cations[†]:

$$
M(\text{standard state}) + \text{solvent} \xrightarrow{-ne^-} M^{n+}(\text{solv}) \text{ in solution}, \Delta X_f^{\ominus} .
$$

The reason for delaying discussion of these parameters until now is their very close relation to redox potentials. Indeed $\Delta G_f^{\ominus}(M^{n+}aq)$ values can be obtained directly from e.m.f. measurements on appropriate cells. Consider a cell with M^+/M^0 and $H^+/\frac{1}{2}H_2$ electrodes, whose reaction is

$$
M^+aq + \frac{1}{2}H_2 = M + H^+aq .
$$

If all the components are in their standard states, then the cell e.m.f., $E^{\ominus}$, is related to the Gibbs free energy for the cell reaction by the usual equation, $\Delta G^{\ominus} = -nFE^{\ominus}$ (here $n = 1$). $\Delta G^{\ominus}$ is the difference between the chemical potentials of the products and the reactants. Of these chemical potentials, those of M and of $\frac{1}{2}H_2$ are by definition zero. If $\Delta G_f^{\ominus}(H^+aq)$ is taken as zero, as is customary, then $\Delta G_f^{\ominus}(M^+aq) = -FE^{\ominus}$. Thus $\Delta G_f^{\ominus}(M^{n+}aq)$ values are directly related to measurable cell e.m.f.s. As mentioned in the previous Section, enthalpies can also be derived from e.m.f. measurements, although in this area they rarely are. $\Delta H_f^{\ominus}(M^{n+}aq)$ values are generally based on

[†] The thermodynamic parameters for the electron must be considered in deriving these single ion quantities. The normal assumption is that the energy of a gaseous electron is, apart from its kinetic energy, zero [51].

measured e.m.f.s at one temperature and on ancillary thermochemical data. Some chemists have worked from enthalpies measured calorimetrically and from free energies derived from solubilities and activity coefficients to $\Delta H_f^\ominus$, $\Delta G_f^\ominus$, and thence $S^\ominus$ and $E^\ominus$ values [8]. This type of approach is, of course, essential where appropriate electrical cells cannot be constructed.

Values for Gibbs free energies of formation of aquo-cations, $\Delta G_f^\ominus(M^{n+}aq)$, are listed in Table 8.14. These values are relative to $\Delta G_f^\ominus(H^+aq) = 0$. This Table, and the following $\Delta H_f^\ominus$ Table, contain in general the most recent values derived either from the standard sources [52] or from subsequently published literature

Table 8.14 Standard Gibbs free energies of formation of aquo-metal cations, $\Delta G_f^\ominus(M^{n+}aq$ $25°C.$

H^+ 0

Li^+ −70.22		
Na^+ −62.593	Be^{2+} −90.75^m	
K^+ −67.70	Mg^{2+} −108.7^m	Zn^{2+} −34.6^a
Rb^+ −67.45	Ca^{2+} −132.30^m	Cd^{2+} −18.6^a
Cs^+ −67.41	Sr^{2+} −133.71^m	Hg^{2+} +39.3$^\cdot$
	Ba^{2+} −134.02^m	
	Ra^{2+} −134.2^m	Hg_2^{2+} +36.7

Ti^{2+} −37.5^e	Cr^{2+} −39.3^e	Mn^{2+} −54.5
Ti^{4+} −84.65^h	Cr^{3+} −49.0^e	

Zr^{4+} −124.7^i

Hf^{4+} −132.6^j

Ce^{2+} −64	Pr^{2+}−92	Nd^{2+}−97	Pm^{2+}−102	Sm^{2+}−123	Eu^{2+}−129	Gd^{2+}−4
Ce^{3+}−160.7	Pr^{3+}−161.9	Nd^{3+}−160.8	Pm^{3+}−160	Sm^{3+}−158	Eu^{3+}−137	Gd^{3+}−1
Ce^{4+}−120.4^k						

a Ref. [3, 8]; *b* refs. [8, 53]; *c* ref. [7]; *d* ref. [16]; *e* ref. [54]; *f* not standard state; *g* ref. [11, 5 in footnote *k* give similar values for lanthanide cations; *m* from Natn. Bur. Stand. Technical Note 270

[53-58]. Earlier estimates can be traced through references cited in recent papers. We have included as many significant figures as the original sources. In some cases, such as the alkali metal cations, this is probably justified. In other less fully studied cases, the number of significant figures given often leaves an over-optimistic impression of the accuracy of the values cited. Two or three estimates of $\Delta G_f^{\ominus}(M^{2+}aq)$ exist for several of the 2+ cations of the first row transition elements. For the case of $Fe^{2+}aq$ these range from -18.9 to -21.8 kcal mol^{-1}. Similarly, $\Delta G_f^{\ominus}(Fe^{3+}aq)$ values range from -1.1 to -4.0 kcal mol^{-1}. Values for such unlikely species as $Au^{3+}aq$ and $Zr^{4+}aq$ have been included

cal mol^{-1}; from reference [7] (for lanthanide cations[i]) or [52] unless otherwise indicated, at

$^{3+}$ -116	—		
$^{3+}$ -140.2	Ga^{3+} -38.0		
$^{3+}$ -165.8	In^{3+} -23.4	Sn^{4+} $+0.7$[e]	
$^{3+}$ -163.8[c]	Tl^{3+} $+51.3$		
	Ga^{2+} (-21)		
	In^{2+} (-12.1)	Sn^{2+} -6.5[f]	
	Tl^{2+} $(+42)$[d]	Pb^{2+} -5.83	Bi^{3+} $+22.07$[g]
	In^{+} -2.9		
	Tl^{+} -7.74		
			Cu^{+} $+11.95$
$^{2+}$ -21.8[a]	Co^{2+} -13.2[a]	Ni^{2+} -10.5[a]	Cu^{2+} $+15.9$[a]
$^{3+}$ -4.0[a]	Co^{3+} $+32$		
		Pd^{2+} $+42.2$	Ag^{+} $+18.433$
	Rh^{3+} $\sim+46$[k]		Ag^{2+} $+64.3$[f]
		Pt^{2+} $(+57)$[k]	
			Au^{3+} $+98$[e]

$^{2+}-77$	$Dy^{2+}-100$	$Ho^{2+}-97$	$Er^{2+}-91$	$Tm^{2+}-110$	$Yb^{2+}-129$
$^{3+}-157$	$Dy^{3+}-160$	$Ho^{3+}-164$	$Er^{3+}-160.9$	$Tm^{3+}-159$	$Yb^{3+}-154$

ref. [56]; i ref. [57]; j ref. [58]; k from Natn. Bur. Stand. Technical Note 270/7 (1973); l ref. cited 1971).

in order to provide as comprehensive a coverage of the Periodic Table as possible.

Values of $\Delta H_f^{\ominus}(M^{n+}aq)$, also with respect to $\Delta H_f^{\ominus}(H^+aq) = 0$, are listed in Table 8.15 [7, 52, 59-72]. Again, where there are several values, the most recent is quoted. Values of $\Delta H_f^{\ominus}$ for aquated oxocations are reported in Table 8.16. A few values of $\Delta G_f^{\ominus}$ and $\Delta H_f^{\ominus}$ for ammono-cations in liquid ammonia have been derived [73, 74].

FURTHER READING

Latimer, W. M., *The Oxidation States of the Elements and Their Potentials in Aqueous Solution, 2nd Edn.*, Prentice-Hall (1952).

Charlot, G., Bézier, D., and Courtot, J., *Oxydo-Reduction Potentials*, Pergamon (1958).

Charlot, G., Collumeau, A., and Marchon, M. J. C., *Selected Constants: Oxida-*

Table 8.15   Standard enthalpies of formation, $\Delta H_f^{\ominus}(M^{n+}aq)/\text{kcal mol}^{-1}$; from reference [52] (la

H^+	0			
Li^+	-66.552^a			
Na^+	-57.433^a	Be^{2+} -91.5^c		
K^+	-60.270^a	Mg^{2+} -111.58^c	Zn^{2+} $-36.3^{d,e,f} -36.66^a$	
Rb^+	-60.019^a	Ca^{2+} -129.74^c	Cd^{2+} $-18.1^{d,e,g}$	
Cs^+	-61.673^a	Sr^{2+} -130.45^c	Hg^{2+} $+40.67^{e,h,i}$	
		Ba^{2+} -128.50^c		
		Ra^{2+} -126.1^c	Hg_2^{2+} $+39.87^h$	

Cr^{2+} -34.3	Mn^{2+} -52.76	Fe^{2+} -22.1^d
Cr^{3+} -60.00^{ff}		Fe^{3+} -12^d

$Eu^{2+} -126^t$

$Ce^{3+} -166.8^l$	$Pr^{3+} -168.0$	$Nd^{3+} -166.8$	$Pm^{3+} -166$	$Sm^{3+} -164.3^l$	$Eu^{3+} -144.6^l$	$Gd^{3+} -16$
$Ce^{4+} -128.4^{gg}$						

$Th^{3+}(-84)^v$ $Pa^{3+}(-104)^v$ $U^{3+}-117.2^v$ $Np^{3+}-125.9^v$ $Pu^{3+}-139.0^v$ $Am^{3+}-147.4^{w,x}$ $Cm^{3+}-147^{x,y}$ $Bk^{3+}-138$

$Th^{4+}-183^{aa,ii}$ $Pa^{4+}-148^{bb}$ $U^{4+}-141^{bb,cc}$ $Np^{4+}-132^{aa}$ $Pu^{4+}-128.2^{aa}$ $Am^{4+}-116^{aa,dd}$

---

a Ref. [60]; *b* not standard state; *c* from Natn. Bur. Stand. Technical Note 270/6 (1971); *d* ref. [8]; *e* see also ref. [61]; *f* see also ref. [62]; *g* see also ref. [63]; *h* refs. [8.53]; *i* ref. [64]; *j* ref. [65]; *k* ref. [7], slightly different earlier estimates in ref [52] and [59]; *l* the most recent values [9] for La^{3+} and some lanthanide(III) cations differ (except for Lu^{3+}) only ver slightly from the values quoted here; *m* an earlier estimate of + 27.7 kcal mol⁻¹ was very different; *n* ref. [16]; *o* in aqueo hydrochloric acid; *p* ref. [66]; *q* see also ref. [67]; *r* see also ref. [68]; *s* ref. [13]; *t* ref. [20]; *u* ref. [52]; and see also re [59]; *v* ref. [59]; *w* currently favoured values, values of -162.7 [59], -162.4 [69], -147.4 [70], and -150 [59] have bee quoted; *x* ref. [71]; *y* ref. [72]; *z* refs. [9. 59]; *aa* ref. [68]; *bb* in M HCl [15], *cc* a range of -139 to -157 is quote *tion-Reduction Potentials of Inorganic Substances in Aqueous Solution,* Butterworths (1971).

Table 8.16   Standard enthalpies of formation of aquated oxocations, $\Delta H_f^{\ominus}/\text{kcal}$ mol^{-1} ; **from reference [68] unless stated otherwise.**

$TiO^{2+}aq$	-164.9^a	$VO^{2+}aq$	-116.3^a
$ZrO^{2+}aq$	-223.1^a	VO_2^+aq	-155.3^a
UO_2^+aq	-247.4^a	NpO_2^+aq	$-230.8, -233.6^b$
$UO_2^{2+}aq$	$-242.2, -250.0$	$NpO_2^{2+}aq$	$-205.5, -208.5^b$
PuO_2^+aq	-219.9^b	AmO_2^+aq	-207.7
$PuO_2^{2+}aq$	-197.8^b	$AmO_2^{2+}aq$	-170.8

a Ref. [52]; *b* non-standard state values given in ref. [52] are similar.

---

ide cations from references [7, 52, 59^{hh}]) unless otherwise stated.

-127.86^j		
-146.8	Ga^{3+} -50.6	
-172.9	In^{3+} -25^{ee}	Sn^{4+} $+7.3^{b,o}$
$-169.4^{k,l}$	Tl^{3+} $+47^m$	
-156^c		Sn^{2+} $-2.1^{b,o}$
	Tl^{2+} $(+45)^n$	Pb^{2+} $+0.27^p$
	Tl^+ $+1.28$	
		Cu^+ $+17.13$
-14.0^q	Ni^{2+} -12.8^b	Cu^{2+} $+15.7^{d,r}$
$+22$		
	Pd^{2+} $(+42)^s$	Ag^+ $+25.275^a$
		Ag^{2+} -64.2^b

-164.5^l $Dy^{3+}-168$ $Ho^{3+}-171.5^l$ $Er^{3+}-169.5$ $Tm^{3+}-167.6$ $Yb^{3+}-161.2^l$ $Lu^{3+}-160.1^{l,u}$

7^z $Cf^{3+}-144$ to -149^z $Es^{3+}-143$ to -150^z $Fm^{3+}-141$ to -149^z $Md^{3+}(\sim+129)^t$ $No^{3+}(\sim+101)^t$ $Lr^{3+}(\sim+145)^t$

---

e more recent reference [68]; *dd* an uncertainty of ±6 is suggested, *ee* surprisingly, a recent publication (Camp-A. N., *Can. J. Chem.*, 54, 703 (1976)) gives a value of -57.8 kcal mol^{-1}, though $\Delta G_f^{\ominus}$ ($In^{3+}aq$) in this reference is to that given in Table 8.14; *ff* this recently (1976) proposed value resolves earlier uncertainties (Dellien, I. and er, L.G., *Can. J. Chem.*, 54, 1383; *gg* from Natn. Bur. Stand. Technical Note 270/7 (1973); *hh* ref. cited in footnote ves similar values; *ii* the recent estimate of -184 kcal mol^{-1}(Morss, L. R. and McCue, M. C., *J. Chem. Engng. Data*, 37 (1976) and refs. therein) is in good agreement.

REFERENCES

[1] West, J. M. *Basic Electrochemistry*, Van Nostrand Reinhold (1973), Appendix A.1.

[2] Latimer, W. M., *Oxidation Potentials, 2nd Ed.*, Prentice-Hall (1952).

[3] Johnson, D. A., *J. Chem. Educ.*, **45**, 236 (1968).

[4] Johnson, D. A., *Some Thermodynamic Aspects of Inorganic Chemistry*, Cambridge University Press (1968) Ch. 4.

[5] Ives, D. J. G., *Principles of the Extraction of Metals*, Royal Institute of Chemistry (1960) pp. 51–52.

[6] Phillips, C. S. G. and Williams, R. J. P., *Inorganic Chemistry, Vol. 2*, Clarendon Press (1966) Appendix I.

[7] Johnson, D. A., *J. Chem. Soc. Dalton*, 1671 (1974).

[8] Larson, J. W., Cerutti, P., Garber, H. K., and Hepler, L. G., *J. Phys. Chem.*, **72**, 2902 (1968); Hepler, L. G. and Olafsson, G., *Chem. Rev.*, **75**, 585 (1975).

[9] Nugent, L. J., *J. Inorg. Nucl. Chem.*, **37**, 1767 (1975).

[10] See p. 59 of reference [1].

[11] Vasil'ev, V. P. and Grechina, N. K., *Russ. J. Inorg. Chem.*, **12**, 315'(1967); Fedorov, V. A., Kalosh, T. N., and Mironov, V. E., *ibid.*, **16**, 1596 (1971).

[12] Izatt, R. M., Eatough, D. J., Morgan, C. E., and Christensen, J. J., *J. Chem. Soc. A*, 2514 (1970).

[13] Goldberg, R. N. and Hepler, L. G., *Chem. Rev.*, **68**, 229 (1968).

[14] Hancock, R. D., Finkelstein, N. P., and Evers, A., *J. Inorg. Nucl. Chem.*, **36**, 2539 (1974).

[15] Fuger, J. and Brown, D., *J. Chem. Soc. Dalton*, 2256 (1975).

[16] Schwarz, H. A., Comstock, D., Yandell, J. K., and Dodson, R. W., *J. Phys. Chem.*, **78**, 488 (1974); and references therein.

[17] Falcinella, B., Felgate, P. D., and Laurence, G. S., *J. Chem. Soc. Dalton*, 1367 (1974).

[18] See chapter 6 of reference [4].

[19] Biedermann, G. and Romano, V., *Acta Chem. Scand.*, **29A**, 615 (1975).

[20] Morss, L. R. and Haug, H. O., *J. Chem. Thermodynamics*, **5**, 513 (1973).

[21] Bidermann, G. and Silver, H. B., *Acta Chem. Scand.*, **27**, 3761 (1973).

[22] Nugent, L. J., Baybarz, R. D., and Burnett, J. L., *J. Phys. Chem.*, **73**, 1177 (1969).

[23] Tendler, Y. and Faraggi, M., *J. Chem. Phys.*, **57**, 1358 (1972).

[24] Nugent, L. J., Baybarz, R. D., Burnett, J. L., and Ryan, J. L., *J. Phys. Chem.*, **77**, 1528 (1973).

[25] Cotton, F. A. and Wilkinson, G., *Advanced Inorganic Chemistry, 3rd Ed.*, Wiley/Interscience (1972) p. 1072.

[26] See p. 1090 of reference [25].

[27] Marcus, Y., Yanir, E., and Givon, M. in *Coordination Chemistry in Solu-*

tion, ed. Högfeldt, E. Swedish Natural Science Research Council (1972) p. 229.

[28] Strehlow, H., in *The Chemistry of Non-aqueous Solvents, Vol. 1,* ed. Lagowski, J. J., Academic Press (1966) Ch. 4.

[29] Constantinescu, E., *Rev. Roumaine Chim.,* **17**, 1819 (1972).

[30] Murgulescu, I. G. and Constantinescu, E., *Rev. Roumaine Chim.,* **12**, 199 (1967).

[31] E.g. Baldwin, J., Gill, J. B., and Prescott, A., *J. Inorg. Nucl. Chem.,* **33**, 2103 (1971).

[32] Navaneethakrishnan, R. and Warf, J. C., *J. Inorg. Nucl. Chem.,* **36**, 1311 (1974).

[33] Dasent, W. E., *Inorganic Energetics,* Penguin Books (1970).

[34] Kilpatrick, M. and Jones, J. G. in *The Chemistry of Non-Aqueous Solvents Vol. 2* ed. Lagowski, J. J., Academic Press (1967) Ch. 2, p. 99.

[35] Senne, J. K. and Kratochvil, B., *Analyt. Chem.,* **44**, 585 (1972).

[36] Senne, J. K. and Dratochvil, B., *Analyt. Chem.,* **43**, 79 (1971).

[37] Kolthoff, I. M. and Coetzee, J. F., *J. Amer. Chem. Soc.,* **79**, 870, 1852 (1957).

[38] Coetzee, J. F. and Simon, J. M., *Analyt. Chem.,* **44**, 1129 (1972).

[39] L'Her, M. and Courtot-Coupez, J., *Bull. Soc. Chim. Fr.,* 3645 (1972).

[40] Larson, R. C. and Iwamoto, R. T., *J. Amer. Chem. Soc.,* **82**, 3239, 3526 (1960).

[41] Schmid, R. and Gutmann, V., *Chemické Zvesti,* **23**, 746 (1969); Gutmann, V. and Schmid, R., *Mh. Chem.,* **100**, 2113 (1969); Gutmann, V., *Allg. Prakt. Chem.,* **21**, 116 (1970).

[42] Duschek, O. and Gutmann, V., *Z. Anorg. Allg. Chem.,* **394**, 243 (1972).

[43] Duschek, O. and Gutmann, V., *Mh. Chem.,* **104**, 1259 (1973).

[44] Gutmann, V., *Pure Appl. Chem.,* **27**, 73 (1971); *Fortschr. Chem. Forsch.,* **27**, 59 (1972).

[45] Kestner, M. O. and Allred, A. L., *J. Amer. Chem. Soc.,* **94**, 7189 (1972).

[46] Fujita, S., Horii, H., and Taniguchi, S., *J. Phys. Chem.,* **77**, 2868 (1973).

[47] Bell, J. T., Friedman, H. A., and Billings, M. R., *J. Inorg. Nucl. Chem.,* **36**, 2563 (1974).

[48] Féher, F. in *The Chemistry of Non-aqueous Solvents, Vol. 3,* ed. Lagowski, J. J., Academic Press (1970) Ch. 4, p. 238.

[49] Waddington, T. C., *Non-aqueous Solvents,* Nelson (1969) Ch. 1.

[50] Lewis, D. in reference [27] p. 59.

[51] Noyes, R. M., *J. Chem. Educ.,* **40**, 2 (1963).

[52] Natn. Bur. Stand. Circular 500 (1952) and Technical Notes 270/3 (1968); 270/4 (1969), 270/5 (1971); and 270/6 (1971).

[53] Vanderzee, C. E. and Swanson, J. A., *J. Chem. Thermodynamics,* **6**, 827 (1974).

[54] See Table A2 of reference [1].

[55] Barton, A. F.M. and Wright, G. A., *Proceedings of the First Australian Conference on Electrochemistry,* ed. Friend, J. A. and Gutmann, F., Pergamon, Oxford (1965) p. 124.

[56] Vasil'ev, V. P., Vorob'ev, P. N., and Khodakovskii, I. L., *Russ. J. Inorg. Chem.,* **19**, 1481 (1974).

[57] Vasil'ev, V. P., Kochergina, L. A., and Lytikin, A. I., *Russ. J. Inorg. Chem.,* **19**, 1640 (1974).

[58] Vasil'ev, V. P., Kochergina, L. A., and Lytikin, A. I., *Russ. J. Inorg. Chem.,* **20**, 9 (1975).

[59] Nugent, L. J., Burnett, J. L., and Morss, L. R., *J. Chem. Thermodynamics,* **5**, 665 (1973).

[60] *J. Chem., Thermodynamics,* **7**, 1 (1975); **8**, 603 (1976).

[61] Wulff, C. A., *J. Chem. Engng. Data.,* **12**, 82 (1967).

[62] Vorob'ev, A. F. and Broier, A. F., *Russ. J. Phys. Chem.,* **45**, 745 (1971); Berg, R. L., and Vanderzee, C. E., *J. Chem. Thermodynamics,* **7**, 229 (1975).

[63] Vorob'ev, A. F. and Broier, A. F., *Russ. J. Phys. Chem.,* **45**, 1358 (1971)

[64] Vanderzee, C. E., Rodenburg, M. L. N., and Berg, R. L., *J. Chem. Thermodynamics,* **6**, 17 (1974).

[65] Krivtsov, N. V., Rosolovskii, V. Ya., and Shirokova, G. N., *Russ. J. Inorg. Chem.,* **16**, 1402 (1971).

[66] Vasil'ev, V. P., Kozlovskii, E. V., and Shitova, V. V., *Russ. J. Phys. Chem.,* **45**, 109 (1971).

[67] Vasil'ev, V. P., Raskova, O. G., Belonogova, A. K., and Vasil'eva, V. N., *Russ. J. Inorg. Chem.,* **19**, 1331 (1974).

[68] Vasil'ev, V. P. and Kunin, B. T., *Russ. J. Inorg. Chem.,* **17**, 1129 (1972).

[69] Fuger, J. in *MTP International Review of Science, Inorganic Chemistry, Series 1, Vol. 7, Lanthanides and Actinides,* ed. Bagnall, K. W., Butterworths, London, and University Park Press, Baltimore (1972) Ch. 5.

[70] Byan, J. L., *J. Chem. Thermodynamics,* **5**, 153 (1973).

[71] Fuger, J., Peterson, J. R., Stevenson, J. N., Noé, M., and Haire, R. G., *J. Inorg. Nucl. Chem.,* **37**, 1725 (1975).

[72] Fuger, J., Reul, J., and Muller, W., *Inorg. Nucl. Chem. Lett.,* **11**, 265 (1975).

[73] Jolly, W. L., *Chem. Rev.,* **50**, 351 (1952).

[74] Lepoutre, G. and Demortier, A., *Ber. Bunsenges. Phys. Chem.,* **75**, 647 (1971).

Chapter 9

HYDROLYSIS

9.1 ACIDITY OF SOLVATED CATIONS

9.1.1 Introduction

Aquo-cations, especially those of 4+, 3+, and small 2+ ions, tend to act as acids in solution:

$$M(OH_2)_x^{n+} \rightleftharpoons M(OH_2)_{x-1}(OH)^{(n-1)+} + H^+aq \ .$$

At the simplest level, this behaviour can be ascribed to the influence of the positive charge on the metal ion facilitating the loss of a water proton (Fig. 9.1). When their solvent molecules contain ionisable protons, solvento-cations in non-aqueous solvents can behave in a similar way. Liquid ammonia or alcohols illustrate this tendency. Such phenomena are special cases of a generally observed characteristic of protic ligands, their greater ease of deprotonation when coordinated to metal cations.

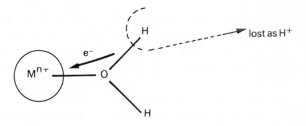

Fig. 9.1 The acidity of an aquo-cation.

9.1.2 Quantitative Aspects

Precise determinations of equilibrium constants characterising the hydrolysis and polymerisation of aquo-metal cations started only some two decades ago, when Sillén began his systematic studies. These have mainly involved accurate e.m.f. measurements at controlled ionic strength over a wide range of concentrations of the reacting species [1].

The acidic properties of solvento-cations in solution are characterised quantitatively by a variety of equilibrium constants, all interrelated and sometimes identical in definition if not in format. The most detailed general compilation of information on the acid behaviour of aquo-cations [2] presents the information in one of two forms, as decadic logarithms of K_1 or $*K_1$ defined as follows[†]:

$$K_1 = \frac{[M(OH)^{(n-1)+}]}{[M^{n+}][OH^-]} \quad \text{and} \quad *K_1 = \frac{[M(OH)^{(n-1)+}][H_3O^+]}{[M^{n+}][H_2O]}.$$

Some authors prefer to use $K_h = 1/*K_1$. The constants K_1 and $*K_1$ are interrelated:

$$\frac{*K_1}{K_1} = \frac{[M(OH)^{(n-1)+}][H_3O^+][M^{n+}][OH^-]}{[M^{n+}][H_2O][M(OH)^{(n-1)+}]} = \frac{[H_3O^+][OH^-]}{[H_2O]} = K_w$$

or
$$\log *K_1 - \log K_1 = \log K_w .$$

Analogous expression apply to successive constants K_n, $*K_n$ referring to the loss of the n th proton from an aquo-cation. In addition, equilibrium constants β_n relating to the overall loss of n protons from an aquo-cation are related to the successive or stepwise constants K_1 to K_n by:

$$\beta_n = \prod_{i=0}^{n} K_i .$$

Determination. Acidity constants for solvento-cations can be measured by any technique able to monitor concentrations of metal ions, hydroxo-complexes (or their equivalents in non-aqueous media), and solvated protons [3]. Reference [2] mentions about forty experimental techniques which have been used in the determination of acidity constants for aquo-cations. These include ultraviolet-visible, infrared-Raman, and n.m.r. spectroscopy, transport properties such as conductivity, e.m.f. measurements, kinetic methods, and a variety of other techniques.

Difficulties. Precise and accurate values are difficult to obtain for many reasons. In the first place, many aquo-cations, including most of the 2+ cations and all of the 1+ ones, are very weak acids. Indeed Rb^+aq and Cs^+aq appear to be so weak that they have so far defied satisfactory measurements of their acid dissociation

[†] Ideally activities would be used, but stability constants are normally defined in terms of concentration units so that they can be calculated from measurable quantities. The concentration of water, $[H_2O]$, is set at unity by arbitrary convention.

constants. Acidities of aquo-cations of higher charge, 3+ or 4+, are readily measurable, but they also often show strong tendencies of form polymeric species in alkaline or even neutral solution (Chapter 10). The large ranges of values (an extreme example is the range of 5.6 to 10.1 for the $p*K_1$ of La^{3+}aq) cited for some aquo-cations can often be attributed at least in part to interference from polymerisation. A related problem is that of 'ageing'. Sometimes systems take a very long time to attain equilibrium, by which time polymeric species are very likely to be present. This area of difficulty has been much investigated, and discussions of the typical systems Al^{3+}aq and Fe^{3+}aq can be found in references [4] and [5] respectively. While this problem of polymerisation causes concern in the derivation of K_1 values, it is much more serious for K_2 and higher constants. See the recent examination and discussion of the case of Cr^{3+}aq [6], for example.

Calculations. The calculation of the required result from experimental observations is often complicated, especially if allowance has to be made for concurrent polymerisation. Fortunately standard computational methods have been developed and published by Sillén and his associates, pioneers in the accurate determination of hydrolysis and polymerisation constants for aquo-metal cations [1]. The basic computational methods, christened LETAGROP, were published in the early sixties [7]. Since then a series of papers has dealt with detailed aspects of the application of this program [8], and with simplification and speeding-up of its operation in certain circumstances [9]. Specific Algol [10] and Fortran [11] versions of LETAGROP have been presented, as well as a Fortran version [11] of the related general program HALTAFALL [12] dealing with the composition of equilibrium mixtures.

Ionic strength. Acidity constants tend to vary with ionic strength. If series of directly comparable values for different aquo-cations are to be obtained, standard conditions must be selected. There are two schools of thought on this matter. For a long time the majority of workers presented results at zero ionic strength, derived from experimental results by extrapolation. More recently, and less attractively, there has been a tendency to obtain and express results under conditions of swamping electrolyte, generally 3M $NaClO_4$. The reasons for this approach, and its attendant disadvantages, have been discussed elsewhere [13]. In practice the effects of ionic strength variation are small compared with uncertainties arising from other sources. The magnitude of variation with ionic strength is illustrated by the results given in Table 9.1. At the ionic strengths cited in this Table, $p*K_1$ values generally increase with increasing ionic strength. The effect is more marked for 4+ and 3+ cations, for example U^{4+} and Ga^{3+}, than for 2+ cations such as Pb^{2+}.

Table 9.1   The effect of ionic strength variation on hydrolysis constants, p^*K_1 (as defined in the text), for aquo-metal cations, at $25°C$ [2].[a]

$Pb^{2+}aq$	$[NaClO_4]/M$	0.3	3.0		
	p^*K_1	7.8	7.9		
$Ni^{2+}aq$	$[KNO_3]/M$	0.03	0.3	1.0	1.5
	p^*K_1	10.22	10.23	10.26	10.18
$Ga^{3+}aq$	$[NaClO_4]/M$	0.1	0.3	0.5	1.0
	$p^*K_1{}^{b}$	2.87	2.48	2.30	1.78
$Tl^{3+}aq$	$[NaClO_4]/M$	1.5	3.0		
	p^*K_1	1.07	1.16		
$V^{3+}aq$	$[NaCl]/M$	1.0	3.0		
	$p^*K_1{}^{c}$	2.85	3.15		
$U^{4+}aq$	$[NaClO_4]/M$	0.19	0.5	1.0	2.0	3.0
	p^*K_1	1.12	1.50	1.56	1.63	2.0

[a] Further relevant results are contained in Table 9.2 ($Zn^{2+}aq$) and in Fig. 9.2 ($Ni^{2+}aq$); [b] ref. [14]; [c] $20°C$.

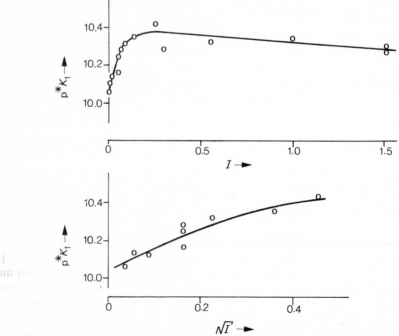

Fig. 9.2   The dependence of the hydrolysis constant p^*K_1 for $Ni^{2+}aq$ on ionic strength and on the square root of ionic strength (at low ionic strength) [15].

Particularly detailed studies of the dependence of $p*K_1$ on ionic strength have been undertaken for some M^{2+}aq cations. The picture for Ni^{2+}aq [15] is shown in Fig. 9.2. The shape of this plot conforms to the pattern expected from extended Debye-Hückel theory and similar plots have been obtained for Mn^{2+}aq [16] and Zn^{2+}aq [17]:

$$p*K_1\,(I=0) \;=\; p*K_1 \;-\; c\,\frac{I^{\frac{1}{2}}}{1 + c'I^{\frac{1}{2}}} \;+\; c''I$$

where I = ionic strength and c, c', and c'' are constants. The fit of the experimental results to this equation is particularly good at low ionic strengths, permitting precise estimates of $p*K_1$ values for these aquo-cations. They are 9.86 ± 0.03 for Ni^{2+}aq [15], 10.59 ± 0.08 for Mn^{2+}aq [16], and 8.96 ± 0.03 for Zn^{2+}aq [17], at 25°C.

Specific ion effects. The nature of the salt used to maintain the ionic strength constant can have some effect on the $p*K$ value for an aquo-cation. This is illustrated, for added cation variation, in Table 9.2. The activity coefficient for Li^+aq is probably similar to that of H^+aq, whereas the activity coefficient of Na^+aq differs somewhat – hence the differences perceived when lithium and sodium perchlorates are variously used to maintain ionic strengths. In view of these activity coefficient considerations, lithium perchlorate would seem a better choice than sodium perchlorate for the maintaining of constant ionic strength. However, it is customary to use the sodium salt.

The nature of the anion present probably has some effect on the determined $p*K$ value. Thus the $p*K_1$ of Pb^{2+} is reported as 7.93 in 2M $NaClO_4$, but 8.84 in 2M $NaNO_3$. However it must be admitted that these two values come from different sources (contrast the cation effects in Table 9.2, where each series is taken from a single source).

Table 9.2 **Specific ion effects on hydrolysis constants, $p*K_1$ (as defined in the text), for aquo-metal cations [2].**

Zn^{2+}aq		2M KCl	2M NaCl	2M NaClO$_4$	3M KCl	3M NaCl
	$p*K_1$	9.02	9.11 to 9.14	9.55	9.26	9.25
Tl^{3+}aq		3M NaClO$_4$		3M LiClO$_4$		
	$p*K_1$	1.14		1.18		

Conclusions. In the light of the foregoing paragraphs, and after an inspection of the ranges of values available for some aquo-cations, the reader will readily agree with authors of textbooks who hold that little reliance can be placed on exact numbers presented for pK values. On the other hand, one can discern the orders

Table 9.3　Hydrolysis constants, $p*K_1$, for sp-block aquo-cations and Ag^+ at 25°C; from [2] except where otherwise indicated.

	$p*K_1(I=0)$	$p*K_1(I>0)$		$p*K_1(I=0)$	$p*K_1(I>0)$
Li^+	13.7 to 14.1	14.2	Ag^+	11.7 to 12.0	9.8 to 10.1
Na^+	14.6 to 14.8		Tl^+	13.3 ± 0.2(6)	
Be^{2+}		5.7 to 6.7	Hg_2^{2+}		3.6 to 5.0[a]
Mg^{2+}	11.4	12.2 to 12.8			
Ca^{2+}	12.6 ± 0.1(7)	13.4	Zn^{2+}	9.5 ± 0.4(6)	8.2 to 9.6
Sr^{2+}	13.1 ± 0.1(3)	13.8	Cd^{2+}	7.9	7.6 to 10.2
Ba^{2+}	13.3 ± 0.1(5)	14.0			10.3 [22]
			Hg^{2+}	2.5	2.4 to 3.7
					3.6 to 3.7 [23]
Al^{3+}	5.0 ± 0.3(11)	4.3 to 5.0			
Sc^{3+}		4.6 to 5.1[d]	Tl^{2+}		4.6 [24]
Y^{3+}		4.7 [18], 4.6 [19]			
		~7 to 9.1	Sn^{2+}	1.7 to 2.1	1.7 to 3.2
La^{3+}		8.0 [19]	Pb^{2+}	7.2	6.5 to 8.4
		8.3 to 10.7			7.9 [25]
		9.1 [19], 10.0 [20]			
			Ga^{3+}	2.6	2.9 to 3.4
			In^{3+}		2.1 to 4.4
			Tl^{3+}		−0.8 to 1.2
					1.5 [26]
			Bi^{3+}	b	1.6 [2], 2.0 [27]
			Sn^{4+}		−0.6 [28]
			Po^{4+}		0.1 to 3.4[c]

a The $p*K_1$ for the transient Hg^+aq cation has been estimated as 5.1 [21]; b a recent estimate of $p*K_1$ for Bi^{3+}aq is 1.56 at zero ionic strength (Antonovich, V. P., Nevskaya, E. M., Shelikhina, E. I. and Nazerenko, V. A., *Russ. J. Inorg. Chem.*, **20**, 1642 (1975)); c the most recent estimate of $p*K_1$ for Po^{4+}aq is 0.5 (Ampelgova, N. I., *Radiokhimiya*, **17**, 68 (1975)); d the most recent review of scandium thermochemistry favours a slightly lower value (Travers, J. G., Dellien, I., and Hepler, L. G., *Thermochim. Acta*, **15**, 89 (1976)).

Table 9.4 Hydrolysis constants, p^*K_1, for transition metal aquo-cations at 25°C; from [2] except where indicated otherwise.

	M²⁺aq		M³⁺aq		M⁴⁺aq
	$p^*K_1(I=0)$	$p^*K_1(I>0)$	$p^*K_1(I=0)$	$p^*K_1(I>0)$	p^*K_1
Ti			2.3	2.6 [2], 1.3 to 2.2 [36] 2.6 [37], 2.8 [38] 1.9 [39], 2.3 [40]	−4
Zr					−0.7 to 0.2 −0.3 or 0.6 [41]
Hf					−0.1 [2], 0 [15,42] 0.2 or 1.1 [41]
V		6.5	2.4 to 2.9	2.4 to 3.2	
Cr		8.7 [29] 9 to 11 [30]	3.8 to 4.0	3.8 to 4.4 4.1 [6,43]	
Mn	10.6	10.5 9.6 [31]		−0.6 to 0.2 0.93 [44], 1.05 [45] 1.0 [46]	
Fe	5.9 to 8.3	3.3 to 9.5^a	2.0 ± 0.3(8)^{d,e}	2.5 to 3.1^d 2.6 to 2.9 [48] 3.1 [49], 3.3 [50]	
Co	7.0 to 12.2	7.6 to 9.9		0.6 or 1.7	
Rh				2.9 to 3.4, 4.1^f	
Ir				4.8^f	
Ni	9.2 to 10.9^b	6.5 to 10.2^b			
Pd	1.0 or 1.6 1.6 [33] >2.5^g				
Pt					
Cu	7.3 to 8.0	6.5 or 6.8 5.4 to 5.8 [34]			
Ag^c		0.5 [35]			

a Values, omitting extremes, average to 7.1 (5 determinations); b but there is a claim that there is no evidence for Ni(OH)⁺, only for Ni(OH)₄⁻ [32]; c see Table 9.3 for Ag⁺aq; d there is an extensive tabulation of hydrolysis constants for Fe³⁺aq in [47]; e the latest estimate is 2.4 (Landry, J. C., Buffle, J., Haerdi, W., Levantal, M. and Nembrini, G., Chimia, 29, 163 (1975)); f recent estimates at 5°C, (Beutler, P and Gamsjäger, H., Chimia, 29, 525 (1975)); g estimated value for the recently characterised Pt²⁺aq cation (Elding, L. I., Inorg. Chim. Acta, 20, 65 (1976)).

of magnitude, together with some trends, and some of these can be explained. The situation and problems are of course the same (or more complicated) for cations in non-aqueous or in mixed aqueous media.

9.2   AQUEOUS SYSTEMS

9.2.1   pK Values for Aquo-cations

Values for hydrolysis constants of aquo-cations in aqueous solution, expressed as p$*K_1$ ($= -\log_{10} *K_1$), are collected together in Tables 9.3 to 9.5 [18-57][†]. Table 9.3 contains cations of the sp-block elements and, for convenience of layout, Ag^+. Table 9.4 contains transition metal cations, while Table 9.5 contains lanthanide and actinide cations. The majority of entries are taken from the exhaustive compilation in [2] (another extensive compilation can be found in [58]) though a few later results are included and referenced. As indicated in the previous Section, hydrolysis constants have been determined under a wide range of conditions. Table 9.3 to 9.5 contain only values at 25°C. They have been split into two groups, those estimated for zero ionic strength and those measured at relatively high ionic strengths, generally within the range 1 to 3M, but sometimes even higher. No single ionic strength with an extensive series of values available for special quotation exists. Where several p$*K_1$ values are available for a given aquo-cation, the range is indicated in Tables 9.3 to 9.5. In those cases where several fairly consistent values exist, their mean and standard deviation (and, in brackets, the number of values) are quoted. Obviously these latter cases are most likely to represent accurate values of hydrolysis constants. By way of contrast, one should emphasise that for cations such as In^{3+}aq, Fe^{2+}aq, Co^{2+}aq, and Ni^{2+}aq, the range of values is over a million-fold. Here, the reader requiring the 'best' value needs to read the original references carefully and critically. It should be said that Tables 9.3 to 9.5, like the compliation of reference [2], contain p$*K_1$ values for a few rather dubious species, for example Bi^{3+}aq, whose existence is far from unequivocally proven.

Despite the uncertainties and misgivings expressed in previous paragraphs, certain trends and exceptional values can be distinguished in the data assembled in these three Tables. There is a general dependence on cation charge, with 4+ cations often more acidic than 3+, 3+ generally more acidic than 2+, and 1+ often so weakly acidic that their pKs verge on the immeasurable. Here as elsewhere, the effect of cation radius is smaller than that of cation charge and often submerged under other effects. However the Group II cations, from Be^{2+}aq down to Ba^{2+}aq, show the electrostatically expected trend of decreasing acidity as the cation radius increases. Other factors must be invoked to explain why many aquo-cation pKs, particularly those of B–Group metals, fit a charge-radius based plot constructed from pK values of 'well-behaved' cations such as those

---

† Hydrolysis and polymerisation of aquo-cations in natural waters, for instance in seawater, are discussed in Stumm, W. and Morgan, J. J., *Aquatic Chemistry*, Wiley/Interscience (1970)

Table 9.5 Hydrolysis constants, $p*K_1$, for lanthanide[a] and actinide aquo-cations at 25°C.

	3+ Cations Various sources	[51][b]	[52]	4+ Cations	
Lanthanides					
La^{3+}	(Table 9.3)	9.33	7.4		
Ce^{3+}	9.3[2]			Ce^{4+}	0.7 or -0.7^c [2]
Pr^{3+}	8.5[2]	8.82	7.1	Pr^{4+}	3.2[56]
Nd^{3+}	8.5[2], 9.4[53]	8.70	7.0		
Pm^{3+}			6.5		
Sm^{3+}		8.61	4.4		
Eu^{3+}	8.0[19]	8.58	4.8		
Gd^{3+}	8.3[19], 8.2[20], 9.2[54]	8.62	7.1		
Tb^{3+}		8.43	5.2		
Dy^{3+}		8.37	5.6		
Ho^{3+}		8.31	5.7		
Er^{3+}	9.0^f	8.26	5.5		
Tm^{3+}		8.22	4.4		
Yb^{3+}	8.0[19]	8.19	4.3		
Lu^{3+}	6.6[2], 7.7[19]	8.17	3.5		
Actinides					
				Th^{4+}	2.4 to 5.0[2], 3.15[19]
				Pa^{4+}	0.1[2]
				U^{4+}	0.7 to 2.0[2], 1.3[57]
Np^{3+}	7.4^e			Np^{4+}	2.3[2]
Pu^{3+}	7.0[2]			Pu^{4+}	$1.5 \pm 0.2(8)^d$ [2]

a An assortment of values for individual lanthanide(III) aquo-cations can be tracked down via references cited in [52]; b values derived from this reference and given in [2] differ by 0.27 due to different choices of pK_w; c the state of cerium(IV) in aqueous solution has been briefly reviewed several times, with current opinion favouring the value of -0.7 for $p*K_1$ [55]; d as in Table 9.3 we quote the mean and standard deviation of eight reasonably concordant reported $p*K_1$ values here, but is should be added that the latest estimate of the $p*K_1$ for Pu^{4+}aq is 0.45 (Guyomont, R., *Radiokhimiya*, **17**, 636 (1975)); e recent estimate by Mefod'eva, M. P., Krot, N. N., Afanas'eva, T. V., and Gelman, A. D., *Izv. Akad. Nauk SSSR, Ser. Khim.*, 2370 (1974); f recent estimate by Burkov, K. A., Lilich, L. S., and Nguyen Dinh Ngo, *Izv. Vyssh. Uchebn. Zaved., Khim. Khim. Tekhnol.*, **18**, 181 (1975) (*Chem. Abs.*, **83**, 49039x (1975)); g see also Heitanen, S. and Sillén, L. G., *Acta Chem. Scand.*, **22**, 265 (1968); Morss, L. R. and McCue, M. C., *J. Chem. Engng. Data*, **21**, 337 (1976).

of Groups I, II(A), and III(A) only poorly. The hard and soft acids and bases concept has been applied (Table 9.6) with some success [59]. 'Soft' cations are very much stronger acids than 'hard' cations of similar size.

Several of the exceptionally acidic aquo-cations, for example Ag^{2+}aq,

Mn^{3+}aq, and Ce^{4+}aq, are strong oxidants, and it has been suggested that these two factors are related [35]. The electron attracting powers of such cations should facilitate loss of H^+ from coordinated water. It is possible that the different geometries, pyramidal or planar, for $M^{n+}-OH_2$ units (see Chapter 2.2.1) may have an effect on the acidity of coordinated water molecules. Other factors have been proposed as influences on the acidity of aquo-cations, for instance the symmetry of the hydrated cation [60] and the hydration number [61].

In the area of transition metal cations, the obvious thing to look for is the crystal field effect and its consequences. The importance of crystal field effects is difficult to assess, because the range of $p*K_1$ values reported for each aquo-transition metal cation is usually so large. The situation is lightly less confused for the M^{3+}aq than for the M^{2+}aq cations. Opinions differ as to whether $p*K_1$ values for such M^{3+}aq cations are significantly affected by crystal field enthalpy contributions [62] or not [63]. The situation also lacks clarity in respect of aquo-lanthanide cations. There may or may not be a discontinuity in the trend of $p*K_1$ values in the middle of the series. The picture for the aquo-lanthanide cations is further complicated by the large discrepancies between different series of reported values (Table 9.5).

Table 9.6 The relation of aquo-cation pK values to the hard and soft acids and bases concept [59].

Cation		Radius/Å	$p*K_1(25°C)$
K^+	hard	1.33	14
Ag^+	soft	1.26	10
Mg^{2+}	hard	0.65	12.2
Cu^{2+}	soft	0.69	7.3
Ca^{2+}	hard	0.99	12.6
Cd^{2+}	soft	0.97	9.0
Sr^{2+}	hard	1.13	13.1
Hg^{2+}	soft	1.10	3.6

pK_n **values.** Despite the difficulties involved, especially the greater chance of the concurrent formation of polynuclear species at relatively high pHs, sets of values for pK_1, pK_2 ⋯ pK_n exist for some aquo-cations. Several such series are included in Table 9.7. In general the acid dissociation constant for the loss of a second proton from an aquo-cation is some 10 to 100 times smaller than that for the loss of the first proton. However there are some odd and unlikely results, for example Hg^{2+} [23], In^{3+}, and Cd^{2+}, where $M(OH)^{(n-1)+}$aq appears to be as strong or stronger an acid than M^{n+}aq! Further proton loss is rarely monitored; one wonders whether such pK_3, pK_4, and so on, values are really meaningful. Such reservations are particularly strong where even the starting aquo-cation, for example Ti^{4+}aq, is none too well characterised.

Table 9.7   Series of $p*K_n$ values for aquo-cations.[a]

	$p*K_1$	$p*K_2$	$p*K_3$	$p*K_4$	$p*K_5$	$p*K_6$	
Sc^{3+}	4.5 to 4.6	5.5 or 8.8	6.3				[18, 19]
Ga^{3+}	2.8 to 2.9	3.5 to 4.4	4.5		10.3	11.7	[2, 14]
V^{3+}	2.4 to 2.9	3.5 to 3.8					[2]
Cr^{3+}	4.1 to 4.2	5.6 to 6.2	6.2				[2, 43]
Fe^{3+}	2.5 to 2.8	4.6 to 4.7					[2, 64]
Ti^{4+}	−4	−3.2	1.5	3.0			[2]
Th^{4+}	3.2	6.6					[19]
U^{4+}	4.2	5.2					[2]

a For conditions of ionic strength and temperature see the references cited.

Protonation of aquo-cations. It seems that it is possible to protonate at least some aquo-cations in sufficiently acidic conditions. The rate laws for water exchange (see Chapter 11) with the Cr^{3+}aq and Ni^{2+}aq cations:

$$\text{rate} = k[M^{n+}\text{aq}] + k'[M^{n+}\text{aq}][H^+]$$

indicate the involvement of $M(OH_2)_5(OH_3)^{(n+1)+}$ species in these reactions [65]. Similar kinetic evidence has been presented for protonation of coordinated water in $Cr(OH_2)_5(N_3)^{2+}$. The likely alternative explanation of protonating the azide ligand in this case was considered and rejected [66]. Unfortunately no estimates of pK values of these protonated species were made.

9.2.2   Oxo-aquo-cations

In a sense these form a special case of the aquo-cations discussed above, for they are formally derived from aquo-cations by the loss of two protons from one coordinated water molecule, for example

$$V(OH_2)_6^{4+} = VO(OH_2)_5^{2+} + 2H^+$$

For actinides, this process occurs twice over, for example

$$U(OH_2)_n^{6+} = UO_2(OH_2)_m^{2+} + 4H^+ + (n-m-2)H_2O.$$

Information on the hydrolysis constants of these oxo-aquo-cations is given in Table 9.8. Values of $p*K_1$ here are considerably lower (in other words these 2+ oxo-aquo-cations are considerably more acidic) than analogous M^{2+}aq cations (Tables 9.3 and 9.4). This state of affairs can presumably be attributed to the high formal charge, 5+ or 6+, on the central metal atom.

Table 9.8 Hydrolysis constants, $p*K_1$, for oxo-aquo-cations [2], at $25°C$.

	$P*K_1 (I = 0)$	$p*K_1 (I > 0)$
VO^{2+}	4.7 to 5.4	6.0
UO_2^{2+}	4.1 to 4.3	4.2 to 6.1
PuO_2^{+}	$\geqslant 9.7$	
PuO_2^{2+}		5.3 to 5.7

Many more oxo-cations have been characterised or postulated than those listed in Table 9.8, for example VO_2^{+}aq, NpO_2^{3+}aq, TcO^{2+}aq, RuO^{2+}aq, and TiO^{2+}aq. A number of hydroxo-oxo-aquo-cations have also been suggested, such as $VOOH^{2+}$aq and PuO_2OH^{+}aq. Unfortunately, quantitative pK data are lacking for these species. The logical extension of this progression from simple aquo-cations through hydroxo-cations and oxo-cations is to what might be inelegantly called poly-oxo-species, that is to say oxoanions consisting of several oxide ligands coordinated to a central metal cation, for example chromate, permanganate, molybdate, and innumerable similar species. The progression may be stated explicitly for the case of vanadium:

$$V^{2+}aq, \quad V^{3+}aq, \quad VO^{2+}aq, \quad VO_2^{+}aq, \quad VO_3^{-} \ .$$

There are many acid dissociation constants known for acids of oxoanions with a central metal cation, but this subject takes us beyond the scope of the present text.

9.2.3 Mixed Aquo-ligand Complexes

A large body of experimental pK values referring to the acidity of coordinated water in complexes of the type $M(OH_2)_x(L)_y^{n\pm}$ exists, and Tables 9.9 to 9.14 present a generous selection of these. Numerous complexes of the cobalt(III)-ammine or amine–water type are included in Table 9.9, while cobalt(III) complexes containing oximato-ligands are covered in Table 9.10. Values for chromium(III) complexes are given in Table 9.11. Table 9.12 presents comparative data for some series of analogous complexes of various transition metal cations. Square-planar complexes are dealt with in Table 9.13. Finally

Table 9.9 Acidity constants for cobalt(III)–ammine or amine–water complexes at 25°C.

	p^*K_1	
$CoL_5(OH_2)^{n\pm}$		
$Co(NH_3)_5(OH_2)^{3+}$	5.7 to 6.6[2,67–71]	
$Co(en)_2(NH_3)(OH_2)^{3+}$	cis 5.8 to 6.3[2,68,72,73]	trans 5.7 to 6.1[2,68,72,74]
$Co(en)_2(amine)(OH_2)^{3+a}$	cis 5.3 to 6.2[73]	
$Co(trien)(Cl)(OH_2)^{2+}$	cis-α7.1[b] cis-β 5.8[b][2]	
$Co(NH_3)_4(Cl)(OH_2)^{2+}$	cis 6.6[b][2]	
$Co(en)_2(Cl)(OH_2)^{2+}$	cis 6.7 to 6.8[c][2]	trans 5.7 to 7.2[d][2]
$Co(NH_3)_4(NO_2)(OH_2)^{2+}$		trans 9.0[e][2]
$Co(en)_2(NO_2)(OH_2)^{2+}$	cis 6.3[2]	trans 6.4[2]
$Co(NH_3)_4(CN)(OH_2)^{2+}$		trans 8.6[e][2]
$Co(en)_2(SO_4)(OH_2)^+$	cis 6.3[f][2]	
$Co(CN)_5(OH_2)^{2-}$	9.7[g][2]	
$CoL_4(OH_2)_2^{n+}$		
$Co(NH_3)_4(OH_2)_2^{3+}$	cis 5.7 to 6.0 [2,75,76][h]	
$Co(en)_2(OH_2)_2^{3+}$	cis 6.0 to 6.1[2,75,77]	trans 4.5 to 4.6[2,77]
$Co(tren)(OH_2)_2^{3+}$	cis 5.4[b][76]	
$Co(trien)(OH_2)_2^{3+}$	cis-α 5.4[b] cis-β 5.3[b][78]	
$Co(trans-14-diene)(OH_2)_2^{3+}$		trans 4.0[79]
$Co(tet-a)(OH_2)_2^{3+}$		trans 2.7[79]
$Co(bipy)_2(OH_2)_2^{3+}$	cis 4.5[75]	
$Co(phen)_2(OH_2)_2^{3+}$	cis 4.7[75]	
$Co(NH_3)_2(NO_2)_2(OH_2)_2^+$	cis 6.3 [2]	trans 6.4[2]
$CoL_3(OH_2)_3^{n+}$		
$Co(NH_3)_3(OH_2)_3^{3+}$	fac 5.3[i][76]	
$Co(tach)(OH_2)_3^{3+j}$	fac 5.5[b][76]	
$Co(bamp)(OH_2)_3^{3+j}$		mer 4.1[b][76]
$CoL_2(OH_2)_4^{n+}$		
$Co(NH_3)_2(OH_2)_4^{3+}$	cis[k] 3.4[i][2]	

a Amine = pyridine, aniline, four aliphatic amines; b 20°C; c p^*K_1 = 7.12 at 20°C [2]; d p^*K_1 = 6.11 at 20°C [2]; e 18°C; f 0°C; g 40°C; h ref. [76] value at 20°C; i 15°C; j see list of abbreviations at back of book for tach, bamp; k this assignment of stereochemistry is only tentative (White, J. D., Sullivan, J. C., and Taube, H., J. Amer. Chem. Soc., 92, 4733 (1970)).

Table 9.14 is devoted to acidity constants for alkylaquomercury(II), alkyl-aquotin(IV), and related cations.

It is impossible to draw together into a succinct paragraph or two all these observations, and the explanations offered for them by various writers. Only a

Table 9.10   Acidity constants for coordinated water molecules in *trans*-Co(dmgH)$_2$L(OH$_2$)$^+$ and related complexes.

	p*K_1		p*K_1		p*K_1
trans-Co(dmgH)$_2$(OH$_2$)$_2^+$	5.14 [80]	*trans*-Co(dmgH)$_2$(X-pyridine)(OH$_2$)$^+$ [83]		*cis*-Co(dmgH)$_2$(OH$_2$)$_2^+$	4.58^a [84]
trans-Co(dmgH)$_2$(NH$_3$)(OH$_2$)$^+$	6.98 [80]	X = 3-Me	5.96	*cis*-Co(bipy)$_2$(OH$_2$)$_2^{3+}$	4.5 [75]
trans-Co(dmgH)$_2$Cl(OH$_2$)	6.15 [81]	3-Br	5.97	*cis*-Co(phen)$_2$(OH$_2$)$_2^{3+}$	4.7 [75]
	6.30 [82]	4-Me	6.05	*cis*-Co(phen)$_2$(OH)(OH$_2$)$^{2+}$	6.8 [86]
trans-Co(dmgH)$_2$Br(OH$_2$)	6.31 [81]	3-PhCH$_2$	6.07	*cis*-Co(phen)$_2$(NO$_2$)(OH$_2$)$^{2+}$	8.4 [86]
	6.44 [82]	H	5.99		
trans-Co(dmgH)$_2$I(OH$_2$)	6.87 [81]				
trans-Co(dmgH)$_2$(NO$_2$)(OH$_2$)	7.28 [80]	*trans*-Co(dmgH)$_2$(X-C$_6$H$_4$NH$_2$)(OH$_2$)$^+$ [80]		Co(porph)(OH$_2$)$_2^{5+b}$	6.0[87]; 5.5[88]
	7.66 [82]		6.69 to 7.09	Co(porph)(py)(OH$_2$)$^{5+b}$	8.1 [89]
trans-Co(dmgH)$_2$(OH)(OH$_2$)	8.05 [82]				

a The cited p*K_1 value for this complex may refer to loss of a proton from one of the dmgH$^-$ ligands, as seems to be the case for *trans*-Co(dmgH)$_2$(tu)(OH$_2$)$^+$ [85]; *b* porph = tetrakis-(4-N-methylpyridyl)porphine.

Table 9.11 *See next page.*

Table 9.12   Comparison of acidity constants for aquo-ligand complexes of some transition metal cations. ^{a}

	Cr(III)	Co(III)	Rh(III)	Fe(III)	Ru(III)	Ru(II)	Pt(IV)
M(CN)$_5$(OH$_2$)$^{n-}$	~9 [95]	9.7 [2,95]		8.4 [95]			
M(NH$_3$)$_5$(OH$_2$)$^{n+}$	4.8 to 5.3	5.7 to 6.6	5.9 to 6.9 [2,70,96-99]		4.1 [101] or 4.2 [2]	13.1 [101]	4 [102]
M(NH$_3$)$_4$(OH$_2$)$_2^{n+}$	5.1 to 5.5	5.7 to 6.0					2.0 [71]
MCl(OH$_2$)$_5^{n+}$	5.2		4.9 [100]				
cf. M(OH$_2$)$_6^{n+}$	3.8 to 4.1	0.6 to 1.7	2.9 to 3.4	2.0 to 3.1			

a Values of p*K_1 taken from Tables 9.9 to 9.11 and/or [2] unless otherwise stated.

few generalisations and trends derived from the data presented in Tables 9.9 to 9.14 will be pointed out here.

Ligand basicity. In general terms, the effects of coordinated ligands on the acidity of coordinated water molecules can to some extent be related to the electron releasing and withdrawing properties, that is to say basicity, of the former. These effects can either be transmitted through or across the metal cation, while the ease of transmission will depend on such properties of the cation as its electron configuration and the availability of suitable orbitals.

The pK values for the coordinated water in certain series of related compounds can be correlated with the base strength of other ligands coordinated to the metal. Examples include the cis-$Co(en)_2L(OH_2)^{3+}$ and trans-$Co(dmgH)_2$ (substituted pyridine)$(OH_2)^+$ series (Tables 9.9, 9.10 respectively). In both these cases, the variation in the p$*K_1$ value for the coordinated water is much less than the variation in the pK value for the coordinated L or substituted pyridine. Here, and usually, the effect of ligand basicity is greatly attenuated by

Table 9.11   Acidity constants for aquo-chromium(III) complexes at 25°C.

	p$*K_1$	
$CrL_5(OH_2)^{n\pm}$		
$Cr(NH_3)_5(OH_2)^{3+}$		4.8 to 5.3 [2,69,70,90]
$Cr(CN)_5(OH_2)^{2-}$		~9 [95]
$Cr(ox)_2(NCS)(OH_2)^{2-}$	cis 7.1[a] [91]	
$CrL_4(OH_2)_2^{n\pm}$		
$Cr(NH_3)_4(OH_2)_2^{3+}$	cis 5.08 [2,77]	trans 5.5 [77]
$Cr(en)_2(OH_2)_2^{3+}$	cis 4.80 [77]	trans 4.1, 4.27 [2,77]
$Cr(en)(tmd)(OH_2)_2^{3+}$		trans 4.13 [77]
$Cr(tmd)_2(OH_2)_2^{3+}$		trans 4.15 [77]
$Cr(bipy)_2(OH_2)_2^{3+}$		trans (?)[b] 4.5 [92]
$Cr(ox)_2(OH_2)_2^-$	cis 7.1 [93]	trans 7.2 [93]
$CrL_2(OH_2)_4^{n\pm}$		
$Cr(NH_3)_2(OH_2)_4^{3+}$		trans[c] 4.11[d] [2]
$CrCl_2(OH_2)_4^+$		trans[e] 5.4 to 5.7 [2]
$Cr(ox)(OH_2)_4^+$	cis 5.6 [43]	
$CrL_5(OH_2)^{n\pm}$		
$CrCl(OH_2)_5^{2+}$		5.2 [2] or ~6 [94]
cf. $Cr(OH_2)_6^{3+}$		3.8 to 4.1[f]

a At 1°C, which low temperature was needed to minimise aquation; b it is usually believed that compounds $M(bipy)_2L_2$ have cis geometry; c assignment of trans geometry only tentative [2]; d at 20°C; e for assignment of trans geometry see Johnson, H. B. and Reynolds, W. L., *Inorg. Chem.*, 2, 468 (1963); f Table 9.4.

the time the protons of the coordinated water have been reached.

Ligand denticity. A feeling exists that the replacement of unidentate ligands by analogous chelating, especially macrocyclic, ligands leads to more pronounced acidity of coordinated water molecules. The values cited in Table 9.9 for complexes of the general formula $CoL_4(OH_2)_2^{3+}$ support this idea. Explanations are offered in terms of solvation differences attributable to the bridging portions of the ligands, which are generally hydrophobic, and of inductive and field effects [111]. However, in some cases this macrocyclic effect is small, and may indeed

Table 9.13 Acidity constants for aquo-ligand complexes of d^8 cations $ML_x(OH_2)_{4-x}^{n+}$.

	$p*K_1$	
$ML_3(OH_2)^{n+}$		
$Pt(NH_3)_3(OH_2)^{2+}$	5.5 [103]	
$Pt(dien)(OH_2)^{2+}$	6.13 [104]	
$Pt(NH_3)_2(CH_2=CH_2)(OH_2)^{2+}$	*trans* 3.33	
$Pt(NH_3)_2(CH_3CH=CH_2)(OH_2)^{2+}$	*trans* 3.60	
$Pt(NH_3)_2(CH_3CMe=CH_2)(OH_2)^{2+}$	*trans* 3.64	[103]
$Pt(NH_3)_2(C_6H_5CH=CH_2)(OH_2)^{2+}$	*trans* 3.28	
$Pt(NH_3)_2(p\text{-}ClC_6H_4CH=CH_2)(OH_2)^{2+}$	*trans* 3.18	
$Pt(NH_3)_2(p\text{-}MeOC_6H_4CH=CH_2)(OH_2)^{2+}$	*trans* 3.42	
$Pt(NH_3)_2Cl(OH_2)^+$	6 [2] [a]	
$Pt(ala)(NH_3)(OH_2)^+$	*cis* 6.3	
$Pt(ala)(py)(OH_2)^+$	*cis* 5.9	[105]
$Pt(ala)(Et_2S)(OH_2)^+$	*cis* 5.7	
$Pt(ala)(DMSO)(OH_2)^+$	*cis* 4.4	
$PtCl_3(OH_2)^-$	7 [33]	
$PdCl_3(OH_2)^-$	7 [33]	
$AuCl_3(OH_2)$	0.6 to 0.8 [33,106]	
$ML_2(OH_2)_2^{n+}$		
$Pt(NH_3)_2(OH_2)_2^{2+}$	*cis* 5.5 to 5.6 [2,107]	*trans* 4.2 to 4.5 [2,107]
$Pt(MeNH_2)_2(OH_2)_2^{2+}$	*cis* 5.6 [107]	*trans* 4.5 [107]
$Pt(EtNH_2)_2(OH_2)_2^{2+}$	*cis* 5.5 [107]	*trans* 4.5 [107]
$Pt(NH_3)(py)(OH_2)_2^{2+}$	*cis* 5.2 [107]	*trans* 2.1 [107] or 4.1 [2]
$Pt(py)_2(OH_2)_2^{2+}$ [b]	*cis* 4.5 [107,108]	*trans* 3.8 to 3.9 [2,107,108]
$Pt(bipy)(OH_2)_2^{2+}$	*cis* 4.7 [2]	
$PdCl_2(OH_2)_2$	2.1 [2], 4.3 [33] [a]	

a stereochemistry unspecified; *b* other complexes $Pt(amine)_2(OH_2)_2^{2+}$ have similar pK values, for instance with amine = β-picoline, piperidine, hydroxylamine, or O-methylhydroxylamine [2, 107, 108].

not be significant (in, for example, aquo-porphine complexes [87, 88]).

Number of coordinated water molecules. The $p*K_1$ of coordinated water in the series $Co(NH_3)_x(OH_2)_{6-x}^{3+}$ increases as x increases, in other words the acidity increases with the number of water molecules. A similar trend is apparent for chromium(III), for example for the series of complexes $Cr(ox)_x(OH_2)_{6-2x}^{(3-2x)+}$ (Table 9.11).

Sterochemistry (octahedral). There is no apparent pattern in the relative importance of *cis* and *trans* effects of ligands.

Square-planar complexes. Here ligand effects on the acidity of coordinated water are considered in terms of both the well-established *trans* influence and the more recently documented *cis* influence. When $p*K_1$ values for *cis* and *trans* compounds in the $Pt(amine)_2(OH_2)_2^{2+}$ series are compared with the parent compounds $Pt(NH_3)_2(OH_2)_2^{2+}$, a bigger variation in $p*K_1$ values is revealed for *cis* compounds than for their *trans* analogues [103]. Similarly, coordinated ethylene enhances the acidity of *trans* much more than of *cis* water ligands

Table 9.14 Acidity constants for aquo-alkyl-[109], aquo-halogeno-[110], and related [110] mercury(II) cations, and for aquo-alkyl-tin(IV) cations [2], at 25°C.

	$p*K_1$		$p*K_1$
$MeHg^+aq$	3.59 to 4.78[a]	$HgOH^+aq$	2.6
$EtHg^+aq$	4.9	$HgCl^+aq$	3.1
$PrHg^+aq$	5.12	$HgSCN^+aq$	3.4
$BuHg^+aq$	5.17	$HgBr^+aq$	3.5
		HgI^+aq	4.0
$PhHg^+aq$	4.1 to 4.9		
CF_3Hg^+aq	3.24		
$C_2F_5Hg^+aq$	3.42	cf. $Hg^{2+}aq$	3.6 to 5.1[b]
$C_3F_7Hg^+aq$	3.50		
$Me_2Sn^{2+}aq$	3.11 to 3.55[a]		
$Et_2Sn^{2+}aq$	2.4 to 3.5[c]		
Me_3Sn^+aq	6.16 to 6.60[c]		
Et_3Sn^+aq	6.81		
Ph_3Sn^+aq	4.8[d]		

a Range of six independent determinations using various experimental techniques; *b* Table 9.3, *c* range of three estimates; *d* at 30°C.

[112]. The cis-Pt(ala)L(OH$_2$)$^+$ series in Table 9.13 demonstrates how π-acceptor properties of coordinated ligands produce a significant increase in the acidity of coordinated water molecules [105].

9.2.4 Enthalpies, Entropies, and Volumes

Enthalpies corresponding to the ionisation of a proton from an aquo-cation can be measured directly by calorimetry, and can also be derived from the temperature variation of the acidity constant:

$$\frac{\partial \ln K_1}{\partial(1/T)} = -\frac{\Delta H_1}{R}.$$

Once the enthalpy for the process has been obtained, the entropy change can be calculated from this and the Gibbs free energy (acidity constant). The volume change for the ionisation of a proton from an aquo-cation can be estimated from the variation of the acidity constant with pressure:

$$\frac{\partial \ln K_1}{\partial P} = -\frac{\Delta V}{RT}.$$

In view of the uncertainties in acidity constants apparent from the preceding Section, quantities derived from their temperature or pressure variation must obviously be approached in a critical manner. Here, as in other circumstances, direct calorimetry is preferable to temperature variation of equilibrium constant

Table 9.15 ΔH_1^* values (kcal mol^{-1}) for the dissociation of a proton from aqu… otherwise indicated [2, 113].

Li$^+$ 16.4[a]		Ca^{2+} 14.6 to 15.4		Zn^{2+} 13.2[a]	
		Sr^{2+} 14.6		Cd^{2+} 13.1[b]	
		Ba^{2+} 14.6 to 15.2		Hg^{2+} 7.2[b] [23,114	

| | V^{2+} 15.5 | | | | Mn^{2+} 13.4 |
| Ti^{3+} ~14[a] | V^{3+} 9.7 or 10 | Cr^{3+} 9.3 [119], 9.4, or 12.4[b] | | | Mn^{3+} 4 |

Ce^{4+} 15.5

| Th^{4+} 5.9 | U^{4+} 10.7 or 11.7 | Pu^{4+} 7.3 |

a ΔH_1^* values computed by the present author from the published temperature dependence p*K_1 values; b direct calorimetric determination; c see also Ref. [116]; d values from the te… perature dependence of p*K_1 range from 8.2 to 12.3 kcal mol^{-1}, direct calorimetric determi…

methods for the estimation of enthalpies. However, in the case of enthalpies of ionisation of aquo-cations even directly calorimetry seems to be somewhat unreliable (see the range of values reported for $Fe^{3+}aq$ below).

Just as there are different ways of presenting acidity constants for aquo-cations, for example as pK, $p*K$, or pK_h (see Chapter 9.1.2), so there is also a choice of ways of reporting enthalpies, entropies, and volumes for loss of a proton from an aquo-cation. Either

$$\Delta H_1 \text{ for } M^{n+}aq + OH^- = MOH^{(n-1)+}$$

or $$\Delta H_1^* \text{ for } M^{n+}aq + H_2O = MOH^{(n-1)+} + H_3O^+$$

These two quantities are interrelated by the enthalpy change for the reaction

$$H_3O^+ + OH^- = 2H_2O \,,$$

which is $-13.41 \pm 0.10 \text{ kcal mol}^{-1}$ at zero ionic strength.[†] This mean value is derived from eleven determinations [2]. It becomes slightly more negative as the ionic strength increases, its numerical value probably increasing by about 0.1 or 0.2 kcal mol^{-1} on going from zero ionic strength to 3M, depending on the nature of the salt used to provide this ionic strength. We have used the value of -13.41 kcal mol^{-1} wherever required to convert ΔH to ΔH^* values, regardless of ionic strength. ΔS_n and ΔS_n^* are similarly interconvertible via ΔS for the

[†]The 'best' 1977 value is quoted as -13.34 kcal mol^{-1}, at 25°C (Olofsson, G. and Olofsson, I. *J. Chem. Thermodynamics*, **9**, 65 (1977)).

ations. The values are derived from the temperature variation of $p*K_1$ values unless

$^{3+}$	19^a			In^{3+}	4.9^b				
				Tl^{3+}	1.7 to 8.3^a				
				Tl^+	13.8		Sn^{2+}	3.4	Bi^{3+}	17^b [115]
e^{2+}	12.0^a			Co^{2+}	7.7^a, 8.2^c	Ni^{2+}	7.8 or 12.4	Cu^{2+}	11 to 12 [117]
e^{3+}	8.2 to $19.7^{d,e}$			Co^{3+}	10				
				Rh^{3+}	(4.3)				

ions have given values between 11.0 [114] and 19.7 kcal mol^{-1}; e values of ΔH_1^* over a range of mperatures can be estimated from the reported variation of $p*K_1$ with temperature over the tensive range of 50 to 350°C [118].

$H_3O^+ + OH^-$ reaction, which is $+19.3$ cal deg^{-1} mol^{-1} [2].

Enthalpies and entropies. Many of the available results for enthalpies and entropies corresponding to acidity constants of aquo-cations are listed in Tables 9.15 and 9.16 respectively. They are given in the form of ΔH_1^* and ΔS_1^*. Where the original results are in the form of ΔH_1 and ΔS_1, these have been converted as stated in the previous paragraph. The ΔH_1 values have been rounded off to one decimal place, which probably still gives an optimistic impression of the precision in some cases. A few sets of pK values over a range of temperatures exist from which the original authors did not derive enthalpies and entropies. The present author has computed enthalpies and entropies for those series which give a tolerably linear plot of pK against reciprocal temperature. This computation used a standard unweighted least-mean-squares program, and such results are indicated in Table 9.15.

Table 9.16 ΔS_1^* values (cal deg^{-1} mol^{-1} at 25°C) for the dissociation of a proton from aquo-cations [2, 113].

Ca^{2+} −7 to −9	Cd^{2+} −3	In^{3+} −4
Sr^{2+} −12	Hg^{2+} +8 [23]	
Ba^{2+} −10 to −12		

Tl^+ −14	Sn^{2+} +4	Bi^{3+} +27 [115]

Ni^{2+} +3

V^{3+} +26 Cr^{3+} +12[119], +14,or +24 Mn^{3+} +16 Fe^{3+} +21 to +50^a Co^{3+} +25

Ce^{4+} +55

Th^{4+} +1	U^{4+} +33 or +36	Pu^{4+} +19

a The most attractive estimates are in the range +21 to +31. Direct calorimetric determination of ΔH_1^* led to a value of +23 cal deg^{-1} mol^{-1} for ΔS_1^* [114].

There seems to be no obvious pattern to the ΔH_1^* values of Table 9.15, not even a grouping of values by cation charge.[†] There is some pattern in the ΔS_1^* values of Table 9.16. At least the large negative ΔS_1^* value for Tl^+aq can be contrasted with the small negative or positive values for M^{2+}aq and the larger positive values for M^{3+}aq and M^{4+}aq. This grouping by charge type presumably arises from the importance of hydration changes in the reaction

$$M^{n+}aq \longrightarrow MOH^{(n-1)+}aq + H^+aq,$$

and the ratio of $(n-1)$ to n.

† Wells [62] has discussed the variation of ΔH_1^* along the series of 3+ transition metal cations (first row).

Enthalpies and entropies for aquo-oxo-cations are almost non-existent. There is a choice of values for UO_2^{2+}aq, 11 or 20.8 kcal mol^{-1}, which is a most unsatisfactory situation. As shown in Table 9.17, there is a little information on mixed aquo-ligand complexes.

Table 9.17 Enthalpies and entropies for the dissociation of a proton from aquo-ligand complexes.

	ΔH_1^*/kcal mol^{-1}	ΔS_1^*/cal deg^{-1} mol^{-1}	
$Co(NH_3)_5(OH_2)^{3+}$	9.0	+3	[69]
$Rh(NH_3)_5(OH_2)^{3+}$	9.4^a	+4^a	[69]
$Cr(NH_3)_5(OH_2)^{3+}$	8.3	+6	[69]
	10.5^b		
$RhCl(OH_2)_5^{2+}$	8.9^c		

a Earlier estimates were $\Delta H_1^* = 6$ kcal mol^{-1} and $\Delta S_1^* = -9$ cal deg^{-1} mol^{-1} [97]; *b* computed from data in [90]; *c* computed from data in [100].

ΔH_2^* and ΔS_2^* values, corresponding to the loss of a second proton from an aquo-cation, have been reported for Hg^{2+}aq, Cr^{3+}aq, and U^{4+}aq [2]. In view of the uncertainities expressed earlier in this Chapter about p*K_2 values and about the derivation of enthalpies and entropies from acid dissociation constants, such ΔH_2^* and ΔS_2^* values should be treated with caution.

Another facet of this area of chemistry which should be treated with at least a degree of scepticism concerns heat capacities of proton loss from aquo-cations. Such quantities are the second derivatives of acid dissociation constants with respect to temperature. Values of +31, +74, and +224 cal deg^{-1} mol^{-1} have been claimed for Sr^{2+}aq, Ba^{2+}aq, and Th^{4+}aq respectively [2].

Volumes. Volume changes corresponding to the dissociation of a proton from a coordinated water molecule are estimated from the variation of the acid dissociation constant with pressure. The usual technique employed is spectrophotometry, and pressures up to nearly 1800 bar have been used. Currently available values of ΔV_1^* are listed in Table 9.18. The main point for comment is the smallness, and indeed the sign, of the experimental results. One can estimate expected values of ΔV_1^* from the Drude-Nernst equation:

$$\Delta V = \frac{-z^2 e^2}{2Dr} \left[\frac{\partial \ln D}{\partial P} \right].$$

This forecasts values for ΔV_1^* of about +10 cm^3 mol^{-1} from solvent electrostriction changes accompanying the proton loss from the aquo-cation. The difference between expected and observed ΔV_1^* values can be rationalised by treating the hydroxo-product as an ion-pair M^{n+}, OH^- rather than a true hydroxo-complex $MOH^{(n-1)+}$, but this is hardly a satisfactory situation. It may well be that a better picture emerges when many more ΔV_1^* values are known.

Table 9.18 ΔV_1^* **values for the dissociation of a proton from coordinated water molecules.**

	$\Delta V_1^*/cm^3\,mol^{-1}$	
$Tl^{3+}aq$	-3.2	[120]
$Cr^{3+}aq$	-3.8	[119]
$Fe^{3+}aq$	-1.2	[120]
$Co(NH_3)_5(OH_2)^{3+}$	-2.3	[120]

9.3 D₂O

Although it has sometimes been claimed, for instance for $Tl^{3+}aq$ [2] and for $Fe^{3+}aq$ [121], that $p^*K_1(H_2O) = p^*K_1(D_2O)$, investigators generally find a small but significant difference between $pK(M^{n+}, H_2O)$ and $pK(M^{n+}, D_2O)$. In all cases the acidity of the solvated metal cation is slightly less in D_2O than in H_2O (Table 9.19).

Values of ΔH_1^* and ΔS_1^* have been reported for the Fe^{3+} cation in D_2O. These are 9.3 ± 0.6 kcal mol^{-1} and $+18 \pm 2$ cal deg^{-1} mol^{-1} respectively [123]. Because of considerable doubt in the analogous values for the $Fe^{3+}aq$ cation in water (see Tables 9.15 and 9.16) comparison of D_2O against H_2O values is worthless.

Table 9.19 Acidity constants for aquo-cations and aquo-ligand complexes in H_2O and in D_2O, at 25°C.

	$p^*K_1(H_2O)$	$p^*K_1(D_2O)$	
Cu^{2+}	7.22	7.71	[122]
Fe^{3+}	2.54	2.85	[123]
La^{3+}	10.04	10.35	[124]
Gd^{3+}	8.20	8.34	[124]
U^{4+}	1.68	1.74	[2]
Np^{4+}	2.3	2.5	[2]
Pu^{4+}	1.73	1.94	[2]
$Cr(NH_3)_5(OH_2)^{3+a}$	5.10	5.26^b or 5.59^c	[70]
$Co(NH_3)_5(OH_2)^{3+a}$	6.18	6.35^b or 6.67^c	[70]
	6.22	6.70	[67]

a p^*K_1 in D_2O refers to $M(NH_3)_5(OD_2)^{3+}$ cations; b uncorrected values (see c); c corrected via acetic acid/acetate buffer measurements in H_2O and D_2O.

Table 9.20 Acidity constants for metal cations in binary aqueous solvent mixtures, mole fraction x_2 organic component, at $25°C$.

Methanol

Fe^{3+}

x_2	0	0.04	0.08	0.12	0.20	0.25	0.31		
$p*K_1$	2.66	2.62	2.52	2.48	2.21	2.06	2.03		[125]
x_2	0	0.15	0.20	0.25	0.395	0.443	0.519		
$p*K_1$	2.78	2.32	2.21	2.06	1.86	1.81	1.72		[126]
x_2	0	0.52	0.69						
$p*K_1$	~2.1	1.77	1.44						[2]

Ethanol[a]

	x_2	0	0.09	0.21		
Sc^{3+}	$p*K_1$	4.55	4.13	3.72		
La^{3+}	$p*K_1$	9.06	8.70	8.17		[19]
Th^{4+}	$p*K_1$	3.15	2.77	2.17		

Dioxan

Na^+

x_2	0	0.14	0.32	
$p*K_1$	~14.5	13.5	11.9	[2]

Cu^{2+}

x_2	0	0.20	
$p*K_1$	8.0	7.6	[2]

Tl^{3+}

x_2	0	0.32	
$p*K_1$	0.8	0.5	[127]

Acetone

In^{3+}

x_2	0	0.046	0.096	0.201	0.360	
$p*K_1$	4.26	4.22	4.05	3.84	3.64	[127]

Dimethyl sulphoxide[c]

In^{3+}

x_2[b]	0	0.002	0.005	0.010	0.016	0.021	0.033	0.046	
$p*K_1$	4.22	4.35	4.55	4.82	5.19	5.89	6.10	6.70	[128]

Tl^{3+}

x_2	0	0.011	0.023	0.049	
$p*K_1$	1.19	1.47	1.88	2.72	[127]

Acetonitrile

In^{3+}

x_2	0	0.007	0.014	0.032	0.064	0.130	0.195	0.346	
$p*K_1$[d]	4.2	3.8	3.6	3.4	3.2	3.0	2.7	2.7	[129]

Tl^{3+}

x_2	0	0.032	0.080	0.195	0.346	
$p*K_1$	1.19	1.10	0.96	0.76	0.55	[127]

a Values for Y^{3+}, Eu^{3+}, Gd^{3+}, Yb^{3+}, and Lu^{3+} at the same mole fractions of ethanol are also given in [19]; *b* approximate mole fractions estimated from the molarities given in [128]; *c* there may be some solvation by the dimethyl sulphoxide as well as by water here; *d* these $p*K_1$ values are means of determinations made by pH-metric and e.m.f. experiments.

9.4   MIXED AQUEOUS SYSTEMS

Acidity constants have been reported for several M^{n+} aq species in a variety of binary aqueous mixtures. Generally the results refer to water-rich solvent mixtures in which the primary coordination sphere of the metal cation will probably contain only water molecules (see Chapter 6). If the primary solvation shell contains molecules of both solvent components, then obviously here one is introducing another variable into the system and making discussion more difficult. If the pK estimates are made electrochemically, the question of junction potentials must always be borne in mind. If the pK estimates are made spectrophotometrically, then one has to worry about the possibility of frequencies of maximum absorption and molar extinction coefficients varying with solvent composition.

Some $p*K_1$ values for metal cations in binary aqueous mixtures are quoted in Table 9.20. It is interesting that in nearly all cases $p*K_1$ values decrease as the proportion of non-aqueous component increases. The only certain exceptions are In^{3+} and Tl^{3+} in water + dimethyl sulphoxide. It may be significant that these are the only TNAN (Chapter 1.1.2. p. 16) solvent mixtures represented in Table 9.20, but see footnote c there. Values of $p*K_1$ for the $Cr(NH_3)_5(OH_2)^{3+}$ cation and its cobalt(III) analogue are higher in aqueous dioxan than in water [70], and this trend runs counter to that exhibited by simple aquo-cations. It would be unwise to attempt any generalisations yet, in view of these apparent oddities and of the small number of results available at present. Nonetheless, the trend of $p*K_1$ values for Fe^{3+} aq in water + methanol mixtures conforms to the prediction of pK dependence of a positively charged acid in such media [130].

There is one report of a ΔH_1^* determination, of 11.6 kcal mol^{-1} for Fe^{3+} in aqueous methanol, mole fraction 0.2 methanol [125]. Because of the great range of values claimed for ΔH_1^* for Fe^{3+} aq in water (Table 9.15), useful comparison is ruled out.

9.5   NON-AQUEOUS SOLVATES

As so often, the amount of information available on cation solvates in non-aqueous solution is very much less than that available for aquo-cations in water. Indeed data on non-aqueous solvento-cations in non-aqueous media are very nearly non-existent. Somewhat more common are data on non-aqueous solvento-cations (for example ammono-cations) in aqueous solution. This area (Chapter 9.5.1) provides a convenient link between earlier Sections of this Chapter and the purely non-aqueous systems which conclude it (Chapter 9.5.2).

9.5.1   In Aqueous Solution

Ammonia is easily the most common solvent ligand encountered here. Table 9.21 gives pK values for coordinated ammonia, measured in aqueous solution, for hexa- and tetra-ammine complexes, while Table 9.22 lists those

Table 9.21 p*K_1 values for coordinated ammonia in ammines in aqueous solution at 25°C.

Cation	p*K_1	
$Au(NH_3)_4^{3+}$	7.5	[131]
$Co(NH_3)_6^{3+}$	14	[102]
$Rh(NH_3)_6^{3+}$	14	[102]
$Ru(NH_3)_6^{3+}$	12.4[a]	[132]
$Pt(NH_3)_6^{4+}$	7.0 to 8.8[b,c]	[2, 71, 102, 143]

a This value may be somewhat in error, as discussed in Ref. [133]; b values are sensitive to the nature of the gegenion present, presumably since a 4+ cation is prone to form ion-pairs in solution [135]; c p*K_2 = 10.5 at zero ionic strength [134].

Table 9.22 p*K_1 values for coordinated ammonia in mixed ligand ammine complexes in aqueous solution, at 25°C.

Cation	p*K_1
Platinum(IV)	
$Pt(NH_3)_5(OH)^{3+}$	9.5 [2]
$Pt(NH_3)_5Cl^{3+}$	8.1 to 8.9[a] [2, 71, 136]
$Pt(NH_3)_5Br^{3+}$	8.3 [2]
$Pt(NH_3)_3(en)Cl^{3+}$	7.7, 8.9[b] [137]
cis-$Pt(NH_3)_4Cl_2^{2+c}$	9.4 to 9.7 [2, 102, 136]
$trans$-$Pt(NH_3)_4Cl_2^{2+}$	11.2 or 11.3 [2, 102]
mer-$Pt(NH_3)_3Cl_3^+$	11.1 [2]
$Pt(NH_3)_2(py)_2Cl_2^{2+}$	9.4, 10.0[b] [107]
$Pt(NH_3)_2(CN)_4$	12.1[d] [2]
$Pt(NH_3)_2(CN)_3Cl$	12.7[d] [2]
$Pt(NH_3)_2(CN)_3Br$	12.8[d] [2]
$Pt(NH_3)_2(CN)_3I$	13.0[d] [2]
Platinum(II)	
$Pt(NH_3)_3(C_2H_4)^{2+}$	8.56 [103]
$Pt(NH_3)_3(DMSO)^{2+}$	9.06 [103]

a p*K_2 ≥ 10.3 [136]; b these values refer to two isomers whose stereochemistry is not specified in the original reference; c p*K_2 = 10.3 or 12.4 [136]; d 20°C.

for mixed ligand ammine complexes. The value of ΔH_1^* for $Pt(NH_3)_6^{4+}$ has been estimated as $20.7\ kcal\ mol^{-1}$ [2]. Values of $p*K_2$ have been reported for $Pt(NH_3)_6^{4+}$, $Pt(NH_3)_5Cl^{3+}$, and cis-$Pt(NH_3)_4Cl_2^{2+}$ [2]. As for coordinated water, the $p*K_1$ for coordinated ammonia in $Pt(NH_3)_6^{4+}$ is higher in D_2O than in H_2O [2].

Acidity constants for some coordinated alcohols are compared with those for coordinated water in analogous complexes in Table 9.23; differences are small. The acidity of coordinated hydrogen sulphide is very different from that of water. The $p*K_1$ of $Ru(NH_3)_5(SH_2)^{2+}$ is 4.0, whereas that of $Ru(NH_3)_5$-$(OH_2)^{2+}$ is 13.1. The $p*K_1$ of $Ru(NH_3)_5(SH_2)^{3+}$ is about -10. Under all conditions so far examined the ligand is thus SH^- rather than SH_2 [101].

Table 9.23 Comparison of $p*K_1$ values for coordinated alcohols with those for coordinated water (latter from Table 9.9).

	$p*K_1$		$p*K_1$
$Co(NH_3)_5(MeOH)^{3+}$	5.6 [138]	$Co(NH_3)_5(OH_2)^{3+}$	5.7 to 6.6
$Co(en)_2(HOCH_2CH_2OH)^{3+}$	2.6, 5.8 [111]	cis-$Co(en)_2(OH_2)_2^{3+}$	6.0 to 6.1
$Co(en)_2(H_2NCH_2CH_2OH)^{3+}$	3.5 to 3.6 [111]	cis-$Co(en)_2(NH_3)(OH_2)^{3+}$	5.8 to 6.3

9.5.2 In Non-Aqueous Solution

The author has been unsuccessful in tracking down any $p*K$ values for non-aqueous solvento-cations in non-aqueous solvents that are analogous to the $p*K$ values discussed above for aquo-cations in aqueous solution.

FURTHER READING

Baes, C. F. and Mesmer, R. E., *The Hydrolysis of Cations,* Wiley (1976).

REFERENCES

[1] *Coordination Chemistry in Solution,* ed. Högfeldt, E., Swedish Natural Science Research Council, Lund (1972); see especially Martell, A. E., p.3.

[2] Sillén, L. G. and Martell, A. E., *Stability Constants of Metal-ion Complexes, Chemical Society Special Publications 17 and 25 (Supplement 1),* The Chemical Society, London (1964, 1971).

[3] Rossotti, F. J. C. and Rossotti, H. S., *The Determination of Stability Constants,* McGraw-Hill (1961).

[4] Cotton, F. A. and Wilkinson, G., *Advanced Inorganic Chemistry, 3rd edn.* Wiley/Interscience (1972) p. 266.

[5] Danesi, P. R., Chiarizia, R., Scibona, G., and Riccardi, R., *Inorg. Chem.,* **12**, 2089 (1973).

[6] Von Meyenburg, U., Široký, O., and Schwarzenbach, G., *Helv. Chim. Acta,* **56**, 1099 (1973).

[7] Sillén, L. G., *Acta Chem. Scand.,* **16**, 159 (1962); Ingri, N. and Sillén, L. G., *ibid.,* **16**, 173 (1962); Sillén, L. G., *ibid.,* **18**, 1085 (1964).

[8] Sillén, L. G. and Warnqvist, B., *Ark. Kemi.,* **31**, 315, 341, 377 (1969); *Acta Chem. Scand.,* **22**, 3032 (1968); Brauner, P., Sillén, L. G., and Whiteker, R., *Ark. Kemi,* **31**, 365 (1969).

[9] Arnek, R., Sillén, L. G., and Wahlberg, O., *Ark. Kemi,* **31**, 353 (1969).

[10] Ingri, N. and Sillén, L. G., *Ark. Kemi,* **23**, 97 (1965).

[11] Ekelund, R., Sillén, L. G., and Wahlberg, O., *Acta Chem. Scand.,* **24**, 3073 (1970).

[12] Ingri, N.,Kakořowicz, W., Sillén, L. G., and Warnqvist, B., *Talanta,* **14**, 1261 (1967).

[13] Bjerrum, J., p. 71 of reference [1].

[14] Biryuk, E. A. and Nazarenko, V. A., *Russ. J. Inorg. Chem.,* **18**, 1576 (1973).

[15] Perrin, D. D., *J. Chem. Soc.,* 3644 (1964).

[16] Perrin, D. D., *J. Chem. Soc.,* 2197 (1962).

[17] Perrin, D. D., *J. Chem. Soc.,* 4500 (1962).

[18] Komissarova, L. N., Prutkova, N. M., and Pushkina, G. Ya., *Russ. J. Inorg. Chem.,* **16**, 954 (1971).

[19] Usherenko, L. N. and Skorik, N. A., *Russ. J. Inorg. Chem.,* **17**, 1533 (1972).

[20] Amaya, T., Kakihana, H., and Maeda, M., *Bull. Chem. Soc. Japan,* **46**, 1720 (1973).

[21] Fujita, S. -I., Horii, H., Mori, T., and Taniguchi, S., *J. Phys. Chem.,* **79**, 960 (1975).

[22] Matsui, H. and Ohtaki, H., *Bull. Chem. Soc. Japan,* **47**, 2603 (1974).

[23] Ciavatta, L., Grimaldi, M., and Palombari, R., *J. Inorg. Nucl. Chem.,* **37**, 1685 (1975); Hepler, L. G. and Olofsson, G., *Chem. Rev.,* **75**, 585 (1975·).

[24] O'Neill, P. and Schulte-Frohlinde, B., *J. Chem. Soc. Chem. Commun.,* 387 (1975).

[25] Pavlović, Z. and Popović, R., *Rev. Roumaine Chim.,* **15**, 313 (1970).

[26] Kul'ba, F. Ya., Yakovlev, Yu. B., and Kopylov, E. A., *Russ. J. Phys. Chem.,* **45**, 408 (1971).

[27] Moeller, T. and O'Conner, R., *Ions in Aqueous Systems,* McGraw-Hill (1972).

[28] Nazarenko, V. A., Antonovich, V. P., and Nevskaya, E. M., *Russ. J. Inorg. Chem.,* **16**, 980 (1971).

[29] Wells, C. F., and Salam, M. A., *J. Chem. Soc. A,* 1568 (1968).

[30] Thamburaj, P. K. and Gould, E. S., *Inorg. Chem.,* **14**, 15 (1975).

[31] Wells, C. F. and Salam, M. A., *J. Inorg. Nucl. Chem.,* **31**, 1083 (1969).

[32] Kawai, T., Otsuka, H., and Ohtaki, H., *Bull. Chem. Soc. Japan,* **46**, 3753

(1973).

[33] Bekker, P. van Z. and Robb, W., *J. Inorg. Nucl. Chem.*, **37**, 829 (1975).

[34] Hay, R. W., Clark, C. R., and Edmonds, J. A. G., *J. Chem. Soc. Dalton*, **9** (1974)

[35] Honig, D. S., Kustin, K., and Martin, J. F., *Inorg. Chem.*, **11**, 1895 (1972).

[36] Pecsok, R. L. and Fletcher, A. N., *Inorg. Chem.*, **1**, 155 (1962); Birk, J. P. and Logan, T. P., *ibid.*, **12**, 580 (1973).

[37] Paris, M. R. and Gregoire, C. L., *Analytica Chim. Acta*, **42**, 439 (1968).

[38] Krentzien, H. and Brio, F., *Ion(Madrid)*, **30**, 14 (1970); *Chem. Abstr.*, **73**, 7865b (1970); Martin, A. H. and Gould, E. S., *Inorg. Chem.*, **14**, 873 (1975).

[39] Birk, J. P., *Inorg. Chem.*, **14**, 1724 (1975).

[40] Orhanović, M. and Earley, J. E., *Inorg. Chem.*, **14**, 1478 (1975).

[41] Noren, B., *Acta Chem. Scand.*, **27**, 1369 (1973).

[42] Nazarenko, V. A., Antonovich, V. P., and Nevskaya, E. M., *Russ. J. Inorg. Chem.*, **16**, 1273 (1971).

[43] Schenk, C., Stieger, H., and Kelm, H., *Z. Anorg. Allg. Chem.*, **391**, 1 (1972).

[44] Wells, C. F. and Davies, G., *J. Chem. Soc. A*, 1858 (1967).

[45] Rosseinsky, D. R., Nicol, M. J., Kite, K., and Hill, R. J., *J. Chem. Soc. Faraday I*, **70**, 2232 (1974).

[46] Goncharik, V. P., Tikhonova, L. P., and Yatsimirskii, K. B., *Russ. J. Inorg. Chem.*, **18**, 658 (1973).

[47] Sylva, R. N., *Rev. Pure Appl. Chem.*, **22**, 115 (1972).

[48] Knudsen, J. M., Larsen, E., Moreira, J. E., and Nielsen, O. F., *Acta Chem. Scand.*, **29A**, 833 (1975); Koren, R. and Perlmutter-Hayman, B., *Inorg. Chem.*, **11**, 3055 (1972).

[49] Ciavatta, L. and Grimaldi, M., *J. Inorg. Nucl. Chem.*, **37**, 163 (1975).

[50] Kolosov, I. V., Intskirveli, L. N., and Varshal, G. M., *Russ. J. Inorg. Chem.*, **20**, 1179 (1975).

[51] Frolova, U. K., Kumok, V. N., and Serebrennikov, V. V., *Izvest. Vyssh. Ucheb. Zaved. SSSR*, **9**, 176 (1966).

[52] Guillaumont, R., Désiré, B., and Galin, M., *Radiochem. Radioanalyt. Lett.*, **8**, 189 (1971).

[53] Kurkov, K. A., Lilich, L. S., Nguen Din Ngo, and Smirnov, A. Yu., *Russ. J. Inorg. Chem.*, **18**, 797 (1973).

[54] Nguen Din Ngo and Burkov, K. A., *Russ. J. Inorg. Chem.*, **19**, 680 (1974).

[55] Adamson, M. G., Dainton, F. S., and Glentworth, P., *Trans. Faraday Soc.*, **61**, 689 (1965); Offner, H. G. and Skoog, D. A., *Analyt. Chem.*, **38**, 1520 (1966); Amjad, Z. and McAuley, A., *J. Chem. Soc. Dalton*, 2521 (1974).

[56] Faraggi, M. and Feder, A., *J. Chem. Phys.*, **56**, 3294 (1972).

[57] Glevob, V. A., Klygin, A. E., Smirnova, I. D., and Kolyada, N. S., *Russ. J. Inorg. Chem.*, **17**, 1740 (1972).

[58] Perrin, D. D., *Dissociation Constants of Inorganic Acids and Bases in Aqueous Solution,* Butterworths (1969).

[59] Gutmann, V., *Allg. Prakt. Chem.,* **21**, 116 (1970); *Fortschr. Chem. Forsch.,* **27**, 59 (1972).

[60] Latysheva, V. A., *Russ. J. Phys. Chem.,* **48**, 773 (1974).

[61] McGowan, J. G., *Nature,* **168**, 601 (1951); *Rec. Trav. Chim.,* **85**, 777 (1966).

[62] Wells, C. F., *Nature,* **205**, 693 (1965).

[63] Hunt, J. P., *Metal Ions in Aqueous Solution,* Benjamin (1963) Ch. 4.

[64] Hemmes, P., Rich, L. D., Cole, D. L., and Eyring, E. M., *J. Phys. Chem.* **75**, 929 (1971).

[65] Swift, T. J. and Stephenson, T. A., *Inorg. Chem.,* **5**, 1100 (1966).

[66] Balahura, R. J. and Jordan, R. B., *Inorg. Chem.,* **9**, 2639 (1970).

[67] Splinter, R. C., Harris, S. J., and Tobias, R. S., *Inorg. Chem.,* **7**, 897 (1968).

[68] Yajima, F., Yamasaki, A., and Fujiwara, S., *Inorg. Chem.,* **10**, 2350 (1971).

[69] Cunningham, A. J., House, D. A., and Powell, H. K. J., *Aust. J. Chem.,* **23**, 2375 (1970).

[70] Chan, S. C. and Hui, K. Y., *Aust. J. Chem.,* **21**, 3061 (1968).

[71] Lagowski, J. J., *Modern Inorganic Chemistry,* Marcel Dekker (1973) p. 653.

[72] Cannon, R. D. and Earley, J. E., *J. Amer. Chem. Soc.,* **88**, 1872 (1966).

[73] Udovenko, V. V., Reiter, L. G., and Shkurman, E. P., *Russ. J. Inorg. Chem.,* **17**, 1008 (1972).

[74] Martin, D. F. and Tobe, M. L., *J. Chem. Soc. A,* 1388 (1962).

[75] Francis, D. J. and Jordan, R. B., *Inorg. Chem.,* **11**, 461 (1972).

[76] Schwarzenbach, G., Boesch, J., and Egli, H., *J. Inorg. Nucl. Chem.,* **33** 2141 (1971).

[77] Couldwell, M. C., House, D. A., and Powell, H. K. J., *Aust. J. Chem.,* **26**, 425 (1973).

[78] Searle, G. H. and Sargeson, A. M., *Inorg. Chem.,* **12**, 1014 (1973).

[79] Liteplo, M. P. and Endicott, J. F., *Inorg. Chem.,* **10**, 1420 (1971).

[80] Ablov, A. V., Bovykin, B. A., and Samus', N. M., *Russ. J. Inorg. Chem.,* **11**, 978 (1966).

[81] Bovykin, B. A., *Russ. J. Inorg. Chem.,* **17**, 89 (1972).

[82] Palade, D. M., Volokh, T. N., Breslavskaya, N. A., and Golyshina, T. N., *Russ. J. Inorg. Chem.,* **15**, 1401 (1970).

[83] Bovykin, B. A. and Ivanov, N. V., *Russ. J. Inorg. Chem.,* **16**, 370 (1971).

[84] Bologa, O. A. and Samus', N. M., *Russ. J. Inorg. Chem.,* **16**, 1730 (1971).

[85] Samus', N. M., Damaskina, O. I., and Ablov, A. V., *Russ. J. Inorg. Chem.,* **19**, 1210 (1974).

[86] Palade, D. M. and Volokh, T. N., *Russ. J. Inorg. Chem.,* **17**, 1016 (1972).

[87] Pasternack, R. F. and Cobb, M. A., *J. Inorg. Nucl. Chem.*, **35**, 4327 (1973).

[88] Ashley, K. R., Berggren, M., and Cheng, M., *J. Amer. Chem. Soc.*, **97**, 1422 (1975).

[89] Pasternack, R. F., Cobb, M. A. and Sutin, N., *Inorg. Chem.*, **14**, 866 (1975).

[90] Banerjea, D. and Sarkar, S., *Z. Anorg. Allg. Chem.*, **393**, 301 (1972).

[91] Ashley, K. R. and Kulprathipanja, S., *Inorg. Chem.*, **11**, 444 (1972).

[92] Bratushko, Yu. I. and Nazarenko, Yu. P., *Russ. J. Inorg. Chem.*, **17**, 241 (1972).

[93] Kelm, H. and Harris, G. M., *Z. Phys. Chem. Frankf. Ausg.*, **66**, 8 (1969).

[94] Wang, R. T. and Espenson, J. H., *J. Amer. Chem. Soc.*, **93**, 1629 (1971).

[95] Espenson, J. H. and Wolenuk, J. G., *Inorg. Chem.*, **11**, 2034 (1972); and references therein.

[96] Brønsted, J. N. and Volqvartz, K., *Z. Phys. Chem.*, **134**, 97 (1928).

[97] Pöe, A. J., Shaw, K., and Wendt, M. J., *Inorg. Chim. Acta*, **1**, 371 (1967).

[98] van Eldik, R., *Z. Anorg. Allg. Chem.*, **416**, 88 (1975).

[99] Ford, P. C. and Petersen, J. D., *Inorg. Chem.*, **14**, 1404 (1975).

[100] Pavelich, M. J. and Harris, G. M., *Inorg. Chem.*, **12**, 423 (1973).

[101] Kuehn, C. G. and Taube, H., *J. Amer. Chem. Soc.*, **98**, 689 (1976).

[102] Humphreys, D. and Staples, P. J., *J. Chem. Soc. Dalton*, 897 (1973).

[103] Gel'fman, M. I., Karpinskaya, N. M., and Razumovskii, V. V., *Russ. J. Inorg. Chem.*, **15**, 1438 (1970).

[104] Alcock, R. M., Hartley, F. R., and Rogers, D. E., *J. Chem. Soc. Dalton*, 1070 (1973).

[105] Kukushkin, Yu, N. and Gur'yanova, G. P., *Russ. J. Inorg. Chem.*, **16**, 580 (1971).

[106] Peshchevitskii, B. I., Belevantsev, V. I., and Kurbatova, N. V., *Russ. J. Inorg. Chem.*, **16**, 1007 (1971); Bekker, P. van Z. and Robb, W., *Inorg. Nucl. Chem. Lett.*, **8**, 849 (1972).

[107] Stetsenko, A. I., *Russ. J. Inorg. Chem.*, **15**, 1608 (1970).

[108] Grinberg, A. A., Stetsenko, A. I., Mitkinova, N. D., and Tikhonova, L. S., *Russ. J. Inorg. Chem.*, **16**, 137 (1971).

[109] Ingman, F. and Liem, D. H., *Acta Chem. Scand.*, **28A**, 947 (1974); Rabenstein, D. L., Evans, C. A., Tourangeau, M. C., and Fairhurst, M. T., *Analyt. Chem.*, **47**, 338 (1975).

[110] Ahlberg, I. and Leden, I., p. 17 of reference [1]; Ciavatta, L., Grimaldi, M., and Palombari, R., *J. Inorg. Nucl. Chem.*, **38**, 823 (1976).

[111] E.g. Bennett, L. E., Lane, R. H., Gilroy, M., Sedor, F. A., and Bennett, J. P., *Inorg. Chem.*, **12**, 1200 (1973).

[112] Gel'fman, M. I., Karpinskaya, N. M., and Razumovskii, V. V., *Russ. J. Inorg. Chem.*, **19**, 1533 (1974).

[113] Christensen, J. J. and Izatt, R. M., *Handbook of Metal Ligand Heats,*

Marcel Dekker (1970).

[114] Arnek, R., *Ark. Kemi,* **32**, 55 (1970).

[115] Olin, Å., *Acta Chem. Scand.,* **29A**, 907 (1975).

[116] Salam, M. A. and Raza, M. A., *Chemy. Ind.,* 601 (1971).

[117] Arnek, R. and Patel, C. C., *Acta Chem. Scand.,* **22**, 1097 (1968).

[118] Lewis, D., p. 59 of reference [1].

[119] Swaddle, T. W. and Pi-Chang Kong, *Can. J. Chem.,* **48**, 3223 (1970).

[120] Stranks, D. R., personal communication.

[121] Hudis, J. and Dodson, R. W., *J. Amer. Chem. Soc.,* **78**, 911 (1956).

[122] Kakihana, H., Amaya, T., and Maeda, M., *Bull. Chem. Soc. Japan,* **43**, 3155 (1970).

[123] Fukushima, S. and Reynolds, W. L., *Talanta,* **11**, 283 (1964); Knight, R. J., and Sylva, R. N., *J. Inorg. Nucl. Chem.,* **37**, 779 (1975).

[124] Amaya, T., Kakihana, H. and Maeda, M., *Bull. Chem. Soc. Japan,* **46**, 2889 (1973).

[125] Wada, G. and Endo, A., *Bull. Chem. Soc. Japan,* **45**, 1073 (1972).

[126] Wada, G. and Kobayashi, Y., *Bull. Chem. Soc. Japan,* **48**, 2451 (1975).

[127] Kul'ba, F. Ya., Kopylov, E. A., Yakovlev, Yu. B., and Kolerova, E. G., *Russ. J. Inorg. Chem.,* **17**, 1364 (1972); Kul'ba, F. Ya., Zenchenko, D. A., and Yakovlev, Yu. B., *ibid.,* **20**, 1314, 1464 (1975).

[128] Kul'ba, F. Ya., Yakovlev, Yu. B., and Zenchenko, D. A., *Russ. Inorg. Chem.,* **19**, 502 (1974).

[129] Kul'ba, F. Ya., Yakovlev, Yu. B., and Zenchenko, D. A., *Russ. J. Inorg. Chem.,* **20**, 994 (1975).

[130] Ohtaki, H., p. 185 of reference [1].

[131] Skibsted, L. H. and Bjerrum, J., *Acta Chem. Scand.,* **28A**, 740 (1974).

[132] Waysbort, D. and Navon, G., *Chem. Commun.,* 1410 (1971).

[133] Armor, J. N., *J. Inorg. Nucl. Chem.,* **35**, 2067 (1973); Evans, I. P., Everett, G. W. and Sargeson, A. M., *J. Chem. Soc. Chem. Commun.,* 139 (1975).

[134] Klein, B. and Heck, L., *Z. Anorg. Allg. Chem.,* **416**, 269 (1975).

[135] Grunwald, E. and Dodd-Wing Fong, *J. Amer. Chem. Soc.,* **94**, 7371 (1972).

[136] Johnson, R. C., Basolo, F., and Pearson, R. G., *J. Inorg. Nucl. Chem.,* **24**, 59 (1962).

[137] Gil'dengershel', Kh. I., Pechenyuk, S. I., Stetsenko, A. I., and Budanova, V. F., *Russ. J. Inorg. Chem.,* **16**, 1084 (1971).

[138] Jordan, R. B., Sargeson, A. M., and Taube, H., *Inorg. Chem.,* **5**, 1091 (1966).

Chapter **10**

POLYMERISATION

10.1 BACKGROUND

The loss of protons from aquo-cations to give hydroxo-complexes was considered in some detail in Chapter 9. Here we shall be considering primarily the polynuclear[†] species obtained, actually or formally, by condensation reactions between these hydroxo-complexes, for example,

$$2M(OH)^{2+} \longrightarrow M_2(OH)_2^{4+} .$$

The polynuclear products thus generated are held together by hydroxo- or oxo-bridges, e.g. $\left[M \begin{smallmatrix} \diagup OH \diagdown \\ \diagdown OH \diagup \end{smallmatrix} M \right]^{4+}$. It is generally difficult to tell whether such a species does indeed contain a double hydroxide bridge, or whether it contains a single oxo-bridge, $[M\text{--}O\text{--}M]^{4+}$. There are a few examples of polynuclear metal cations containing metal–metal bonds, and these will be considered separately.

Almost all cations of charge 3+ or higher give rise to polynuclear species in aqueous solution over an appropriate pH range. Here and elsewhere the appropriate pH range lies between low pHs, where the cation exists as the simple aquo-ion M^{n+}aq, and high pHs, where it precipitates as hydroxide or hydrated oxide. Many 2+ cations also give rise to polynuclear species. This is true not only of the particularly small Be^{2+}, which so often behaves similarly to a 3+ cation, but also of several larger M^{2+} cations. These include several transition metal(II) species, for example the much-studied Ni^{2+} cation. There is no fixed limit to the number of metal atoms per polymeric unit. Examples containing up to six metal atoms are common, while an Al_{13} unit has been proposed for aluminium with some confidence. Increasing the size of the polymeric unit eventually leads to the formation of insoluble hydroxides or hydrated oxides as the limiting form. The full sequence has been examined in detail for such cations as those of iron(III) and aluminium(III) (see below).

† Throughout this chapter polynuclear cations are designated according to the number of metal atoms they contain, $M_2(OH)_2^{4+}$ or M_2O^{4+} is thus referred to as a binuclear species.

A number of experimental techniques have been used to demonstrate the presence, and investigate the chemistry, of polynuclear cations in aqueous solution. Perhaps the most important are those of electrochemistry, particularly for monitoring hydrogen ion concentrations, X-ray diffraction, and Raman spectroscopy. Other techniques, including ultraviolet-visible spectroscopy, polarography, cyclic voltammetry, and exotica such as Tyndallometry, have also been of use from time to time. X-ray diffraction investigations have proved particularly useful in two respects. First, it is usually possible to determine the metal–metal distances with some precision from the radial distribution curves. Secondly, X-ray studies of appropriate solids, some of them actually obtained from hydrolysed and polymerised solutions, give a good idea of the structures that polynuclear cations may well adopt in solution. Another technique giving valuable information is ion-exchange. Here the determination of the charge per metal atom is possible, as the example of molybdenum shows (see below).

A fair amount of quantitative data are available for many polynuclear species [1]. These data are usually presented in the form of stability constants, $*\beta_{pq}$, for the general (hypothetical) reaction

$$q\,M^{n+} + p\,H_2O = M_q(OH)_p^{(nq-p)+} + p\,H^+ .$$

Sometimes, as in reference [2], the data are presented as equilibrium constants (K) for reactions of the type

$$n\,MOH^{m+} = M_n(OH)_n^{nm+} .$$

One needs to tread warily in this area of quantitative chemistry. Accurate and precise experimental results are required for the input to those computer programs which search for the most plausible species distribution to fit the facts provided. A small change in the input data can have a large effect on the species proposed by the computer as giving the best fit, or on the stability constants estimated for the various species present. Another difficulty is that the ideal inert supporting electrolyte does not exist; results will vary, sometimes negligibly but sometimes significantly. This occurs, for instance, when sodium perchlorate is replaced by lithium perchlorate. This is especially important at high ionic strengths, such as the recently much favoured 3M. The time taken to reach equilibrium is a further source of uncertainity, and this may be important in some systems. It is particularly relevant for such kinetically-inert cations as chromium(III), and also where solution-solid equilibria are involved. In the latter case the 'ageing' of precipitates may cause trouble, as for example for aluminium(III) or chromium(III) [3].

Several attempts to estimate the enthalpy and entropy changes associated with polynuclear complex formation have been made. As is always the case, those which depend on direct calorimetric measurement are more likely to merit some measure of confidence than those derived from the temperature dependence of stability constants. Here the difference in levels of confidence is perhaps more important than usual, in view of all the complications associated with the determination of equilibrium constants in this area of chemistry. Methods and precautions involved in determining these enthalpy and entropy changes have been discussed in a recent paper dealing with a range of cations [4]. The interested reader should consult this extensive discussion by an experienced worker in a particularly difficult field.

In view of the misgivings with respect to quantitative data mentioned above, the main body of this Chapter has been divided into qualitative and quantitative sections. Much of the information available on polynuclear cations refers to aqueous media. The descriptive material available on proved and postulated species is presented in Chapters 10.2.1 to 10.2.4, arranged according to the Periodic Table. Some of the available quantitative results, equilibrium constants and enthalpies and entropies, are presented in Chapter 10.2.5. A few facts refer to heavy water (Chapter 10.3) and to mixed aqueous solvents (Chapter 10.4). Very little is known about this type of chemistry in non-aqueous solvents (Chapter 10.5). Polynuclear species containing metal–metal bonds rather than hydroxo- or oxo- bridges are discussed in the following section (Chapter 10.6), and polynuclear entities containing more than one type of cation in the closing paragraphs (Chapter 10.7). Polynuclear species containing ligands other than aquo-, hydroxo-, and oxo- groups have occasionally been described [5]. These species are analogous to the mononuclear mixed ligand hydroxo-species dealt with in Chapter 9.2.3.

10.2 AQUEOUS SYSTEMS

10.2.1 sp-Elements

Group II. Polynuclear species, like hydroxo-species, are most prevalent for cations with high charge to radius ratios. They can therefore be expected for the small beryllium(II) cation in aqueous media. Species containing up to five beryllium atoms have been proposed by various authors on various grounds. Potentiometric [6, 7] and recent ion-exchange and infrared and Raman [8] investigations all suggest that $Be_3(OH)_3^{3+}$ is the predominant species in weakly acidic media. Eight estimates for the decadic logarithm of the stability constant, $*\beta_{33}$, defined for the following equation:

$$3\,Be^{2+} + 3\,H_2O = Be_3(OH)_3^{3+} + 3\,H^+,$$

range between -8.66 and -8.91 ($25°C$, various ionic strengths) [1, 9, 10]. This $Be_3(OH)_3^{3+}$ species is thus tolerably well established and quantitatively characterised. Other polynuclear beryllium(II) species are less firmly established.

Several reports claim the existence of $Be_2(OH)^{3+}$ under limited ranges of conditions, and one report suggests the existence of a pentanuclear species $Be_5(OH)_7^{3+}$ while disfavouring any higher entity [7].

Evidence is sparse for the existence of polynuclear species of the remaining Group IIA elements and for the Group IIB elements zinc and cadmium. Values of $*\beta_{44}$ for the formation of tetranuclear species are quoted for magnesium(II) and for cadmium(II) [1] [†], but the latter figure has been disputed [11]. A binuclear zinc(II) species [1] has been tentively proposed, and evidence presented for the existence of $Cd_2(OH)^{3+}$ in strong solutions of cadmium(II) [12].

At the bottom of Group IIB there is a lot of information, especially from X-ray and Raman experiments, concerning polynuclear species in solutions containing mercury(II). In fact the possibility of the existence of such species in aqueous media first seems to have been mentioned as long ago as 1839. Early studies of mercury(II) solutions were interpreted as indicating the presence of binuclear species. In 1926 a report on an extensive potentiometric study using both a glass electrode to monitor pH and a redox electrode to monitor mercury(II) concentration showed that it was necessary to postulate both binuclear and tetranuclear species if the best fit to the experimental results were to be obtained [13]. Radial distribution curves for strong solutions of mercury(II) in perchlorate media are complicated. A careful analysis indicates the presence of several polynuclear species containing only a few mercury atoms each, the principle component appearing to be a binuclear species. The mercury-mercury distance in these presumably hydroxo-bridged entities is 3.64Å [14]. Raman spectra of mercury(II) solutions in nitrate media also indicate the presence of polynuclear cations [15]. From such solutions salts containing $Hg_3O_2^{2+}$ units have been isolated in the form of crystalline nitrates [16], and these salts are analogous to $Hg_3O_2^{2+}$ salts earlier prepared by a different route [17]. Recently the crystal structure of a salt of this type, formulated as $Hg(OH)_2.2HgSO_4.H_2O$, indicated the existence of a zig-zag trimercury(II) unit with hydroxo-bridges (Fig. 10.1) [18]. All of the mercury species mentioned so far have the mercury atoms linked by hydroxo- or oxo- bridges; the special case of the mercury–mercury bonded Hg_2^{2+} cation will be discussed in the appropriate Section (Chapter 10.6).

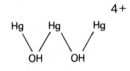

Fig. 10.1　The hydroxo-mercury(II) cation $Hg_3(OH)_2^{4+}$ in $Hg(OH)_2.2HgSO_4.H_2O$.

† Also, very recently, for zinc(II) (Zinevich, N. I. and Garmash, L. A., *Russ. J. Inorg. Chem.*, **20**, 1571 (1975)).

Group III. Polynuclear aquo-hydroxo-derivatives of aluminium(III) have been studied by a variety of techniques, including potentiometry, vibrational spectroscopy (infrared and Raman), and n.m.r. (^{1}H and ^{27}Al) spectroscopy. There has been considerable disagreement about the results of such studies and their interpretation, although the most recent ^{27}Al n.m.r. investigation [19] resolves some of the earlier arguments. It seems that highly polymeric species only form slowly at relatively high pHs, so that results obtained are liable to depend on the experimental methods used. What is claimed to be reliable evidence for the existence of species $Al_2(OH)_2^{4+}$ and $Al_{13}O_4(OH)_{24}^{7+}$ has been presented. The latter may contain both 4- and 6- coordinated aluminium [5], while some aluminium seems to be present in another form, probably an Al_8 unit, perhaps $Al_8(OH)_{20}^{4+}$. At high concentrations of aluminium(III) and high pHs, infrared, Raman, and ^{27}Al n.m.r. spectroscopies all suggest the coexistence of $Al_2O(OH)_6^{2-}$ with $Al(OH)_4^-$ [20]. There is evidence that Al_2 and Al_3 units, thought to be $Al_2(OH)_2^{4+}$ and $Al_3(OH)_4^{5+}$ respectively, are important components at higher temperatures (62.5°C and upwards) in the presence of potassium chloride (1M), along with an Al_n species with $n = 14 \pm 1$ (cf. $Al_{13}O_4(OH)_{24}^{7+}$ above) [21]. The logical extension of the examination of Al_n units to higher and higher n values, that is to the precipitation of the hydroxide, has been investigated for aluminium(III) solutions at high pHs [22]. The environmental relevance of the hydrolysis, polymerisation, and precipitation of aluminium(III) species has been reviewed [23].

In perchlorate media indium(III) can exist as $In(OH)^{2+}$, $In(OH)_2^+$, (Chapter 9) and in the form of $In(In(OH)_2)_x^{(3+x)+}$ [24, 25]. Of the latter species, $In_2(OH)_2^{4+}$ is thought to predominate [24]. In choride solution an analogous di-indium species $In_2(OH)Cl^{4+}$ is claimed [26].

Group IV. Cations $Ge(GeOH)_n^{(4+3n)+}$, with $0 < n < 10$, have been tentatively suggested to exist in solutions of germanium(IV) in strong acid ($0 < pH < 2$) [27]. Polynuclear units $Sn_3(OH)_4^{2+}$ (the major component) and $Sn_2(OH)_2^{2+}$ exist in hydrolysed tin(II) solutions [28]. A recent X-ray examination of tin(II) perchlorate solutions indicated a considerable degree of polynuclear complex formation, with each tin linked to 1.5 other tin atoms on average. Triangular Sn_3 units, as found in crystals of $Sn_3O(OH)_2(SO_4)$ [29], $Sn_2O(SO_4)$, and $Sn_6O_4(OH)_4$ [30], are perhaps present in these hydrolysed tin(II) solutions. The average tin-tin distance is around 3.6Å. A value for the formation constant of $Sn_3(OH)_4^{2+}$ has been determined [31].

In the present polymerisation context the most studied Group IV cation is lead(II). Units such as $Pb_2(OH)^{3+}$, $Pb_3(OH)_4^{2+}$, $Pb_4(OH)_4^{4+}$ and $Pb_6(OH)_8^{4+}$ have been variously suggested [30, 32], and equilibrium constants for reactions interconnecting these cations estimated [33]. X-ray investigations of both hydrolysed and polymerised solutions containing lead(II) and of analogous crystals, yield some information on the species present in the solutions. Thus the $Pb_4(OH)_4^{4+}$

4+

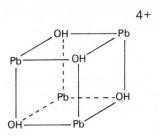

Fig. 10.2   The structure of the $Pb_4(OH)_4^{4+}$ cation.

unit suggested in solution may have the same almost cubic (tetrahedral Pb_4)
structure (Fig. 10.2) [34] as the discrete $Pb_4(OH)_4^{4+}$ cation found in crystals of
of $[Pb_4(OH)_4]_3(CO_3)(ClO_4)_{10}.6H_2O$ and in $[Pb_4(OH)_4](ClO_4)_4.2H_2O$. However
the lead to lead distance in the former solid is 3.78Å, but it is slightly larger,
3.85Å, in the solutions studied [35]. There has been similar X-ray work on the
$Pb_6O(OH)_6^{4+}$ cation (Fig. 10.3) [36]. A kinetic study of the depolymerisation of
the $Pb_4(OH)_4^{4+}$ cation provides evidence for an 'open' $Pb_4(OH)_2^{6+}$ cation *en route*
to the binuclear species $Pb_2(OH)^{3+}$ [37].

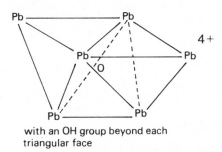

4+

with an OH group beyond each
triangular face

Fig. 10.3   The structure of the $Pb_6O(OH)_6^{4+}$ cation.

Group V. There is a little evidence for the existence of oxo- and hydroxo-
bridged polynuclear derivatives of antimony(III), though $Sb_2(OH)_2^{4+}$ has been
proposed to exist in perchloric acid solution [38]. A crystal structure determination of $Sb_4O_5(OH)(ClO_4)$ revealed chains of SbO_3 and SbO_4 polyhedra, a type
of structure which could also persist in solution [38].

Although there is scant evidence for the existence of the simple Bi^{3+}aq
cation in aqueous solution (Chapter 1.2.1), there are several fairly well established polynuclear bismuth(III) species in aqueous media. The cations $Bi_2(OH)_4^{2+}$,
$Bi_2(OH)_5^+$, $Bi_4O_3(OH)_4^{2+}$, $Bi_4O_4^{4+}$, $Bi_6(OH)_{12}^{6+}$, $Bi_6O_6^{6+}$, $Bi_8O_{10}^{4+}$, $Bi_9(OH)_{20}^{7+}$,
$Bi_9(OH)_{21}^{6+}$, $Bi_9(OH)_{22}^{5+}$ have all been proposed, amongst others [39]! The
$Bi_6(OH)_{12}^{6+}$ cation is the best established and characterised of these. Its structure
was established from infrared and Raman spectra, both of solutions and of
solids [40], and from X-ray diffraction studies of concentrated solutions con-

taining bismuth(III) [41]. A recent X-ray reexamination of the $Bi_6(OH)_{12}^{6+}$ unit in a crystal of its perchlorate salt [42] showed that the bismuth-bismuth distances along the edges of the Bi_6 octahedra (Fig. 10.4) are between 3.64 and 3.77Å, while the shortest bismuth-bismuth distances between octahedra are 8.78Å. The Bi_6 unit is thus established beyond doubt as a separate and distinct entity. A normal coordinate analysis of Raman spectroscopic data indicated some weak bismuth-bismuth interaction between nearest bismuth atoms in the Bi_6 octahedra, despite their separation by hydroxo-groups [40]. Various equilibrium constants have been published linking the various polynuclear species [39], and the enthalpy and entropy of generation of the $Bi_6(OH)_{12}^{6+}$ species have been estimated calorimetrically [43]. Some kinetic results are available for the formation and dissociation of $Bi_6(OH)_{12}^{6+}$ [44] (Chapter 14).

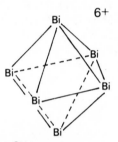

with an OH group on each
of the twelve Bi-Bi edges

Fig. 10.4   The structure of the $Bi_6(OH)_{12}^{6+}$ cation.

10.2.2 Transition Metals

Group III. No consistent picture emerges from the published information on scandium(III). The most recent discussion reviews earlier work and presents potentiometric and spectrophotometric evidence for the existence of $Sc_2(OH)_2^{4+}$ in the pH range 3.0 to 4.5 [45]. In sulphate medium, polynuclear species are thought to contain double hydroxo-bridges, and the onset of scandium hydroxide precipitation occurs at a pH of 6.4 [46]. Yttrium hydroxide does not start to precipitate until a considerably higher pH, because $Y_2(OH)_2^{4+}$ appears to exist in aqueous media at pHs above about 7.8 [47]. Polynuclear species derived from La^{3+} are listed, for convenience, with the lanthanides (see below).

Group IV. Titanium(III) solutions are more susceptible to oxidation at pHs above about 3.5 than at lower pHs. This observation is explained if more readily oxidisable polynuclear species are formed at pHs above 3.5. However, no guesses as to the composition of such species are available [48], except for a kinetic demonstration of the presence of low concentrations of a dimer [49]. The position for titanium(IV) is uncertain. Infrared spectra have been used to provide evidence for the presence of Ti–O–Ti units in weak acid [50], but dialysis and ion-exchange experiments were held to prove that titanium(IV) in dilute hydrochloric acid existed either as a monomeric cation or as colloidal TiO_2aq [51]. Such confusion may well stem from the very long time, a matter of days, required to established equilibrium [52]. The latest (1973) opinion is that polynuclear species are present when the pH is above 1 [53].

There is little or no convincing evidence for the existence of Zr^{4+}aq or ZrO^{2+}aq in moderately concentrated aqueous acids, but considerable evidence for polynuclear species. Cations proposed include $Zr_2(OH)_2^{6+}$ [54], $Zr_3(OH)_4^{8+}$, and $Zr_4(OH)_8^{8+}$ [55], the second and third thought to be linear and cyclic respectively. Qualitative observations and stability constants have been cited for Hf_4 as well as Zr_4 species [1], and $Hf_2(OH)_2^{6+}$ has also been postulated [54].

Group V. The vanadium(III) binuclear species VOV^{4+} is thought to be a transient intermediate in the reduction of VO^{2+}aq by V^{2+}aq. Small concentrations of VOV^{4+} can be detected in solutions of vanadium(III) at pHs close to 2.7 [56]. There have been hints that vanadium(IV) solutions may undergo polymerisation under certain conditions [57]. Many polynuclear species containing vanadium(V) exist, but all of these are anionic [58].

Group VI. The binuclear chromium(III) species $CrOCr^{4+}$ or $Cr_2(OH)_2^{4+}$ has been invoked many times to explain various observations in chromium chemistry [56, 59]. The most recent study has established its pK, of coordinated water rather than of bridging hydroxide of course, as 3.5 at $5°C$. Thus this binuclear species is a stronger acid than Cr^{3+}aq, whose pK is approximately 4.6 under similar conditions [60].

Bi- or poly- nuclear species are known for oxidation states II, III, IV, V, and VI of molybdenum. Molybdenum(II) exists in aqueous solution as Mo_2^{4+}aq (Chapter 10.6.2) [61]. There was some disagreement over the characterisation of the Mo^{3+}aq cation, part of it probably arising from the presence of some dimeric species in several of the samples studied. This binuclear species seems more likely to be $(H_2O)_4Mo\!\!\begin{smallmatrix}/OH\backslash\\\backslash OH/\end{smallmatrix}\!\!Mo(OH_2)_4^{4+}$ than $(H_2O)_5Mo\text{–}O\text{–}Mo(OH_2)_5^{4+}$ [62]. The proposed molybdenum(IV) dimer is thought to contain a double oxo-bridge, $(H_2O)_4Mo\!\!\begin{smallmatrix}/O\backslash\\\backslash O/\end{smallmatrix}\!\!Mo(OH_2)_4^{4+}$ [63]. Reduction of acid (perchloric) solutions of molybdenum(VI), or oxidation of solutions containing molybdenum(II) or molybdenum(III), yields solutions containing molybdenum(V). This exists as a binuclear species of total charge 2+. Thus the structure suggested for this binuclear cation is $(H_2O)_3Mo\!\!\begin{smallmatrix}O\ \ \ \ O\\ \|\ /O\backslash\ \|\\ \backslash O/\end{smallmatrix}\!\!Mo(OH_2)_3^{2+}$ [64]. There has been a suggestion that higher polynuclear cationic species of molybdenum(V) can be generated at a pH of about 1 [65]. There have also been reports of a cation Mo_2^{5+}aq, but as this is stable in sulphuric acid but not in p-toluene sulphonic acid it is presumably a binuclear sulphato-complex [66]. Much of this aqueous solution chemistry of molybdenum has been established only recently. Ion-exchange techniques have been particularly important in this area, in establishing the charge per molybdenum atom in the various species. Polynuclear molybdenum(VI) species are all anionic.

Group VII. Some dimer formation takes place in manganese(II) solutions at pHs higher than 3.5, with significant concentrations of dimer at pH > 5. Under these mildly acidic conditions the dimer may be an aquo-bridge species $Mn(OH_2)Mn^{4+}$, while at pHs above 7 hydroxo-bridging is said to become important [67]. Aged solutions of manganese(III) are thought to contain some 15 to 25% of the manganese(III) in the form of di- or poly- meric species [68]. Doubtless the species $Mn_2(OH)_2^{4+}$ of familar general formula figures among these.

Group VIII. Iron(III) solutions have been extensively studied in respect of hydrolysis and polynuclear species formation [69]. A great deal of evidence points to the existence of a dinuclear species. This may be $Fe_2(OH)_2^{4+}$ [70, 71] or, as suggested recently on the basis of vibrational and Mössbauer spectra, a μ-oxo-species Fe_2O^{4+}aq [72]. It is difficult to be certain of the formation constant value for this dinuclear species, or of the composition and stability of higher polynuclear cations, because neutral and alkaline solutions of iron(III) are very reluctant to reach equilibrium. Indeed the time-ageing of one such solution has been monitored over a period of 15 years [73]! There has been much discussion of the slow formation of precipitates, including such species as FeOOH in its α- (goethite), β-, and γ- (lepidocrocite) forms, from polynuclear species in solution [70, 74]. For iron(II), the existence of $Fe(OH_2)Fe^{4+}$ has been suggested [67].

The aqueous solution chemistry of ruthenium(III) and ruthenium(IV) is not well characterised. This is readily understandable in the case of ruthenium(III), which is particularly keen to be oxidised or to form complexes. Ruthenium(III) solutions in non-complexing acids may contain much Ru^{3+}aq, a little $RuOH^{2+}$aq, and some bi- or poly- nuclear species. Ruthenium(IV) solutions appear to contain RuO^{2+}aq (and/or $Ru(OH)_2^{2+}$aq) and a tetranuclear species, which may be $Ru_4(OH)_{12}^{4+}$ or $Ru_4O_6^{4+}$. Persistent rumours also hint at the existence of condensed ruthenium species which must, from the determined charge per ruthenium atom, contain a mixture of oxidation states. Such species can be generated by coulometry or cyclic voltammetry, and can themselves be reduced to a transient then a stable polynuclear ruthenium(III) species. The mixed oxidation state species are said to be Ru_2O^+, which contains $Ru^{3.5+}$, and another species containing $Ru^{3.75+}$ [75].

Cobalt(II) solutions may contain $Co(OH_2)Co^{4+}$ dimers (see Mn^{2+} above) in very weakly acidic solution [76]. There is still argument over whether or not significant concentrations of dimeric species exist in aqueous acidic solutions of cobalt(III) [77]. Aqueous solutions of rhodium(II) contain the rhodium wholly in the form of the dimer Rh_2^{4+}aq [78] (Chapter 10.6.2).

There is much evidence for the existence of polynuclear species in nickel(II) solutions at pHs around 7. The predominant species is $Ni_4(OH)_4^{4+}$, in water and in binary aqueous mixtures. The structure of $Ni_4(OH)_4^{4+}$ is thought, by analogy with $Pb_4(OH)_4^{4+}$, to be that shown in Fig. 10.5. There are also indications that small concentrations of a binuclear species $Ni_2(OH)^{3+}$ are present [79].

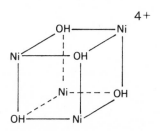

Fig. 10.5   The structure of the $Ni_4(OH)_4^{4+}$ cation.

Group IB. Stability constants, estimated by e.m.f. methods, have been reported for a variety of polynuclear copper(II) cations, including $Cu_2(OH)^{3+}$, $Cu_2(OH)_2^{2+}$, $Cu_3(OH)_2^{4+}$, and $Cu_3(OH)_4^{2+}$ [1, 80]. An enthalpy of formation has been obtained for the binuclear $Cu_2(OH)_2^{2+}$ cation [81].

10.2.3 Lanthanides

Relatively little is known about polynuclear hydroxo- and oxo- cations derived from these elements, which are indeed rather reluctant to form such entities in aqueous solution. Several polynuclear cations in solutions containing lanthanum(III) under approximately neutral conditions have been proposed. These include the $La_2(OH)^{5+}$, $La_5(OH)_9^{6+}$ and $La_6(OH)_{10}^{8+}$ cations [82]. Equilibrium constants characterising the equilibria established between these species and La^{3+}aq and $LaOH^{2+}$aq have been calculated [82], as have equilibrium constants relating to the formation of $Nd_2(OH)_2^{4+}$ [83], $Er_2(OH)_2^{4+}$ [47], and $Gd_2(OH)_2^{4+}$ [84] (see Chapter 10.2.5). For the special case of cerium(IV), both evidence and stability constants have been presented for bi-, ter-, tetra-, and dodeca- cerium units, and for species with molecular weights up to about 40,000 [85].

10.2.4 Actinides

Because these cations tend to contain the metals in relatively high oxidation states, they are prone to undergo hydrolysis (Chapter 9) and polymerisation despite their large size. This is true particularly for the 4+ state, which is both common and stable at the start of this series of elements, and also for the oxo-cations of the 5+ and 6+ oxidation states.

The solution chemistry of thorium(IV) is very complicated, except in strong acid, where $Th^{4+}aq$ predominates. There are many experimental results for other conditions, but a variety of interpretations [86, 87]. X-ray results suggest that the polynuclear entities are built up from $Th\begin{smallmatrix}/OH\backslash\\\backslash OH/\end{smallmatrix}Th$ units, in which the thorium--thorium distance is 3.95Å [38, 88]. Similar comments regarding the quantity of results and the diversity of interpretation apply to the much studied case of uranium(IV) [89]. Again X-ray results are consistent with $U\begin{smallmatrix}/OH\backslash\\\backslash OH/\end{smallmatrix}U$ bridging, with a uranium--uranium distance of 4.0Å [90]. Solutions containing uranium(VI) contain polynuclear species at pHs above about 2 [91], though it is not clear whether the mode of bridging is again $O_2U\begin{smallmatrix}/OH\backslash\\\backslash OH/\end{smallmatrix}UO_2$ or $O_2U-O-UO_2$ [92]. More than a score of independent determinations have been made of the stability constant for the formation of the uranium(VI) dimer (next section). There are indications of bi- and ter- nuclear cations in solutions containing neptunium(VI) or plutonium(VI) [93], but the study of the plutonium(VI) system is particularly difficult. Radiolysis of the solution leads to reduction, and plutonium(III), (IV), and (V) can all coexist in equilibrium with plutonium(VI) [94].

10.2.5 Formation Constants

The stability of a polynuclear cation is usually expressed in terms of the equilibrium constant $*\beta_{pq}$ for the equilibrium:

$$q\,M^{n+} + p\,H_2O = M_q(OH)_p^{(nq-p)+} + p\,H^+ .$$

Values have been estimated for a variety of polynuclear species, up to one (for aluminium(III)) containing thirteen metal atoms. To give some idea of the dependence of $*\beta_{pq}$ values on the nature of the metal and the number of metal atoms per polynuclear unit, a selection of logarithms of $*\beta_{pq}$ ($p*\beta_{pq} = -\log_{10}*\beta_{pq}$) values is presented in Table 10.1. In fact the variation of $*\beta_{pq}$ with the nature of the metal can be ascribed almost entirely to the differences in pK values for the mononuclear $M^{n+}aq$. This is because the equilibrium constant K_{mm} for such reactions varies relatively little with the nature of the metal

Table 10.1   Formation constants $*\beta_{pq}$, as defined in the text, for polynuclear hydroxo-cations $M_q(OH)_p^{(nq-p)+}$ derived from $M^{n+}aq$ cations. All values refer to $25°C$ and have been taken from reference [1] unless otherwise stated.

	$\log_{10} *\beta_{22}$	conditions	
Sn^{2+}	-2.7 to -4.6	various	
Cu^{2+}	-10.5 to -10.9^a	0–3M $NaClO_4$	
Hg^{2+}	-4.95 to -5.16	3M $NaClO_4$, 3M Ca or $Mg(ClO_4)_2$	[13]
Al^{3+}	-6.9 to -7.6	various	[1, 21, 95]
Sc^{3+}	-5.35 to -6.17	0.01 to 1M $NaClO_4$	
Y^{3+}	-14.04 to -14.30	3M $LiClO_4$	[1, 47]
Ti^{3+}	-3.30	3M KBr	
V^{3+}	-4.0^a	3M KCl	
Cr^{3+}	-3.4	1 or 2M $NaClO_4$	
Fe^{3+}	-2.14 to -2.96	0 to 3M $NaClO_4$	[1, 96]
Nd^{3+}	-13.93	3M $NaClO_4$	[83]
Er^{3+}	-13.72	3M $LiClO_4^c$	[47]
Gd^{3+}	-14.23	3M $NaClO_4$	[84]
Th^{4+}	-4.4 to -5.5	various	
UO_2^{2+}	-5.8 to -6.3^b	various	[1, 93]
NpO_2^{2+}	-6.68	M $NaClO_4$	[93]
PuO_2^{2+}	-8.51	M $NaClO_4$	[93]
	$\log_{10} *\beta_{33}$		
Be^{2+}	-8.66 to -8.91	various	[1, 9, 10]
	$\log_{10} *\beta_{44}$		
Mg^{2+}	-39.8	3M NaCl	
Cd^{2+}	-31.8		
Pb^{2+}	-18.7 to -21.7	various	
Ni^{2+}	-27.4 to -28.4	various	

a Values available at $18°C$, $20°C$ are consistent with these values for $25°C$; b this range excludes four erratic values, but covers 21 concordant results; c an identical value has been reported for 3M $NaClO_4$ (Burkov, K. A., Lilich, L. S., and Nguyen Dinh Ngo, *Izv. Vyssh. Uchebn. Zaved., Khim. Khim. Tekhnol.*, **18**, 181 (1975)).

if the formation of polynuclear units is expressed in a form from which proton loss has been eliminated, for example

$$m \, MOH^{n+} \rightleftharpoons M_m(OH)_m^{nm+} .$$

The series of K_{22} values presented in Table 10.2 illustrates this.

Table 10.2 Equilibrium constants K_{22} as defined in the text for the generation of $M_2(OH)_2^{4+}$ from $2MOH^{2+}$ [2], at 25°C.

Binuclear cation	$\log_{10}K_{22}$
$In_2(OH)_2^{4+}$	3.6
$Sc_2(OH)_2{}^{4+}$	3.5^a
$Y_2(OH)_2^{4+}$	3.9
$Cr_2(OH)_2^{4+}$	3.1
$Fe_2(OH)_2^{4+}$	3.2

a Values of 3.82, 3.90 have also been quoted [1].

Large uncertainties are likely in enthalpies and entropies corresponding to the constants $*\beta_{pq}$ and K_{nm} and this discourages the presentation of a Table of such values here. In the case of the best studied cation, Fe^{3+}, the seven estimates of the enthalpy change $(\Delta*H_{22})$ corresponding to $*\beta_{22}$ range from 8.4 to 13.5 kcal mol^{-1}. As p and q increase, $\Delta*H_{pq}$ increases. Thus for Al^{3+}, $\Delta*H_{22}$ is 18.3, $\Delta*H_{33}$ is 30.9, and $\Delta*H$ for formation of the Al_{13} (Al_{14} ?) unit is 262.8 kcal mol^{-1} [21]. This continuing increase arises from the addition of a further enthalpy for each new hydroxo- or oxo- linkage formed. The value of the enthalpy per metal atom is probably very roughly constant in such series.

10.3   POLYNUCLEAR CATIONS IN D_2O

Few data are available here. $Log_{10}*\beta_{22}$ values for copper(II), yttrium(III), erbium(III), dioxouranium(VI), and iron(III) are all slightly more negative, by about 0.6, in D_2O than in H_2O [97].

10.4   MIXED AQUEOUS SOLUTIONS

Again data are in very short supply, and refer mainly to aqueous dioxan. Facts are available for beryllium(II), nickel(II), and copper(II). The addition of dioxan, even up to a mole fraction of 0.5, has remarkably little effect on the composition or distribution of polynuclear species. There is a suggestion that higher polynuclear species for copper(II), though apparently not for beryllium(II) or nickel(II), become somewhat disfavoured as the mole fraction of dioxan increases [80].

10.5   NON-AQUEOUS SYSTEMS

Here the chemistry involved becomes rather different, for example with NH_2, NH, or N bridges if liquid ammonia is the solvent. There has been almost no investigation of this area of chemistry.

10.6 METAL–METAL BONDED CATIONS

In previous sections of this Chapter we have dealt with polynuclear cations in which the metal atoms have been linked by hydroxo- or oxo- bridges. Now we deal with the much rarer species in which the metal atoms are either known or thought to be directly bonded to each other [98].

10.6.1 Mercury

The longest known, best established, most studied, and most stable metal-metal bonded cation in solution is Hg_2^{2+}aq. Indeed, this is the only such cation that can be considered stable in aqueous solution with respect both to dissociation and to redox decomposition. This Hg_2^{2+}aq cation behaves, and can be treated, just like any ordinary M^{n+} cation, as it has been treated elsewhere in this book.

Further catenation of mercury atoms in cations has been established only recently, and that in non-aqueous media only. This has developed from work and ideas in the chemistry of non-metallic polynuclear ions such as I_2^+ and Te_4^{2+}, and of novel cations such as XeF^+ and XeF_5^+. Anions like AsF_6^- and $Sb_2F_{11}^-$ seem to be good at stabilising such non-metallic cations. When mercury is dissolved in liquid sulphur dioxide containing arsenic pentafluoride, solutions are obtained that are first red, then yellow, then eventually colourless. This last simply contains Hg_2^{2+} as its AsF_6^- salt, but from the yellow solution yellow crystals of $Hg_3(AsF_6)_2$ can be obtained, and from the red solution red crystals of $Hg_4(AsF_6)_2$. Thus one can postulate the presence of Hg_3^{2+} and of Hg_4^{2+} in these solutions. Higher members of the series Hg_n^{2+} with $n > 5$ have not been found yet, but a strikingly gold solid $Hg_{2.86}(AsF_6)$ corresponding to $Hg_{5.72}^{2+}$ has been obtained and seems to represent the limit for the series. While the Hg_2^{2+}, Hg_3^{2+}, and Hg_4^{2+} salts contain discrete 2-, 3-, and 4-mercury units in their respective AsF_6^- salts (Fig. 10.6), the $Hg_{5.72}(AsF_6)_2$ contains infinite linear chains of mercury atoms along the a and b axes. Because the chains are independent, one has effectively a one-dimensional metal here. The stoichiometric composition of this limiting compound is thus fixed simply by the relative sizes of the mercury atoms and the AsF_6 units. The mercury–mercury distance is 2.64Å, slightly greater than in Hg_4^{2+} but markedly less than in metallic mercury (3.00Å) [99].

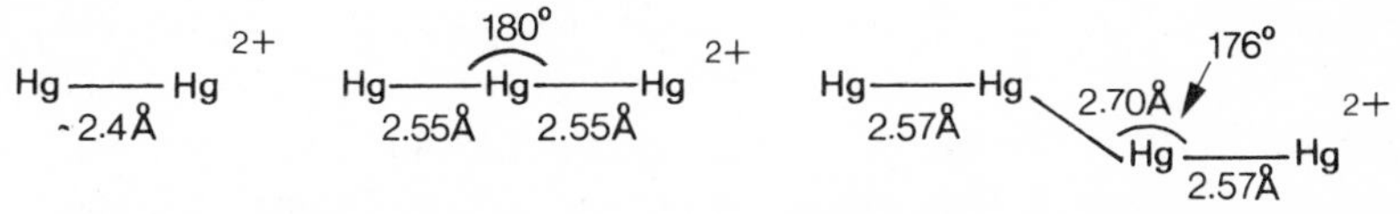

Fig. 10.6 The geometry and dimensions of Hg_n^{2+} cations.

Thus an Hg_n^{2+} chemistry, for $n > 2$, exists in sulphur dioxide solution, but these ions and their salts are extremely water-sensitive. The situation contrasts strongly with that of the water-stable Hg_2^{2+} cation. As an interesting historical footnote, the Hg_3^{2+} cation may well have been generated in the 1930s; a yellow colour, ascribed to Hg_2^{2+}, was reported when metallic mercury had been treated with fluorosulphonic acid [100].

10.6.2 Other Elements

The two nearest neighbours to mercury, cadmium and zinc, do not form analogous stable M_2^{2+}aq cations. Raman spectroscopy gives some indication of M_2^{2+} entities in melts [101], with Zn–Zn and Cd–Cd stretching force constants of 0.6 and 1.1 mdynes $Å^{-1}$ (contrast Hg–Hg at about 2.5 mdynes $Å^{-1}$). Solid salts containing Cd_2^{2+} or Zn_2^{2+} also exist, but the existence of Cd_2^{2+}aq or Zn_2^{2+}aq seems unlikely.† Similarly there is reason to believe that gallium-gallium bonded units in the solid state are present in such compounds as those of $Ga_2X_6^{2-}$, formally halide complexes of gallium(II). While it is possible to replace one or two of the ligands X by solvent molecules such as ethers or amines, it is not possible to generate a solvated Ga_2^{4+} cation in solution.

Molybdenum(II) exists in aqueous solution as Mo_2^{4+}aq [61], with a claimed bond order of 4 for the molybdenum–molybdenum bond [102]. However, despite determined attempts in more than one laboratory, it has so far proved impossible to precipitate this cation by the addition of a non-complexing anion. The molybdenum–molybdenum distance has not been estimated and so the bond order in this species cannot be confirmed [102].

Rhodium(II) in aqueous solution also exists as a dimer, Rh_2^{4+}aq [78, 103]. In this case no information exists concerning the nature of the rhodium-rhodium bond. Indeed the presence of such a bond in Rh_2^{4+}aq is implied, from the crystal structure of the rhodium(II) acetate dimer, rather than explicity stated [78, 102, 103].

10.7   MIXED POLYNUCLEAR CATIONS

This is an area more of potential than of actual importance and interest, both from the point of view of pure and of applied chemistry. There is industrial relevance in such areas as the extraction of titanium (Ti/Fe and Ti/Al cations), production of paints (Ti-containing cations again), and corrosion of steel (Fe^{II}/ Fe^{III} cations).

Polynuclear cations containg titanium(IV) and iron(II), vanadium(IV), or chromium(VI) [49, 104], titanium(IV) or zirconium(IV) with chromium(III) or aluminium(III) [105], and tin(IV) with iron(III) [106] have been described. Special cases of such cations containing one metal in two oxidation states have also been mentioned, for example titanium(III) and (IV) [104, 107]. In the case

† Since this paragraph was written, evidence has been presented for cations Hg_n^{2+}, Cd_n^{2+}, and Zn_n^{2+}, with n = 2, 3, and 4 (Cutforth, B. D., Gillespie, R . J., and Ummat, P. K., *Revue Chim. Minér.*, **31**, 119 (1976).

of iron, the stoichiometry of such a cation has been established as $1Fe^{II}:3Fe^{III}$, with the proposed structure shown in Fig. 10.7. A $1Fe^{II}:1Fe^{III}$ cation has also been detected [108].

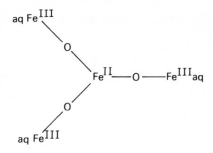

Fig. 10.7 The proposed structure of a mixed iron(II)–iron(III) cation [107].

REFERENCES

[1] Sillén, L. G. and Martell, A. E., *Stability Constants of Metal Ion Complexes, Chemical Society Special Publications 17 and 25 (Supplement 1)*, The Chemical Society, London (1964, 1971).

[2] Schwarzenbach, G., Boesch, J., and Egli, H., *J. Inorg. Nucl. Chem.*, **33**, 2141 (1971).

[3] Smyshlyaev, S. I. and Simonova, L. A., *Russ. J. Phys. Chem.*, **48**, 287 (1974).

[4] Arnek, R., *Ark. Kemi*, **32**, 55 (1970).

[5] Martell, A. E. in *Coordination Chemistry in Solution*, ed. Högfeldt, E., Swedish Natural Science Research Council, Lund (1972) p. 3.

[6] Kakihana, H. and Sillén, L. G., *Acta Chem. Scand.*, **10**, 985 (1956).

[7] Mesmer, R. E. and Baes, C. F., *Inorg. Chem.*, **6**, 1951 (1967).

[8] Cooper, M. K., Garman, D. E. J., and Yaniuk, D. W., *J. Chem. Soc. Dalton*, 1282 (1974); Bertin, F. and Derouault, J., *C. R. Hebd. Séanc. Acad. Sci., Paris*, **280C**, 973 (1975).

[9] Vanni, A., Gennaro, M. C., and Ostacoli, G., *J. Inorg. Nucl. Chem.*, **37**, 1443 (1975).

[10] Tsukuda, H., Kawai, T., Maeda, M., and Ohtaki, H., *Bull. Chem. Soc., Japan*, **48**, 691 (1975).

[11] Matsui, H. and Ohtaki, H., *Bull. Chem. Soc. Japan*, **47**, 2603 (1974).

[12] Cotton, F. A. and Wilkinson, G., *Advanced Inorganic Chemistry, 3rd edn.*, Wiley/Interscience (1972) p. 513.

[13] Ahlberg, I., *Acta Chem. Scand.*, **16**, 887 (1962).

[14] Johansson, G., *Acta Chem. Scand.*, **25**, 2787, 2799 (1971).

[15] Cooney, R. P. J. and Hall, J. R., *Aust. J. Chem.*, **22**, 337 (1969).

[16] Cooney, R. P. J. and Hall, J. R., *Aust. J. Chem.*, **25**, 1159 (1972).

[17] Bernstein, R. B., Pars, H. G., and Blumenthal, D. C., *J. Amer. Chem. Soc.*, **79**, 1579 (1957).

[18] Björnland, G., *Acta Chem. Scand.*, **28A**, 169 (1974).

[19] Akitt, J. W., Greenwood, N. N., Khandelwal, B. L., and Lester, G. D., *J. Chem. Soc. Dalton*, 604 (1972).

[20] Moolenaar, R. J., Evans, J. C., and McKeever, L. D., *J. Phys. Chem.*, **74**, 3629 (1970).

[21] Mesmer R. E. and Baes, C. F., *Inorg. Chem.*, **10**, 2290 (1971).

[22] Macdonald, D. D., Butler, P., and Owen, D., *J. Phys. Chem.*, **77**, 2474 (1973).

[23] *Aqueous-Environmental Chemistry of Metals*, ed. Rubin, A. J., Wiley (1975).

[24] Biedermann, G., *Ark. Kemi*, **9**, 277 (1956).

[25] Kudryavskii, Yu. P. and Kazantsev, E. I., *Russ. J. Phys. Chem.*, **48**, 226 (1974).

[26] Ferri, D., *Acta Chem. Scand.*, **26**, 747 (1972).

[27] Alekseeva, I. I. and Nemzer, I. I., *Russ. J. Inorg. Chem.*, **16**, 987 (1971).

[28] Tobias, R. S., *Acta Chem. Scand.*, **12**, 198 (1958).

[29] Grimvall, S., *Acta Chem. Scand.*, **27**, 1447 (1973); **29A**, 590 (1975).

[30] Johansson, G. and Ohtaki, H., *Acta Chem. Scand.*, **27**, 643 (1973).

[31] See p. 331 of reference [12].

[32] Olin, Å., *Acta Chem. Scand.*, **14**, 126, 814 (1960); Pokrić, B. and Pučar, Z. *J. Inorg. Nucl. Chem.*, **35**, 1987 (1973).

[33] Pavlović, Z. and Popović, R., *Revue Roum. Chim.*, **15**, 313 (1970).

[34] Maroni, V. A. and Spiro, T. G., *Inorg. Chem.*, **7**, 188 (1968).

[35] Hong, S. -H. and Olin, Å., *Acta Chem. Scand.*, **27**, 2309 (1973); **28A**, 233 (1974).

[36] Olin, Å. and Söderquist, R., *Acta Chem. Scand.*, **26**, 3505 (1972).

[37] Frei, V. and Wendt, H., *Z. Phys. Chem. Frankf. Ausg.*, **88**, 59 (1974).

[38] Bovin, J. -O., *Acta Chem. Scand.*, **28A**, 723 (1974).

[39] Drăgulescu, C., Nimară, A., and Julean, I., *Revue Roum. Chim.*, **17**, 1181 (1972); and references therein.

[40] Maroni, V. A. and Spiro, T. G., *Inorg. Chem.*, **7**, 183 (1968).

[41] Levy, H. A., Danforth, M. D., and Agron, P. A., *J. Chem. Phys.*, **31**, 1458 (1959).

[42] Sundvall, B., *Acta Chem. Scand.*, **28A**, 1036 (1974).

[43] Olin, Å., *Acta Chem. Scand.*, **29A**, 907 (1975).

[44] Pokrić, B. and Pučar, Z., *J. Inorg. Nucl. Chem.*, **35**, 3287 (1973).

[45] Akalin, S. and Özer, U. Y., *J. Inorg. Nucl. Chem.*, **33**, 4171 (1971).

[46] Shatskii, V. M., Bashkov, B. I., Komissarova, L. N., and Grevtsev, A. M., *Russ. J. Inorg. Chem.*, **19**, 1103 (1974).

[47] Blewett, F. McC. and Watts, P., *J. Chem. Soc. B*, 881 (1971).

[48] Pecsok, R. L. and Fletcher, A. N., *Inorg. Chem.*, **1**, 155 (1962).

[49] Birk, J. P. and Logan, T. P., *Inorg. Chem.*, **12**, 580 (1973).

[50] Bragina, M. I. and Bobyrenko, Yu. Ya., *Russ. J. Inorg. Chem.*, **17**, 61 (1972).

[51] Babko, A. K., Gridchina, G. I., and Nabivanets, B. I., *Russ. J. Inorg. Chem.* **7**, 66 (1962).

[52] Sheytanov, C. E. and Rizov, N., *Inorg. Nucl. Chem. Lett.*, **6**, 785 (1970).

[53] Ellis, J. D. and Sykes, A. G., *J. Chem. Soc. Dalton*, 537 (1973).

[54] Alekseeva, I. I., Nemzer, I. I., and Yuranova, L. I., *Russ. J. Inorg. Chem.*, **20**, 50 (1975).

[55] Peshkova, V. M. and P'ēng Ang, *Russ. J. Inorg. Chem.*, **7**, 1091 (1962); Fratiello, A., Vidulich, G. A., and Mako, F., *Inorg. Chem.*, **12**, 470 (1973); Claude, R. and Vivien, D., *Bull. Soc. Chim. Fr.*, 65 (1974); Baglin, F. G. and Breger, D., *Inorg. Nucl. Chem. Lett.*, **12**, 173 (1976). and references therein.

[56] Newton, T. W. and Baker, F. B., *Inorg. Chem.*, **3**, 569 (1964).

[57] Henry, R. P., Mitchell, P. C. H., and Prue, J. E., *J. Chem. Soc. Dalton*, 1156 (1973); Francavilla, J. and Chasteen, N. D., *Inorg. Chem.*, **14**, 2860 (1975).

[58] Yatsimirskii, K. B. and Zhelyazkova, B. G., *Russ. J. Inorg. Chem.*, **17**, 970 (1972).

[59] Ardon, M. and Plane, R. A., *J. Amer. Chem. Soc.*, **81**, 3197 (1959); Ardon, M. and Linenberg, A. *J. Phys. Chem.*, **65**, 1443 (1961).

[60] Von Meyenburg, U., Široký, O., and Schwarzenbach, G., *Helv. Chim. Acta* **56**, 1099 (1973).

[61] Bowen, A. R. and Taube, H., *J. Amer. Chem. Soc.*, **93**, 3287 (1971); *Inorg. Chem.*, **13**, 2245 (1974).

[62] Ardon, M. and Pernick, A., *Inorg. Chem.*, **13**, 2275 (1974).

[63] Ardon, M. and Pernick, A., *J. Amer. Chem. Soc.*, **95**, 6871 (1973); Ardon, M., Bino, A., and Yahav, G., *ibid.*, **98**, 2338 (1976).

[64] Ardon, M. and Pernick, A., *Inorg. Chem.*, **12**, 2484 (1973).

[65] Viossat, B. and Lamache, M., *Bull. Soc. Chim. Fr. A*, 1570 (1975).

[66] Pernick, A. and Ardon, M., *J. Amer. Chem. Soc.*, **97**, 1255 (1975).

[67] Wells, C. F. and Salam, M. A., *J. Inorg. Nucl. Chem.*, **31**, 1083 (1969).

[68] Rosseinsky, D. R., Nicol, M. J., Kite, K., and Hill, R. J., *J. Chem. Soc. Faraday I*, **70**, 2232 (1974).

[69] Sylva, R. N., *Rev. Pure Appl. Chem.*, **22**, 115 (1972); Kolosov, I. V., Intskirveli, L. N., and Varshal, G. M., *Russ. J. Inorg. Chem.*, **20**, 1179 (1975).

[70] Feitknecht, W., Giovanoli, R., Michaelis, W., and Müller, M., *Helv. Chim. Acta*, **56**, 2847 (1973); Knight, R. J. and Sylva, R. N., *J. Inorg. Nucl. Chem.*, **36**, 591 (1974).

[71] Ciavatta, L. and Grimaldi, M., *J. Inorg. Nucl. Chem.*, **37**, 163 (1975).

[72] Knudson, J. M., Larsen, E., Moreira, J. E., and Nielsen, O. F., *Acta Chem Scand.*, **29A**, 833 (1975).

[73] Feitknecht, W., Giovanoli, R., Michaelis, W., and Müller, M., *Z. Anorg. Allg. Chem.*, **417**, 114 (1975).

[74] Sommer, B. A., Margerum, D. W., Renner, J., Saltman, P., and Spiro, T. G., *Bioinorg. Chem.*, **2**, 295 (1973).

[75] Wallace, R. M. and Propst, R. C., *J. Amer. Chem. Soc.*, **91**, 3779 (1969); Bremard, C., Nowogrocki, G., and Tridot, G., *Bull. Soc. Chim. Fr.*, 110, 392 (1974); and references therein.

[76] Salam, M. A. and Raza, M. A., *Chemy Ind.*, 601 (1971).

[77] Davies, G. and Warnqvist, B., *J. Chem. Soc. Dalton*, 900 (1973); Wells, C. F. and Fox, D., *J. Inorg. Nucl. Chem.*, **38**, 107 (1976).

[78] Maspero, F. and Taube, H., *J. Amer. Chem. Soc.*, **90**, 7361 (1968).

[79] Ohtaki, H. and Biedermann, G., *Bull. Chem. Soc. Japan*, **44**, 1822 (1971); Burkov, K. A., Zinevich, N. I., and Lilich, L. S., *Russ. J. Inorg. Chem.*, **16**, 926 (1971); Kawai, T., Otsuka, H., and Ohtaki, H., *Bull. Chem. Soc. Japan*, **46**, 3753 (1973).

[80] Ohtaki, H. and Kawai, T., *Bull. Chem. Soc. Japan*, **45**, 1735 (1972).

[81] Arnek, R. and Patel, C. C., *Acta Chem. Scand.*, **22**, 1097 (1968).

[82] Biedermann, G. and Ciavatta, L., *Acta Chem. Scand.*, **15**, 1347 (1961).

[83] Kurkov, K. A., Lilich, L. S., Nguen Din Ngo, and Smirnov, A. Yu., *Russ. J. Inorg. Chem.*, **18**, 797 (1973).

[84] Nguen Din Ngo and Burkov, K. A., *Russ. J. Inorg. Chem.*, **19**, 680 (1974).

[85] Hardwick, T. J., and Robertson, E., *Can. J. Chem.*, **29**, 818 (1951); Danesi, P. R., *Acta Chem. Scand.*, **21**, 143 (1967); Louwrier, K. P. and Steemers, T., *Inorg. Nucl. Chem. Lett.*, **12**, 185 (1976).

[86] Katz, J. J. and Seaborg, G. T., *The Chemistry of the Actinide Elements*, Methuen (1957) pp. 52-56.

[87] Milić, N. B., *Acta Chem. Scand.*, **25**, 2487 (1971).

[88] E.g. Johansson, G., *Acta Chem. Scand.*, **22**, 399 (1968).

[89] See pp. 171-179 of reference [86].

[90] Pocev, S. and Johansson, G., *Acta Chem. Scand.*, **27**, 2146 (1973).

[91] E.g. Rabinowitch, E. and Belford, R. L., *Spectroscopy and Photochemistry of Uranyl Compounds*, Pergamon (1964) ch. 2.

[92] Mavrodin-Tărăbîc, M., *Revue Roum. Chim.*, **18**, 73, 609 (1973); **19**, 1461 (1974).

[93] Schedin, U., *Acta Chem. Scand.*, **25**, 747 (1971); **29A**, 333 (1975); Cassol, A., Magon, L., Tomat, G., and Portanova, R., *Inorg. Chem.*, **11** 515 (1972).

[94] Costanzo, D. A., Biggers, R. E., and Bell, J. T., *J. Inorg. Nucl. Chem.*, **35**, 609 (1973); and see pp. 292-304 of reference [86].

[95] Turner, R. C., *Can. J. Chem.*, **53**, 2811 (1975).

[96] Vertes A., Ranogajec-Komor, M., and Gelencser, P., *Magy. Kém. Foly.*,

References
309

78, 639 (1972); Landry, J. C., Buffle, J., Haerdi, W., Levental, M., and Nembrini, G., *Chimia*, **29**, 263 (1975).

[97] Amaya, T., Kakihana, H., and Maeda, M., *Bull. Chem. Soc. Japan*, **46**, 2889 (1973); Knight, R. J. and Sylva, R. N., *J. Inorg. Nucl. Chem.*, **37**, 779 (1975).

[98] Taylor, M. J., *Metal-to-Metal Bonded States of the Main Group Elements*, Academic Press (1976).

[99] Cutforth, B. D., Gillespie, R. J., and Ireland, P. R., *J. Chem. Soc. Chem. Commun.*, 723 (1973); Cutforth, B. D., Davies, C. G., Dean, P. A. W., Gillespie, R. J., Ireland, P. R., and Ummat, P. K., *Inorg. Chem.*, **12**, 1343 (1973).

[100] Meyer, J. and Schram, G., *Z. Anorg. Allg. Chem.*, **206**, 24 (1932).

[101] See p. 507 of reference [12].

[102] Cotton, F. A., *Chem. Soc. Rev.*, **4**, 27 (1975).

[103] Wilson, C. R. and Taube, H., *Inorg. Chem.*, **14**, 405, 2276 (1975).

[104] Reynolds, M. L., *J. Chem. Soc.*, 2991, 2993 (1965).

[105] Kolpachkova, N. M., Maiskaya, T. Z., Nekhamkin, L. G., and Zaitsev, L. M., *Russ. J. Inorg. Chem.*, **20**, 53 (1975).

[106] Morgenstern-Badarau, I. and Michel, A., *J. Inorg. Nucl. Chem.*, **38**, 1400 (1976)

[107] Jørgenson, C. K., *Acta Chem. Scand.*, **11**, 73 (1957); Goroshchenko, Ya. G. and Godneva, M. M., *Russ. J. Inorg. Chem.*, **6**, 744 (1961).

[108] Misawa, T., Hashimoto, K., Suëtaka, W., and Shimodaira, S., *J. Inorg. Nucl. Chem.*, **35**, 4159 (1973); Misawa, T., Hashimoto, K., and Shimodaira, S., *ibid.*, **35**, 4167 (1973).

Chapter **11**

KINETICS AND MECHANISMS: SOLVENT EXCHANGE

11.1 INTRODUCTION

The exchange of solvent molecules between the primary solvation shell of a cation and 'bulk' solvent may be considered the fundamental reaction for metal ions in solution. Solvent exchange is closely linked with the formation of complexes, as will become apparent in Chapter 12, and can also be significant in the mechanisms of some redox reactions (Chapter 13).

Rates of solvent exchange at metal ions vary enormously with the nature of the cation. The range of observed first-order rate constants for aqueous solutions at 25°C, is from almost 10^{10} sec^{-1} for Cu^{2+} or for Cr^{2+} right down to less than 10^{-7} sec^{-1} for Rh^{3+}. The exchange rate for a given cation is strongly solvent dependent. For example, rate constants for solvent exchange at Fe^{3+}, again at 25°C, range from 2×10^4 sec^{-1} in ethanol down to less than 40 sec^{-1} in acetonitrile or dimethylformamide. Our concern will be to examine reasons for these reactivity ranges, and to ascertain likely mechanisms for solvent exchange at various cations. The very simplicity of these systems has drawbacks, particularly because it is impossible to vary the concentration of the reactant in a pure solvent in order to discover its exponent in the rate law. Attempts to establish the order with respect to solvent by having a dilute solution of that solvent in an 'inert' cosolvent, lead only to the confused and controversial field of mixed solvents where new problems are more likely to be generated than the original ones solved.

11.2 METHODS

Only two experimental methods seem suited to the determination of rates of solvent exchange. These are isotopic labelling for slow exchange, and n.m.r. spectroscopic studies for those few cations where solvent exchange rates are of the same order of magnitude as the reciprocal of the n.m.r. signal frequency.

11.2.1 Isotopic Labelling

Either the hydrated cation or the bulk solvent may be labelled, for example with $H_2^{18}O$. Then the aquo-cation can be precipitated at intervals and its content of the label monitored by mass spectrometry. This permits monitoring of either the rate of release of labelled water from the primary coordination sphere, or the reverse process. The long time-scale required for this procedure is obvious, and to date the only cations for which this method has been used have been $Cr^{3+}aq$ and $Rh^{3+}aq$.

11.2.2 N.m.r. Spectroscopy

This method has proved to be rather more widely applicable, as solvent exchange at several cations is within the range appropriate to n.m.r. investigation. Moreover this method can be used just as easily for non-aqueous solvents as for water. In systems where the frequency of solvent exchange is comparable with the frequency of the n.m.r. signal within the experimentally accessible temperature range, the number and shape of the lines in the n.m.r. spectra will show a marked variation with temperature. At low temperatures solvent exchange between the two distinguishable environments of the cation primary solvation shell and 'bulk' (that is anion primary, secondary, and bulk solvent, see Chapter 1) is slow, and the n.m.r. spectrum for a given solvent nucleus will consist of two peaks (see Chapter 2.2.1). On the other hand, at high temperatures the rate of solvent exchange between these two situations is fast. Here the environment of each solvent nucleus is time-averaged on the n.m.r. time-scale and only a single peak will be observed. Spectra will vary smoothly between these two extremes. As the temperature rises, the two separate coordinated and 'bulk' solvent peaks first broaden and then coalesce, and finally the merged peak sharpens again. Figure 11.1 illustrates this [1]. At the coalescence point the frequency of solvent exchange at this particular temperature can be simply and directly derived from the known n.m.r. frequency. The estimation of solvent exchange rates at other temperatures, and from these the determination of the activation enthalpy and entropy, is a more complicated matter [2].

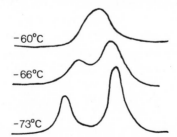

Fig. 11.1 The variation of ^1H n.m.r. spectra with temperature for the Mg^{2+} cation in magnesium perchlorate solution in aqueous acetone [1].

In the case of paramagnetic cations it is often impossible to detect the signal from the coordinated solvent. This situation prevails when correlation times for electronic relaxation are long, as for V^{2+}, Mn^{2+}, and Cu^{2+}. Relaxation times for Co^{2+} and Ni^{2+} are more suitable, and both bulk and coordinated solvent nucleus resonances can be detected in n.m.r. spectra of salts of these cations in solution. However, even when the coordinated solvent resonance is broadened beyond detection, it is still possible to extract some kinetic results from the linewidth of the bulk solvent signal as it coalesces with the (invisible) line due to the coordinated solvent [3, 4].

For non-aqueous solvents there is often a choice of nuclei whose n.m.r. resonances can be monitored. In the case of ethanol, one can look at the hydroxyl protons, the methylene protons, or the methyl protons. For acetonitrile there is a choice between ^{1}H, ^{13}C, and ^{14}N n.m.r. Two factors determine the actual choice of nucleus. The first is the relative ease of working with particular nuclei, the second the advantages offered by working with the nucleus coordinated directly to the cation. Of these, one advantage is the certainty of following exchange of the whole solvent molecule. The hydroxyl-proton n.m.r spectra of an alcoholic solution of a salt may reveal the kinetics of proton exchange with coordinated alcohol molecules rather than exchange of whole alcohol molecules at the cation (Chapter 14). Admittedly this point can easily be cleared up for alcohols by monitoring the resonance of protons in methyl or methylene groups. There is another advantage in monitoring n.m.r. signals of nuclei directly coordinated to cations. Because the ligating nucleus is closest to the cation, the difference of its chemical shift from that in bulk solvent is likely to be greatest.

Activation parameters for solvent exchange can be obtained from the variation of linewidth with temperature in the region of the coalescence temperature. The mathematical procedures for this are well-established [2], but their accurate application to actual experimental results has been the subject of much controversy. The difficulties lie in the complicated nature of the dependence of linewidth on temperature. Although only one section of a linewidth against temperature plot is concerned with solvent exchange, the linewidth-temperature behaviour in other temperature ranges needs to be well-defined if accurate rate constants and thence activation parameters are to be extracted from the relevant experimental observations. This is not a minor point, as can most dramatically be illustrated by quoting activation parameters for dimethyl sulphoxide exchange at Ni^{2+} determined by several different groups of workers (Table 11.1) [5–13]. A similar scatter afflicts determinations of the activation parameters for methanol exchange at Ni^{2+}. Variations between results from different sources are rather smaller for solvent exchange at Co^{2+}, which is not too surprising because the nuclear properties of Co^{2+} are more suitable than those of Ni^{2+} for the observation of good n.m.r. spectra of solvent nuclei.

If the difficulties which beset the estimation of activation parameters for solvent exchange from the temperature variation of n.m.r. spectra are to be

Table 11.1 Activation parameters determined by n.m.r. spectroscopy for dimethyl sulphoxide exchange at Ni^{2+}.

$\Delta H^{\ddagger}$/kcal mol^{-1}	$\Delta S^{\ddagger}$/cal deg^{-1} mol^{-1}	
6.2	−20	[5]
8	−16	[6]
8.1	−14	[7]
8.7	−11	[8]
11.7	−1	[9]
12.1	−1	[10]
12.1	+1	[11]
13.0	+3	[12, 13]

appreciated the usual linewidth-temperature dependence must be briefly considered. Such a dependence for a typical system is illustrated diagrammatically in Fig. 11.2 [14]. The linewidth is determined by chemical exchange in the region B to C, but other relaxation effects in the regions of slow exchange (A to B) and fast exchange (D to E) have to be allowed for in calculating solvent exchange rates and activation parameters from the information in the first region. Poor definition of the linewidth-temperature dependence in regions A to B or D to E can have 1 or 2 kcal mol^{-1} effect on the estimated activation parameters. Often the temperature range B to C corresponds to a temperature range which is inconveniently short for the precise determination of activation parameters. In principle, salvation here lies in using an n.m.r. machine operating at a higher frequency.

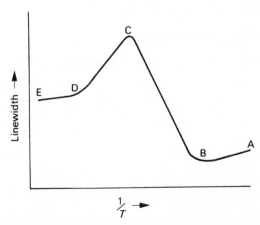

Fig. 11.2 Typical dependence of linewidth on temperature for a solution in which the rate of solvent exchange at the cation varies from much slower to much faster than that corresponding to the n.m.r. frequency.

The derivation of kinetic parameters discussed in the preceding paragraphs is from the spin-spin relaxation time T_2. Only recently has the derivation of kinetic parameters from the spin-lattice relaxation time T_1 been described, with particular reference to water exchange at $Ni^{2+}aq$ [15].

The temperature dependence of chemical shifts for bulk and for coordinated solvent molecules rather resembles that described above for linewidths. Kinetic parameters can also be extracted from spectra run in the region where solvent exchange is responsible for determining the chemical shifts.

Some of the difficulties encountered in obtaining kinetic information by n.m.r. spectroscopy can be avoided if the magnetic field as well as the temperature is varied. When this suggestion was first made, n.m.r. technology was insufficiently advanced for it to be practicable. However, there are now n.m.r. spectrometers able to cope with this approach [16].[†]

Table 11.2   Kinetic parameters for solvent exchange at the Ni^{2+} cation.[a]

Solvent	$\log_{10} k_{25}$[b]	$\Delta H^{\ddagger}$ /kcal mol^{-1}	$\Delta S^{\ddagger}$ /cal deg^{-1} mol^{-1}
Water[c]	4.4 to 4.6	9.9 to 12.1	-5.2 to $+2.9$
Ammonia	5.0	9.9 to 11	-2 to $+2$
Methanol	2.3 to 3.0	13.1 to 15.8	-4 to $+8$
Ethanol	4.0	10.8	-4
Dimethylformamide	3.6 to 3.9	9.4 to 15	-9 to $+8$
Acetonitrile[d]	3.3 to 4.2	6.2 to 16.4	-20 to $+12$
Dimethyl sulphoxide[e]	3.5 to 4.0	6.2 to 13.0	-20 to $+3.2$

a Values taken from ref. [17], except where otherwise indicated; *b* units of k are sec^{-1}; *c* also refs. [15] and [18]; *d* also ref. [19]; *e* from Table 11.1.

11.3   KINETIC PARAMETERS

Rate constants[††] and activation parameters for solvent exchange in both aqueous and non-aqueous media are collected together in Tables 11.2 to 11.5

---

† An alternative experimental method, using a stopped-flow device ingeniously incorporated in a pulse Fourier transform n.m.r. spectrometer, has recently been described (Couch, D. A., Howarth, O. W., and Moore, P., *J. Phys. E:Sci. Instrum.*, **8**, 831 (1975)). Its first application has been to the exchange of $(CD_3)_2SO$ with $Al[(CH_3)_2SO]_6^{3+}$ (Brown, A. J., Couch, D. A., Howarth, O. W., and Moore, P., *J. Magnetic Resonance*, **21**, 503 (1976)). The log k vs $1/T$ plot for these results is collinear with that obtained at higher temperatures from the ordinary n.m.r. line-broadening method. A combination of the two techniques can thus cover an unusually extended temperature range, promising accurate $\Delta H^{\ddagger}$ and $\Delta S^{\ddagger}$ values for solvent exchange.

†† It is sometimes difficult, or even impossible, to tell whether published rate constants for solvent exchange (or mean residence times for coordinated solvent molecules) refer to one solvent molecule or to the solvento-cation. The difference in $\log_{10} k$ values is 0.8 for hexasolventocations.

[17–48]. These Tables include a generous selection of published results, but are far from exhaustive compilations of kinetic data for solvent exchange at cations. Rate constants are all quoted as observed first-order constants. Activation entropies are derived from these and therefore strictly apply to dissociative exchange. However, assuming a second-order process makes a difference of only about 4 cal deg^{-1} mol^{-1} to the calculated activation entropy. Kinetic parameters for solvent exchange at the two most widely studied cations, Ni^{2+} and Co^{2+}, are listed in Tables 11.2 and 11.3. These Tables provide some idea of the dependence of rates and activation parameters on the nature of the exchanging solvent. Table 11.4 includes kinetic parameters for water exchange at a selection of 2+ cations, both of transition metals and of sp-block elements. It may be added that for many cations of the latter type, for example Ba^{2+}, Hg^{2+}, and Sn^{2+}, rate constants for water exchange of greater than 10^4 sec^{-1} have been demonstrated. Table 11.5 presents a selection of kinetic results illustrating solvent exchange at 3+ cations. In cases where several different results are available for a given reaction, the range of published values is indicated. In the case of $M^{3+}aq$ some of the variation in reported rate constants may derive from observations at different pHs, some $M(OH)^{2+}aq$ for instance may be present at times [49].

Table 11.3 **Kinetic parameters for solvent exchange at the Co^{2+} cation.**[a]

Solvent	$\log_{10} k_{25}$[b]	$\Delta H^{\ddagger}$ /kcal mol^{-1}	$\Delta S^{\ddagger}$ /cal deg^{-1} mol^{-1}
Water[c]	6.1 to 6.4	8.0 to 11.9	−4.1 to +10.6
Ammonia	6.9	11.2	+10.2
Methanol	3.6 to 4.3	11.5 to 13.8	−4 to +7.2
Dimethylformamide	5.4 to 5.6	7.1 to 13.4	−10 to +12.9
Acetonitrile	5.1 to 5.5	8.1 to 11.4	−7.5 to +5.2
Dimethyl sulphoxide	5.2	6.9 to 9.6	−14.5 to −0.4

a Sources as Table 11.2; b units of k are sec^{-1}; c see also refs. [20, 21].

All the kinetic parameters quoted in Tables 11.1 to 11.5 refer to 'normal' thermochemical solvent exchange. Very little attention has been paid to solvent exchange under photochemical conditions. The only quantitative studies have been of $Cr^{3+}aq$ and of the mixed ligand complex $Rh(NH_3)_5(OH_2)^{3+}$. In both cases a marked increase in reactivity occurs when the systems are irradiated at appropriate frequencies [50].

Table 11.4　Kinetic parameters for water, methanol, and ethanol exchange at dipositive cations.[a]

	Water			Methanol			Ethanol		
	$\log_{10} k_{25}$ [b]	$\Delta H^{\ddagger}$ /kcal mol⁻¹	$\Delta S^{\ddagger}$ /cal deg⁻¹ mol⁻¹	$\log_{10} k_{25}$ [b]	$\Delta H^{\ddagger}$ /kcal mol⁻¹	$\Delta S^{\ddagger}$ /cal deg⁻¹ mol⁻¹	$\log_{10} k_{25}$ [b]	$\Delta H^{\ddagger}$ /kcal mol⁻¹	$\Delta S^{\ddagger}$ /cal deg⁻¹ mol⁻¹
sp-**Block tetrahedral**									
Be²⁺	~2.5[c] or 4.5[d]								
sp-**Block octahedral**									
Mg²⁺	5.7	10.2	+2	3.7	16.7	+14	6.4	17.7	+30
Zn²⁺	7.5				14	+15			
Cd²⁺	8.2								
TM: First row									
V²⁺	1.9	16.4	+5						
Cr²⁺	8.5 or 9.9	3	−3						
Mn²⁺	7.5	8.1	+2.9	5.6	6.2	−12			
Fe²⁺	6.5	7.7	−3.0	4.7	12.0	+3			
Co²⁺	6.1 to 6.4	8.0 to 11.9	−4.1 to +5.3						
Ni²⁺	4.4 to 4.6	10.3 to 12.1	−5.2 to +2.9				4.0	10.8	−4
Cu²⁺	9.3 or 9.9[e]	5.6	+6	~8	10	+13			
TM: Second row									
Ru²⁺	~−1[f]								

a Values from ref. [17] for water and from ref. [22] for alcohols except where otherwise indicated; *b* units of *k* are sec⁻¹; *c* ref. [23]; *d* ref. [24]; *e* ref. [25]; *f* ref. [26];

Table 11.5 Kinetic parameters for solvent exchange at tripositive cations.

Solvent	$\log_{10} k_{25}$[a]	$\Delta H^{\ddagger}$ /kcal mol^{-1}	$\Delta S^{\ddagger}$ /cal deg^{-1} mol^{-1}	
sp-Elements				
Al^{3+} Water	-0.8	27	$+28.$	[27]
Methanol	3.6	3.9	-29	[27]
Dimethylformamide	-0.8	17.7	$+5$	[27]
Dimethyl sulphoxide	-1.2	20	$+4$	[28][b]
Ga^{3+} Water	3.3	6.3	-22	[27]
Methanol	4.0	4.2	-26	[27]
Dimethylformamide	0.2	14.6	-8	[27]
In^{3+} Water	~4.3	~4		[29]
Tl^{3+} Water	$\geqslant 9.5$			[30]
Bi^{3+} Water	>4			[24]
TM: First row				
Ti^{3+} Water	4.8	6.2	-15	[31]
Methanol	5.1	3.3	-24	[31]
V^{3+} Methanol	3.1	6.2	-23	[32]
Cr^{3+} Water	-6.3	26	0	[33,34]
Ammonia	-5.1	32		[35]
Dimethylformamide	-7.3	24.2	-10	[36]
Dimethyl sulphoxide	-7.5	23.1	-11.8	[37]
Fe^{3+} Water	3.5	8.9	-13	[38]
Methanol	3.4 to 3.7	10.1 to 10.7	-8 to -7	[39,40]
Ethanol	4.3	6.2	-18	[40]
Dimethylformamide	1.5	10.1 to 12.5	-16.5 to -10	[40,41]
Dimethyl sulphoxide	1.7	~10		[40,42]
Acetonitrile	<1.6			[40]
TM: Second row				
Ru^{3+} Water	<-6			[43
Rh^{3+} Water	-7.5	32	$+14$	[44
Lanthanides				
La^{3+} Water	7.3			[45
Gd^{3+} Water	9.0 or 9.3			[46,47
Dy^{3+} Water	7.8			[4ε

[a] Units of k are sec^{-1}; [b] see also footnote on page 314.

11.4   MECHANISMS

Solvent exchange (this Chapter) and complex formation (next Chapter) are special cases of nucleophilic substitution reactions of metal complexes. Several standard textbooks have discussed the mechanisms of such substitution reactions at length [24, 51, 52]. This section will merely outline the basic pattern and indicate the general approaches used in establishing mechanisms of substitution at inorganic centres.

11.4.1 Mechanisms of Nucleophilic Substitution

The basic classification of nucleophilic substitutions is into dissociative and associative, or unimolecular and bimolecular, processes. These are the $S_N 1$ and $S_N 2$ mechanisms of organic chemistry [53]. Many inorganic chemists use a slightly more detailed classification, and subdivide the range of possible mechanisms into the following four groups.

D **mechanism.** In this limiting dissociative mechanism a transient intermediate of reduced coordination number is generated. It persists long enough to be able to discriminate between potential nucleophiles in its vicinity. In particular it can choose between reacting with the added nucleophile or recombining with the group which has just been lost. The latter alternative gives, of course, no nett reaction.

$$ML_6^{n+} \rightleftharpoons ML_5^{n+} + L$$
$$\downarrow L'$$
$$ML_5L'^{n+}$$

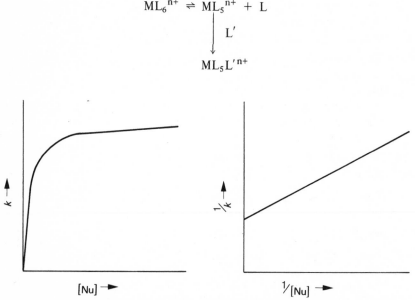

Fig. 11.3 The dependence of rate constant on incoming nucleophile concentration for *D* substitution.

The dependence of observed rate constant on the concentration of the incoming ligand has the characteristic shape shown in Fig. 11.3. The limiting rate will be the same for all incoming nucleophiles, but the curvature at lower incoming nucleophile concentrations reflects the competition between incoming and out-going groups. This situation is best illustrated for substitution at complexes of the type $Fe(CN)_5(OH_2)^{3-}$, discussed in Chapter 12.7.4.

I_d **mechanism.** This is the dissociative interchange mechanism. Here transition state formation involves considerable extension of the metal to leaving group bond, but very little interaction between the metal ion and the incoming group in the transition state. Nonetheless this mechanism does involve outer-sphere association between the starting complex and the incoming nucleophile. Because of this the incoming group is suitably placed to enter the primary coordination sphere of the metal ion as soon as the out-going group has left (Fig. 12.1).

$$ML_6^{n+} + L' \rightleftharpoons ML_6^{n+},L'$$

$$\downarrow \text{ rate-determing step}$$

$$[L_5M...L,L']^{\ddagger n+}$$

$$\downarrow$$

$$ML_5L'^{n+} + L$$

The outer-sphere associated species ML_6^{n+},L' is simply an ion-pair if the complex and incoming group have opposite charges. However it is not necessary for the incoming group to be charged for this mechanism to operate. In particular, when solvolysis or solvent exchange reactions take place the incoming L' in the associated species ML_6^{n+},L' will be present in the secondary solvation shell of the complex ML_6^{n+} and so suitably situated for interchange. This I_d mechanism features prominently in the discussion of formation reactions of metal complexes in Chapter 12.

I_a **mechanism.** Here, just as in the I_d mechanism, there is interchange of ligands between the primary and secondary coordination regions. However, this time there is significant interaction between the incoming group and the metal ion in the transition state.

A **mechanism.** At this extreme the main feature is the formation of an intermediate, rather than a transition state, of increased coordination number:

$$ML_6^{n+} + L' \longrightarrow ML_6L'^{n+} \longrightarrow ML_5L' + L.$$

As indicated at the beginning of this section, there is a continuous gradation of mechanisms from D via I_d and I_a to A. The extent of interaction between the metal cation and the incoming group in the transition state corresponds to this state of affairs. It may be anything between nil (D) and predominant (A), and there are no hard and fast boundaries here.

11.4.2 Diagnosis of Mechanism

A wide variety of methods are used in attempting to establish the mechanisms of reactions. The choice of method depends on circumstances. In certain cases, for instance solvent exchange, the very nature of the system rules out some methods of diagnosis.

Rate law determination. A first-order rate law is often indicative of a unimolecular (dissociative) reaction, a second-order rate law of a bimolecular (associative) reaction. However this simple criterion must be applied with caution, because both D and I_d mechanisms may give second-order rate laws under certain conditions. As is evident from Fig. 11.3, the curve characteristic of a D mechanism approximates to a straight line at low concentrations of the incoming nucleophile. If this nucleophile is weak the straight portion may extend to the highest accessible nucleophile concentration, giving the misleading impression that the reaction in question is first-order in incoming nucleophile and second-order overall. Similarly, if the association between the complex and the incoming ligand is weak (as it usually is), then an I_d process also appears to follow a second-order rate law (see Chapter 12).

Rate comparisons. If a complex reacts with various nucleophiles at different rates, an associative mechanism is indicated. However, if it reacts with various nucleophiles at the same rate then a dissociative mechanism seems more probable. This approach, using comparisons of rates for solvent exchange with those for reaction with a range of incoming ligands, has proved valuable in the assignment of mechanisms both for solvent exchange and for complex formation.

Activation parameters. As in the case of rate comparisons, comparisons of activation enthalpies and entropies for series of related reactions may reveal whether or not they share a common dissociative rate-determining step. Diagnosis of mechanism from the enthalply of activation for a single reaction is rarely attempted, but diagnosis from the entropy of activation is more common. This approach is based on the assumptions that a dissociative process will have a positive activation entropy and that an associative process will have a negative activation entropy. It is useful in situations where solvation changes on forming the transition state from the initial state are relatively small, but it can be dangerous in other circumstances.

Activation enthalpies and entropies are readily obtainable from the temperature variation of reaction rates,

$$\left. \frac{\partial \ln k}{\partial T} \right|_P = \frac{E_a}{RT^2}$$

and thence $\Delta H^{\ddagger}$ and $\Delta S^{\ddagger}$. Establishing the variation of rates with pressure is technically more difficult, but the volumes of activation that can thereby be established,

$$\left. \frac{\partial \ln k}{\partial P} \right|_T = \frac{\Delta V^{\ddagger}}{RT},$$

are easier to visualise and so easier to use in the diagnosis of substitution mechanisms than enthalpies and entropies of activation. A positive activation volume suggests a dissociative process, a negative activation volume an associative process. However here, as in the case of activation entropies, it is essential to take account of volume changes both of the reactants and of solvating solvent in generating the transition state. Activation volumes can be valuable guides to mechanism for reactions involving negligible separation or cancellation of charge in transition state formation.

The first differentials of rate with respect to temperature and pressure give activation enthalpies, entropies, and volumes, and are widely used in establishing reaction mechanisms. The second differentials with respect to temperature give the heat capacity of activation and with respect to pressure the compressibility of activation. They have recently found limited use in mechanism diagnosis.

Other methods. Information on the mechanism operating in a given reaction may also be obtained from one or more other methods. The variation of rate with ionic strength indicates changes of charge involved in transition state formation, and can help in deciding between dissociative and associative mechanisms. The variation of rate with ligand substituent can prove informative when a series of compounds differing only in this respect can be synthesised and studied. The variation of reaction rate with solvent composition, long used in the diagnosis of mechanism in organic chemistry, has been used in probing mechanisms of inorganic reactions. Here however the results are sometimes controversial.

Thus the problem of establishing the mechanism of an inorganic reaction can be approached in many ways. The application of a number of these methods to solvent exchange, complex formation, and redox reactions will be detailed later in this Chapter and in Chapters 12 and 13 respectively.

11.4.3 Mechanisms of Solvent Exchange

Two fundamental questions have to be answered, firstly whether solvent exchange is dissociative or associative, and secondly whether the mechanism varies according to the cation's size, charge, and position in the Periodic Table. One would also like to know if there is any variation in mechanism with variation in solvent for a given cation.

As already mentioned, the very simplicity of the solvent exchange reaction brings problems in ascertaining its mechanism. One advantage is that there are no worries about ion-association and similar complications (contrast formation kinetics, in the next Chapter). On the other hand, the solvent is the outgoing group, the incoming group, and, necessarily, the bulk solvent medium, and this means that there are fewer potential variables to press into service in elucidating the mechanism. In particular, because the solvent is present in large and constant excess, the order of reaction cannot be determined with respect to the solvent. Mixed solvents are potentially useful here, though solvent mixtures in which there is much intercomponent or intracomponent interaction must be avoided. Consequently it is difficult to establish a reaction order for water by this means, but informative results have been obtained for exchange of less complicated solvents in appropriate dilute solutions in inert solvents. For example, the rate of trimethyl phosphate exchange at Be^{2+} in trimethyl phosphate+dichloromethane solvent mixtures does not fluctuate significantly with variation of the trimethyl phosphate concentration [23]. Similar observations and deductions have been made for Co^{2+} and Ni^{2+} in dimethyl sulphoxide+dichloromethane mixtures [12, 13], and for these two cations and Cr^{3+} in dimethyl sulphoxide+nitromethane mixtures [12, 54]. The results obtained for the Cr^{3+} system do not accord with the determined activation volume for dimethyl sulphoxide exchange, which strongly suggests an associative mechanism (see below).

What guidance there is to establishing mechanisms of solvent exchange, especially in pure (that is, single) solvents, comes principally from activation parameters. The actual magnitude of the activation enthalpy for solvent exchange may not be diagnostic of mechanism, but useful information may be yielded by comparison of values for related cations. This point is most strikingly illustrated by comparing activation enthalpies for water exchange at Al^{3+} and at Ga^{3+}. These are 27 and 6.3 kcal mol^{-1} respectively. Such a large difference indicates a difference of mechanism. The higher activation enthalpy for exchange at the aluminium suggests a dissociative mode of activation at that centre, and the much lower activation enthalpy for exchange at gallium is consistent with an associative mechanism. In this case, bond formation between the gallium and the incoming water assists in the separation of the outgoing water in generating the transition state. These conclusions are consistent with the observed activation entropies (see below).

Although the activation enthalpy for solvent exchange at one particular cation has no diagnostic value, individual activation entropies may have. A broad distinction can be made, with markedly positive activation entropies reflecting the greater freedom in the transition state for a dissociative process, and markedly negative values indicating an associative process, where the solvent molecule incorporated into the transition state undergoes some loss of freedom. In this context one should remember that an assignment of mechanism is implicit in the derivation of these activation entropies. The choice between expressing solvent exchange rate constants in first-order or second-order units will be reflected in the numerical value of the activation entropy obtained. The difference is not large. For water exchange at nickel(II), the respective activation entropies are -2 cal deg^{-1} mol^{-1} assuming a first-order process, -6 cal deg^{-1} mol^{-1} assuming a second-order process [17].

The most striking example of the assignment of mechanism from activation entropies is provided by water exchange at the Group IIIA cations Al^{3+} and Ga^{3+}. The value of $+28$ cal deg^{-1} mol^{-1} for the former strongly suggests dissociative activation, while the value of -22 for the latter equally strongly suggests associative activation. These conclusions are consistent with tentative conclusions derived from the respective activation enthalpies. They are also consistent with the greater ease of formation of a transition state of increased coordination number expected for the larger cation, Ga^{3+}.[†] These conclusions are strengthened when rates for water exchange are compared with those for complex formation. Al^{3+}aq reacts with SO_4^{2-} at a rate comparable with that for water exchange under analogous conditions, indicating a common, dissociative, rate-determining step. The rate constant for the reaction of Ga^{3+}aq with SO_4^{2-} is very different from that for water exchange, suggesting an associative mechanism for one or both of the processes [55].

Of all the activation parameters perhaps activation volume, $\Delta V^{\ddagger}$ is the most easily visualised when applied to the diagnosis of mechanism. $\Delta V^{\ddagger}$ values are more revealing guides to mechanism in solvent exchange reactions than in some others. When a reaction involves charge separation or charge cancellation, the measured activation volume represents the resultant of the volume change of the actual reactant species together with any volume change arising from solvating solvent in transition state formation. Consequently unequivocal deduction of mechanism is ruled out by the the need to separate the measured activation volume into its two contributions, sometimes in a fairly arbitrary manner. When solvent exchange takes place at a cation, the outgoing and incoming groups are uncharged, so there is no reason for a significant volume change of solvating solvent in transition state formation. Moreover, because the reactants and products are identical, $\Delta V^{\ominus}$ for the reaction is zero, and there is no need to worry how far $\Delta V^{\ominus}$ should be reflected in $\Delta V^{\ddagger}$. So in these solvent exchange

† Similar entropies of activation for methanol exchange at Al^{3+} and at Ga^{3+} have recently been reported, at -29 and -26 cal deg^{-1} mol^{-1} respectively [27]. This suggests an associative mechanism for both cations, which is in uncomfortable contrast with the situation in aqueous solution. The activation parameters for methanol exchange at Al^{3+} are very different from those for exchange of water, dimethylformamide, or dimethyl sulphoxide (see Table 11.5).

reactions, the determined activation volume corresponds simply to the change in volume of the reactants going from the initial state to the transition state.

$$\Delta V^{\ddagger} \; = \; \Delta V^{\ominus} \, (\text{transition state}) \; - \; \Delta V^{\ominus} \, (\text{initial state}).$$

A positive value for $\Delta V^{\ddagger}$ indicates reactant expansion in forming the transition state, the result expected for a dissociative process (Fig. 11.4(a) shows this). A negative volume of activation indicates contraction during transition state generation, which would be expected for an associative process, as shown in Fig. 11.4(b). The number of determinations of $\Delta V^{\ddagger}$ for solvent exchange, indeed for any type of reaction, is relatively small. This reflects the greater degree of technical difficulty inherent in measuring rates as a function of pressure. However, in the last few years the determination of activation volumes for inorganic reactions has become considerably more popular, both for 'slow' reactions and for such 'fast' reactions as can conveniently be monitored by the temperature-jump method. Sufficient results for solvent exchange are now available for some pattern to be apparent, though if the picture is to become clear solvent exchange at mixed ligand-solvent complexes as well as at pure solvento-complexes must be considered.

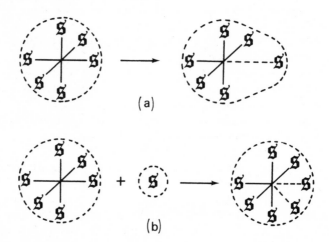

Fig. 11.4   Volume changes on transition state formation for (a) dissociative and (b) associative solvent exchange.

The one hexa-aquo-cation to have its volume of activation for water exchange determined is $Cr(OH_2)_6^{3+}$, which is almost uniquely inert and stable. In this case, $\Delta V^{\ddagger} = -9.3 \pm 0.3 \; \text{cm}^3 \, \text{mol}^{-1}$ [34]. This value approaches that of about $-15 \; \text{cm}^3 \, \text{mol}^{-1}$ which is the extreme that can readily be visualised for an associative mechanism, assuming that the incoming water molecule is accommodated into the transition state without any increase in volume accompanying the conversion of the octahedral initial state into the seven-coordinate transition state. Interestingly, $\Delta S^{\ddagger}$ for water exchange at $Cr(OH_2)_6^{3+}$ is $+1$ cal deg^{-1} mol^{-1} [33]; the activation entropy appears of less diagnostic value than the activation volume here.

The activation volumes for dimethylformamide and for dimethyl sulphoxide exchange at $Cr(dmf)_6^{3+}$ and at $Cr(dmso)_6^{3+}$ are also negative, -6.3 [37] and -11.3 [54] cm^3 mol^{-1} respectively. Again an associative interchange mechanism is indicated, though other evidence does not always support this assignment [55]. Mechanisms of substitution at chromium(III) are still the subject of controversy [56]. The strong indications of associative mechanisms for solvent exchange obtained from activation volumes provide a valuable piece of evidence to include in such discussions.

It would be useful to have, for comparison, the activation volume for water exchange by a dissociative mechanism. Unfortunately, studies of water exchange at Co^{3+}aq (the most likely candidate for solvent exchange by a dissociative pathway) are almost impossible because of the relatively rapid cobalt(III) oxidation of water. However, many mixed aquo-ammine complexes of cobalt(III), and particularly those containing a high proportion of ammonia, are stable with respect to oxidising water as well as being substitution-inert. Thus it is a straightforward matter to determine their activation volumes for water exchange. The activation volume for water exchange at $Co(NH_3)_5(OH_2)^{3+}$ can be compared directly with those for water exchange at its chromium(III), rhodium(III), and iridium(III) analogues, as in Table 11.6 [57, 58]. Here the contrast between the positive activation volume for water exchange at the cobalt(III) complex and the negative activation volume for water exchange at the analogous chromium(III) complex indicates dissociative and associative mechanisms respectively. The negative value for the rhodium(III) complex suggests considerable associative character here too, as also for the iridium(III) analogue. Associative character to water exchange at chromium(III) can be ascribed to the lower electron density (d^3) in the t_{2g} orbitals compared with that for the d^6 cobalt(III). Hence nucleophilic approach of water to the metal centre is less impeded in the former case. Associative attack at rhodium(III) and at iridium(III), despite the d^6 electron configuration, can be rationalised in steric

Table 11.6 Activation volumes for water exchange with aquo-ammine complexes.

	$\Delta V^{\ddagger}$/cm^3 mol^{-1}	
$Co(NH_3)_5(OH_2)^{3+}$	$+1.2$ $(25, 35°C)$	
$Cr(NH_3)_5(OH_2)^{3+}$	-5.8 $(25°C)$	[57]
$Rh(NH_3)_5(OH_2)^{3+}$	-4.1 $(35°C)$	
$Ir(NH_3)_5(OH_2)^{3+}$	-3.2 $(70.5°C)$	[58]

terms. The formation of a transition state of increased coordination number should be easier at the larger centres (ionic radii: Co^{3+} 0.665, Rh^{3+} 0.805, Ir^{3+} 0.87Å [59]).

A word of caution is required in connection with activation volumes for water exchange at cobalt(III) complexes. A surprisingly large difference exists between the value for $Co(NH_3)_5(OH_2)^{3+}$ (Table 11.6) published in 1958 [60], and that for water exchange at the very similar complex $trans\text{-}Co(en)_2(OH_2)_2^{3+}$, which is $+14$ cm^3 mol^{-1} [57]. Water exchange at $trans\text{-}Co(en)_2(SeO_3)(OH_2)^+$ and its conjugate acid $trans\text{-}Co(en)_2(SeO_3H)(OH_2)^{2+}$ has an activation volume of about $+8$ cm^3 mol^{-1} in both cases [57]. Technical difficulties may have led to a low value in the early work, but at any rate the activation volumes in all cases are positive and so qualitatively consistent with the operation of a mechanism of predominantly dissociative character.

11.5   REACTIVITIES

One would like to be able to explain fully the wide variations of kinetic parameters, k, $\Delta H^{\ddagger}$, and $\Delta S^{\ddagger}$, for solvent exchange as cations and solvents are varied. The ideal state of full and consistent comprehension seems far away at the time of writing. In this section it is only possible to point out some of the factors involved in determining reactivities.

11.5.1 Cation Variation

sp-**Elements.** The important factors controlling solvent exchange rates for non-transition metal cations in solution are the charge and size of the cation. This is as would be expected from an electrostatic model. Information about solvent exchange rates at 'soft' cations such as Cd^{2+} and Hg^{2+} is insufficient to determine how far covalent interactions in such cases affect the general picture.

Water exchange at the alkali metal cations, whose charge is only 1+, is very fast. It is also fast at the larger cations of charge 2+, such as the alkaline earth series Ca^{2+} to Ba^{2+}. For small cations of charge +2, Mg^{2+} and especially Be^{2+}, and for cations of charge +3, rates of water exchange are considerably less. At these latter cations rates of water exchange, and of exchange of other solvents, are often in the range which can be monitored by n.m.r. techniques.

Transition metals. Cation charge and size are still important in determining rates of solvent exchange at transition metal cations, but here another factor is also operative, one of comparable importance. This additional factor is crystal field stabilisation, which varies with and reflects the distribution of d electrons in these cations. Despite their name, crystal field effects are just as relevant to co-ordinated solvents as to any other coordinated ligands (see Chapter 3.1.1). Cations which have large crystal field stabilisation energies (CFSE) are likely to have large crystal field activation energies (CFAE) with respect to substitution processes:

$$CFAE = CFSE(\text{transition state}) - CFSE(\text{initial state}).$$

Particularly for a dissociative process, the cation with the biggest CFSE has the most to lose when one ligand is partially removed in the transition state. The electronic configurations with the most important crystal field stabilisation energies are d^3, d^6 (low-spin) and d^8. Water exchange which is relatively or very slow has been established for such cations as V^{2+}aq and Cr^{3+}aq (both d^3), Rh^{3+}aq (d^6), and Ni^{2+} aq (d^8).

For the case of solvent exchange at a series of first row transition metal 2+ cations, Table 11.7 illustrates the qualitative correlation of both rates and activation enthalpies with crystal field effects [13]. Detailed inspection of solvent exchange rates and of activation enthalpies shows that it would be incautious to rely totally on crystal field effects as an explanation for reactivity patterns in the area of transition metal cations. In any case, difficulties arise as soon as one tries to calculate CFAE values. Of the four cations in Table 11.7, two (Fe^{2+} and Co^{2+}) should have *negative* CFAE values for a dissociative process. The ratio of solvent exchange rates at Co^{2+} and Ni^{2+} shows a considerable variation between the solvents water, methanol, and dimethylformamide, despite the fact that these three solvents all have the same Dq value (Chapter 3.1.1), 850 cm^{-1} for the Ni^{2+} cation. The variation of rate ratios between these three solvents is comparable with the difference between ratios for water and ammonia exchange (a pair of solvents with Dq values as different as 850 and 1080 cm^{-1}) for Ni^{2+} [61].

The importance of cation charge is immediately obvious from comparisons of rates of solvent exchange at analogous 2+ and 3+ ions. For example, water

Table 11.7 Correlation of rate constants (k) and activation enthalpies ($\Delta H^{\ddagger}$) for solvent exchange at some first row transition metal 2+ cations with crystal field stabilisation energies (CFSE) [17, 19].

		Mn^{2+}	Fe^{2+}	Co^{2+}	Ni^{2+}
Electron configuration		d^5	d^6	d^7	d^8
CFSE/Dq		0	4	8	12
$\log_{10} k_{25}{}^a$	water	7.5	6.5	6.1 to 6.4	4.4 to 4.6
	acetonitrile	7.1	5.7	5.5	3.5
$\Delta H^{\ddagger}$/kcal mol^{-1}	water	8.1	7.7	8.0 to 11.9	10.3 to 12.1
	acetonitrile	7.2	9.6	11.4	15.3

a Units of k are sec^{-1}; values of k from Tables 11.2, 11.3, ref. [17].

exchange at $Cr^{3+}aq$ (d^3) is about 10^8 times slower than at $V^{2+}aq$ (also d^3). A combination of 3+ charge and maximum crystal field stabilisation for the low-spin d^6 configuration results in the slowest rates of exchange, for instance at $Rh^{3+}aq$ (Table 11.5).

The special cases of solvento-complexes with marked tetragonal distortion, for example $Cr^{2+}(d^4)$ and $Cu^{2+}(d^9)$, undergo particularly rapid solvent exchange (Table 11.4). Presumably the two distant solvent molecules are loosely held and thus readily and easily available for exchange with bulk-secondary solvent. The importance of this Jahn-Teller distortion for the hexasolvento-complexes is neatly illustrated by kinetics of acetonitrile exchange at $Cu(MeCN)_6^{2+}$ and at mixed ligand–acetonitrile complexes of Cu^{2+}. As soon as the non-solvent ligands prevent the facile Jahn-Teller distortion pathway, the exchange rate drops dramatically (Chapter 11.6.4) [62].

Lanthanides. Water exchange with lanthanide aquo-cations is very fast, which might be considered surprising in view of their 3+ charge. Thus the water exchange rate constant for $Gd^{3+}aq$ at $25°C$ is 2×10^9 sec^{-1}. Rate constants for water exchange at $Tb^{3+}aq$, $Dy^{3+}aq$, $Ho^{3+}aq$, $Er^{3+}aq$, and $Tm^{3+}aq$ lie between 3×10^6 and 1.4×10^8 sec^{-1} (at $24°C$) with faster exchange at the larger cations here [63]. The activation enthalpy for water exchange at $Dy^{3+}aq$ is less than 5 kcal mol^{-1}. Such large rate constants are usually accounted for in terms of the high coordination number of these cations in solution. Coordination numbers of eight or nine are thought to involve this number of water molecules at two or more somewhat different distances from the central lanthanide cation. Here, as in the Jahn-Teller distorted transition metal solvento-cations mentioned above, the more distant water molecules are fairly loosely bound and susceptible to exchange. However, it should be remembered that a hydration number of six (with no apparent reason for deviation from octahedral symmetry) has been proposed for some of the lanthanide(III) cations from n.m.r. studies of their solutions in mixed aqueous solvents at low temperatures (see Chapter 5.3.8).

11.5.2 Solvent Variation

Before embarking on explanations of observed variations of solvent exchange rates or activation parameters, it would seem sensible to see if there is a pattern of behaviour common to various different cations. Rates of solvent exchange at the 2+ cations of the first row of transition metals fall into a reasonably coherent pattern, as far as one can tell from the modest number of results available. This is illustrated in Fig. 11.5, where cobalt(II) and manganese(II) values have been plotted against the respective values for nickel(II). Both comparisons indicate approximately linear correlations (at least of logarithms of rate constants) with the relative rates of solvent exchange in the order

$$NH_3 > OH_2 > DMSO, MeCN, DMF > MeOH.$$

This order bears little resemblance to solvent E_T values, Gutmann donor numbers, or any of the other more or less empirical solvent parameters listed in Chapter 1.3. Attempts to establish a correlation with the Gutmann solvent donor numbers have recently been resumed, with some limited success in certain restricted areas [64].

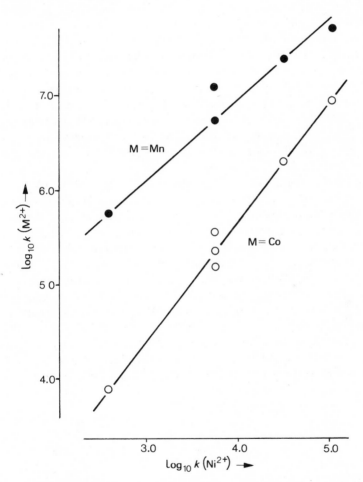

Fig. 11.5 Correlation of solvent exchange rates at 2+ cations of some first row transition metals.

Attempts to extend correlations of the type shown in Fig. 11.5 to 3+ cations meet with disappointment. A plot of logarithms of rate constants for solvent exchange at Fe^{3+} versus analogous values for Ni^{2+} shows no correlation.

There are too few rate constants for solvent exchange at Fe^{3+}, or at any other M^{3+} for that matter, to tell whether any correlation with E_T values or Gutmann solvent numbers exists.

The reason for the lack of success in these correlations may lie in the number of variables involved. Solvent exchange depends on the nature of the solvent in several ways. The nucleophilicity of the solvent will obviously affect the ease of departure of a coordinated solvent molecule from the cation. Equally the nucleophilicity of the solvent may affect the ease of approach of the in-coming solvent molecule to the metal ion. These two factors are opposite in effect, but the importance of the latter will depend greatly on the mechanism, be it D, I_d, I_a, or A. Moreover, rates of solvent exchange are likely to be affected by steric factors. Many of the relatively simple organic solvents are much bulkier than water. As a result, there may be strain in the initial state, which could be relieved by the generation of a dissociative transition state. On the other hand, it may be very difficult to pack an extra solvent molecule around the cation to generate the transition state for an associative solvent exchange process. All these steric factors are likely to be more important for M^{3+} cations than for M^{2+}, since the former are generally smaller.

The comparison of solvent exchange rates at Ni^{2+} and at Fe^{3+} lends some support to the idea of steric effects playing an important role in determining reactivities. The solvents concerned can be grouped into two classes. In the first are the alcohols, with rates of solvent exchange at Fe^{3+} about twice those at Ni^{2+}. In the second are the remainder, whose rates of exchange at Fe^{3+} are only about one two-hundredth of those at Ni^{2+}. There are two groups or atoms bonded to the ligating atom of the solvent in the alcohols, alkyl-carbon and hydrogen. In the other solvents the ligating atom is bonded directly to only one other atom of the solvent molecule, O only to S in DMSO, N only to C in MeCN. Hence there may be some strain in packing six solvent molecules around the small Fe^{3+} cation where the alcohols are concerned, strain which may decrease or disappear when the same molecules are packed around the larger Ni^{2+}, or when solvents such as DMF or DMSO are packed around Fe^{3+}. This theory is illustrated diagrammatically in Fig. 11.6 [40].

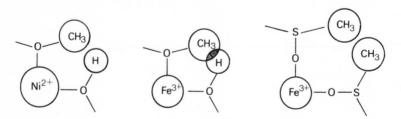

Fig. 11.6 Diagrams illustrating a possible source of steric strain in Fe^{3+}-alcohol solvates, and its absence from Ni^{2+}-alcohol and Fe^{3+}-dimethyl sulphoxide solvates (not to scale).

Consequently it would seem over-optimistic to expect a simple correlation between solvent exchange rates and any single solvent parameter[†] to apply to solvento-cations in general. The discussion in the previous paragraphs suggests that any generalised correlation would have to be cast in the form of a multi-parameter equation.

11.6 MIXED SOLVENT-LIGAND COMPLEXES

In a complex of the general type $ML_xS_{y-x}^{n+}$, the presence of the ligands L may be expected to have at least some effect on the rate of exchange of the remaining solvento-ligands S, in comparison with the solvent exchange rate for the simple solvate MS_y^{m+}. Solvent exchange rates in mixed complexes have been extensively studied for nickel(II), less so for cobalt(II), manganese(II), and the special case of oxovanadium(IV). Reactivities in these and related systems will be described in this section. In fact non-leaving ligand effects have been even more thoroughly studied for 'ordinary' substitution reactions, which are outside the scope of this book, than they have been for solvent exchange or complex formation.

11.6.1 Aquo-ligand Complexes of Transition Metal(II) Cations

Selected rates of water exchange at Ni^{2+}, Co^{2+}, and Mn^{2+} are listed in Table 11.8 [65–71]. The division of this Table into two groups of ligands reflects the two types of rate trend observed. The first group contains simple ligands such as ammonia, chloride, and thiocyanate, in fact ligands in which σ-bonding between ligand and cation is of predominant importance. Replacement of coordinated water by ligands in this group results in markedly faster water exchange rates for the remaining coordinated water molecules. Indeed there is a marked correlation of water exchange rate constants with the number of water molecules remaining coordinated, as shown in Fig. 11.8 for Ni^{2+}. This correlation operates regardless of the nature of the replacing ligand, or even of its charge. This effect of ligands can be rationalised in terms of the σ-electron donation from the ligand to the central cation reducing the effective charge on this metal(II) centre. The remaining metal(II) to water bonds will thereby be weakened, and water exchange with bulk solvent made easier and faster. It is surprising that ligands as different as ammonia and the halides have such similar effects.

The second group of ligands includes diimines of the bipyridyl and phenan-

---

† Several authors have considered the role of solvent structural features in solvent exchange reactions, especially with M^{2+} cations. The relationship between the kinetics and mechanism of the solvent exchange process and solvent viscosity and vaporisation has been discussed in detail. Readers interested in the role of solvent structure in these reactions are referred to the original Bennetto and Caldin papers (Bennetto, H. P. and Caldin, E. F., *J. Chem., Soc., A*, 2198 (1971); *J. Solution Chem.*, 2, 217 (1973)) and to subsequent discussions, such as Chattopadhyay, P. K. and Coetzee, J. F., *Inorg. Chem.*, 12, 113 (1973) and Tanaka, M., *ibid.*, 15, 2325 (1976).

Table 11.8 Rate constants for water exchange at mixed ligand-aquo-complexes of some transition metal(II) cations. The cited rate constants are per water molecule. Values for Ni^{2+} ($25°C$) are from reference [65], for Co^{2+} ($25°C$) from reference [21], and for Mn^{2+} ($0°C$) from reference [66], except where otherwise indicated.

	$10^{-5} k/\text{sec}^{-1}$			$10^{-5} k/\text{sec}^{-1}$	
	Ni^{2+}	Co^{2+}		Ni^{2+}	Mn^{2+}
$M(OH_2)_6^{2+}$	0.32	22.4	$M(OH_2)_6^{2+}$	0.32	59
$M(OH_2)_5(NH_3)^{2+}$	2.5	155			
$M(OH_2)_5(NCS)^+$		95^d			
$M(OH_2)_5Cl^+$	1.4	170			
$M(OH_2)_4(NH_3)_2^{2+}$	6.1	650	$M(OH_2)_4(bipy)^{2+}$	0.49	
$M(OH_2)_4(en)^{2+}$	4.4		$M(OH_2)_4(phen)^{2+}$		130
$M(OH_2)_4(NCS)_2$		3000^d			
$M(OH_2)_4(mal)$		2200			
$M(OH_2)_3(NH_3)_3^{2+}$	25		$M(OH_2)_3(terpy)^{2+}$	0.52	
$M(OH_2)_3(dien)^{2+}$	18^a		$M(OH_2)_3(sb)^{2+}$	0.38^e	
$M(OH_2)_3(NCS)_3^-$	11	$>5000^d$			
$M(OH_2)_2(en)_2^{2+}$	54		$M(OH_2)_2(bipy)_2^{2+}$	0.66	
$M(OH_2)_2(tren)^{2+}$	$8.2, \sim90^b$		$M(OH_2)_2(phen)_2^{2+}$		310
$M(OH_2)(edta)^{2-}$	7^c				

a Ref. [67]; *b* ref. [68]; *c* ref. [69]; *d* ref. [70], $27°C$; *e* ref. [71], the ligand sb is shown in Fig. 11.7.

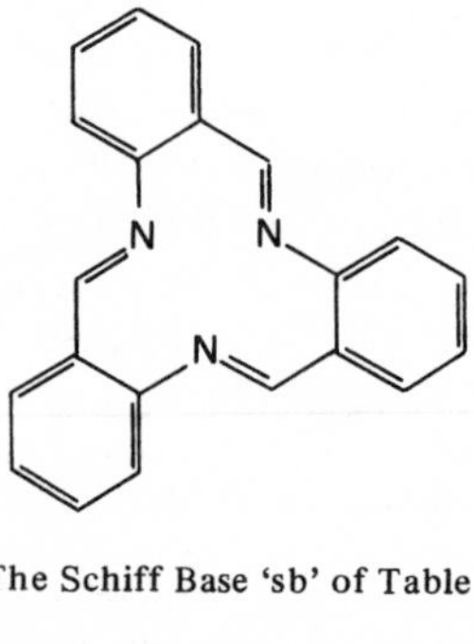

Fig. 11.7 The Schiff Base 'sb' of Table 11.8.

throline type. Here the replacement of coordinated water has very little effect on rates of water exchange for the remaining coordinated water molecules. This may be a consequence of the strong π-bonding which exists between metal

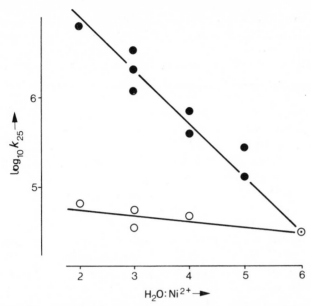

Fig. 11.8   The variation of rate constant for water exchange (at $25°C$, k_{25}) with the number of water ligands per Ni^{2+}; ○ diimine ligands, ● other ligands. The kinetic data are taken from Table 11.8.

atom and ligand in these cases. The effect of σ-electron donation by the ligating nitrogen is counterbalanced by π-electron donation from metal cation back to the ligand (Fig. 11.9). In consequence the effective charge on the central metal atom is not altered overmuch by coordination of this type of ligand. The metal–water bonds remain as strong as those in the hexa-aquo-complex, and the rate of water exchange is little affected.

Fig. 11.9   Metal ion to ligand σ- and π-bonding.

Whether the effect is small or large, introduction of other ligands seems in practically every case to increase the lability of the remaining water molecules. This general pattern extends, of course, to biochemical systems. The coordination of adenosine triphosphate leads to an increased rate constant for water exchange of $5.0 \times 10^7 \ sec^{-1}$ at $25°C$ for the $Mn(atp)aq^{2-}$ anion [72]. The rate constant for water exchange with manganese(II) bound to rabbit muscle pyruvate kinase is appreciably faster than water exchange with $Mn^{2+}aq$ [73]. The only exception to the general pattern appears to be Mn(pyruvate carboxylate)-

(OH_2). Here the rate constant for water exchange, 1.0×10^6 sec^{-1} at $25°C$ [74], is less than that for $Mn^{2+}aq$ (2.2×10^7 sec^{-1} [3]). One is tempted to attribute this to some exceptional property of the complex, in which the biological ligand exerts some specific retarding effect on the coordinated water, probably through hydrogen bonding.

In virtually all the compounds cited in this section so far, only one rate constant for water exchange has been reported, even when there are two or more differently situated water molecules in the complex. Consequently only one rate constant has been reported for water exchange at each complex $M(OH_2)_5(NH_3)^{n+}$, despite the fact that such complexes have one water molecule *trans-* to the ammonia, and four waters *cis-* to the ammonia. It is not usually clear whether the reporting of a single rate constant means that water exchange takes place at the same rate at the different sites or, at the other extreme, that it takes place at such widely different rates that exchange is only monitored at one stereochemical type of site. In contrast, in the case of the aquo-$\beta\beta'\beta''$-triaminotriethylaminenickel(II) cation, $Ni(tren)(OH_2)_2^{2+}$ (Fig. 11.10), two rate constants have been reported for water exchange, of 8.2 and 90×10^5 sec^{-1} [68]. Comparison of these values with others in Table 11.8, and consideration of the relative *trans-*effects of $-NH_2$ and $>N$ on acetonitrile exchange at Ni^{2+} (see below), suggest that the fast rate applies to the water *trans-* to the $-NH_2$ of the coordinated tren, the slower rate to water *trans-* to the tertiary nitrogen (Fig. 11.10).

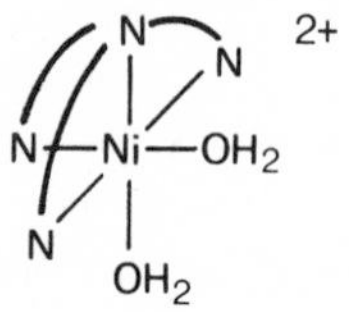

Fig. 11.10 Schematic representation of the geometry of the $Ni(tren)(OH_2)_2^{2+}$ cation.

Activation parameters have been estimated for a variety of mixed aquoligand complexes of this type. A selection of values is given in Table 11.9. Generally the uncertainties in the values cited are sufficiently large to prevent the confident establishment of trends, though there seems to be a correlation between increasing rate and decreasing activation enthalpy, at least for manganese(II) [73]. All that can be said with confidence is that the activation parameters cover only a relatively small range for each metal centre.

It seems likely that copper(II) complexes will not conform to the pattern set out here for the other transition metal(II) cations. This may be for reasons associated with the operation of the Jahn-Teller distortion in the symmetrical $Cu(OH_2)_6^{2+}$ cation, but the non-operation of this effect in unsymmetrical mixed complexes. In practice this effect is better illustrated by acetonitrile complexes than by aquo-complexes, as discussed in Chapter 11.6.4 below.

Table 11.9 Activation parameters for water exchange with mixed aquo-ligand complexes.

	$\Delta H^{\ddagger}$ /kcal mol^{-1}	$\Delta S^{\ddagger}$ /cal deg^{-1} mol^{-1}	
$Ni(OH_2)_6^{2+}$	9.9 to 12.1	-5 to $+3$	Table 11.2
$Ni(OH_2)_5(NH_3)^{2+}$	10.6	$+1.8$	
$Ni(OH_2)_5Cl^+$	8		
$Ni(OH_2)_4(NH_3)_2^{2+}$	7.8	-6.0	
$Ni(OH_2)_4(en)^{2+}$	10.0	$+10$	
$Ni(OH_2)_4(bipy)^{2+}$	12.6	$+5.1$	[65]
$Ni(OH_2)_3(NH_3)_3^{2+}$	10.2	$+5.0$	
$Ni(OH_2)_3(terpy)^{2+}$	10.7	-1.0	
$Ni(OH_2)_3(NCS)_3^-$	6	-11	
$Ni(OH_2)_2(en)_2^{2+}$	9.1	$+2.6$	
$Ni(OH_2)_2(bipy)_2^{2+}$	13.7	$+9.2$	
$Co(OH_2)_6^{2+}$	8.0 to 11.9	-4 to $+11$	Table 11.3
$Co(OH_2)_5(NH_3)^{2+}$	12.6	$+16.7$	[21]
$Co(OH_2)_5Cl^+$	13.8	$+21$	
$Co(OH_2)_5(NCS)^+$	10.1		[70]
$Co(OH_2)_4(NH_3)_2^{2+}$	9.4	$+7$	[21]
$Co(OH_2)_4(mal)$	12.9	$+18$	
$Mn(OH_2)_6^{2+}$	7.8 or 8.8	$+5$	[66,75]
$Mn(OH_2)_4(phen)^{2+}$	9	$+6$	[66]
$Mn(OH_2)_2(phen)_2^{2+}$	9	$+8$	
$Mn(OH_2)_2(nta)^-$	6.6	$+6$	[76]
$Mn(OH_2)(edta)^{2-a}$	7.7	$+7$	
$Mn(aq)(pyruvate\ kinase)^{2-}$	6.6		[73]

a Seven-coordinate manganese.

11.6.2 Aquo-ligand Complexes of Transition Metal(III) Cations

Water exchange at mixed aquo-ligand complexes of transition metal 3+ cations has been much less extensively studied than that at 2+ cations. Some rate constants for water exchange are listed in Table 11.10 [60, 77–79]. The contrast between the very small *cis*-effects and the large *trans*-effects of halide ligands in $Cr(OH_2)_5X^{2+}$ is striking. The *trans*-labilisation series $I^- > Br^- > Cl^- > NCS^- > OH_2$ is the same as that observed for rhodium(III). It is difficult to make comparisons with cobalt(III) or platinum(IV) because of parallel iso-merisation and redox processes at these centres. The NO ligand has a particularly large *trans*-effect. This has been rationalised by the hypothesis that $Cr^{III}NO^-$ is equivalent to $Cr^{I}NO^+$. The latter contains a labile d^5 rather than an inert d^3 centre. Water exchange at the nitrosyl and iodo-complexes may well go by a D mechanism, under the influence of the strong *trans*-effects of these ligands (cf. ligand substitution at chromium(III) [80]). Methyl ligands are also very effective at labilising coordinated water molecules, for example in the transition metal(III) cation Me_2Au^+aq and in related species such as $Me_3Pt(OH_2)_3^+$, Me_2Ga^+aq, and $Me_2Sn^{2+}aq$ [29].

Table 11.10 Rates constants for water exchange at mixed aquo-ligand complexes of chromium(III)^b and cobalt(III), at 25°C.

	$10^6 k(cis)/\text{sec}^{-1}$	$10^6 k(trans)/\text{sec}^{-1}$	
$Co(NH_3)_5(OH_2)^{3+}$	5.9		[60]
$Co(en)_2(OH_2)_2^{3+}$	7.5	11.3	[77]
$Cr(OH_2)_5Cl^{2+}$	2.8	33	[78]
$Cr(OH_2)_5Br^{2+}$	3.0		[78]
$Cr(OH_2)_5I^{2+}$	3.1	270	
$Cr(OH_2)_5(NCS)^{2+}$	3.0	17	[78]
$Cr(OH_2)_5(NO)^{2+}$	290^a		[79]

a This value assumes that all five waters are exchanging at the same rate; the comparable value for $Cr(OH_2)_5Cl^{2+}$ is 3.8; *b* the rate of water exchange at $Cr(OH_2)_5(ONO_2)^{2+}$ is thought to be about 10^5 times that for $Cr(OH_2)_6^{3+}$, see Mitchell, M. L., Montag, T., Espenson, J. H., and King, E. L., *Inorg. Chem.*, 14, 2862 (1975).

Water exchange at $FeOH^{2+}aq$ takes place much more rapidly than at $Fe^{3+}aq$. The respective rate constants are approximately 10^4 and 150 sec^{-1} [81]. The rate constant for water exchange at the iron(III)-aquo-pyridine-haemin complex sketched in Fig. 11.11 is similar to that for exchange at $FeOH^{2+}aq$. The rate constant for the biochemical complex is 3.6×10^3 sec^{-1}. Activation parameters have also been determined for water exchange at this haemin derivative. They are $\Delta H^{\ddagger} = 7.4$ kcal mol^{-1} and $\Delta S^{\ddagger} = -16$ cal deg^{-1}mol^{-1}. Interestingly, the kinetic parameters for pyridine exchange at this same complex are very similar to these water exchange parameters [82].

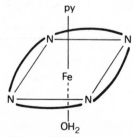

Fig. 11.11 The environment of the iron(III) in the haemin derivative of reference [82].

11.6.3 Oxocation Complexes

Oxovanadium(IV), $VO^{2+}aq$, and oxoactinide(VI), $ActO_2^{2+}aq$, complexes are special cases of mixed ligand-aquo-complexes with one or two oxo-ligands and five or (probably) four water ligands respectively. In the case of $VO^{2+}aq$, some doubt is possible as to whether the determined rate constant applies to the *cis*- or the *trans*- water molecules, or even to both. The variable temperature n.m.r. spectra are ambiguous on this point. The determined rate constants are more likely to apply to exchange between the coordinated water *cis*- to the oxo-ligand and bulk solvent (but see VO^{2+} in dimethylformamide below).

Some kinetic data for water exchange at these oxo-cations are given in Table 11.11 [83–85]. Water exchange at vanadium(IV) is thought to be dissociative in character. Water exchange at $UO_2^{2+}aq$ is fast, considering the uranium has a formal change of 6+, so it is obviously the charge of 2+ on the dioxouranium unit which counts here. The large size of the uranium may also permit some associative character to the exchange mechanism.

Table 11.11 Kinetic parameters for water exchange with oxo-aquo-cations.

	$\log_{10} k_{25}{}^{a}$	$\Delta H^{\ddagger}$/kcal mol^{-1}	$\Delta S^{\ddagger}$/cal deg^{-1} mol^{-1}	
$VO^{2+}aq(H_2O)^{b}$	3.89	12	0	[83]
$(D_2O)^{b}$	3.52	14	+2	[83]
$UO_2^{2+}aq (H_2O)$	~5			[85]

a Units of k are sec^{-1}; *b* at pH = 7; *c* other estimates of the kinetic parameters for water exchange at $VO^{2+}aq$ can be found in ref. [84].

11.6.4 Mixed Complexes in Non-aqueous Solvents

Some work has been done on the kinetics of exchange of non-aqueous solvents with mixed ligand complexes, particularly of nickel(II) and cobalt(II). We shall concentrate on solutions in methanol, acetonitrile, and dimethyl-formamide, since the majority of kinetic studies in this area have been conducted in one of these three solvents.

Methanol. Rate constants for methanol exchange with a variety of mixed complexes are collected together in Table 11.12 [86–93]. It does not usually seem possible to obtain separate rate constants for the *cis*- and *trans*- methanol molecules in complexes $M(MeOH)_5X^{n+}$, though this has been achieved for the case of $Co(MeOH)_5(py)^{2+}$ [91]. In these mixed ligand methanol complexes, as in analogous mixed ligand aquo-complexes, the introduction of a chloride or thiocyanate ligand has a marked labilising effect on the remaining coordinated solvent molecules.

Table 11.12 Rate constants (per methanol molecule) for methanol exchange with mixed complexes. Rate constants for methanol exchange with hexamethanol solvates under analogous conditions are included for comparison.

Compound	Temp./°C	$\log_{10} k^a$	$(\log_{10}k(M(MeOH)_6^{2+}))^a$	
$Mn(MeOH)_5Cl^+$	-17	5.0	(4.8)	[86]
$Co(MeOH)_5Cl^+$	-63	2.9	(-0.5)	[87]
$Co(MeOH)_5(NCS)^+$	25	6.4^b	(4.3)	[88]
$Co(MeOH)_5(OAc)^+$	-70	3.0^c		[89]
$Co(MeOH)_5(OH_2)^{2+}$	-30	2.8	(2.0)	[90]
$Co(MeOH)_5(py)^{2+}$	-30	$2.6^d, 3.3^e$	(2.0)	[91]
$Ni(MeOH)_5Cl^+$	25	5.4	(3.0)	[92]
$Ni(MeOH)_5(OH_2)^{2+}$	25	3.3	(2.9)	[93]

a Units of k are sec^{-1}; b rate constant for exchange of methanol cis- and trans- to the NCS$^-$ ligand; c believed to apply to exchange of methanol molecules cis- to the OAc$^-$ ligand; d for exchange of methanol cis- to py; e for exchange of methanol trans- to py.

Activation parameters for methanol exchange are listed in Table 11.13. Here a pattern of slightly decreasing activation enthalpies seems to emerge when one methanol ligand is replaced by another group. As one would expect from the higher activation enthalpy for methanol exchange at $Ni(MeOH)_6^{2+}$ than for water exchange at $Ni(OH_2)_6^{2+}$, activation enthalpies for methanol exchange with these mixed complexes are larger than those for water exchange with aquo-analogues.

Table 11.13 Activation parameters for methanol exchange with mixed complexes.

	Co^{2+}			Ni^{2+}		
	$\Delta H^{\ddagger}$ /kcal mol^{-1}	$\Delta S^{\ddagger}$ /cal deg^{-1} mol^{-1}		$\Delta H^{\ddagger}$ /kcal mol^{-1}	$\Delta S^{\ddagger}$ /cal deg^{-1} mol^{-1}	
$M(MeOH)_6^{2+}$	13.8	$+7$	[89]	15.8^a	$+8^a$	[92]
$M(MeOH)_5(OH_2)^{2+}$	12.8	$+7$	[90]			
$M(MeOH)_5Cl^+$	11.4	$+10$	[89]	15.2	$+7$	[92]
$M(MeOH)_5(NCS)^+$	11.7	$+10$	[88]			
$M(MeOH)_5(py)^{2+}$ cis	11.9	$+3$	[91]			
trans	11.2	$+2$				

a These values differ from those given earlier. Here we cite only the values which were obtained in the same investigation as those for $Ni(MeOH)_5Cl^+$.

Acetonitrile. Table 11.14 lists rate constants for acetonitrile exchange, and Table 11.15 some activation parameters [62, 94]. From these kinetic parameters for the Co^{2+} and Ni^{2+} series a labilising effect series is derived:

$$-NH_2 > \gg N > -OH \sim MeCN.$$

This series is rationalised in terms of relative degrees of electron donation from ligand to metal, and the consequent variation in charge reduction at the metal(II) centre [94].

Table 11.14 Rate constants (per single acetonitrile molecule) for acetonitrile exchange with mixed complexes. Rate constants for the respective hexaacetonitrile solvates are included for comparison. Values are at $25°C$ except where otherwise indicated. Ligand formulae are shown in Fig. 11.12 [62, 94].

	$10^{-3} k/sec^{-1}$		
	Co^{2+}	Ni^{2+}	Cu^{2+}
Tridentate ligands			
triam (NNN)		555	
diap (NNO)	4500	180	
triol (OOO)	230	>1	
Tetradentate ligands			
trien (NNNN)	4000^c		
tren (NNNN)	$2000^{d,e}$	$165, >2000^h$	$5.1^{e,g}$
$Me_6 tren^e$ (NNNN)	0.1^f	0.1^f	0.1^g
cyclam (NNNN)	8000^a		
trenol (NNNO)	2000^b	40, 2	80^g
cf. $M(MeCN)_6^{2+}$	620	2.0	16000^g

a At $-7°C$; *b* at $-50°C$; *c* at $-15°C$, at which temperature the rate constant for exchange of MeCN with $Co(MeCN)_6^{2+}$ is 1.5×10^3 sec^{-1}; *d* at $-40°C$; *e* gives five coordinate compounds, i.e. there is only one molecule of coordinated acetonitrile; *f* at $80°C$; *g* at $-45°C$; *h* these two rate constants refer to the two different solvation sites.

There is a very marked difference in kinetic behaviour between these cobalt(II) and nickel(II) complexes and their copper(II) analogues. Whereas tren and trenol both labilise acetonitrile coordinated to Co^{2+} or Ni^{2+}, which is the normal behaviour, the presence of either of these ligands coordinated to Cu^{2+} has a large retarding effect on rates of acetonitrile exchange. This is because the Jahn-Teller effect has a strong labilising effect on $Cu(MeCN)_6^{2+}$, but it no longer operates for the mixed ligand complexes. These therefore undergo acetonitrile exchange at rates comparable with those for analogous complexes of other transition metal(II) cations [62].

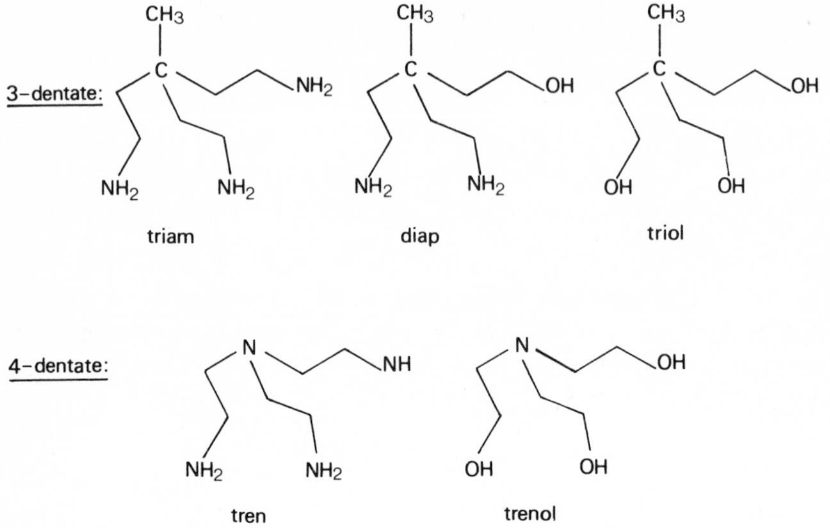

Fig. 11.12 The polydentate amino-/hydroxo- ligands of Tables 11.13 and 11.14.

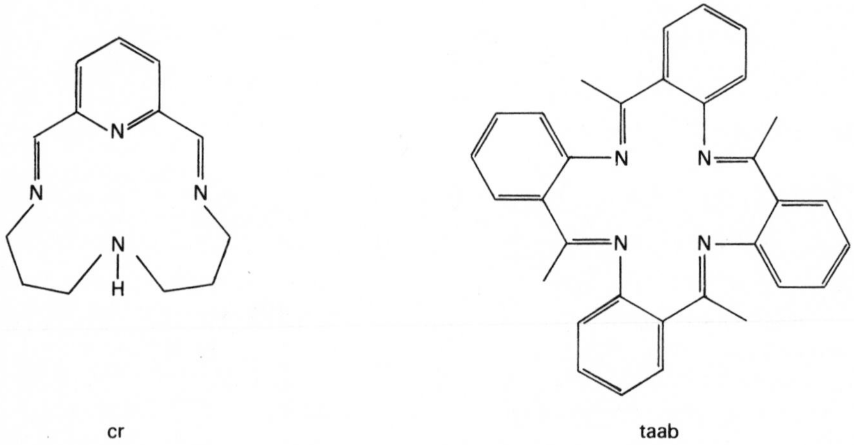

Fig. 11.13 The Schiff base ligands 'cr' and 'taab' of Table 11.16.

Table 11.15 Activation parameters for acetonitrile exchange with mixed complexes; ligand formulae in Fig. 11.12 [62, 94].

	Co²⁺		Ni²⁺		Cu²⁺	
	$\Delta H^{\ddagger}$ /kcal mol^{-1}	$\Delta S^{\ddagger}$ /cal deg^{-1} mol^{-1}	$\Delta H^{\ddagger}$ /kcal mol^{-1}	$\Delta S^{\ddagger}$ /cal deg^{-1} mol^{-1}	$\Delta H^{\ddagger}$ /kcal mol^{-1}	$\Delta S^{\ddagger}$ /cal deg^{-1} mol^{-1}
Tridentate ligands						
triam			9.3 ± 1.0	-0.9 ± 3.5		
diap	7.4 ± 0.6	-3.3 ± 1.4	12.0 ± 1.1	$+5.4 \pm 3.5$		
triol	9.6 ± 0.5	-2.3 ± 1.5	15.9 ± 0.8	$+8.7 \pm 2.5$		
Tetradentate ligands						
tren			10.8 ± 1.5	$+1.4 \pm 5.0$	10.8 ± 1.0	$+6.2 \pm 3.8$
trenol			12.0 ± 1.2	$+2.8 \pm 4.0$	6.2 ± 1.9	-8.1 ± 7.6
cf. M(MeCN)$_6^{2+}$	11.4 ± 0.5	$+5.2 \pm 2.0$	16.4 ± 0.5	$+12.0 \pm 2.0$		

Acetonitrile will exchange with the cis-$M(NO)_2(MeCN)_4^{2+}$ cations with $M = Mo$ or W, but not at an appreciable rate with the $trans$-$Cr(NO)_2(MeCN)_4^{2+}$ cation. This difference in behaviour may arise from $trans$-labilisation of MeCN by NO, which is only possible for the Mo and W compounds (cf. water exchange at $Cr(OH_2)_5(NO)^{2+}$ above). The mechanism of acetonitrile exchange for the molybdenum and tungsten compounds is thought to be dissociative, since the rate of exchange is independent of acetonitrile concentration in a series of acetonitrile+nitromethane mixtures [95].

Table 11.16 Kinetic parameters for dimethylformamide exchange at porphyrin-metal cation complexes, at 25°C.

	k/sec^{-1}	$\Delta H^{\ddagger}$ /kcal mol^{-1}	$\Delta S^{\ddagger}$ /cal deg^{-1}mol^{-1}	
$\alpha\beta\gamma\delta$-tetraphenylporphineiron(III)	10^{6} [a]	9.4 ± 0.4	3.8 ± 0.7	[41]
cf. $Fe(dmf)_6^{3+}$	10	10.1 ± 1.0	-16.5 ± 3.0	[41]
$Ni(cr)(dmf)_2^{2+}$ [b]	1.9×10^{6} [c]	9.5	2.2	[97]
$Ni(taab)(dmf)_2^{2+}$ [b]	7.3×10^{4}	11.5	2.5	[98]
cf. $Ni(dmf)_6^{2+}$	3.8×10^{3}	15.0	8.0	[99]

[a] Water, methanol exchange are also about 10^{5} times faster with the porphine complex than with the respective hexa-solvento-cations [100]; [b] ligands cr and taab are depicted in Fig. 11.13; [c] a similar rate enhancement has recently been reported for exchange of methanol with $Ni(cr)(MeOH)_2^{2+}$ [101].

Dimethylformamide. Dimethylformamide exchange kinetics have been studied for a small and varied group of mixed complexes of iron(III) [41] (Table 11.16), manganese(III) [96], and nickel(II) [97-101] (Table 11.16). It seems that solvent exchange rates are generally faster in solvento-metalloporphyrins and analogous Schiff base model compounds than in the corresponding hexa-solvento-cations. Again this is normal behaviour, and parallel to that observed in similar situations in water, methanol, or acetonitrile.

Table 11.17 Kinetic parameters for solvent exchange at oxovanadium(IV), at 25°C.

	$\log_{10}k$ [a]	$\Delta H^{\ddagger}$ /kcal mol^{-1}	$\Delta S^{\ddagger}$ /cal deg^{-1}mol^{-1}	
Water	3.89	12	0	[83]
Methanol	2.76	9.5	-14	[102]
Acetonitrile	3.45	7.1	-20	[103]
Dimethylformamide [b]	2.76	7.3	-16	[104]

[a] Units of k are sec^{-1}; [b] but see also Table 11.18.

Table 11.17 compares kinetic parameters for the exchange of dimethyl-formamide, methanol, and acetonitrile at VO^{2+} (taken from one series of investigations) with those for water exchange [102-104]. In each case the activation enthalpy for exchange for the non-aqueous solvent is less than that for water exchange. Unfortunately this simple picture has recently been upset by a reinvestigation of the kinetics of dimethylformamide exchange at VO^{2+}. It is now claimed that the activation enthalpy for dimethylformamide exchange is higher than that for water exchange. The most interesting feature of this recent work is the claim that it provides the first unambiguous demonstration of axial (*trans*) and equatorial (*cis*) solvent exchange occurring in parallel. The relevant kinetic parameters [105] are reproduced in Table 11.18. The standard errors of the activation parameters have been included here so that the reader may assess for himself whether the reported differences between axial and equatorial values are significant at a convincing level of confidence. It is disturbing that these new results are so different from the previous values. The rate constants indicate a marked *trans*-labilising effect of the oxo-ligand, contrary to some earlier ideas. Interestingly, the difference between the rate constants for axial and equatorial exchange seems to be determined by the activation entropies rather than by the activation enthalpies.

Table 11.18   Kinetic parameters for dimethylformamide exchange at the axial (*trans*-) and equatorial (*cis*-) positions of oxovanadium(IV); k and $\Delta S^{\ddagger}$ values refer to 25°C [105].

	k/sec^{-1}	$\Delta H^{\ddagger}/\text{kcal mol}^{-1}$	$\Delta S^{\ddagger}/\text{cal deg}^{-1}\text{mol}^{-1}$
Equatorial	2.0×10^{2}	13.1 ± 0.6	-4.1 ± 1.8
Axial	4.6×10^{4}	15.3 ± 2.5	$+13.9 \pm 9.0$

General. The activation parameters $\Delta H^{\ddagger}$ and $\Delta S^{\ddagger}$ vary in a compensatory manner with the solvent. This pattern can be ascribed to solvation and solvent structural changes, as described by Bennetto and Caldin (see Chapter 12.9 and footnote in Chapter 11.5.2).

11.7   MIXED SOLVENTS

In mixtures of coordinating solvents several exchange processes, involving several mixed complexes, are likely to be taking place concurrently. In aqueous ammoniacal cobalt(II) solutions exchange takes place with $Co(NH_3)_6^{2+}$, $Co(NH_3)_5(OH_2)^{2+}$, and $Co(NH_3)_4(OH_2)_2^{2+}$ simultaneously. It is a difficult and complicated matter to analyse the kinetic observations so as to obtain individual rate constants. However, in this cobalt(II)-ammonia–water system it has at least been established that rates of ammonia exchange are similar, at between 3 and 5×10^{5} sec^{-1} at $-23°$C, for the three complexes named [106].

Mixed solvents consisting of one coordinating and one non- (or poorly) coordinating solvent are more amenable to kinetic study and interpretation. In particular, it is often possible to establish how the rate of solvent exchange depends on the concentration of the coordinating solvent, and from this to make a guess at the mechanism of this solvent exchange. This has already been mentioned in the previous section, for acetonitrile exchange at cis-$M(NO)_2(MeCN)_4^{2+}$ cation with $M = Mo$ or W for example. There a small dependence of solvent exchange rate on its concentration was taken as evidence for a dissociative mechanism. Similar results and a similar conclusion have been obtained for dimethyl sulphoxide exchange at cobalt(II) and at nickel(II) in dimethyl sulphoxide-nitromethane and in dimethyl sulphoxide-methylene chloride mixtures [12, 13]. Dimethylformamide exchange at nickel(II) in dimethylformamide-nitromethane mixtures [107], trimethyl phosphate exchange at magnesium(II) in trimethyl phosphate-acetone or -methylene chloride mixtures [108], at aluminium(III) in trimethyl phosphate-nitromethane mixtures [109], and water exchange at $Co(NH_3)_5(OH_2)^{3+}$ in methanol-water mixtures [60] again appear to be dissociative. In some cases, for instance dimethyl sulphoxide and dimethylformamide exchange at nickel(II) and trimethyl phosphate at aluminium(III), subtle details of the kinetic pattern have been used as evidence to favour a D rather than an I_d mechanism. The kinetic parameters (particularly $\Delta S^{\ddagger}$) for solvent exchange at hexasolvento-aluminium(III) cations, solvent = trimethyl phosphate or dimethyl methylphosphonate, in nitromethane solution, indicate a dissociative mechanism. At the tetrasolvento-aluminium(III) cation $Al(hmpa)_4^{3+}$ however, the mechanism of exchange is associative [110].

REFERENCES

[1] Matwiyoff, N. A. and Taube, H., *J. Amer. Chem. Soc.*, **90**, 2796 (1968).

[2] E.g. Caldin, E. F., *Fast Reactions in Solution*, Blackwell (1964) Ch.11.

[3] Swift, T. J. and Connick, R. E., *J. Chem. Phys.*, **37**, 307 (1962).

[4] Olson, M. V., Kanazawa, Y,, and Taube, H., *J. Chem. Phys.*, **51**, 289 (1969).

[5] Cock, P. A., Cottrell, C. E., and Boyd, R. K., *Can. J. Chem.*, **50**, 402 (1972).

[6] Thomas, S. and Reynolds, W. J., *J. Chem. Phys.*, **46**, 4164 (1967).

[7] Vigee, G. S. and Ng, P., *J. Inorg. Nucl. Chem.*, **33**, 2477 (1971).

[8] Blackstaffe, S. and Dwek, R. A., *Molec. Phys.*, **15**, 279 (1968).

[9] Matwiyoff, N. A., personal communication quoted in reference [10].

[10] Angerman, N. S. and Jordan, R. B., *Inorg. Chem.*, **8**, 2579 (1969).

[11] Boubel, C. and Delpuech, J. J., *J. Chim. Phys.*, **70**, 578 (1973).

[12] Frankel, L. S., *Chem. Commun.*, 1254 (1969).

[13] Frankel, L. S., *Inorg. Chem.*, **10**, 814 (1971).

[14] Ravage, D. K., Stengle, T. R., and Langford, C. H., *Inorg. Chem.*, **6**, 1252 (1967).

[15] Strehlow, H. and Frahm, J., *Ber. Bunsenges. Phys. Chem.*, **79**, 57 (1975).

[16] Gutowsky, H. S. and Cheng, H. N., *J. Chem. Phys.*, **63**, 2439 (1975).

[17] Bennetto, H. P. and Caldin, E. F., *J. Chem. Soc. A.*, 2198 (1971); Wilkins, R. G., *The Study of Kinetics and Mechanism of Reactions of Transition Metal Complexes*, Allyn and Bacon (1974) p. 219.

[18] Granot, J., Achlama (Chmelnik), A. M., and Fiat, D., *J. Chem. Phys.*, **61**, 3043 (1974).

[19] Kapur, V. K and Wayland, B. B., *J. Phys. Chem.*, **77**, 634 (1973); Lincoln, S. F. and West, R. J., *Aust. J. Chem.*, **26**, 255 (1973).

[20] Zeltmann, A. H., Matwiyoff, N. A., and Morgan, L. O., *J. Phys. Chem.*, **73**, 2689 (1969).

[21] Hoggard, P. E., Dodgen, H. W., and Hunt, J. P., *Inorg. Chem.*, **10**, 959 (1971).

[22] Burgess, J., *Inorganic Reaction Mechanisms – Chemical Society Specialist Periodical Report*, **1**, 137 (1971).

[23] Connick, R. E. and Fiat, D. N., *J. Chem. Phys.*, **39**, 1349 (1963); Crea, J. and Lincoln, S. F., *J. Chem. Soc. Dalton*, 2075 (1973).

[24] Basolo, F. and Pearson, R. G., *Mechanisms of Inorganic Reactions, 2nd Edn.*, Wiley/Interscience (1967) p. 152.

[25] Poupko, R. and Luz, Z., *J. Chem., Phys.*, **57**, 3311 (1972).

[26] Kallen, T. W. and Earley, J. E., *Inorg. Chem.*, **10**, 1149 (1971).

[27] Movius, W. G. and Matwiyoff, N. A., *Inorg. Chem.*, **6**, 847 (1967); **8**, 925 (1969); Richardson, D. and Alger, T. D., *J. Phys. Chem.*, **79**, 1733 (1975).

[28] Thomas, S. and Reynolds, W. L., *J. Chem. Phys.*, **44**, 3148 (1966).

[29] Glass, G. E., Schwabacher, W. B., and Tobias, R. S., *Inorg. Chem.*, **7**, 2471 (1968).

[30] Sutin, N., *A. Rev. Phys. Chem.*, **17**, 119 (1966).

[31] Chmelnik, A. and Fiat, D., *J. Chem. Phys.*, **51**, 4238 (1969).

[32] Chmelnik, A. M. and Fiat, D., *J. Magnetic Resonance*, **8**, 325 (1972).

[33] Hunt, J. P. and Plane, R. A., *J. Amer. Chem. Soc.*, **76**, 5960 (1954).

[34] Stranks, D. R. and Swaddle, T. W., *J. Amer. Chem. Soc.*, **93**, 2783 (1971).

[35] Swaddle, T. W., Coleman, L. F., and Hunt, J. P., *Inorg. Chem.*, **2**, 950 (1963).

[36] Swaddle, T. W. and Lo, S. T. D., *Inorg. Chem.*, **14**, 1878 (1975).

[37] Carle, D. L. and Swaddle, T. W., *Can. J. Chem.*, **51**, 3795 (1973).

[38] Connick, R. L., *Symposium on Relaxation Techniques*, Buffalo (1965).

[39] Breivogel, F. W., *J. Chem. Phys.*, **51**, 445 (1969).

[40] Breivogel, F. W., *J. Phys. Chem.*, **73**, 4203 (1969).

[41] Hodgkinson, J. and Jordan, R. B., *J. Amer. Chem. Soc.*, **95**, 763 (1973).

[42] Langford, C. H. and Chung, F. M., *J. Amer. Chem. Soc.*, **90**, 4485 (1968).

[43] Sutin, N., *Chemical Society Autumn Meeting*, York (1971).

[44] Plumb, W. and Harris, G. M., *Inorg. Chem.*, **3**, 542 (1964).

[45] Akitt, J. W., *J. Chem. Soc. A*, 2347 (1971).

[46] Morgan, L. O. and Nolle, A. W., *J. Chem. Phys.*, **31**, 365 (1959).

[47] Koenig, S. H. and Epstein, M., *J. Chem. Phys.*, **63**, 2279 (1975).

[48] Reuben, J. and Fiat, D., *Chem. Commun.*, 729 (1967).

[49] Sasaki, Y. and Sykes, A. G., *J. Chem. Soc. Dalton*, 1048 (1975).

[50] Plane, R. A. and Hunt, J. P., *J. Amer. Chem. Soc.*, **79**, 3343 (1957); Ford, P. C. and Petersen, J. D., *Inorg. Chem.*, **14**, 1404 (1975).

[51] Tobe, M. L., *Inorganic Reaction Mechanisms*, Nelson (1972).

[52] Langford, C. H. and Gray, H. B., *Ligand Substitution Processes*, Benjamin (1966).

[53] Ingold, C. K., *Structure and Mechanism in Organic Chemistry*, Bell (1953, second edition 1969).

[54] Langford, C. H., Scharfe, R. and Jackson, R., *Inorg. Nucl. Chem. Lett.*, **9**, 1033 (1973).

[55] Miceli, J. and Stuehr, J., *J. Amer. Chem. Soc.*, **90**, 6967 (1968).

[56] Baltisberger, R. J. and King, E. L., *J. Amer. Chem. Soc.*, **86**, 795 (1964); Ardon, M., *Inorg. Chem.*, **4**, 372 (1965); Carey, L. R., Jones, W. E., and Swaddle, T. W., *ibid.*, **10**, 1566 (1971).

[57] Stranks, D. R., *Pure Appl. Chem.*, *15th Int. Conf. Coord. Chem.*, **38**, 303 (1974).

[58] Tong, S. B. and Swaddle, T. W., *Inorg. Chem.*, **13**, 1538 (1974).

[59] Adams, D. M., *Inorganic Solids*, Wiley/Interscience (1974) pp. 32–33.

[60] Hunt, H. R. and Taube, H., *J. Amer. Chem. Soc.*, **80**, 2642 (1958).

[61] Babiec, J. S., Langford, C. H., and Stengle, T. R., *Inorg. Chem.*, **5**, 1362 (1966).

[62] West, R. J. and Lincoln, S. F., *J. Chem. Soc. Dalton*, 281 (1974).

[63] Reuben, J. and Fiat, D., *J. Chem. Phys.*, **51**, 4918 (1969).

[64] Gutmann, V. and Schmid, R., *Coord. Chem. Rev.*, **12**, 263 (1974).

[65] See p. 140 of Reference [22].

[66] Grant, M., Dodgen, H. W., and Hunt, J. P., *Inorg. Chem.*, **10**, 71 (1971).

[67] Melvin, W. S., Rablen, D. P., and Gordon, G., *Inorg. Chem.*, **11**, 488 (1972).

[68] Rablen, D. P., Dodgen, H. W., and Hunt, J. P., *J. Amer. Chem. Soc.*, **94**, 1771 (1972).

[69] Grant, M. W., Dodgen, H. W., and Hunt, J. P., *J. Amer. Chem. Soc.*, **93**, 6828 (1971).

[70] Zeltmann, A. H. and Morgan, L. O., *Inorg. Chem.*, **9**, 2522 (1970).

[71] Letter, J. E. and Jordan, R. B., *J. Amer. Chem. Soc.*, **93**, 864 (1971).

[72] Zetter, M. S., Dodgen, H. W., and Hunt, J. P., *Biochemistry N. Y.*, **12**, 778 (1973).

[73] Reuben, J. and Cohn, M., *J. Biol. Chem.*, **245**, 6539 (1970).

[74] Mildvan, A. S. and Scrutton, M. C., *Biochemistry N. Y.*, **6**, 2978 (1967).

[75] Hague, D. N., *Inorganic Reaction Mechanisms – Chemical Society Specialist Periodical Report*, **2**, 206, 233 (1972).

[76] Zetter, M. S., Grant, M. W., Wood, E. J., Dodgen, H. W., and Hunt, J. P., *Inorg. Chem.*, **11**, 2701 (1972).

[77] Kruse, W. and Taube, H., *J. Amer. Chem. Soc.*, **83**, 1280 (1961).

[78] Bracken, D. E. and Baldwin, H. W., *Inorg. Chem.*, **13**, 1325 (1974).

[79] Moore, P., Basolo, F., and Pearson, R. G., *Inorg. Chem.*, **5**, 223 (1966).

[80] Burgess, J., *Inorganic Reaction Mechanisms – Chemical Society Specialist Periodical Report*, **3**, 163 (1974).

[81] Judkins, M. R., *PhD Thesis*, University of California (Lawrence Radiation Laboratory Report UCRL 17561).

[82] Degani, H. A. and Fiat, D., *J. Amer. Chem. Soc.*, **93**, 4281 (1971).

[83] Reuben, J. and Fiat, D., *J. Amer. Chem. Soc.*, **91**, 4652 (1969).

[84] Wüthrich, K. and Connick, R. E., *Inorg. Chem.*, **6**, 583 (1967); **7**, 1377 (1968); Reuben, J. and Fiat, D., *ibid.*, **6**, 579 (1968); Zeltmann, A. H. and Morgan, L. O., *ibid.*, **10**, 2739 (1971).

[85] Frei, V. and Wendt, H., *Ber. Bunsenges. Phys. Chem.*, **74**, 593 (1970).

[86] Levanon, H. and Luz, Z., *J. Chem. Phys.*, **49**, 2031 (1968).

[87] Luz, Z., *J. Chem. Phys.*, **41**, 1756 (1964).

[88] Vriesenga, J. R., *Inorg. Chem.*, **11**, 2724 (1972).

[89] Kurtz, J. and Vriesenga, J. R., *J. Chem. Soc. Chem. Commun.*, 711 (1974).

[90] Vriesenga, J. R. and Gronner, R., *Inorg. Chem.*, **12**, 1112 (1973).

[91] Plotkin, K., Copes, J. and Vriesenga, J. R., *Inorg. Chem.*, **12**, 1494 (1973).

[92] Luz, Z., *J. Chem. Phys.*, **51**, 1206 (1969).

[93] Luz, Z. and Meiboom, S., *J. Chem. Phys.*, **40**, 1066 (1964).

[94] Lincoln, S. F. and West, R. J., *Aust. J. Chem.*, **25**, 469 (1972); **27**, 97 (1974); *Inorg. Chem.*, **12**, 494 (1973); *J. Amer. Chem. Soc.*, **96**, 400 (1974); Lincoln, S. F., *Aust. J. Chem.*, **27**, 899 (1974).

[95] Johnson, B. F. G., Khair, A., Savory, C. G., and Walter, R. H., *J. Chem. Soc. Chem. Commun.*, 744 (1974).

[96] Rusnak, L. and Jordan, R. B., *Inorg. Chem.*, **11**, 196 (1972).

[97] Rusnak, L. L. and Jordan, R. B., *Inorg. Chem.*, **10**, 2686 (1971).

[98] Rusnak, L. L., Letter, J. E., and Jordan, R. B., *Inorg. Chem.*, **11**, 199 (1972).

[99] Matwiyoff, N. A., *Inorg. Chem.*, **5**, 788 (1966).

[100] Jordan, R. B., *Chemical Society Meeting*, Swansea, Wales (1973).

[101] Rusnak, L. L. and Jordan, R. B., *Inorg. Chem.*, **14**, 988 (1975).

[102] Angerman, N. S. and Jordan, R. B., *Inorg. Chem.*, **8**, 1824 (1969).

[103] Angerman, N. S. and Jordan, R. B., *Inorg. Chem.*, **8**, 65 (1969).

[104] Angerman, N. S. and Jordan, R. B., *J. Chem. Phys.*, **48**, 3983 (1968).

[105] Miller, G. A. and McClung, R. E. D., *J. Chem. Phys.*, **58**, 4358 (1973).

[106] Murray, R., Lincoln, S. F., Glaeser, H. H., Dodgen, H. W., and Hunt, J.
P., *Inorg. Chem.,* **8**, 554 (1969).
[107] Frankel, L. S., *Inorg. Chem.,* **10**, 2360 (1971).
[108] Crea, J., Lincoln, S. F., and West, R. J., *Aust. J. Chem.,* **26**, 1227
(1973).
[109] Frankel, L. S. and Danielson, E. R., *Inorg. Chem.,* **11**, 1964 (1972).
[110] Delpuech, J. -J., Khaddar, M. R., Peguy, A. A., and Rubini, P. R., *J. Amer.
Chem. Soc.,* **97**, 3373 (1975).

Chapter **12**

KINETICS AND MECHANISMS: COMPLEX FORMATION

12.1 INTRODUCTION

The reactions discussed in this Chapter represent an extension and generalisation of those of the previous Chapter. In both cases a solvent molecule in the starting solvento-cation is replaced:

$$MS_6^{n+} + S^* \longrightarrow MS_5S^{*n+} + S$$

$$MS_6^{n+} + L \longrightarrow MS_5L^{n+} + S$$

There are therefore important connections between the kinetics and mechanisms (for a general discussion of substitution mechanisms see Chapter 11.4.1) of the two types of reaction, especially in the many cases where both are dissociative.

The experimental technique normally used to monitor kinetics for solvent exchange, n.m.r. spectroscopy, is equally applicable to aqueous and to non-aqueous media. The techniques most commonly used in the case of complex formation, stopped-flow and temperature-jump, are easier to use for aqueous solutions. This technical encouragement reinforces the inorganic kineticist's strong preference for working in the most readily available, most extensively documented, but most complicated solvent, and it has resulted in a preponderance of kinetic results for complex formation in aqueous solution. We shall therefore deal with aqueous media first, and then compare results in non- and mixed aqueous solvents with those in water.

The basic mechanism for the formation of complexes from aquo-cations is that developed by Eigen, Wilkins, and their associates some fifteen years ago, though the ideas involved can be traced right back to Werner. The Eigen-Wilkins mechanism [1, 2] derives from observations that rates and activation parameters for complex formation (particularly for 2+ cations reacting with unidentate, uncharged, ligands) are generally similar for a given metal ion. Moreover they are also similar to the kinetic parameters for water exchange at the said metal ion. These observations suggest a dissociative mechanism. However, because the formation rates depend somewhat both on the nature and on the concentration of the incoming ligand, a mechanism involving only rate-determining water loss must be an oversimplification. When rapid, equilibrium, association between the reactants previous to the interchange step is considered, a formation mechanism consistent with the majority of the kinetic results emerges. Interestingly, the reaction sequence for formation reactions of solvento-transition metal cations built up by Eigen, Wilkins and their associates had been foreshadowed earlier in studies of salt effects on acetolysis of benzenesulphonates [3].

Returning to our general picture of a solvento-cation in solution (Fig. 1.2), the incoming solvated ligand has to make its way from the bulk solvent region (D) through the secondary solvation shell to its final position in the primary coordination sphere (A) of the cation. The kinetic pattern does not reflect the early stages in this process, when there is little interpenetration of the solvation shells of the cation and the potential ligand. Reaction rates are determined by two steps, the rapid equilibrium formation of an ion-pair (or outer-sphere association complex if the ligand is uncharged or positively charged) and the subsequent rate-determining interchange step. Here the ligand bonds to the cation after release of a water molecule from its primary coordination shell (Fig. 12.1).

Fig. 12.1 Simplified picture of metal complex formation, showing the kinetically important steps of association and interchange.

The above description and the pictorial representation of Fig. 12.1 can be expressed in chemical symbols[†] as:

$$M(OH_2)_x^{n+} + L \overset{K_{os}}{\rightleftharpoons} M(OH_2)_x^{n+}, L \overset{k_i}{\longrightarrow} M(OH_2)_{x-1}L^{n+} + H_2O .$$

The actual ligand interchange step may be dissociative or associative in character, though it will be seen below that the former is more common.

The rate law corresponding to such a mechanism is, omitting coordinated water molecules and taking the case of an uncharged ligand:

$$+ \frac{d[ML^{n+}]}{dt} = \frac{K_{os}k_i[M^{n+}][L]}{1 + K_{os}[L]} .$$

Under the usual experimental conditions of a small value for K_{os} and a relatively low concentration of incoming ligand, this reduces to:

[†] The particular case of an uncharged ligand L is shown for the sake of clarity.

$$+ \frac{d[ML^{n+}]}{dt} = K_{os}k_i[M^{n+}][L] \ .$$

Thus the observed dependence of rates of complex formation on the nature and concentration of the incoming ligand can arise from the proposed mechanism even when the rate-determining interchange step is dissociative in character. Observed first-order rate constants for complex formation, k_f, are given by:

$$k_f = K_{os}k_i \ .$$

There is good evidence to support this mechanism, especially from ultra-sonic experiments, which can be operated on a sufficiently short time scale to detect the fast pre-association of reactants. Early ultrasonic relaxation studies [2] of solutions of metal(II) sulphates were interpreted by a three reaction sequence. However, more recent investigations, for example of manganese(II) sulphate solutions [4], have shown that two relaxations attributable to association and interchange are sufficient to account for the experimental observations. Experiments on beryllium(II) sulphate solutions in a particularly sophisticated temperature-jump apparatus also indicate that formation of the $Be(SO_4)$ complex proceeds by two kinetically distinguishable steps [5]. Kinetic studies of slow formation reactions of chromium(III) and cobalt(III) complexes have also shown a mechanism of fast association followed by rate-determining interchange (see Chapter 12.7.4 below). Aquo-cations associate with potential ligand ions at very fast rates, approaching the diffusion controlled limiting rate. These rates have been determined in some cases, for example for several $M^{2+}aq$ cations with SO_4^{2-} [6], and (by e.s.r. techniques) for transition metal aquo-cations plus nitrosyldisulphonate [7]. The changes in enthalpy [8] and in volume [9] attendant on ion-pair formation between $Mg^{2+}aq$ and sulphate have also been estimated.

Both for the rapidly reacting systems whose kinetics can be monitored by ultrasonic or temperature-jump techniques, and for the slowly reacting systems where there is time to study the ion association equilibrium by leisurely conventional methods, values of K_{os} can be estimated and thus k_f values separated into their K_{os} and k_i components. Occasionally such a separation is possible for nickel(II) systems, for example with thiocyanate [10] or with methylphosphate [11]. However, the only way to separate k_f into K_{os} and k_i for the majority of formation reactions is to estimate K_{os} [12]. This is usually done using the Fuoss equation [13].

$$K_{os} = \frac{4\pi N_A a^3}{3000} \exp\left(-U/k_B T\right)$$

where U is the Coulomb energy for interaction between oppositely charged ions, k_B is Boltzmann's constant, N_A is Avogadro's number, and a the distance apart of the ions in the ion-pair (often taken as 5Å for reactions of aquo-cations with ordinary ligands). A similar equation has been developed for estimating K_{os}

values for the association of solvated cations with uncharged, or even positively charged, dipolar ligands [14]. K_{os} values predicted by these equations and measured K_{os} values agree tolerably in the very few cases where directly measured K_{os} values are available. The predicted value of K_{os} for Ni^{2+}aq and methylphosphate is 14 M^{-1}, the observed value 40 M^{-1}, while for Ni^{2+}aq and ammonia the predicted and observed values are both 0.15 M^{-1}. However the estimation of K_{os} values often involves uncertainties and dubious assumptions, particularly for incoming groups of irregular geometry [15].

Once K_{os} has been measured or calculated, k_f can be divided thereby to obtain k_i, which can be compared with k_i values for other formation reactions and with k_{ex} for solvent exchange in order to diagnose or confirm the nature of the interchange step. For instance, the measured k_f value for the formation of $Ni(CH_3PO_4)$ is 2.9×10^5 M^{-1}sec^{-1} while the K_{os} value is 40 M^{-1}, and from these one can calculate that k_i here is 7×10^3 sec^{-1}. A direct ultrasonic estimate of k_i for Ni^{2+}aq,SO_4^{2-} is 15×10^3 sec^{-1} [2]. The rate constant for water exchange, k_{ex}, is similar, at 30×10^3 sec^{-1} (Chapter 11). This similarity of interchange and solvent exchange rate constants indicates a dissociative mechanism. Here and in many other systems k_i is slightly smaller in value than k_{ex}, but this can be attributed to statistical factors. Although the reacting Ni^{2+}aq cation is surrounded by water, it has only one immediately adjacent ligand molecule or ion. Hence a water molecule is certain to be in the right position for interchange, but the ligand may not be suitably placed with respect to the leaving water molecule to replace it. Another reason for small differences between k_i and k_{ex} is that the latter is measured for a free solvento-cation, while k_{ex} for the same solvento-cation in an ion-pair will be somewhat different. The effect of ion-pairing on solvent exchange rates has been examined only recently. Methanol exchange with the $Co(MeOH)_5(OAc)^+$ cation almost doubles in the presence of added acetate. There is n.m.r. evidence for ion-pairing under the conditions of these kinetic experiments [16]. On the other hand, evidence has been presented to show that k_{ex} for dimethylformamide with the $Cr(dmf)_6^{3+}$, ClO_4^- ion-pair is only about half that for dimethylformamide exchange with the free $Cr(dmf)_6^{3+}$ cation [17].

12.2   NICKEL(II) IN AQUEOUS SOLUTION

12.2.1   General

The kinetics of more formation reactions have been studied for nickel(II) than for any other metal cation. Amongst possible reasons for this state of affairs is the purely technical one that rates of formation are in a range that is particularly convenient for monitoring by stopped-flow techniques. Rates are neither so fast as to require apparatus too sophisticated and expensive for general availability, nor so slow as to tax the patience of the research worker.

Consequently the mechanistic pattern for formation of metal complexes has to a large extent been built up from observations on nickel(II) reactions. The pattern of fast equilibrium association followed by rate-determining dissociative interchange is illustrated by the results contained in Table 12.1 [18-28]. When divided by appropriate K_{os} values, a wide range of k_f values gives a much smaller range of interchange rate constants k_i. The k_i values are close to the rate constant for water exchange at nickel(II), which is about 30×10^3 sec^{-1} at 25°C (Chapter 11). A dissociative interchange (I_d) mechanism operates here.

Table 12.1 Rate constants and association constants (defined in the text) for formation reactions of Ni^{2+}aq in aqueous solution, at 25°C.

Ligand	Measured $10^{-3} k_f/M^{-1} sec^{-1}$	Calculated K_{os}/M^{-1}	Estimated $10^{-3} k_i/sec^{-1}$	
tetrenH$_3^{3+}$	0.0035	a		[18]
tetrenH$_2^{2+}$	0.32	a		[18]
trienH$_2^{2+}$	0.10	a		[18]
NNN-Me$_3$en$^+$	0.5	0.02	25	[19]
ImidazoleH$^+$	0.4	0.02	20	[20]
Ammonia	5	0.15	33	[21]
Hydrogen fluoride	3	0.15	20	[22]
Pyridine	~4	0.15	30	[23]
Imidazole	2 to 7	0.15	13 to 47	[20, 24]
Histidine	2.2^b	0.15	15	[20]
Diglycine	21	0.17	12	[15]
Fluoride$^-$	8	1	8	[22]
Acetate$^-$	100	3	30	[25]
OxalateH$^-$	5	2	3	[26]
Oxalate^{2-}	75	13	6	[26]
Malonate^{2-}	450	95	5	[27]
Methylphosphate^{2-}	290	40^c	7	[11]
Sulphate^{2-}			10^d	[13]
Pyrophosphate^{3-e}	2100	88	24	[28]
Tripolyphosphate^{4-f}	6800	570	12	[28]

a Values of K_{os} were not estimated for these ligands; b at 23.7°C; c K_{os} estimated from the kinetic results; d at 20°C; e P$_2$O$_7$H^{3-}; f P$_3$O$_{10}$H^{4-}.

Activation parameters for complex formation from hexa-aquo-nickel(II) are also consistent with a dissociative mechanism. Values of $\Delta H_i^{\ddagger}$ and of $\Delta S_i^{\ddagger}$ for the interchange step (Table 12.2 [26-29]) are similar to the activation parameters for water exchange. Activation volume values for formation reactions of such ligands as ammonia, pada (Fig. 12.2), and glycine[†] with hexa-aquo-nickel(II) (Table 12.3 [30-32]) are markedly positive and do not correlate with $\Delta V^{\ominus}$ values. Thus they too are consistent with dissociative activation. In particular, observed activation volumes do not reflect the very different molar volumes of the incoming ligands. The increase in volume in going to the transition state corresponds to an extension about 20% in the nickel-water bond. Interestingly, the activation volumes for these complex formation reactions are similar to those for $S_N 1$ reactions of organic cations such as alkylsulphonium cations in aqueous solution.

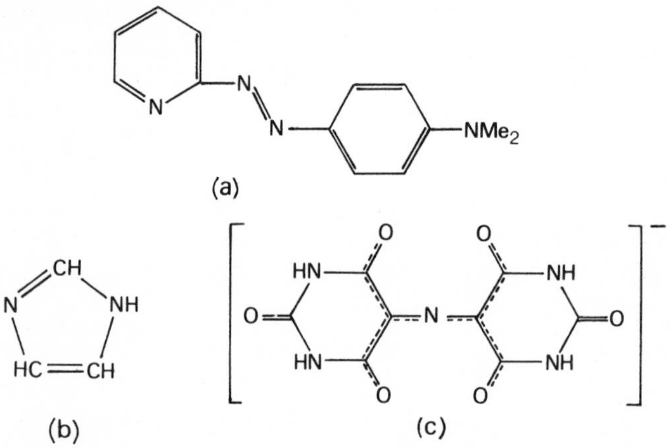

(a)

(b)   (c)

Fig. 12.2 (a) Pyridine-2-azo-p-dimethylaniline (pada); (b) imidazole; (c) murexide.

Table 12.2 Enthalpies and entropies of activation for the ligand interchange step in complex formation reactions of Ni^{2+}aq.

Ligand	$\Delta H_i^{\ddagger}/\text{kcal mol}^{-1}$	$\Delta S_i^{\ddagger}/\text{cal deg}^{-1}\text{mol}^{-1}$	
Succinate^{2-}	11.7	+0.4	[29]
Malonate^{2-}	11	−2.8	[29]
Oxalate^{2-}	11	−3	[26]
OxalateH$^-$	12	−0.8	[26]

† A correction of 3 cm³ mol⁻¹ must be applied to the observed activation volume for this reaction involving a charged incoming group in order to estimate the activation volume for the ligand interchange step. Electrostriction effects will be much smaller for the association of nickel(II) with the uncharged ammonia or pada ligands, so for these reactions the observed activation volume can be taken as the activation volume for the interchange step.

Table 12.3 Volumes of activation and of reaction for complex formation reactions of $Ni^{2+}aq$.

Ligand	$\Delta V^{\ddagger}/cm^3 mol^{-1}$	$\Delta V^{\ominus}/cm^3 mol^{-1}$	
Ammonia	7.1 ± 1.0	−2.3	[30]
Glycinate⁻	8 ± 2	+2.1	[31]
pada [a]	8.2 ± 0.2	+0.9	[30]
Imidazole [b]	11 ± 1.6		[32]
Murexide [b]	12.2 ± 1.5	+22.6	[32]

[a] Pyridine-2-azo-p-dimethylaniline, Fig. 12.2; [b] Fig. 12.2.

A change of solvent from water to heavy water (D_2O) will affect K_{os} values as estimated from the Fuoss equation. For $Ni^{2+}aq$ plus malonate, K_{os} values are 95 M^{-1} in H_2O and 99 M^{-1} in D_2O. From these estimates and observed second-order rate constants for formation, k_i values come out as 4.6×10^3 sec^{-1} in H_2O and 4.0×10^3 sec^{-1} in D_2O (at 25°C) [27]. A more dramatic example of the effect of solvent change from H_2O to D_2O can be found in the subsequent section on magnesium(II).

So far the standard mechanism for complex formation reactions from $Ni^{2+}aq$ is seen to be fast ion-association followed by rate-determining dissociative interchange. We shall now consider several special types of ligands whose formation reactions show more or less exceptional kinetic features.

12.2.2 Multidentate Ligands

A three step reaction sequence must be considered when a bidentate ligand reacts with an aquo-cation:

$$Ni(OH_2)_6^{2+} + LL \xrightarrow{K_{os}} Ni(OH_2)_6^{2+}, LL \qquad (12.1)$$

$$Ni(OH_2)_6^{2+}, LL \underset{k_d}{\overset{k_i}{\rightleftharpoons}} Ni(OH_2)_5(L-L)^{2+} + H_2O \qquad (12.2)$$

$$Ni(OH_2)_5(L-L)^{2+} \underset{k_{ro}}{\overset{k_{rc}}{\rightleftharpoons}} Ni(OH_2)_4(\overset{L}{\underset{L}{<}|)^{2+} + H_2O \qquad (12.3)$$

If $k_{rc} \gg k_d$ and k_i then the same kinetic pattern will be observed as for reaction with a monodentate ligand, with

$$+ \, d\left[\mathrm{Ni(OH_2)_4}\!\!\left(\begin{smallmatrix}\mathrm{L}\\ |\\ \mathrm{L}\end{smallmatrix}\right)^{2+}\right]\!/dt \; = \; K_{os}k_i \; .$$

If, however, the chelation (ring closure) step of (12.3) is difficult to achieve and k_{rc} is thus $\sim$ or $< k_d$ and k_i, the overall rate constant for formation (k_f) of the chelate product will be less than $K_{os}k_i$. This latter state of affairs, which has been labelled **sterically controlled substitution**, or SCS, was first described and explained for the reactions of Ni^{2+} and Co^{2+}aq with α- and with β- alanine (Fig. 12.3) [33]. The rate of formation of the β-alanine chelate of Co^{2+}, which entails closing of a six-membered ring, is some ten times slower than that of the α-alanine (five-membered ring) complex. The difficulty attendant on closing the larger ring makes this process rather than the water loss from Co^{2+}aq rate-limiting. Because rates of solvent loss from Ni^{2+} are slower, rates of ring closure are rarely markedly less. In the case of the Ni^{2+}-alanine reactions, the rate of formation of the β-complex is only twice as slow as that for the α-complex. Rate constants can sometimes be obtained both for formation of the mono-dentate intermediate ($K_{os}k_i$) and for the chelation step (k_{rc}) from the analysis of relaxation (pressure-jump usually) experiments. Estimated values of k_{rc} ($25°C$) include $1.8 \times 10^3 \, sec^{-1}$ for the α-aminopropionate complex and $0.8 \times 10^3 \, sec^{-1}$ for its β-analogue [25]; $1.9 \times 10^3 \, sec^{-1}$ for lactate [34]; 1.9 [35] or 2.5 [25] $\times 10^3 \, sec^{-1}$ for glycolate; and $0.45 \times 10^3 \, sec^{-1}$ for 2-hydroxyglutarate (Fig. 12.3(c)) and $0.3 \times 10^3 \, sec^{-1}$ for its 3-hydroxy ana-logue [36]. In the case of the hydroxyglutarates, it has been shown that a carb-oxylate-oxygen bonds to the nickel first and that the hydroxy-oxygen becomes bonded to the nickel in the subsequent chelation step. Authors disagree on whether rates of ring closure are significantly less than rates of formation of the respective monodentate intermediates in formation of malonato- and oxalato-

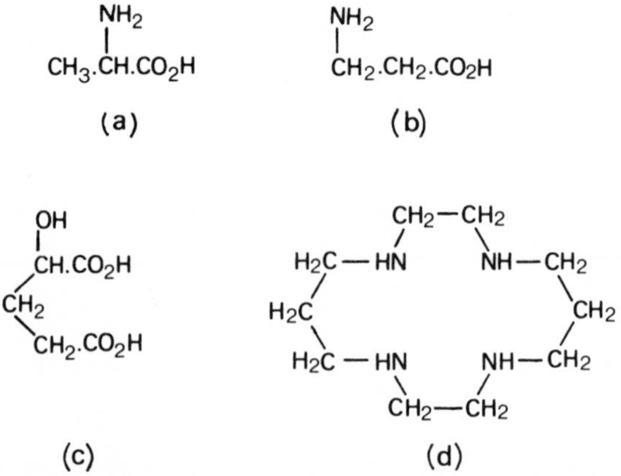

$$\begin{array}{cc} \mathrm{NH_2} & \mathrm{NH_2} \\ | & | \\ \mathrm{CH_3.CH.CO_2H} & \mathrm{CH_2.CH_2.CO_2H} \\ \\ \text{(a)} & \text{(b)} \end{array}$$

Fig. 12.3   (a) α-Alanine; (b) β-alanine; (c) 2-hydroxylgutaric acid; (d) cyclam.

complexes [37]. The formation of a seven- instead of a six- membered chelate ring, as in the reaction with phthalate [38], has no marked effect on k_{rc}. Consequently rates of ring closure are only slightly (about ten times) slower than interchange rates at nickel(II), at any rate in aqueous solution. The ring closure step may become kinetically much more important in mixed and in non-aqueous media, as discussed in Chapter 12.8 below. The kinetic effects of the SCS mechanism are much more important for aquo-cations where interchange rates are faster. As substitution rates speed up, for example in going to Co^{2+}aq or, faster still, to Mn^{2+}aq or alkali metal cations, ring closing difficulties are much more likely to become obtrusive, and examples of the SCS mechanism are more frequently encountered.

If the incoming ligand is a macrocycle rather than a monodentate or flexible acyclic polydentate ligand, steric factors produce a much more marked effect on formation rates, even for nickel(II). The cyclam (Fig. 12.3(d)) complex forms some 3×10^4 times more slowly than the complex of the analogous open-chain (garland) ligand trien [39]. This difference is probably due to steric (bulk) factors and to interference from the intramolecular hydrogen bonding within the cyclam rather than simply to the obvious garland versus cyclic distinction [40]. Methyl substitution on the nitrogen atoms of cyclam further slows complex formation with Ni^{2+}aq [41]. Porphyrins are more rigid than cyclam, yet the mechanism of their reactions with many metal(II) aquocations seems to be normal dissociative interchange [42]. Here probably a second cation is involved, to facilitate the conformational changes required in the ligand during the process of complexation [43]. Reactions of porphyrins with iron(III) or with magnesium(II) may well differ and be examples of the SCS mechanism [42].

12.2.3 The Internal Conjugate Base Mechanism[†]

A few polydentate amines react more rapidly than would be expected from the known rates of formation of complexes from analogous monodentate amines or ammonia. The ethylenediamine and NN'-diethylethylenediamine complexes of nickel(II) form about a hundred times more quickly than the mono-ammonia complex, and so their rate constants are amongst the largest known for substitution at Ni^{2+}aq. As there is no obvious reason why k_i alone should be greater for these polydentate amines, the faster rates must be attributed to some special feature of the cation–ligand association. The exceptional behaviour of these ligands is associated with strongly basic properties (ethylenediamine has pK =

---

† Since this Chapter was written, a letter has been published (Jordan, R. B., *Inorg. Chem.*, 5, 748 (1976)) challenging the need to invoke this special mechanism. The experimental results upon which the internal conjugate base mechanism rests have been re-analysed giving due consideration to reactions involving monoprotonated forms of the diamines involved. Thus, it is claimed, these results can be accommodated satisfactorily within the normal Eigen-Wilkins mechanism.

10.00, the monoprotonated form of triethylenetetramine has $pK = 9.2)^\dagger$ and hydrogen-bonding tendencies. Hydrogen-bonding between ligand and non-leaving water on the cation will lead to an increased value of K_{os}, and such hydrogen-bonding may also have an effect on k_i. The situation is illustrated diagrammatically in Fig. 12.4. Clearly this internal conjugate base (ICB) mechanism can operate only for bi- (poly-) dentate incoming groups, because the hydrogen bonding between ligand and non-leaving water must be retained throughout the rate-determining step [21, 44].

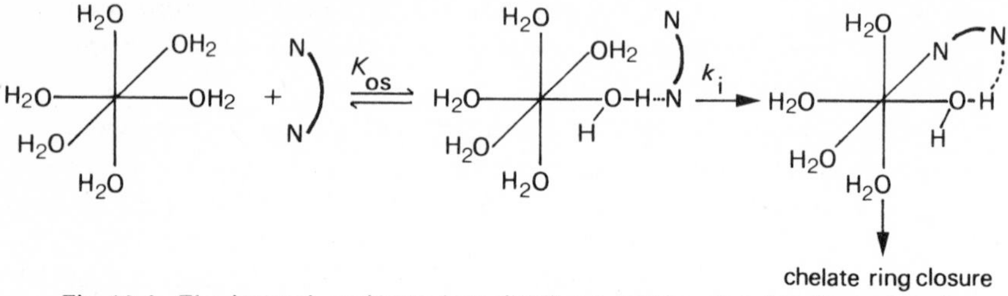

chelate ring closure

Fig. 12.4 The internal conjugate base (ICB) mechanism, showing the proposed role of hydrogen-bonding in facilitating the formation of a chelate complex.

For reactions of polydentate nitrogen bases containing N-alkyl substituents (see NN'-diethylethylenediamine above) there is a possibility that the steric effects of the alkyl groups may interfere with the favourable association between the incoming ligand and the Ni^{2+}aq cation. The balance between the opposing steric and ICB effects is illustrated by the rate constants given in the left-hand column of Table 12.4 [45, 46]. A steady increase in the number of N-methyl substituents causes the formation rate constant to decrease from the exceptionally high value for ethylenediamine itself to values which are lower than that for reaction with ammonia. The effects of steric factors alone are shown by the rate constants in the right-hand column of Table 12.4.

Table 12.4   The effect of steric constraints imposed by N-alkyl substituents on the rates of formation of amine complexes of nickel(II) in aqueous solution, both by the internal conjugate base (ICB) and the normal modes of substitution (25°C).

Ligand	$10^{-3} k_f/M^{-1} sec^{-1}$ [45]	Ligand	$10^{-3} k_f/M^{-1} sec^{-1}$ [46]
NH_3	4.48	NH_3	4.48
en	350	$MeNH_2$	1.31
NN–Me$_2$en	32	Me_2NH	0.332
NN'–Me$_2$en	43		
NNN'–Me$_3$en	2.9	$EtNH_2$	0.865
NNN'N'–Me$_4$en	0.36	Pr^iNH_2	0.605

† Significantly, diglycine (whose pK is 8.2) reacts only slightly faster than expected, and so probably represents the borderline between the ICB and normal mechanisms.

12.2.4 Other Exceptional Cases

In the previous section situations in which hydrogen-bonding between the incoming liquid and cation-solvating water led to anomalously high formation rates were discussed. The present section deals with the complementary situation where hydrogen-bonding effects lead to slower formation rates. The rate constant for the reaction of hydroxyproline with Ni^{2+}aq is 1.2×10^4 M^{-1} sec^{-1}, for proline 3.4×10^4 M^{-1}sec^{-1} (at $25°C$). The marginally slower formation with hydroxyproline is attributed to hydrogen-bonding between the OH group of the ligand and water coordinated to the nickel, which makes the orientation of the ligand within the ion-pair relatively unfavourable for the interchange step [47]. A similar explanation has been offered for the slow rate of formation of the L-dopa complex of nickel(II), and indeed of cobalt(II) as well [48]. The explanation for the slow reactions of Ni^{2+}aq with Hnta^{2-}, Hida$^-$, [49] or the purine base theophylline (Fig. 12.5(c)) [50] is subtly different. Here the interchange step is thought to be more difficult because of unfavourable hydrogen-bonding within the ligand (see also Cu^{2+}aq or La^{3+}aq plus par, Chapter 12.3.3 and 12.6.3).

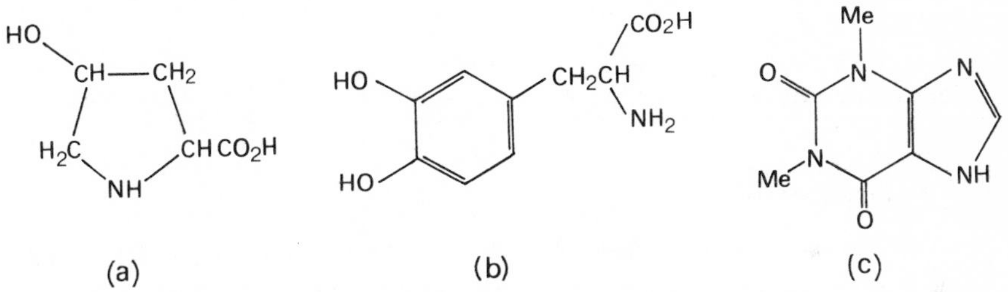

Fig. 12.5   (a) Hydroxyproline; (b) L-dopa; (c) theophylline.

12.3   OTHER TRANSITION METAL(II) CATIONS

Formation reactions of these cations will be discussed in the order dictated by their closeness of kinetic behaviour to nickel(II) and by the amount of information available, rather than by the Periodic Table.

12.3.1 Cobalt(II)

The situation here is very similar to that for nickel(II). Rate constant patterns and activation parameters (enthalpies, entropies, and volumes) suggest as normal behaviour the same mechanism of fast association followed by rate-determining dissociative interchange.

One marked difference between cobalt(II) and nickel(II) is the greater importance of rate-determining ring-closure (SCS mechanism) for formation of chelates from Co^{2+}aq. This difference has already been discussed in the nickel(II) section, with particular reference to reactions with α- and β- alanine. Rate constants for these reactions, and the closely related reactions with the α- and β- aminobutyrate anions [51], are quoted in Table 12.5. The results in this Table also show that rates of formation of cobalt(II) complexes are some twenty to thirty times faster than those for nickel(II) in normal circumstances.

The internal conjugate base mechanism has been suggested to operate at $Co^{2+}aq$ as it does at $Ni^{2+}aq$, for example with poly(aminoalcohol) ligands [52]. The opposite situation, of hydrogen-bonding leading to slow formation rates, has also been demonstrated for the cobalt(II) analogues of the systems discussed earlier in the nickel(II) section.

Table 12.5 Formation rate constants (k_f, at 20°C) for chelates of cobalt(II) and nickel(II), in aqueous solution [51].

| Ligand | $10^4 k_f/M^{-1} sec^{-1}$ | |
	$Co^{2+}aq$	$Ni^{2+}aq$
α-Alanine	60	2.0
β-Alanine	7.5	1.0
α-Aminobutyrate	25	1.0
β-Aminobutyrate	2.0	0.4

12.3.2 Manganese(II)

The same mechanisms apply here as to nickel(II) and cobalt(II), though much less kinetic information is published concerning substitution at $Mn^{2+}aq$. Rate constants (k_f) for the reactions of $Mn^{2+}aq$ with chloride, 8-hydroxyquinolinate (oxine$^-$), and nitrilotriacetate (nta^{3-}) are 1.5×10^7 (25°C, [53]), 1.1×10^8 (16°C, [54]), and 5×10^8 (25°C, [55]) $M^{-1} sec^{-1}$ respectively. The interchange rate constant, k_i, for the formation of the sulphate complex has been estimated from ultrasonic measurements as $4 \times 10^6 sec^{-1}$ at 20°C [2]. These rate constants are all in the same region as rate constants for water exchange at $Mn^{2+}aq$ at the respective temperatures. E.s.r. measurements have shown that the rates of reaction of $Mn^{2+}aq$ with halide ions are all similar, with chloride $k_f = 1.4 \times 10^{10}$, with bromide 1.1×10^{10}, and with iodide 0.8×10^{10} $M^{-1} sec^{-1}$ (at 160°C) [16]. All of these kinetic observations are consistent with the operation of the normal Eigen-Wilkins mechanism, with dissociative interchange rate-determining. Moreover the activation energy for the interchange step in the reaction of $Mn^{2+}aq$ with chloride has been estimated as about 6 kcal mol^{-1} [53], which agrees tolerably well with the value of 8.1 kcal mol^{-1} for water exchange (Chapter 11). The fast rates and low activation energies for substitution at manganese(II) are consistent with the absence of any crystal field stabilisation for this high-spin d^5 ion. In consequence crystal field activation energies here are zero.

As water loss from Mn^{2+}aq can take place so rapidly, the chance is high that ring closure is rate-limiting in formation reactions with polydentate ligands (cf. nickel(II) above). As an example, the formation rate constant for reaction with β-alanine is only 5×10^4 $M^{-1}sec^{-1}$ at 20°C [33]. The rate of formation of $Mn(terpy)^{2+}$ is also extraordinarily slow, due to the stringent steric demands of making the planar double chelate ring system in the product. Indeed the half-life for this formation reaction is a matter of hours [56]. Rate constants for the formation of the 2,2'-bipyridyl and 1,10-phenanthroline complexes are about ten times slower than expected, so ring closure may be just starting to be rate-limiting for these ligands (and perhaps for the oxine complex mentioned above, too) [57].

12.3.3 Copper(II)

The kinetic pattern here is very similar to that for complex formation from Mn^{2+}aq. Again water loss from Cu^{2+}aq is very rapid, this time because the Jahn-Teller distortion of the geometry around this d^9 cation results in two relatively distant and weakly bonded water molecules in the primary hydration sphere. Consequently rates of formation with 'normal' ligands are very fast, for example with ammonia k_f is 2.0×10^8 and with imidazole 5.7×10^8 M^{-1} sec^{-1} respectively, at 25°C [58]. Similarly, rate constants (k_f) for reactions with a variety of bio-chemical ligands, including glycine, serine, sarcosine, leucine [59], valine [60], glycylglycine [61], and phenylalanine [62], all lie within the range 3.5 to 40×10^8 M^{-1} sec^{-1}. Rate constants at the lower end of this range correspond to reaction of Cu^{2+}aq with the bulkier amino-acids [63]. The interchange rate constant within the Cu^{2+}aq,SO_4^{2-} ion-pair is greater than $10^7 sec^{-1}$ [2]. There is some correlation between rate constants for substitution at Cu^{2+}aq and at Ni^{2+}aq; k_{Cu}/k_{Ni} is about 10^5.

Activation enthalpies between 4.3 and 6.0 kcal mol^{-1} have been reported for reactions with several nitrogen bases, though here it must be admitted that awkward values such as $\Delta H^{\ddagger} = -6$ kcal mol^{-1} reported for the reaction with monoprotonated diethylenetriamine tend to introduce some degree of confusion [64]. An activation volume of $+12$ cm^3 mol^{-1} has been found for the reaction of Cu^{2+}aq with glycine [31]. The kinetic behaviour detailed in these two paragraphs, and its relation to the known kinetics of water exchange at Cu^{2+}aq, are once again consistent with the normal Eigen-Wilkins mechanism applying to the reactions of Cu^{2+}aq with normal ligands.

The rate of water loss from Cu^{2+}aq is so fast that in reactions with species that must lose a proton prior to complex formation there is the possibility that such proton loss may be rate-determing. The rate constants (k_f) for the reactions with protonated ethylenediamine, 1.4×10^5 M^{-1} sec^{-1} at 25°C [65], and with other similar protonated species [64], are much less than those for reactions with 'normal' ligands such as ammonia (see above). Relatively slow proton loss also appears to result in relatively slow rates of formation with some ligands of biological importance, for example in the reaction of $Cu^{2+}aq$ with deutero-porphyrin IX disulphonic acid dimethyl ester [66].

The rapidity of water loss from $Cu^{2+}aq$ also makes rate-determining ring closure likely in formation reactions which produce chelates. For example, the rate constant for formation of $Cu(acac)^+$ from $Cu^{2+}aq$ and the enol form of acetylacetone is only 2×10^4 $M^{-1}sec^{-1}$ at $25°C$ [67]. Proton loss as well as ring closure may contribute to this low rate. The rate constant for the reaction of $Cu^{2+}aq$ with L-carnosine (β-alanyl-L-histidine, Fig. 12.6(a)) in its unchanged form is $k_f = 3.5 \times 10^6$ $M^{-1}sec^{-1}$ at $25°C$. Formation of the seven-membered chelate ring is thus rate-determining here, whereas ring closure is not rate-determining in the formation of L-carnosine complexes from $Ni^{2+}aq$ or from $Co^{2+}aq$ [68].

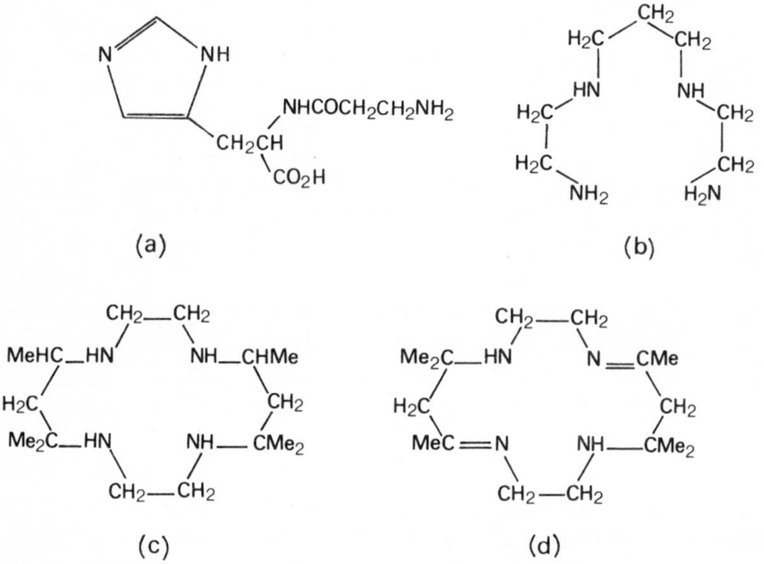

Figure 12.6 (a) Carnosine; (b) 2,3,2-tet; (c) tet-*a* and tet-*b* (stereoisomers); (d) *trans*-[14]-diene.

The reactions of $Cu^{2+}aq$ with macrocyclic ligands take place much more slowly than those with comparable open-chain ligands. Reactions with rigid macrocycles take place even more slowly than those with flexible macrocycles (Table 12.6 [69, 70]). It is not yet established whether the slowness of these formation reactions is entirely a configurational or rigidity effect, or whether ligand solvation is important [40, 41, 71]. In any case, these effects are more important in determining dissociation rates than formation rates. In the reaction with tet-a, the formation of a blue species thought to contain the ligand bonded through two nitrogen atoms to the copper can first be observed, and then the formation of the final product (mauve) containing tetradentate tet-a. Under suitable conditions, the latter process has a half-life of a couple of days [72]. It is interesting that even for reactions of M^{2+}aq with sterically rigid and demanding macrocyles, relative rates of complex formation for different metal cations are not entirely divorced from solvent exchange rates. Reactions with tetramethylcyclam have rates a thousand or more times slower than analogous reactions with, say, ammonia, although the order of formation rates is $Cu^{2+} > Zn^{2+} > Co^{2+} > Ni^{2+}$, just as for solvent exchange [73].

Table 12.6 Rate constants for reactions of Cu^{2+}aq with macrocyclic ligands (formulae in Fig. 12.6) at 25°C in 0.5M NaOH.

Ligand		$k_f/M^{-1}sec^{-1}$	
2,3,2–tet	open chain	10^7	
tet-a	somewhat flexible	1.6×10^3	[69]
tet-b		3.6×10^3	
trans-[14]-diene		5.6×10^3	
Haematoporphyrin IX	rigid	2.0×10^{-2}	
$\alpha\beta\gamma\delta$-Tetra(4-N-methylpyridyl)porphine		8.7×10^{-2}	[70]

Hydrogen bonding effects can be important in determining reactivities for Cu^{2+}aq as for the other M^{2+}aq discussed already in this Chapter. The reaction of Cu^{2+}aq with ethylenediamine occurs more rapidly ($k_f = 3.8 \times 10^9$ $M^{-1}sec^{-1}$ at 25°C [65]) than normal substitution at Cu^{2+}aq. The obvious explanation is that the ICB mechanism is responsible for this fast rate, but it has been suggested that the mechanism may be associative in character. Intraligand hydrogen-bonding can result in slow complex formation, as in the reaction with hydroxyproline (Fig. 12.5(a)) [47] or with par (see Chapter 12.6.3).

12.3.4 Other First Row Transition Metal(II) Cations

For normal monodentate ligands, the Eigen-Wilkins mechanism with rate-determining dissociative interchange probably applies to all of these. The rate constant for the reaction of V^{2+}aq with thiocyanate has a value between 9 and 28 $M^{-1}sec^{-1}$ [74, 75, 76], while that for reaction with 1,10-phenanthroline is 3.0 $M^{-1}sec^{-1}$ [76]. The rate of water exchange at V^{2+}aq is 80 sec^{-1} (Chapter 11). Activation parameters for the reaction of V^{2+}aq with thiocyanate are $\Delta H^{\ddagger} = 13.5$ kcal mol^{-1} and $\Delta S^{\ddagger} = -2$ cal deg^{-1} mol^{-1} [74]; for water exchange they are 16.4 kcal mol^{-1} and $+5$ cal deg^{-1}mol^{-1}. The slow formation rates for V^{2+}aq are reasonable in the light of the high crystal field stabilisation of, and therefore high activation energies for, this d^3 cation.

Rate constants (k_f) at 25°C for formation reactions of Fe^{2+}aq with nitric oxide, 2,2'-bipyridyl, and 2,4,6-tripyridyl-1,3,5-triazine are 5×10^5 [77], 1.5×10^5 [23], and 1.3×10^5 [78] M^{-1}sec^{-1} respectively. The interchange rate constant (k_i) for the reaction of Fe^{2+}aq with sulphate is 1×10^6 sec^{-1} at 20°C [2].

Formation reactions of Cr^{2+}aq are very fast. This d^4 cation, like the d^9 Cu^{2+}aq cation, is Jahn-Teller distorted and contains two distant and very easily lost water molecules in the primary coordination shell. The reaction of Cr^{2+}aq with dioxygen, whose first step is substitution to give CrO$_2^{2+}$, has $k_f = 1.6 \times 10^8$ M^{-1}sec^{-1} [79]. The rate constant for the reaction of Cr^{2+}aq with 2,2'-bipyridyl, 3.5×10^7 M^{-1}sec^{-1} at 24°C, probably represents rate-determining ring closure subsequent to somewhat faster water loss [80]. Rate constants for reactions of Cr^{2+}aq with alkyl radicals are in the range 3.4×10^7 to 3.5×10^8 M^{-1}sec^{-1} [81]. Here, as in the reaction with dioxygen, substitution precedes oxidation of the chromium.

12.3.5 Second and Third Row Transition Metal(II) Cations

The number of aquo-metal(II) cations in the second and third rows of the transition series is small, and the kinetic information in this area correspondingly limited. In fact kinetic studies have been concerned with only the more unusual cations, for example square-planar Pd^{2+}aq and dimeric Mo$_2^{4+}$aq.

Reactions of Pd^{2+}aq with chloride, bromide [82], or substituted anilines [83] follow a second-order rate law. Some kinetic results are given in Table 12.7. The most likely mechanism is associative attack of the incoming ligand at the palladium. This mechanism, which differs from that demonstrated for all the (octahedral) M^{2+}aq cations discussed so far, is proposed in the light of the overwhelming body of evidence that second-order rate terms in square-planar substitution correspond to bimolecular processes. The small negative activation

Table 12.7   Kinetic parameters for formation reactions of Pd^{2+}aq.

Ligand	k_{25}/M^{-1}sec^{-1}	$\Delta H^{\ddagger}$/kcal mol^{-1}	$\Delta S^{\ddagger}$/cal deg mol^{-1}	
Chloride	$1.83 \times 10^{4\,a}$	10	-6	} [82]
Bromide	$9.2 \ \times 10^4$	10	-3	
p-MeO.C$_6$H$_4$.NH$_2$	$1.8 \ \times 10^4$	11.3	$+16$	}
C$_6$H$_5$.NH$_2$	$1.1 \ \times 10^4$	10.2	$+12$	} [83]
p-O$_2$N.C$_6$H$_4$.NH$_2$	$9.8 \ \times 10^2$	7.9	-23	}

a $k_{25} = 1.03 \times 10^4$ M^{-1}sec^{-1} [84].

entropies reported for formation of the chloro- and bromo-complexes from Pd^{2+}aq (Table 12.7) are consistent with an associative reaction. This is especially so when the release of solvating water from the reacting ions as they form the transition state of charge only 1− is taken into account.

12.4   METAL(II) CATIONS OF THE *sp*-ELEMENTS

Two types of kinetic behaviour can be distinguished. The first is the pattern described above for first row transition metal(II) cations, where fast equilibrium pre-association is followed by rate-determining interchange. This pattern applies to, for instance, Mg^{2+}aq. In the second, cations such as the alkaline earth cations Ca^{2+}aq, Sr^{2+}aq, and Ba^{2+}aq have a rate of solvent loss from the cation so fast that it is comparable with rates of association between the aquo-cation and the ligand, and two distinct steps cannot be seen. Formation reactions will be very fast unless some factor such as difficult ring closure interferes.

12.4.1 Magnesium(II)

The rate of water exchange at Mg^{2+}aq is relatively modest (Chapter 11). Hence the reaction sequence is of rapid pre-association followed by rate-determining interchange, just as described at length for nickel(II) at the beginning of this Chapter. The results listed in Table 12.8 [85–88] show this to be the case. The similarity of all the k_i values to each other and to the rate constant for water exchange at Mg^{2+}aq indicates that the interchange process is dissociative. Activation parameters for a few formation reactions are compared with the

Table 12.8   Rate constants and association constants for formation reactions of Mg^{2+}aq in aqueous solution.

	Temp.	$k_f/M^{-1}sec^{-1}$	K_{os}/M^{-1} [a]	k_i/sec^{-1} [b]	
$HP_2O_7^{3-}$	15°C	3.9×10^6	34	1.1×10^5	[85]
$HP_3O_{10}^{4-}$	15°C	8.5×10^6	80	1.1×10^5	[85]
CTP^{4-}	15°C	8.7×10^6	80	1.1×10^5	[85]
Oxine H	16°C	1.1×10^4	0.15^c	0.7×10^5	[86]
Oxine$^-$	16°C	3.8×10^5	2^c	1.4×10^5	[86]
HF	20°C	$\sim 4 \times 10^4$	0.4	$\sim 10^5$	[87]
F^-	20°C	3.7×10^4	1.6	0.2×10^5	[87]
Acetate$^-$	20°C			1×10^5	[87, 88]
SO_4^{2-}	20°C			1×10^5	[2]
$S_2O_3^{2-}$	20°C			1×10^5	[2]
CrO_4^{2-}	20°C			1×10^5	[2]

a Estimated by the authors cited, except where otherwise indicated; *b* first seven entries calculated from k_f and K_{os}, last four entries from direct ultrasonic measurements; *c* K_{os} values here are taken from typical nickel(II) cases (Table 12.1).

values for water exchange in Table 12.9 [89]. The reactions of $Mg^{2+}aq$ with malonate and with tartrate are on the verge of rate-determining ring closure (SCS mechanism). Estimated interchange rate constants are about twice the estimated ring closure rate constants here [90]. The reaction of Mg^{2+} with pyrophosphate, $P_2O_7H^{3-}$, is about 2.4 times slower in D_2O than in H_2O (at $15^\circ C$). The difference is attributed to differences in the association constant, K_{os} [91].

Table 12.9   Activation parameters for substitution at $Mg^{2+}aq$.

	$\Delta H^\ddagger$/kcal mol^{-1}	$\Delta S^\ddagger$/cal deg^{-1}mol^{-1}	
Formation, + ATP	11.3	+9	[89]
+ ADP	12.0	+9	[89]
+ oxine	12.3	+15	[89]
Water exchange	10.2	+2	Chapter 11

12.4.2 Calcium(II), Strontium(II), and Barium(II)

Ultrasonic studies of solutions of acetates and of aminocarboxylates of these cations have led to the rate constants listed in Table 12.10. As ion-pair formation, interchange, and ring closure are thought to have comparable rates for reactions at these cations, the rate constants in Table 12.10 are composite quantities. The rate constants for the reactions of $Ca^{2+}aq$ with the majority of the ligands are tolerably similar, and rate constants for a given incoming group show the expected trend, $Ca^{2+}aq < Sr^{2+}aq < Ba^{2+}aq$. As one would expect, rates of reaction of $Ca^{2+}aq$ with multidentate macrocyclic cryptand ligands are much slower than with simple ligands. The mechanism here is thought to be dissociative, with rate-determining substitution rather than rate-determining ligand conformational change [93].

Table 12.10   Rate constants, sec^{-1}, for the formation of complexes of the alkaline earth cations, in aqueous solution at $20^\circ C$.

Ligand	$Ca^{2+}aq$	$Sr^{2+}aq$	$Ba^{2+}aq$	
Glycinate^{-1}	4×10^8			[87]
Acetate$^-$	$\sim 10^8$			[88]
nta^{3-}				
edta^{4-}	3.0×10^8	3.5×10^8	7.2×10^8	[92]
ida^{2-}				
SO_4^{2-}	$\sim 10^7$			[2]
CrO_4^{2-}	2×10^7			[2]
Cryptands^a	10^2 to 10^4			[93]

a Fig. 12.9(b) and analogous ligands.

12.4.3 Beryllium(II)

This, the first member of the Group IIA cations, is atypical in two ways. First, it is of unique interest as the only aquo-cation of tetrahedral geometry. Secondly it is acidic, so that formation reactions may occur with Be^{2+}aq or with $Be(OH)^+$aq (or even with polynuclear species). In this latter respect beryllium(II) resembles the tripositive cations of Chapter 12.6 below.

Several investigators have suggested that the mechanism of formation, for example with fluoride [94] or with sulphate [2, 5], includes a dissociative interchange step. The difficulty with this postulate is reconciling the experimentally derived activation enthalpies for formation of about 12 kcal mol^{-1} with the estimated enthalpy of removal of a water molecule from Be^{2+}aq of about 100 kcal mol^{-1}. This difficulty has recently and ingeniously been resolved by Strehlow [95]. He has suggested the mechanism depicted in Fig. 12.7. Here, the beryllium atom first moves from the centre of its tetrahedron of primary sovating water molecules towards the centre of one face of this tetrahedron and the associated ligand beyond this face, Fig. 12.7(a). This gives a transition state, Fig. 12.7(b), in which the beryllium atom has squeezed between the coordinated water molecules to become coplanar with three of them. Because the beryllium is so small, it does not yet interact with the incoming group, and so the succeeding interchange step to give to the product, Fig. 12.7(c), is still effectively dissociative in character. Calculations using this model predict an activation enthalpy of about 19 kcal mol^{-1}, which is not impossibly far from the experimentally determined values.

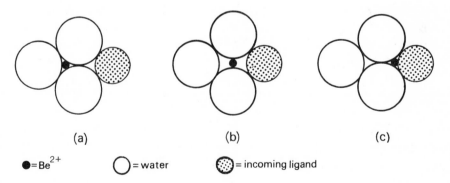

(a)   (b)   (c)

● = Be^{2+} ○ = water   ⊙ = incoming ligand

Fig. 12.7   Mechanism of formation of a complex from Be^{2+}aq.

12.4.4 Zinc(II), Cadmium(II), Mercury(II), Lead(II)

Zinc(II), like magnesium(II), is important in biochemistry and has as a consequence attracted some attention. Ultrasonic studies of the formation of the sulphato-complex are consistent with the reaction sequence of the Eigen-Wilkins mechanism [2, 96]. The interchange rate constant was originally estimated as $> 10^7$ sec^{-1}, at 20°C [2]. Later work demonstrated three relaxation frequencies for zinc sulphate solution, but did not provide an estimate for the interchange rate constant [96]. The interchange rate constant in the formation of the acetate complex is about 10^8 sec^{-1} at 25°C [88]. Formation reactions of Zn^{2+}aq with several ligands are around 10^7 to 10^8 M^{-1}sec^{-1} (Table 12.11 [97, 98]). The reactions of Zn^{2+}aq with bipy or terpy are somewhat slower, with rate constants of about 10^6 M^{-1}sec^{-1}, while the reaction of Zn^{2+}aq with carbonic anhydrase has a rate constant of 10^4 M^{-1}sec^{-1} [99]. Reactions with macrocyclic ligands such as porphyrins are even slower. The rate constant for the reaction with a water soluble derivative of protoporphyrin IX dimethyl ester is as low as 0.24 sec^{-1} at 30°C [100]. The activation volume for the reaction of Zn^{2+}aq with glycine is $+7$ cm^{3}mol^{-1} [31], which is similar to the values for the reactions of Ni^{2+}aq or Co^{2+}aq with this ligand. In fact the value for the Zn^{2+}aq reaction is slightly smaller than the others, perhaps reflecting the tendency of zinc(II) occasionally to be only five-coordinate, in mixed aquo-ligand complexes if not in Zn^{2+}aq itself [101].

Table 12.11 Decadic logarithms of rate constants (in M^{-1}sec^{-1}) for the formation of complexes of the Group IIB cations Zn^{2+} and Cd^{2+} (at 25°C).

Ligand	Zn^{2+}aq	Cd^{2+}aq	
Ammonia	6.6		[97]
1,10-Phenanthroline	6.8	>7.0	[97]
Acetate$^-$	6.9^a	7.3–7.5^a	[88]
Glycinate$^-$	7.8^b		[31]
Dithizonate$^-$	6.8		[98]
ntaH^{2-}		5.9	[55]
edtaH^{3-}	~9	9.0–9.6	[55, 97]
nta^{3-}		10.3	[55]
cf. water	7.5	8.2	Chapter 11

a These are logarithms of relaxation frequencies, in sec^{-1}; b estimated from rate constant measured at 10°C.

As one would expect, the rate constants for formation of mono-ligand complexes of cadmium(II) are somewhat greater than those for the formation of the analogous complexes of zinc(II) (Table 12.11).

The biochemically important formation reaction of alkylmercury cations, RHg$^+$, from Hg^{2+}aq and alkylcobalamins (Fig. 12.8) has been studied by conventional and stopped-flow methods and by 220 MHz n.m.r. spectroscopy. While the observed kinetic pattern looks complicated at the cobalamin end, as far as the Hg^{2+}aq and this book are concerned the reaction is simply heterolysis of the cobalt to alkyl–carbon bond by electrophilic attack by the mercury(II) [102].

Formations reaction of the Pb^{2+}aq cation are probably generally faster than those of Zn^{2+}aq or Cd^{2+}aq. For example the formation rate constant for the acetate complex is 7.5×10^9 $M^{-1}sec^{-1}$ at $25°C$ [103], and for the nitrilotriacetate complex (from $ntaH^{2-}$) is either 4.4×10^6 [104] or 2×10^8 [55] $M^{-1}sec^{-1}$ at $25°C$. The relatively slow rates of reaction of these three cations with $ntaH^{2-}$ may indicate that deprotonation of the anion is involved in the rate-determining step.

Plotting logarithms of rates of metal deposition from solutions containing M^{2+}aq onto the respective metal electrodes or onto mercury against logarithms of rate constants for substitution into the first coordination sphere produces tolerably linear results. This is the case for Pb^{2+}, Cd^{2+}, Zn^{2+}, and several first row transition metal M^{2+} ions. It indicates that, at least for these cations, loss of a water molecule from the aquo-ion is the kinetically important step in electrodeposition [105].

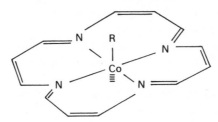

Fig. 12.8   The cobalt atom and its environment in alkylcobalamins.

12.5   METAL(I) CATIONS

12.5.1   Copper(I), Silver(I), and Thallium(I)

Pulse radiolysis studies of the reactions of Cu^+aq with the maleate and fumarate anions indicate rate constants of 2.0×10^9 and 1.7×10^9 $M^{-1}sec^{-1}$ respectively at $22°C$ [106]. These rates are very fast, as expected for a fairly large 1+ cation with a d^{10} configuration (that is, no crystal field stabilisation energy). Formation reactions of Ag^+aq, for example with thiosulphate or with 1,10-phenanthroline, are also very fast [107]. By way of contrast, formation reactions from Tl^+aq seem to be surprisingly slow. The rate constant for the reaction with tri-iodide is only 1.75×10^4 $M^{-1}sec^{-1}$ at $25°C$, while the activation parameters are $\Delta H^{\ddagger} = 7$ $kcal\,mol^{-1}$ and $\Delta S^{\ddagger} = -15$ $cal\,deg^{-1}mol^{-1}$. The slow formation rate is attributed to 'drastic solvent reorganisation' in the course of transition state formation [108].

12.5.2 Alkali Metal Cations

Formation reactions of the alkali metal cations are important both for their intrinsic interest as very fast reactions and for their relevance to the transport of these cations across cell membranes. In view of the reluctance of the alkali metal cations to form complexes, kinetic studies of their formation reactions [109] have to be carried out with strong complexing agents such as ethylene-diaminetetraacetate, nitrilotriacetate, and related biochemical agents (adenosine triphosphate, uramildiacetate, Fig. 12.9(a)). The very fast reaction rates have usually been measured by ultrasonic absorption techniques. There is a marked dependence of rate on ligand nature for a given cation. This is consistent with an Eigen-Wilkins mechanism coupled with chelate ring formation with several steps of comparable rate. Even such a complicated molecule as nonactin (see Fig. 12.17(b) on page 393) reacts very rapidly, though some cryptands react rather less quickly [110, 111]. The cryptand shown in Fig. 12.9(b) reacts with Li^+aq at about the same rate as it does with $Ca^{2+}aq$ (see above) [111]. Nonactin and the cryptands or crown ethers are flexible molecules, and can accommodate themselves to the stepwise loss of water molecules from M^+aq by an appropriate sequence of molecular contortions. This type of ligand and behaviour contrasts with complex formation from porphyrins, for example. Their rigid structure does not permit stepwise displacement of waters of hydration. As usual, rates of complex formation increase as one descends the Periodic Table (Table 12.12).

Table 12.12   Kinetic data for the formation of complexes of alkali metal cations in aqueous solution (20°C) [92].

	$10^{-7}\tau^{-1}$/sec^{-1}				$10^{-7}k_f$/M^{-1}sec^{-1}
	edta^{4-}	nta^{3-}	ida^{2-}	tpp^{5-}	uda^{3-}
Li^+	4.8	4.7	25	90	10
Na^+	4.7	8.8	28	>200	50
K^+	7.5	15		>500	100
Rb^+	14	23		>500	
Cs^+	21	35		>500	

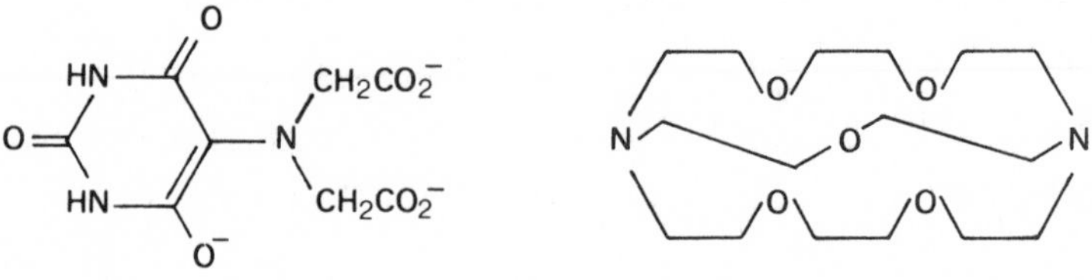

(a)   (b)

Fig. 12.9   (a) Uramildiacetate; (b) cryptand (cf. text).

12.6   METAL(III) AND METAL(IV) CATIONS

Kinetic patterns and mechanistic interpretations are more complicated for formation reactions of metal(III) and metal(IV) aquo-cations. For these cations, and for the very small beryllium(II) cation, such equilibria as

$$M(OH_2)_x^{n+} \rightleftharpoons M(OH_2)_{x-1}(OH)^{(n-1)+} + H^+$$

exist in solution. Both the aquo-ion and its hydroxo-conjugate base may react with potential ligands, so two parallel reaction paths may have to be distinguished. Worse, there is an ambiguity for reactions with ligands derived from weak acids. A distinction between the two conceivable paths

$$M(OH_2)_x^{n+} + L \longrightarrow M(OH_2)_{x-1}(L)^{n+} + H_2O$$

and $$M(OH_2)_{x-1}(OH)^{(n-1)+} + HL^+ \longrightarrow M(OH_2)_{x-1}(L)^{n+} + H_2O$$

cannot be made on the basis of kinetic evidence alone, because

$$+ \, d[complex]/dt \; = \; k[M(OH_2)_x^{n+}][L] \; = \; k'[M(OH_2)_{x-1}(OH)^{(n-1)+}][HL^+].$$

We shall therefore treat formation reactions of metal(III) and metal(IV) aquocations much more briefly than we have those of metal(II) cations, and will concentrate on Fe^{3+}aq, the most studied member of this group.

12.6.1   Iron(III)

Reactions of iron(III) with chloride, bromide, or thiocyanate must involve Fe^{3+}aq plus L^- rather than $FeOH^{2+}$aq plus HL. Consequently the reported rate constants of 4, 4, and 30 sec^{-1} for the ligand exchange step (k_i, estimated from observed k_f values by using an estimated K_{os} value of 5 M^{-1}) must refer to a process involving Fe^{3+}aq itself. At the other extreme, if one tries to interpret the kinetics of reaction of iron(III) with phenol, salicylate, and related ligands in terms of an Fe^{3+}aq plus L^- path, the reaction rates estimated exceed the diffusion controlled limit. Thus rate law terms of the type

$$+ \, d[complex]/dt \; = \; k[Fe^{3+}][salH][H^+]^{-1},$$

which can be taken as

$$+ \, d[complex]/dt \; = \; k'[Fe^{3+}][sal^-]$$

or $$+ \, d[complex]/dt \; = \; k''[Fe(OH)^{2+}][salH],$$

must be interpreted as involving the latter reaction of the hydroxoiron(III) species with salicyclic acid, salH.

Intuition and reasonableness must guide assignment of mechanism for the large number of reactions which fall into neither of these limiting categories. If the Fe^{3+}aq plus L^- mechanism is assumed, interchange rate constants fall into a range from 400 sec^{-1} for the reaction with sulphate to 10^4 to 10^5 for reaction with various carboxylates, fluoride, and chromate. All these values look surprisingly high in comparison with the k_i values given above for the reactions of Fe^{3+}aq with chloride, bromide, and thiocyanate. However, this range of k_i values could be interpreted in terms of associative interchange. On the other hand, if one assumes that reaction occurs between $FeOH^{2+}$aq and HL, wherever this is possible, interchange rate constants are all around unity and so tolerably consistent with dissociative interchange. Accepting that the $FeOH^{2+}$ plus HL path fails to operate only when L^- is too weak a base is perhaps a slightly more attractive alternative than accepting the Fe^{3+}aq plus L^- path until it is abruptly ruled out by the limiting diffusion rate [112]. No experimental results extant are inconsistent with an I_d mechanism, and an I_d mechanism has been demonstrated for complex formation in dimethyl sulphoxide solution, where there is no kinetic ambiguity [113]. A dissociative mechanism also operates for formation reactions of such biochemical species as ferric myoglobin [114].

Enthalpies of activation for complex formation at iron(III) are claimed to be consistent with dissociative interchange despite a disconcertingly wide range of values. $\Delta H^{\ddagger}$ ranges from 6 kcal mol^{-1} for the reaction of $FeOH^{2+}$aq with mandelic acid [115] to 16 kcal mol^{-1} for the reaction of $FeOH^{2+}$aq with sulphate [116] via, for example, $11.1 \text{ kcal mol}^{-1}$ for the reaction with acetic acid [117] and $9.6 \text{ kcal mol}^{-1}$ for the reaction with catechol [118].†

It has been postulated that the internal conjugate base mechanism can operate for iron(III), for instance in the reaction with the diprotonated form of nitrilo-triacetate (ntaH_2^-), which reacts twenty times faster than thiocyanate [119].

12.6.2 Other Transition Metal(III) Cations

Titanium(III). The low electron density in the t_{2g} orbitals of this d^1 cation, and its 3+ charge, suggest that formation reactions of this cation may have some associative character. The interchange rate constant for reaction of Ti^{3+}aq with thiocyanate is about 10^4 sec^{-1}. Indirect arguments favour associative interchange here [120].

Vanadium(III). Here, as for titanium(III), the low electron density in the t_{2g} orbitals (d^2 configuration this time) and the +3 charge should be conducive to

---

† Last year's report of the determination of activation volumes for the reaction of iron(III) with chloride suggests that the reaction of $Fe(OH)^{2+}$ with Cl^- is I_d ($\Delta V^{\ddagger} = +7.8 \text{ cm}^3 \text{ mol}^{-1}$), but that reaction of Fe^{3+}aq with Cl^- is I_a ($\Delta V^{\ddagger} = -4.5 \text{ cm}^3 \text{ mol}^{-1}$) (Hasinoff, B. B., *Canad. J. Chem.*, **54**, 1820 (1976)).

associative activation. The wide range of rate constants reported for formation reactions with various incoming ligands (Table 12.13) suggests a mechanism of appreciably associative character [121]. So does the markedly negative activation entropy (-23 cal deg^{-1} mol^{-1}) for the reaction with thiocyanate [122], together with comparisons of reaction rates of azide with vanadium(III) and with other cations [123]. However the kinetic characteristics· of formation reactions of $V(dmso)_6^{3+}$, where similar rates for different incoming ligands indicate dissociative interchange, suggest caution in interpreting the kinetic results in aqueous solution [124].

In contrast with the behaviour of metal(III) ions later in the transition series, the reactivity of $V(OH)^{2+}$aq seems to be less than that of V^{3+}aq. This is also thought to apply to titanium(III). The difference in reactivity patterns may arise from the difference, associative as against dissociative, between formation mechanisms [75].

Table 12.13 Rate constants (k_f) for the formation of complexes from aquo-vanadium(III) (V^{3+}aq) in aqueous solution ($25°$C) [121].

Ligand	k_f/M^{-1} sec^{-1}
Oxalate[a]	1300
Thiocyanate	110[b]
Hydrazoic acid	0.4
Chloride	3
Bromide	10

a $HC_2O_4^-$; b or 104 M^{-1} sec^{-1} [75].

Chromium(III). Slow substitution rates at this inert d^3 centre encouraged early investigation of the kinetics of complex formation from Cr^{3+}aq. In 1955 it was reported that the rate law for the formation of the monothiocyanato-chromium(III) complex was

$$+ \, d[CrNCS^{2+}]/dt \; = \; k_f[Cr^{3+}aq][NCS^-] \; .$$

If, as for metal(II) cation reactions, k_f is considered the product of an ion-association constant K_{os} and an interchange rate constant k_i, a value of about 2.7×10^{-6} sec^{-1} for k_i results from an estimated value of 7 for K_{os}. Such a value for k_i, close to the rate constant for water exchange with Cr^{3+}aq, strongly suggested a dissociative interchange mechanism for this formation reaction [125]. Three years later a report that formation rates of a variety of carboxylato-complexes depended only slightly on the nature of the incoming anion lent support to the dissociative hypothesis [126].

More recently the question of mechanisms of substitution at chromium(III) has occasioned much argument, with evidence suggesting both dissociative and associative substitution at this centre. These arguments now seem to be in the process of resolution, with the operative mechanism apparently determined by the ligands present on the metal. Thus ligands with a powerful *trans*-effect, such as iodide or methyl, promote dissociative substitution [127]. As far as the particular case of formation reactions of Cr^{3+}aq is concerned, the current interim picture is as outlined in the next paragraph.

The greater range of rate constants for formation from Cr^{3+}aq as opposed to formation from $Cr(OH)^{2+}$aq† has been explained in terms of associative character for the former and dissociative for the latter. The π-bonding propensity of coordinated hydroxide could favour a dissociative path [128]. The observed correlation between free energies of activation and free energies of reaction also suggests an associative mechanism for Cr^{3+}aq formation reactions [129]. The rate of formation of the monoglycinato-complex is ten times that for water exchange, again indicating associative reaction [130]. The activation volume for the reaction of Cr^{3+}aq with oxalate is -2.2 cm^3 mol^{-1}, a moderately negative value suggesting but hardly proving associative character here [131]. Activation enthalpies for the formation of the mono-oxalato-complex (26.6 kcal mol^{-1} [132]) and of the unidentate monoglycol complex (25.2 kcal mol^{-1} [133]) can be compared with that for water exchange at Cr^{3+}aq (26 kcal mol^{-1}, Chapter 11). Interestingly, the activation enthalpy for the closure of the glycol chelate ring,

$$Cr(glycol)(OH_2)_5^{3+} \longrightarrow Cr(glycol)(OH_2)_4^{3+} + H_2O ,$$

is only 15.3 kcal mol^{-1} [133]. The proximity of the entering glycol-hydroxy group to the chromium seems to facilitate reaction, presumably by an associative mechanism.

All the formation reactions mentioned so far take place at the sedate rate normal for Cr^{3+} substitution. In contrast, the tetra-(p–sulphonatophenyl)-porphine complex undergoes formation reactions with such alacrity that the kinetics have to be monitored by stopped-flow techniques. A comparable situation for cobalt(III) was rationalised by invoking a redox mechanism for substitution. Such a mechanism seems less attractive here in the light of the difficulty of reducing chromium(III) to chromium(II) [134]. Yet there must be some special aid to reactivity, for reactions of solvento-cations with rigid porphyrins are usually much slower than those with 'normal' ligands.

For readers with a special interest in kinetics of formation reactions of

---

† There is the same kinetic ambiguity, M^{3+}aq + L$^-$ being equivalent to, and indistinguishable from, $M(OH)^{2+}$aq + HL, here as for iron(III), see above.

chromium(III), extensive compilations both of rate constants [129, 135] and of activation parameters [135] exist.

Molybdenum(III). The incidence of associative substitution seems greater in complexes and in organometallic derivatives of the second and third row transition elements than in those of their first row brethren. In the previous paragraphs several examples of associative mechanisms of substitution at chromium(III) have been cited. One might well, therefore, expect the mechanism of formation of complexes from Mo^{3+}aq to be associative. Due to difficulties in generating and characterising aqueous solutions of Mo^{3+}aq, kinetic studies in this area got off to a bad start. Recent kinetic results (Table 12.14) for the formation of chloro- and of thiocyanato- molybdenum(III) complexes do indeed, by their strong dependence on the nature of the incoming ligand, indicate an associative formation mechanism [136]. As in associative formation reactions of Ti^{3+}aq [120] and of V^{3+}aq [75], but in contrast to dissociative formation from other M^{3+}aq cations, rates of complex formation from Mo^{3+}aq show little dependence on pH.

Table 12.14   Kinetic parameters for formation reactions of Mo^{3+}aq [136].

Ligand	$k_{25}/M^{-1}sec^{-1}$	$\Delta H^{\ddagger}/kcal\ mol^{-1}$	$\Delta S^{\ddagger}/cal\ deg^{-1}mol^{-1}$
Chloride	0.0046	23.5	+9.6
Thiocyanate	0.27	16.3	−6.4

Cobalt(III) and Rhodium(III). The combination of the inertness of Co^{3+}aq towards substitution with its enthusiasm for oxidising water makes studying the kinetics of complex formation at this cation difficult. Nonetheless some results have been achieved, notably with salicylate [137] and chloride [138] as incoming groups. I_d mechanisms seem likely. Substitution at Rh^{3+}aq, as at chloro-aquo-complexes of rhodium(III), is currently thought to take place by a D mechanism [139].

12.6.3 Lanthanide(III) Cations

Complex formation, like solvent exchange, is very fast for these cations. It is therefore usually studied by relaxation methods, either single perturbation methods such as temperature-jump, or periodic perturbation (i.e. ultrasonics). Difficulties of execution and of subsequent interpretation of the results have resulted in a certain amount of controversy in this area, which has not been fully resolved at the time of writing. At least for these relatively non-acidic aquo-cations (Table 9.5, Chapter 9.2.1) the M^{3+}aq against $M(OH)^{2+}$aq problem is not important.

Discussion of kinetic results is usually in terms of the familiar Eigen-Wilkins mechanism. Complications may arise both from uncertainties about the co-ordination number of the lanthanide cation in the initial aquo-complex and in the product, and about the denticity of polydentate ligands. Some caution is also recommended when considering results of experiments in which perchlorate has been used to maintain ionic strength, because the kinetics of complex formation between perchlorate and Er^{3+}aq have recently been described, albeit in aqueous methanol rather than in pure water [140]. Caution is also required in interpreting kinetics of formation of nitrate complexes [141] and of sulphate complexes [142], because only one relaxation is observed by ultrasonic tech-niques. It is not clear whether this indicates a single step rather than a two step Eigen-Wilkins mechanism, or just that one (or more) relaxations are difficult to detect [141]. Two relaxations reported [143] for nitrate complex formation in strong solutions of lanthanide(III) nitrates have been said to relate to the form-ation of mono- and of di- nitrato-complexes [141]. The kinetics of reaction of Dy^{3+}aq with acetate are also complicated by the formation of di- as well as mono- acetate complexes [144]. Comparisons between rates measured in D_2O and in H_2O have been held to indicate that the Eigen-Wilkins mechanism does not apply [145], but this interpretation is not universally accepted.

Despite the difficulties and uncertainties outlined in the previous para-graph, it seems likely that the Eigen-Wilkins mechanism does operate for lanthanide(III) complexation reactions. Rates of formation of complexes are tolerably independent of the nature of the incoming ligand (Table 12.15) [146-148], and are in the same region as solvent exchange rates. For most Ln^{3+}aq the latter can only be estimated, but a rate constant of $9 \times 10^8 \ sec^{-1}$ has been measured by n.m.r. spectroscopy for water exchange at Gd^{3+}aq. Rates of water exchange at other Ln^{3+}aq cations will be similar. The activation energies for several complexation reactions, for example the formation of sulphato-samarium(III), anthranilatodysprosium(III), and the murexide (Fig. 12.2(c)) complex of samarium(III), lie in the same range (2 to 5 kcal mol^{-1}) as the activation energy for water exchange at gadolinium(III) [146].

Table 12.15 Rate constants $(10^{-8}k_f/M^{-1}sec^{-1})$ for the formation of samarium(III) complexes in aqueous solution [146].

Temperature	12 to 12.5°C			25°C
Ionic strength	$\to 0$	0.1M	0.2M	$\to 0$
Murexide	5.5	0.96		8.2
Anthranilate			0.63	
Oxalate				0.82
Malonate				7
Sulphate				$2.1^{a, b}$

a An earlier determination indicated $k_i \cong 10^9 \ sec^{-1}$ [147]; b a later re-evaluation of the observations produced only a small change in this value [148].

The dependence of formation rate constants on atomic number is not regular as one goes across the series from La^{3+}aq to Lu^{3+}aq. Even the patterns for different ligands, say murexide [149] and anthranilate [150], show different irregularities. There is some sort of discontinuity around gadolinium, which may or may not arise from a change in the coordination number of the lanthanide cation in the Ln^{3+}aq cation, in the complex produced, or both.

In view of the rapid loss of water from Ln^{3+}aq cations, one might expect that the rate-limiting step in complex formation reactions need not always be water loss from the aquo-cation (compare for example copper(II) in Chapter 12.3.3). Indeed such is the case in the reaction of La^{3+} with par, shown in Fig. 12.10(a). Here the rate-determining step is the formation of the second chelate ring, as in Fig. 12.10 (b) → (c). This involves breaking the internal hydrogen bond within the ligand so that the oxygen atom is free to coordinate to the lanthanum [151]. Similarly a comparison of the kinetics of reaction of Sm^{3+}aq with sulphate and with malonate indicates that ring closure is rate-determining (SCS mechanism) in the reaction with the latter ligand [146]. For this reaction with malonate, $k_i = 7 \times 10^8$ sec^{-1} and $k_{rc} \cong 6 \times 10^7$ sec^{-1}, at 25°C. On the other hand, the reaction of Tm^{3+}aq with edtaH^{3-} is very fast, $k_f = 1.0 \times 10^{10}$ M^{-1}sec^{-1} at 25°C [152]. The large charge product will favour fast reaction by making K_{os} large; presumably ring closure is of less importance here.

Fig. 12.10 The ligand 4–(2–pyridylazo)resorcinol, par (a), and the final stage in its complexation of La^{3+}, (b) → (c).

12.6.4 *sp*-Element(III) Cations

There have been a few studies of complex formation of the Group IIIA cations Al^{3+}aq, Ga^{3+}aq, and In^{3+}aq, and even an unsatisfactorily incomplete study of the reaction of Tl^{3+}aq with sulphate [96].

There is the usual ambiguity of pathway M^{3+}aq plus L$^-$ as against M(OH)$^{2+}$aq plus HL for a basic incoming ligand L$^-$. This ambiguity flourishes for the ligand semi-xylenol orange (sxo), because the rate law has been interpreted in terms of most of the conceivable combinations of M^{3+}aq, M(OH)$^{2+}$aq, and M(OH)$_2^+$aq with the series sxoH$_6^{2+}$ . . . . . . sxoH$_2^{2-}$ [153] !

Activation energies for the interchange formation step in the generation of AlSO$_4^+$ and of GaSO$_4^+$ are 21.5 and 12.5 kcal mol^{-1} respectively [154]. This parallels the activation energies of 27 and 6.3 kcal mol^{-1} respectively for water exchange at the Al^{3+}aq and Ga^{3+}aq cations, and suggests that the dissociative-associative mechanistic difference in the solvent exchange processes also exists for the formation reactions, making them I_d for aluminium and I_a for gallium. Moreover the interchange rate constant (~ 1 sec^{-1} for sulphate) is similar to the water exchange rate constant (0.13 sec^{-1} at 25°C) for Al^{3+}aq, whereas the analogous rate constants differ greatly for Ga^{3+}aq. Supporting evidence for a dissociative mechanism for formation reactions from Al^{3+}aq is supplied by a study of the reaction of this aquo-cation with methylthymol blue [155].

12.6.5 Metal(IV) Cations

Simple aquo-cations of this type are rare, and many such M^{4+}aq cations are sufficiently strong oxidants to oxidise most potential ligands. But, as will be seen in Chapter 13, such ligand oxidation may be slower than, and kinetically distinguishable from, complex formation. So some kinetic results are available for formation reactions of M^{4+}aq cations, for example of Pr^{4+}aq (or $Pr(OH)^{3+}$aq) with bisulphate, nitrite, bromide, and hydrogen peroxide [156], of Pu^{4+}aq with hydrogen peroxide [157], and of Ce^{4+}aq with diethylenetriaminepenta-acetate [158].

12.7 FORMATION OF TERNARY COMPLEXES

This section deals with formation reactions from aquo-species which already have at least one non-solvent ligand bonded to the cation, ML^{n+}aq + L' $\rightarrow MLL'^{n+}$aq. The kinetic interest here, as in the comparable section in the solvent exchange Chapter, is to examine how the bound ligand L changes the reactivity of M^{n+}aq. There is much activity in this area, especially involving ligands and complexes of biochemical interest, but we shall be able to present only a small selection of results.

The complementary reaction, ML^{n+}aq + M'^{m+}aq $\rightarrow MLM'^{(n+m)+}$aq, is a much rarer subject of study. However, the kinetics of formation of Ni_2(histidine) have been briefly described [20].

12.7.1 Nickel(II)

Tables 12.16 and 12.17 give a selection of rate constants for the formation of ternary complexes of nickel(II) [159-163]. Several reactivity trends can be discerned in these results, while others can be extracted from the now copious literature on this subject. Indeed the quantity of information available has permitted generalisation and extension to environmental systems [164].

The first and vaguest generalisation to be made is that formation rate constants (k_f) for ordinary complexes of nickel(II) reacting with ordinary ligands are of similar magnitude to those for Ni^{2+}aq itself. The importance of

Table 12.16   Rate constants ($k_f/M^{-1}\,sec^{-1}$) for the reaction of substituted 1,10-phenanthroline complexes of nickel(II), $Ni(X\text{-}phen)^{2+}aq$, with variously charged entering groups, at 25°C [159].

X	nta^{3-}	$dienH^+$	$dienH_2^{2+}$
5,6-Me$_2$	4.15×10^6	426	1.30
5-Me	3.78×10^6	408	0.91
H	2.82×10^6	359	0.42
5-Cl	2.34×10^6	327	0.43
5-NO$_2$	1.15×10^6	381	0.42

electrostatics in determining k_f values (via K_{os} of course) is immediately apparent from the Table 12.16 and 12.17 values for $Ni(phen)^{2+}$ reacting with nta^{3-}, NH_3, $dienH^+$, and $dienH_2^{2+}$. Here the difference in charge product of the reactants between 6− and 4+ is reflected in a million-fold difference in second-order rate constants. The variation of ligand bound to the nickel in the starting complex has a less dramatic effect, but one that can at times be readily rationalised. Thus for the $Ni(X\text{-}phen)^{2+}$ series of complexes reacting with nta^{3-} or $dienH^+$, rate constants correlate with Hammett σ values for the phen substituents (Fig. 12.11) [59].

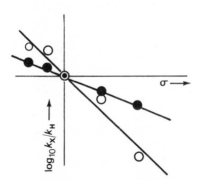

Fig. 12.11   Correlation of logarithms of ratios of rate constants for the reactions of ligands with $Ni(X\text{-}phen)(OH_2)_4^{2+}$ and with $Ni(phen)(OH_2)_4^{2+}$, $\log_{10} k_X/k_H$, with Hammett σ constants for the substituents X. Incoming ligands: ● $dienH^+$, ○ nta^{3-} [159].

The effects of already-bound ligands on the ease of water replacement are, as would be expected, similar to those reported for water exchange at comparable complexes (see Chapter 11). Thus coordinated water lability increases markedly in polyamine complexes, with increasing lability correlating with increasing denticity (that is, increasing number of ligating nitrogen atoms) of the polyamine [160].[†] Similarly thiocyanate ligands also greatly increase the lability of bound water. In consequence the complex $Ni(NCS)_4^{2-}$ aq reacts with thiocyanate with a rate constant of 5×10^5 $M^{-1} sec^{-1}$, despite the unfavourable ion-association contribution [165]. Conversely, heteroaromatic ligands such as 2,2'-bipyridyl or 1,10-phenanthroline have little effect on bound water lability [161]. The one exception to this parallelism between complex formation and solvent exchange rates is the $Ni(terpy)^{2+}$ aq complex. This reacts exceptionally rapidly with a further molecule of terpyridyl (Table 12.17) but undergoes water exchange at a normal rate [166]. Some special interaction between the complex and the incoming terpyridyl making a large contribution to K_{os} was suggested to explain the abnormal formation rate [161]. This special interaction has been more fully investigated for reactions of $Ni(phen)^{2+}$ aq [167]. If one heteroaromatic ligand can stack (Fig. 12.12) above the coordinated phen, this gives a high K_{os}. If it also has a nitrogen atom suitably placed for interchange, it keeps the usual k_i, and k_f will be much larger than for a normal entering ligand (Table 12.18). The favourable stacking interaction has been attributed to hydrophobic interaction, with a polarisation contribution when electron-releasing methyl substituents are present in the incoming ligand.

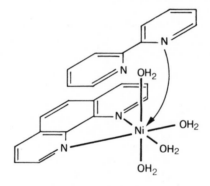

Fig. 12.12   Stacking interaction in the reaction of $Ni(phen)(OH_2)_4^{2+}$ with phen [167].

The special case of ternary complex formation where the bound and incoming ligands are identical fits comfortably into this picture. Examples involving such biologically significant ligands as glycine, tyrosine, and leucine have been reported, with copper(II) [168] as well as with nickel(II) [169]. The presence of such ligands coordinated to nickel(II) has little effect on the lability

† This statement also applies to rate constants for the actual interchange step. k_i values in Table 12.17 have been obtained from k_f values not only by allowing for K_{os} but also by applying a statistical factor to allow for the differing numbers of water molecules available for replacement in the different complexes.

Table 12.17 Rate constants for formation reactions from aquo-polyamine and from aquo-phenanthroline type nickel(II) complexes, at 25°C [160, 161].

		$10^{-3}k_f$ $/M^{-1}sec^{-1}$	$10^{-3}k_i$ $/sec^{-1}$		$10^{-3}k_f$ $/M^{-1}sec^{-1}$
$Ni(en)^{2+}$	$+ NH_3$[a]	12[b]	180	$Ni(phen)^{2+} + NH_3$	1.5
$Ni(dien)^{2+}$	,,	43	860	$Ni(bipy)^{2+} + bipy$	4.7
$Ni(trien)^{2+}$	,,	120	3600	$Ni(bipy)^{2+} + terpy$	5
$Ni(tren)^{2+}$	,,	260	7800	$Ni(terpy)^{2+} + terpy$	220
$Ni(tetren)^{2+}$	,,	v. fast[c]	—		

a The pattern of reactivities with pada is similar, and seems to be determined by the activation enthalpy variation [162]; b rate constants for the reaction of $Ni(en)^{2+}$ with bipy or phen are similar, at 5.0 and 9.5 X 10^3 $M^{-1}sec^{-1}$ respectively [163]; c $Ni(NH_3)_5^{2+}$ plus NH_3 has a rate constant of 720 X 10^3 $M^{-1}sec^{-1}$ [160].

Table 12.18 Rate enhancement arising from 'stacking' interactions in nickel(II) formation reactions [167]; all reactions with $Ni(phen)^{2+}aq$ at 25°C.

Entering ligand	$10^{-3}k_f/M^{-1}sec^{-1}$	
NH_3	1.5	No stacking
phen	3	Stacking, but interchange
bipy	37	difficult
Me_2bipy	158	Stacking promoted by methyls
pada	96	
terpy	100	Stacking, and suitably placed nitrogen ready for interchange
tptz	129	

of coordinated water, and so the rates of formation of the bis-complexes are similar to those of the mono-complexes. For instance, the reaction of $Ni^{2+}aq$ with tyrosine has $k_f = 1.4 \times 10^4$ $M^{-1}sec^{-1}$, and of the mono-complex with a second tyrosine molecule 2.4×10^4 $M^{-1}sec^{-1}$, at 25°C [169].

The rate constant for the reaction of ammonia with the complex shown in Fig. 12.13, $k_f = 7 \times 10^3$ $M^{-1}sec^{-1}$ at 25°C, is very similar to that of phen or bipy with the simpler analogue $Ni(en)_2^{2+}$ ($k_f = 9.3$ and 8.4×10^3 $M^{-1}sec^{-1}$ respectively, again at 25°C). However there is an extremely slow chelation reaction for the reaction of 12.13(a) with the bidentate ligands glycine or DL-alanine. As these ligands must span cis-positions at the metal atom, considerable reorganisation of the complex is required before chelate ring closure can take place [170]. Similar examples of geometrical reorganisation as a barrier to ternary complex formation will be found in the subsequent copper(II) section.

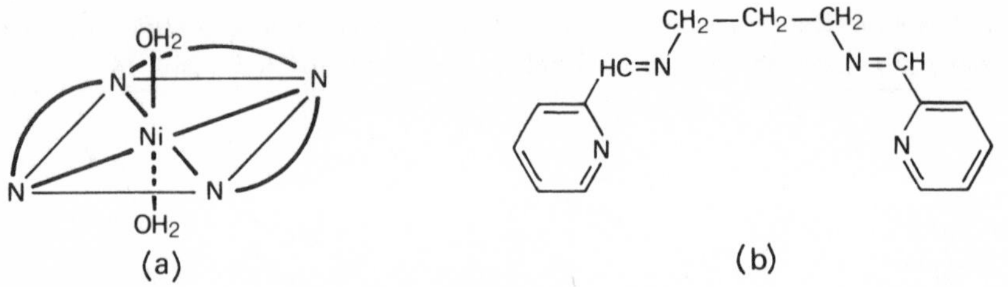

Fig. 12.13 (a) The complex $Ni(NNNN)(OH_2)_2^{2+}$, with $NNNN = NN'$-bis-(2-pyridylmethylene)–1,3–diaminopropane, (b).

Formation reactions of aminocarboxylate-aquo-nickel(II) complexes will be mentioned in Chapter 12.7.3 when reactivities of metal(II) complexes are compared. The reactivities of these nickel(II) complexes fall into the pattern described above, with the number of (non-aromatic) nitrogen atoms bonded to the nickel having a large effect on reactivity, as in the case of the polyamine complexes (see above) [171].

12.7.2 Copper(II)

Rate constants for the formation of some ternary complexes of copper(II) in aqueous solution are listed in Table 12.19 [60, 168, 172, 173]. Most of these are of the same order of magnitude as the rate of water exchange at Cu^{2+}aq, which is 10^8 sec^{-1} at 15°C (Chapter 11). When water loss is so fast, the possibility arises that some proton transfer step to or from ligand or solvent may be rate-determining. The rate constant for the reactions of $Cu(en)^{2+}$ and $Cu(bipy)^{2+}$ with protonated ethylenediamine, enH$^+$, are only 6×10^4 M^{-1}sec^{-1} (37°C) [173] and 2.2×10^4 M^{-1} sec^{-1} (25°C) [174] respectively. The slowness of these rates is attributed to the comparative difficulty of deprotonation of enH$^+$ [65]. Conversely, the apparent rates for the reaction of $Cu(en)^{2+}$ and of $Cu(hm)^{2+}$ with en are exceptionally fast. At 3.6×10^9 and 1.2×10^{10} M^{-1} sec^{-1} respectively (at 37°C), these rate constants are thought to be artefacts, in that the actual reaction is more likely to involve $Cu(OH)(ligand)^+$ plus a protonated incoming ligand [172].

Different reactivities for different complexes CuL^{2+}aq, or CuL^+aq, have been discussed in terms of steric effects, statistical factors, charge effects, and effects of bound ligands L on the lability of the remaining coordinated water molecules, as well as the possible operation of an associative mechanism in some cases. The relatively small differences in reactivity do not seem to permit an unravelling of these factors with any degree of conclusiveness. At least the effect of charge in reactions of CuL^{2+} or of CuL^+ with L$^-$ (Table 12.19, sections A and B, [168, 172]) can readily be explained by an electrostatic contribution to ion association reflected in the formation rate constant. It is interesting that rates of complex formation of ML_2^{2+} from $ML^{2+} + L$ are slower than those of ML^{2+} from

$M^{2+}aq + L$ for $M = Cu$, whereas for $M = Ni$ or Co the converse is the case (Table 12.19, section C). The relative slowness of bis-complex formation is exaggerated when the incoming ligand is bulky, for example bic (NN-dihydroxy-ethylglycine) [60].

Table 12.19 Rate constants, k_f, for the formation of ternary complexes of copper(II) in aqueous solution.

		Temp/°C	$k_f/M^{-1}sec^{-1}$	
A	$Cu(hm)^{2+} + hm$	37	5.6×10^8	
	$Cu(en)^{2+} + hm$	37	3.6×10^8	
	$Cu(en)^{2+} + ser^-$	37	6.2×10^8	[172]
	$Cu(ser)^+ + ser^-$	37	2.8×10^8	
B	$Cu(bipy)^{2+} + gly^-$	25	1.6×10^9	
	$Cu(gly)^+ + gly^-$	25	4×10^8	[168]
C	$Cu^{2+} + val$	25	1.1×10^9	
	$Cu(val)^{2+} + val$	25	2.3×10^8	
	$Cu^{2+} + bic$	25	0.95×10^9	[60]
	$Cu(bic)^{2+} + bic$	25	3.2×10^7	
D	$Cu(en)^{2+} + pada$	25	$\sim 5 \times 10^8$	
	$Cu(bipy)^{2+} + pada$	25	$> 10^8$	
	$Cu(dien)^{2+} + pada$	25	2.0×10^8	
	$Cu(bipy)_2^{2+} + pada$	25	$\sim 3 \times 10^7$	[173]
	$Cu(en)_2^{2+} + pada$	25	3.9×10^5	
	$Cu(gly)_2^{2+} + pada$	25	1.5×10^6	

The requirement of an incoming ligand to displace a *cis*-pair of water molecules from an aquo-cation is strikingly reflected in the kinetics of reaction of bis-aquo-copper(II) complexes with pada. The first four entries in section D of Table 12.19 refer to complexes containing at least one pair of water molecules *cis* to each other. Rates of reaction in these systems are comparable with the other normal rate constants in sections A to C of this Table. However, while $Cu(bipy)_2^{2+}$ contains *cis*- water ligands, $Cu(en)_2^{2+}$ and $Cu(gly)_2^{2+}$ contains *trans*-water ligands, and these last two complexes react much more slowly with pada. Once one end of the pada has coordinated to the copper, a major geometrical reorganisation is required as well as water loss before the pada chelate ring can be closed [173].

12.7.3 Other Metal(II) Cations

Patterns of reactivity for formation reactions of mixed aquo-ligand complexes of other metal(II) cations are less fully estabished, though the basic Eigen-Wilkins mechanism applies as usual, together with all the modifications detailed elsewhere in this Chapter. For example, the kinetics of reaction of the aquo-manganese(II) complex of pyruvate carboxylase with pyruvate can be analysed into K_{os} and k_i components, and the estimated rate constant for ligand interchange, $k_i = 1.5 \times 10^6$ sec^{-1}, is the same as the rate constant for water exchange at the aquo-manganese(II)-pyruvate carboxylate complex under similar conditions [175]. Some idea of the variation of rate constants with the nature of the bound ligand and with the nature of the cation can be obtained from the results collected together in Table 12.20 [57, 176–179]. In this Table the complexes have been arranged in order of overall charge, but it is obvious from the values in the various columns that overall charge is not the dominant factor in determining reactivitiy. Local charge densities and the number of ligating atoms on the ligand are also relevant. Correlations between pairs of cations are none too good, though it is possible to make vague generalisations of the sort that the magnesium(II) and manganese(II) series show less sensitivity of rate than the nickel(II) series to variation in nature of the bound ligand. Reaction rates in the zinc(II) series are very insensitive to the nature of the bound ligand.

Some of the irregularities may arise from interference from ring-closure (bipy) or slow proton loss (oxineH). In extreme cases, the ligand rather than the metal determines the rate of reaction. For example, the rate constants for the reactions of adenosine diphosphate complexes of Mn^{2+}, Mg^{2+}, and Ca^{2+} with creatine kinase all lie between 1.7 and 7.4×10^6 M^{-1}sec^{-1} at 11°C [180].

12.7.4 Cobalt(III), Rhodium(III), Chromium(III), and Iron(II)

Cobalt(III), rhodium(III), chromium(III), and low-spin iron(II) are all substitution inert centres with slow substitution rates. Formation reactions of complexes of the type $ML_5(OH_2)^{3+}$, with M = Co or Cr, have been studied for several decades, but the overall mechanistic picture is still not clear [181]. For dissociative activation there are two distinct routes (see Chapter 11.4.1), the interchange (I_d) mechanism of Fig. 12.14(a) and the limiting S_N1 (D) mechanism of Fig. 12.14(b). The former is the Eigen-Wilkins mechanism which has cropped up so often in this Chapter. The latter involves the generation of a transient five-coordinate intermediate whose lifetime is sufficient for it to discriminate between reacting with a molecule of the potential incoming ligand L$'$ or with water. Variation of incoming ligand concentration will affect observed rates for either mechanism. A higher concentration will give more of the pre-associated species in the Eigen-Wilkins pathway, or compete more effectively with water in the other pathway. So by either mechanism observed rate constants will increase as the ligand concentration increases, and the dependence of

Table 12.20 Rate constants (k_f/M^{-1} sec^{-1}) for complex formation from mixed aquo-ligand complexes of manganese(II), cobalt(II), zinc(II), magnesium(II), and nickel(II) with 2,2′-bipyridyl (bipy), 8-hydroxyquinoline (oxineH) and its anion (oxine$^-$), 5-nitrosalicylate (nsa^{2-}), and pyridine-2-azo-p-dimethylaniline (pada), all at 25°C.

	N^a	Mn(II) bipy $10^{-5}k_f$ [57]	Mn(II) oxineH $10^{-5}k_f$ [176]	Mn(II) oxine$^-$ $10^{-7}k_f$ [176]	Co(II) pada $10^{-4}k_f$ [177]	Zn(II) pada $10^{-6}k_f$ [178]	Mg(II) oxineH $10^{-4}k_f$ [176]	Mg(II) oxine$^-$ $10^{-4}k_f$ [176]	Mg(II) nsa^{2-} $10^{-4}k_f$ [179]	Ni(II) nsa^{2-} $10^{-3}k_f$ [57]
M^{2+}		2.8	4.9	33	7.6	6.8	1.3	60	70	1.02
M(dien)$^{2+}$	3	5.3				2.3				15
M(trien)$^{2+}$	4	2.2				1.1				46
M(tren)$^{2+}$	4	2.1								52
M(edda)^0	4	2.6				0.9				0.26
M(ida)^0	3				65	15				0.63
M(mida)^0	3	6.8								
M(nta)$^-$	4	14	17	2.3	370	2	2.5	15	12	0.15
M(uda)$^-$	4		14	2.2			9.3	5.5		
M(adp)$^{2-}$	2		14	8.5			2.5	25		
M(atp)$^{2-}$	3		93	2.5			2.4	12	12	
M(tpp)$^{3-}$	3	0.94	40	9.1	8.0	1.4	4.5	5.6	23	0.15

a N is the number of ligand atoms thought to be bonded to the metal cation.

rate on concentration is unfortunately the same for either mechanism. In both cases a plot of reciprocal rate constant against reciprocal ligand concentration is linear, at least under conditions relevant here.

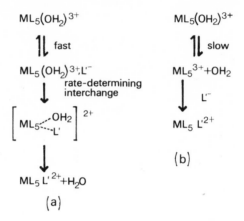

Fig. 12.14   Dissociative interchange, I_d (a), and limiting dissociative, S_N1(lim) or D (b), mechanisms for ternary complex formation.

There is evidence for the formation of ion-pairs in reactions of $Co(NH_3)_5(OH_2)^{3+}$ with ligands L', but this does not necessarily mean that the formation mechanism is via ion-pairs. Indeed in some cases, as when $L' =$ sulphate, reaction rates continue to increase on the addition of further ligand even after all of the cobalt(III) complex is in the ion-paired form. Consequently the I_d mechanism does not look particularly attractive, but it is difficult to gain definitive evidence for the operation of the alternative D mechanism. The inter-mediacy of $Co(NH_3)_5^{3+}$ in aquation of chloro-cobalt(III) complexes catalysed by mercury(II), or of azido-cobalt(III) complexes catalysed by nitrous acid, is fairly well established, so its intermediacy in formation reactions is not an unreasonable hypothesis. In principle I_d and D mechanisms can be distinguished by comparing the kinetics of formation with the kinetics of solvent exchange, but the differences are subtle [182]. Currently the I_d mechanism is favoured for cobalt(III) complexes. Recent kinetic investigations with results which seem best accommodated by an I_d mechanism include those of the reactions of $Co(NH_3)_5(OH_2)^{3+}$ with formate [183] and with other carboxylates [184], and also of $Rh(NH_3)_5(OH_2)^{3+}$ with oxalate [185]. The last-named should be contrasted with the proposal that an I_a mechanism operates in the reactions of $Rh(NH_3)_5(OH_2)^{3+}$ with simple inorganic anionic ligands such as chloride, bro-mide, azide, and sulphate [185]. The D mechanism has been suggested for reac-tions of the chloro- analogue of the preceding ammine complex, $RhCl_5(OH_2)^{2-}$, and of $RhCl(OH_2)_5^{2+}$ and Rh^{3+}aq with chloride [139, 186].

The situation is clearer for reactions of $Co(CN)_5(OH_2)^{2-}$ with ligands L' than for the ammine complex. The pentacyanoaquocation is not going to ion-pair with an incoming anionic ligand, so any marked dependence of formation rate on ligand concentration will indicate a D mechanism. In addition, cyanide is good at stabilising transition metal cations in relatively low coordination numbers, for example in the $Co(CN)_5^{3-}$, $Ni(CN)_4^{2-}$, and $Ni(CN)_5^{3-}$ anions. Formation reactions of dimethylglyoximato-complexes (cobaloximes) can also be more straightforward than those of $Co(NH_3)_5(OH_2)^{3+}$, as $Co(dmgH)_2(OH_2)L$ will be uncharged and so not eager to ion-pair when L has a charge of $1-$. Because L and OH_2 are *trans*- to each other in these complexes, they form a good system for investigating the *trans*-effect of ligands on formation reaction kinetics in octahedral systems. For instance the very strong *trans*-effect of alkyl groups (R) is illustrated by rates of formation from compounds (a) of Fig. 12.15, which are about 10^5 times faster than those from analogous compounds where the alkyl group has been replaced by an iodide or nitro-ligand [187]. As the *trans*-effect of alkyl groups is so strong, one can be sure that a D mechanism operates for compounds of the Fig. 12.15(a) type. Strong *trans*- labilisation by alkyl groups has also been demonstrated in compounds of the type Fig. 12.15(b) [188], and in the closely related and biochemically important alkylcobalamins (corrinoids), whose cobalt environment is shown in Fig. 12.8 [189]. The reaction of vitamin B_{12} with iodide has an activation volume of $+5.5\ cm^3 mol^{-1}$.

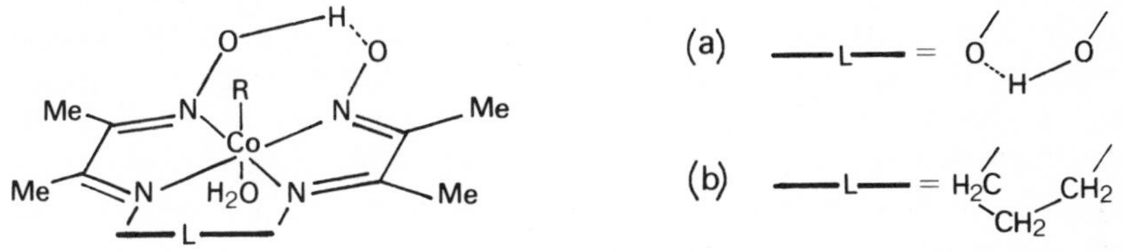

Fig. 12.15 Alkylcobaloximes.

This small positive value is consistent with a dissociative mechanism. The smallness of the value may stem from an unusually long cobalt-water bond length in the vitamin B_{12}. A long bond here would be consistent with the known rapidity of formation reactions, and with their low activation enthalpies, as compared with ordinary aquo-cobalt(III) complexes [190]. Bis-aquo-porphyrin complexes of cobalt(III) contain labile water ligands, as has been demonstrated by kinetic studies of their reactions with pyridine and with thiocyanate. The reaction of bis-aquo-tetrakis(4-N-methylpyridyl)porphinecobalt(III) with thiocyanate is a million times faster than that of $Co(NH_3)_5(OH_2)^{3+}$ with thiocyanate. The fast formation reactions of the aquo-porphyrin complex are thought to go by a D mechanism [191], and the reactions of bis-aquo-haematoporphyrincobalt(III) with thiocyanate and with cyanide show a kinetic pattern typical of D substitution [192].

Reactions of iron(II) complexes of the type $Fe(CN)_5L^{3-} + L'$ generally proceed by a D mechanism (Chapter 11.4.1). Substitution at $Fe(CN)_5(OH_2)^{3-}$ is no exception to this general pattern [193]. The reaction of haem with carbon monoxide has a small positive activation volume [194]. The reactions of haemoglobin and of myoglobin with oxygen have activation volumes of +5 and +8 $cm^3\,mol^{-1}$ respectively, but when the same two compounds react with carbon monoxide they have negative activation volumes. This marked difference between two incoming ligands whose similarities are often noted may be due to differences in the placing of these small molecules in the haem pocket. The activation volume of $-9\ cm^3\,mol^{-1}$ for the reaction of carbon monoxide with myoglobin is close to the value one would expect if the sixth coordination position of the iron in this compound is vacant, as seems likely. However one should be careful of thinking in too simplistic terms, for these activation volumes must also reflect the configurational changes in the haemoglobin and myoglobin molecules which accompany their reactions [195].

Ternary complex formation at cobalt(III) and at low-spin iron(II) thus seems to take place by dissociative mechanisms, either I_d or D. The situation with respect to ternary complex formation at chromium(III), particularly from $Cr(NH_3)_5(OH_2)^{3+}$ and related compounds, is more complicated. Here the additional question of whether there is significant associative character to the mechanism arises (see Chapter 12.6.2). There is both kinetic and spectroscopic evidence for an I_d mechanism for the reaction of $Cr(OH_2)_4(mal)^+$ with the hydrogen malonate anion [196]. In the special case of organo-aquo-chromium(III) compounds, replacement of coordinated water is sometimes very easy, as under the strong cis-labilising influence of such ligands as $-CH_3$, $-CH_2C(CO_2H)Me_2$, and (to a lesser extent) $-CF_3$ [197]. However, the mechanism of formation of many other ternary chromium(III) complexes may well involve significant associative character.

12.8   COMPLEX FORMATION IN NON-AQUEOUS MEDIA

There are indications that the Eigen-Wilkins mechanism for complex formation also applies to a variety of formation reactions in non-aqueous solvents. For the especially favourable case of Ni^{2+} reacting in methanol with trithiocarbonate, CS_3^{2-}, where association between the reactants is particularly extensive due to the 2− charge and the fairly small size of the CS_3^{2-}, the full equation can be tested for the Eigen-Wilkins mechanism (Chapter 12.1). The observed kinetic pattern fits the full equation over a wide range of concentrations, and the value of K_{os} estimated from the kinetics, 610 M^{-1}, corresponds ·to a sensible inter-reactant distance if this is estimated from the Fuoss equation for ion association. Moreover the interchange rate constant, k_i, is 300 sec^{-1} [198], which is just right in relation to the rate constant of 1000 sec^{-1} [199] established for solvent exchange under comparable conditions. As a further test of the applicability of the Eigen-Wilkins mechanism, this time for several M^{2+} cations, rate constants for formation reactions with chloride, when divided by a reasonable estimate for K_{os} of 170 M^{-1} give interchange rate constants which compare well with methanol exchange rate constants at the respective cations (Table 12.21) [220].

Table 12.21 Rate constants for formation of chloro-compléxes of transition metal(II) cations in methanol, at 20°C [200].

	Measured $k_f/M^{-1}\sec^{-1}$	Estimated $k_i/\sec^{-1}$	Compare $k_{ex}/\sec^{-1}$
Ni^{2+}	7×10^4	4×10^2	5×10^2
Co^{2+}	1.2×10^6	7×10^3	1.0×10^4
Fe^{2+}	3.5×10^6	2×10^4	3.0×10^4
Mn^{2+}	2.7×10^7	2×10^5	3.1×10^5

12.8.1 Simple Ligands

Nickel(II). Rate constants for the formation of complexes from solvento-nickel(II) cations in non-aqueous media are given in Table 12.22 [201-205].

Table 12.22 Kinetic parameters for the formation of nickel(II) complexes in non-aqueous media.

	Measured $10^{-3}k_f/M^{-1}\sec^{-1}$	Calculated K_{os}/M^{-1}	Estimated $10^{-3}k_i/\sec^{-1}$	
Methanol (0°C)				
acetate	26	170	0.15	
chloroacetate	21	170	0.12	
dichloroacetate	12	170	0.07	[201]
trichloroacetate	9	170	0.05	
cf. solvent exchange			0.08	
Acetonitrile (20°)				
nitrate	190	73	2.6	
trifluoroacetate	150	73	2.1	
toluene-p-sulphonate	130	73	1.8	[202]
cf. solvent exchange			1.2 to 3.9	
Dimethyl sulphoxide				
chloride (20°C)	70	45	1.6	[203]
thiocyanate (20°C)	80	45	1.8	
thiocyanate (25°C)	89^a	30	3^a	[204]
cf. solvent exchange			3 to 10	

a Very similar results were obtained in a recent reinvestigation of this system [205].

For each solvent in this Table, estimated interchange rate constants compare satisfactorily with solvent exchange rate constants.

Activation enthalpies, derived from k_f values, for the reaction of Ni^{2+} in acetonitrile with nitrate and with p-toluenesulphonate are 17.5 and 17.0 kcal mol^{-1} respectively. From these values and from estimated enthalpies of ion association (derived from the temperature dependence of ion-pairing constants) an interchange activation enthalpy of 15 kcal mol^{-1} has been estimated. This, and an interchange activation entropy of $+6.5$ cal $deg^{-1} mol^{-1}$, are within the rather wide ranges of values reported for acetonitrile exchange at nickel(II) and so consistent with a dissociative interchange mechanism here as well [202].

There has been some argument over the mechanism of the reaction of Ni^{2+} with thiocyanate in dimethyl sulphoxide. The current view is that this is probably the expected I_d [205].

Other metal(II) cations. Rate constants for a small and arbitrary selection of M^{2+} plus ligand formation reactions are listed in Table 12.23. Here again similarity of interchange and solvent exchange rate constants is consistent with dissociative interchange in the formation reactions. Such unlikely anions as perchlorate and p-toluenesulphonate readily form complexes with many M^{2+} cations in non-aqueous solvents, though not in water. Perchlorate complexes with Cu^{2+} or Zn^{2+} in acetonitrile, but not with Ni^{2+} [206].

For the formation of the trifluoroacetate complex of Mg^{2+} in methanol an activation enthalpy of 21 kcal mol^{-1} has been reported. The activation enthalpy for methanol exchange at Mg^{2+} is 17 kcal mol^{-1} [199].

Table 12.23 Rate constants for some formation reactions of metal(II) cations in non-aqueous solvents, at 25°C except where indicated otherwise.

		$k_f/M^{-1} sec^{-1}$	k_i/sec^{-1}	compare k_{ex}/sec^{-1}	
$Cu^{2+} + ClO_4^-$	methanol		9.4×10^7	7.4×10^7	[204]
$Cu^{2+} + ClO_4^-$	acetonitrile	2×10^{10}	5×10^8		[206]
$Zn^{2+} + ClO_4^-$	acetonitrile	5×10^9	7×10^7		[206]
$Mg^{2+} + CF_3CO_2^-$	methanol[a]		4×10^3	2.5×10^3	[200]

[a] At 20°C.

Metal(III) cations. Dimethyl sulphoxide is a popular solvent for kinetic studies of complex formation. In this polar but aprotic (in the ionisation sense) solvent there is no awkwardness of the M^{3+}aq against $M(OH)^{2+}$aq type that complicates kinetics in aqueous solution. $V(dmso)_6^{3+}$ reacts with a variety of different ligands at similar rates, suggesting a dissociative mechanism [124]. On the other hand, $Cr(dmso)_6^{3+}$ reacts with different ligands at very different rates, a pattern consistent with associative activation, especially as the formation rate constants are all greater than the rate constant for dimethyl sulphoxide exchange at chromium(III) [207]. A different approach to the diagnosis of mechanism has been used for the reaction of $Fe(dmso)_6^{3+}$ with thiocyanate. If this goes by an associative mechanism, one would expect the rate to be much faster in dimethyl sulphoxide than in water, because the chemical potential of the small anion will be higher in the dimethyl sulphoxide. In fact the rate of formation is rather slower in dimethyl sulphoxide than in water, which accords much better with a dissociative mechanism [208], though the activation parameters for this formation reaction are none too close to those for solvent exchange at iron(III).

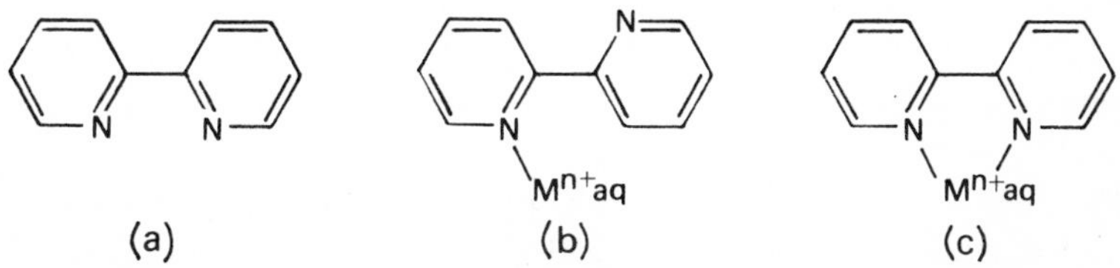

Fig. 12.16 The reaction of 2,2'-bipyridyl (a) with M^{n+} to give a unidentate intermediate (b) and thence the chelate product (c).

12.8.2 Abnormal Ligands

2,2'-Bipyridyl, Fig. 12.16(a), and closely related compounds such as 2,2',2"-terpyridyl and 1,10-phenanthroline have long been popular ligands for use in kinetic studies of complex formation, in both water and non-aqueous and mixed aqueous solvents. Formation reactions in water conform to the standard Eigen-Wilkins pattern, with rate-determining interchange to give a unidentate intermediate, Fig. 12.16(b), followed by rapid ring closure to give the chelate product, Fig. 12.16(c). Some years ago Bennetto and Caldin, in a stimulating series of papers on formation reactions of transition metal(II) cations, used the reactions of these cations with 2,2'-bipyridyl to probe solvent structural effects on kinetic behaviour in non-aqueous and mixed aqueous solvents, assuming the Eigen-Wilkins mechanism to operate throughout [209, 210]. Unfortunately, as has since gradually become apparent, in at least some non-aqueous solvents the rate-determining step shifts to being the ring closure step. This is clearly shown in the comparison of kinetics in dimethyl sulphoxide and in water in Table 12.24 [211–213]. In dimethyl sulphoxide the unidentate intermediate of Fig. 12.16(b) is apparently readier to dissociate back to solvated cation plus ligand than to undergo ring closure. In other words, the solvent competes successfully for the nickel(II) with the non-ligated nitrogen in the intermediate. Ethanol is more like water than dimethyl sulphoxide, so the operation of the Eigen-Wilkins mechanism with its rate-determining interchange is not surprising for the reaction of Ni^{2+} with 2,2'-bipyridyl in this solvent. However the reaction of Ni^{2+} with the rigid 1,10-phenanthroline in ethanol shows deviations from the Eigen-Wilkins kinetic pattern [214]. Even in methanol, reactions of Mn^{2+} with 2,2'-bipyridyl, 2,2'2''-terpyridyl, or 1,10-phenanthroline proceed at rates determined by chelate ring closure. This is understandable in the light of much faster solvent loss and interchange at Mn^{2+} compared with Ni^{2+} (see Chapter 12.2.2, 12.3.2) [215].

Table 12.24 Rate constants (k_f) for reactions of nickel(II) with pyridine, 2,2'-bipyridyl, and related compounds in water and in dimethyl sulphoxide, at 25°C.

Ligand	Water $k_f/M^{-1}sec^{-1}$	Dimethyl sulphoxide $k_f/M^{-1}sec^{-1}$
Pyridine	~4000 [23]	~20 [212]
4-Phenyl pyridine		16 [211]
2,2'-Bipyridyl	1500 [211]	0.7 [209]
2,2',2''-Terpyridyl	1400 [211]	0.3 [209, 213]
cf. Solvent exchange[a]	~30000[b]	32 to 93 [213]

a The units here are sec^{-1}; b Chapter 11.3.

Anomalous behaviour for bidentate ligands is not restricted to 2,2'-bipyridyl and its close analogues. Dithiocarbamates also react about a hundred times slower with M^{2+} cations in dimethyl sulphoxide than would be expected for rate-determining interchange, again indicating rate-limiting ring closure [216]. The formation of $Cu(acac)^+$ in methanol, as in water, also involves rate-determining ring closure [67].

Rates of reaction of porphyrins with M^{2+} cations, often studied in glacial acetic acid solution for solubility reasons, are very slow. The fairly rigid macrocyclic nature of these ligands is sufficient explanation for this. Some years ago a reasonable picture of reactivities was emerging in this area, for reactions with tetrapyridylporphyrin [217] and with haematoporphyrin IX [218, 219]. Unfortunately more recent kinetic studies, particularly on the reactions of the dimethyl ester of protoporphyrin IX, have shown that the kinetics of these reactions approximate less closely than originally thought to rate laws with integral orders of reaction [220].

Complex formation between alkali metal cations and flexible macrocylic ligands takes place rapidly. Rate constants for the reaction of Na^+ with dibenzo-18–crown–6, Fig. 12.17(a), are on the range 2 to 30×10^7 M^{-1} sec^{-1} at 25°C in methanol, dimethylformamide, and dimethoxyethane [221]. For the reaction of K^+, Rb^+, Cs^+, and Tl^+ with dibenzo–30–crown–10 in methanol rate constants are between 60 and 80×10^7 $M^{-1}sec^{-1}$ at 25°C [222], while for the reactions of Na^+ and K^+ with nonactin, Fig. 12.17(b), they are about 30×10^7 $M^{-1}sec^{-1}$ in methanol [87]. These rates are similar to those for the reactions of the alkali metal cations with simple ligands in aqueous solution (Chapter 12.5). Rates of reaction of these cations with more sterically demanding ligands such as cryptands, Fig. 12.17(c), are considerably less. In fact kinetic results here are for cation exchange, where rate constants are about $100 \ sec^{-1}$ at $25°C$ in D_2O and about $250 \ sec^{-1}$ in ethylenediamine [110, 223]. It seems likely that the rate-determining step in these exchange reactions is dissociation, but formation rate constants will be of the same order of magnitude, extremely small for alkali metal cations.

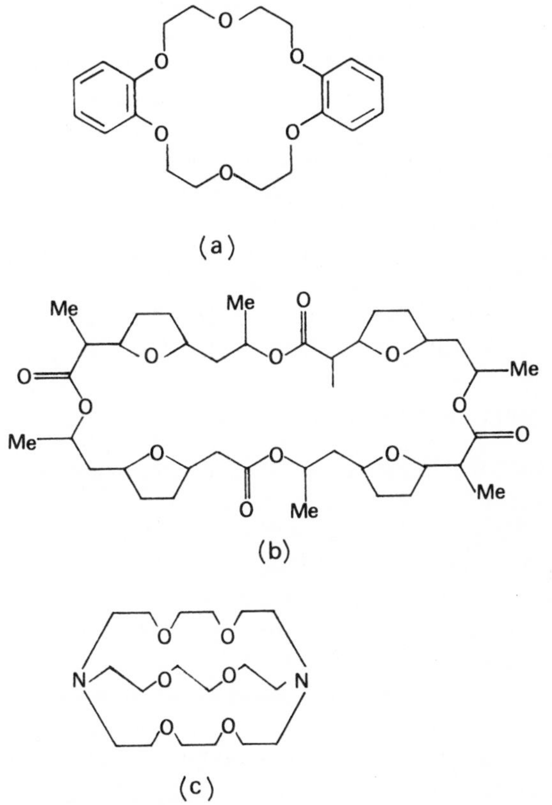

(a)

(b)

(c)

Fig. 12.17   (a) Dibenzo-18-crown-6, (b) nonactin, (c) cryptand.

12.8.3 The Role of the Solvent

Apart from a mention of the competition between solvent dimethyl sulphoxide and bipyridyl-nitrogen for Ni^{2+} in the SCS formation reaction of $Ni(bipy)^{2+}$ in dimethyl sulphoxide solution, the role of the solvent in determining reactivities and mechanism has not been discussed in the previous two sections. Because this topic is still a matter of controversy, only a few recent lines of approach to the subject will be indicated here. It would obviously be pleasing to the orderly mind if some correlation between kinetic parameters and some of the numerous empirical solvent parameters (Chapter 1.3) could be found. Such a correlation was sought in, for example, the reaction of Ni^{2+} with thiocyanate. There is a correlation of dissociation rates of $Ni(NCS)^+$ with Gutmann donor numbers of several solvents, but no correlation of the formation rate [224].

The role of solvent structure in determining reactivities is currently under discussion [209, 225], as is the closely related topic of viscosity [226]. There is a dramatic change in the kinetic pattern for the formation of pada complexes of cobalt(II), nickel(II), copper(II), and zinc(II) when the solvent is changed from water to the much more viscous glycerol. In water, formation rates of these complexes vary greatly with the nature of the cation, but in glycerol rates of formation are very similar for the four cations named. In water rates are activation controlled, determined by the ease of cation–water bond breaking. but in glycerol they are diffusion controlled, dominated by the rigid structure of this viscous solvent. On warming, or applying pressure to, the glycerol systems, the kinetic behaviour starts to become more like that in aqueous systems as the viscosity of the glycerol decreases with increasing temperature or pressure.

12.9   COMPLEX FORMATION IN MIXED AQUEOUS SOLVENTS

First solvent mixtures containing only 'good' or 'good' and 'bad' (in this solvating sense) co-solvents can be differentiated. In the former type of mixture there will be competition between the two solvent components for the primary coordination shell of the cation. Hence the reactivity of M^{n+} plus L varies as much by reason of variation of primary solvent shell composition and of differing reactivities of compounds $M(OH_2)_{6-x}(solvent)_x^{n+}$ as from any other cause. So here the problem is one of the relative lability of the various solvated cation species, that is it is one of ternary complex formation, a topic already discussed in Chapter 12.7 above. Excellent case studies to illustrate this type of system include those of Ni^{2+} with ammonia in methanol+water mixtures [227] and of Cr^{3+} with thiocyanate, again in methanol+water mixtures [228].

In aqueous mixtures containing a poorly solvating co-solvent such as acetone, the actual formation reaction remains the same, M^{n+} aq plus ligand, throughout the solvent composition range. Now the variation of rates or of activation parameters with solvent composition must be attributed to properties of the mixed solvent. That bulk solvent properties may influence kinetics has been recognised for water itself, for instance in the reactions of Mg^{2+} with oxalate [229] and of ruthenium(IV) with halides [230]. Consequently effects of added organic solvents on the structure of water and on reactant solvation are likely to be reflected in reactivity variations. There are two ways in which a formation reaction proceeding by the usual Eigen-Wilkins mechanism might be affected, through K_{os} or through k_i. One would expect the former to be the more important, because it involves transfer of reactant ligand from bulk solvent into the secondary solvation shell, whereas the interchange step is more localised and less under the influence of the bulk solvent. Indeed, there are instances where it is known that the interchange rate constant is insensitive to bulk solvent composition. Ultrasonic studies of the reaction of Zn^{2+}aq with sulphate indicate that k_i is 3×10^7 sec^{-1} in glycol+water mixtures throughout the composition range 0–77% by weight of glycol. Neither the eightfold change in viscosity[†] nor the 1½ times change in dielectric constant over this range is reflected in k_i [231].

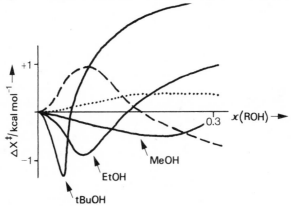

Fig. 12.18 The dependence of activation parameters, $\Delta X^{\ddagger}$, for the reaction of Ni^{2+}aq with bipy on the mole fraction of alcohol, $x(ROH)$, in alcohol+water solvent mixtures [180, 182]. ——— $\Delta H^{\ddagger}$; -------- $T\Delta S^{\ddagger}$; ·········· $\Delta G^{\ddagger}$; all referred to zero in pure water ($T\Delta S^{\ddagger}$ and $\Delta G^{\ddagger}$ for ethanol+water only).

It must be admitted that k_f (i.e. K_{os}) is little more sensitive than k_i to solvent composition variation. This is witnessed by the change of only about 40% in k_f for the reaction of nickel(II) with 2,2′-bipyridyl in going from mole fraction 0 to 0.3 methanol, or by the slightly larger change from 0 to 0.3 mole fraction t-butyl alcohol [210, 212]. Similarly small variations have been described for such diverse formation reactions as those of Mn^{2+}aq with sulphate [232], of lanthanide cations with murexide [233], and of Er^{3+}aq with chloride or nitrate [234]. Although $k_f(\Delta G_f^{\ddagger})$ is insensitive to solvent composition, the same is not always true for $\Delta H_f^{\ddagger}$ and $\Delta S_f^{\ddagger}$. These may show marked, though of course compensatory, changes with solvent composition (Fig. 12.18) [210, 212]. Figure 12.18[††] both shows extrema in $\Delta H_f^{\ddagger}$ and in $T\Delta S_f^{\ddagger}$ (but not in $\Delta G_f^{\ddagger}$) and indicates

---

[†] Such a viscosity change is small compared with the water to glycerol change which causes such a marked change in kinetics (Chapter 12.8.3).

[††] Figure 12.18 has been restricted to water-rich mixtures in order to avoid solvent composition regions where there may be a mixed primary solvation shell around the nickel, or where the SCS mechanism might operate.

that extrema in $\Delta H_f^{\ddagger}$ are affected in size and position by the nature of the cosolvent. Effects increase in the order methanol < ethanol < t-butyl alcohol. A similarly large effect has been reported for cosolvent t-butyl alcohol on the activation parameters for Ni^{2+}aq plus pada [235]. It is interesting that the sizes of the extrema in Fig. 12.18 follow the relative magnitudes of the effects of these alcohol cosolvents on solvent structure, and that the positions of the extrema tie in with extrema of several other spectroscopic and thermodynamic properties of these solvent mixtures (Chapter 1.3). The reaction of Ni^{2+}aq with bipyridyl is not unique in its kinetic pattern; similar extrema in $\Delta X^{\ddagger}$ against x_2 plots to those shown in Fig. 12.18 have been reported for several other reactions, such as solvolysis of organic halides.

As in the case of formation reactions in non-aqueous media (see above), theories of the role of the solvent in formation reactions in mixed aqueous media are still in their infancy. It is apparent that solvent composition affects reactivity in a variety of ways, whose relative importance varies with system. Important factors include dielectric properties [19, 236], structural properties, effects on the initial association equilibrium, and, for chelate formation, the relative rates of solvent loss from the solvento-cation and of ring closure [225].

Almost all kinetic studies of formation reactions of M^{n+} cations in mixed solvents have been in aqueous mixtures. To close this section the pioneering work of Bennetto should be mentioned. He has conducted a full kinetic study of the Ni^{2+} plus 2,2'-bipyridyl reaction in mixed non-aqueous solvents, that is methanol + acetonitrile [237].

12.10   CATALYSIS BY METAL IONS

12.10.1 Introduction

Metal ion catalysis in a range of organic and inorganic reactions is closely related to the formation reactions discussed in the preceding sections of this Chapter. This important class of reactions includes catalysed aquation of transition metal complexes, catalysed hydrolysis of such organic compounds as esters and peptides, and conversely catalysis of the formation of certain types of organic molecules such as Schiff bases. It extends from simple reactions through to the complexities of biochemistry and enzymes [238, 239]. In many cases the formation of a complex between the solvated metal ion and the substrate(s) has been shown to be an important first step.

12.10.2 Inorganic Complexes

The best documented type of reaction here is the catalysed hydrolysis (aquation) of halogeno-cobalt(III) complexes in the presence of mercury(II) cations, for example

$$Co(NH_3)_5Cl^{2+} + Hg^{2+} \xrightarrow{aq} Co(NH_3)_5(OH_2)^{3+} + HgCl^+ .$$

This type of reaction has also been described for related complexes of other metals, such as chromium(III), rhodium(III), and rhenium(IV), for other leaving ligands, such as thiocyanate, and for other 'soft' metal cations, thallium(III), silver(I), and mercury as Hg_2^{2+} among them [240]. Of these cations, thallium(III) is generally less effective than mercury(II), and silver(I) catalysis is complicated by the precipitation of silver halide product. These reactions are formally S_N1 with respect to the transition metal ($HgCl^+$ is a good leaving group), S_E2 with respect to the 'soft' cation. However the mechanism may be either a simple one-step process or involve a ligand-bridged binuclear complex as a transient intermediate,

$$M{-}Cl + M' \rightleftharpoons M{-}Cl{-}M' \longrightarrow M + M'{-}Cl .$$

Of course, as the stability of the binuclear intermediate decreases, the two-step mechanism changes into a one-step mechanism without a sharp dividing line. The balance between the operation of a one- or two- step mechanism may be fine. Mercury(II) catalysed aquation of cis-$Co(en)_2Cl_2^+$ is believed to go by the latter, but mercury(II) catalysed aquation of trans-$Co(en)_2Cl_2^+$ by the former [241]. In special cases, for instance the mercury(II) assisted aquation of some thiocyanato-chromium(III) complexes, the binuclear intermediate is particularly stable and its properties and breakdown can be studied [242]. Metal ion catalysis of dealkylation has also been described, for example for alkylcobalamins and alkylcobaloximes. Here mercury(II) catalysis is S_E2, and involves stereochemical inversion at the alkyl-carbon. There is no indication of a transient Co–alkyl–Hg intermediate [243].

The importance of complex formation in this type of reaction has been demonstrated in several reaction series where rate constants for metal ion catalysis have been shown to parallel stability constants for formation of complexes between the incoming metal ions and potential bridging and leaving ligands. A variety of observations in this area can be systematised according to the 'hard and soft acids and bases' picture [244].

This type of metal ion catalysis applies not only to simple ligands such as halides and thiocyanate, but also to multidentate ligand complexes. Metal ion dechelation of oxalato-complexes provides a simple example here, while metal ion dechelation of a range of aminocarboxylato-complexes (for example those of edta) offers more complicated instances in which a variety of binuclear intermediates are possible as the multidentate ligand is 'unwrapped' from its original host cation [245].

Metal ion catalysis of aquation of inorganic complexes bears a close resemblance to analogous acid catalysis, for example of aquation of azide or cyanide complexes,

$$M-N_3^{n+} + H^+ \overset{fast}{\rightleftharpoons} M-N_3H^{(n+1)+} \xrightarrow[aq]{slow} M-OH_2^{(n+1)+} + HN_3$$

It also bears a formal resemblance to redox catalysis of aquation, as in chromium(II)-promoted aquation of substitution inert chromium(III) complexes (see Chapter 13).

12.10.3 Organic Compounds: Hydrolysis and Related Reactions

Alkyl halides. Metal ion catalysis of solvolysis of alkyl halides forms a link between the previous section, dealing mainly with inorganic halide complexes, and this section, devoted to organic compounds. As for inorganic substrates, carbon–halide solvolysis is catalysed by such 'soft' metal cations as Hg^{2+}, Cd^{2+}, and Tl^{3+} when the halide is chloride or bromide, but by 'hard' cations such as Al^{3+} when the halide is fluoride. The solvolysis of t-butyl halides in the presence of metal ion catalysts follows the rate law

$$-d[\text{t-BuX}]/dt = \left(k_1 + k_2 [M^{n+}] \right) [\text{t-BuX}]$$

for X = F, Cl, or Br. The rate constants k_1 can be assigned to uncatalysed solvolysis. The rate constants k_2 can be correlated with the stability constants of complexes $MX^{(n-1)+}$, except in those few cases such as Cr^{3+} where substitution is very slow [246]. Catalysis by Ag^+ does not fit into this picture. While solvolysis rates for catalysis by the majority of metal cations are first-order in metal ion concentration, rate laws for catalysis by Ag^+ contain a term in $[Ag^+]^n$ where n may be anywhere from about $+\frac{1}{2}$ to $+2$ [247].

Esters. Metal ion catalysis of ester hydrolysis has been much studied over the past few decades. Recent examples include hydrolysis of oxalate esters, where the relative effectiveness of a series of M^{2+} cations was established [248], of ethyl thio-oxalate [249], of ethyl glycinate [250], and of several dipeptide esters, including ethyl glycylglycinate [251]. In the last mentioned case the peptide linkage remains intact, and only the ester linkage is broken. Cu^{2+} is about a thousand times more effective than H^+ here [251].

A closely related field of study is that of the hydrolysis of esters which are coordinated to an inert metal centre, usually cobalt(III). For example, ethyl glycinate in the cationic complex $cis\text{-Co(en)}_2(\text{NH}_2\text{CH}_2\text{CO}_2\text{Et})\text{Cl}^{2+}$ hydrolyses about a hundred times faster than the free ester [252]. Many other similar examples have been documented [253]. This latter type of reaction provides a model for the metal ion catalysed hydrolyses mentioned in the previous paragraph. In both types of reaction, the metal cation acts by activating some appropriate point in the organic molecule to nucleophilic attack by, say, water or hydroxide.

Metal ions can act equally well as catalysts for hydrolysis of esters of inorganic acids, for instance in the Mg^{2+} or Zn^{2+} catalysis of hydrolysis of phosphate esters [254].

Peptides. Metal-ion catalysis of hydrolytic cleavage of peptide linkages is of considerable biochemical significance. Acid catalysis of dipeptide hydrolysis has long been known, with rates [255] and activation parameters [256] established for several dipeptides. The first recognitition that metal ions may also be effective catalysts dates from 1951, when the catalysis of glycylglycine hydrolysis by Co^{2+}aq was briefly reported [256]. Cu^{2+}aq was shown to be an effective catalyst under acid conditions [257], but was found to retard hydrolysis in alkali [258]. Rate law determinations for hydrolysis of catalysis of glycylglycine in the presence of Mn^{2+}aq, Co^{2+}aq, Ni^{2+}aq, or Zn^{2+}aq showed a first-order dependence on the concentration of M^{2+}aq [259]. The particularly important role of Zn^{2+}aq has been further probed for a range of dipeptides [260], and in enzymic peptide hydrolysis [261].

As in the case of ester hydrolysis, there is considerable interest in the peptide hydrolysis of such ligands bound to an inert metal centre [253], as in the reported $> 10^4$ times acceleration for glycylglycine in β-Co(trien)(glyglyH)$^{2+}$ [262]. Again the positively charged metal cation activates the coordinated ligand to nucleophilic attack by water or hydroxide.

Miscellaneous. Another biochemically significant reaction that shows marked catalysis by metal ions is the decarboxylation of oxaloacetate to pyruvate,

$$^-O_2CCOCH_2CO_2^- + H^+ \longrightarrow CH_3COCO_2^- + CO_2 .$$

This reaction occurs about a hundred times faster in the presence of 2+ or 3+ metal ions at concentrations in the region of 10^{-3} M [263, 264]. Catalysis by Cu^{2+} is complicated, as this cation catalyses both keto $\rightleftharpoons$ enol interconversion for oxaloacetate and the decomposition to pyruvate plus carbon dioxide [265]. The importance of ease of complex formation between substrate and metal ion is neatly illustrated by decarboxylation of acetonedicarboxylate, $^-O_2CCH_2COCH_2$-CO_2^-. The dianionic form, L^{2-}, is much more susceptible to metal-ion catalysis than the monoprotonated form HL^-, while the uncharged form H_2L is again much less susceptible than HL^- [266].

To give some impression of the range of organic reactions which can be catalysed by metal ions, the above examples are supplemented by a random selection of systems in Table 12.25 [267-276]. The iodination of 2-acetylpyridine [271], the decomposition of tetramethyl-1,2-dioxetan, Fig. 12.19(c) [272], and the metal-ion induced activation of methylene groups in edta [276] show the relative catalytic effectiveness of different metal ions. In the first case the effectiveness increases from a 5×10^3 times acceleration by Zn^{2+} to 2×10^5 times acceleration by Cu^{2+}, with Ni^{2+} intermediate. In the dioxetan case, the catalytic effect parallels the stability constants of malonate complexes of the

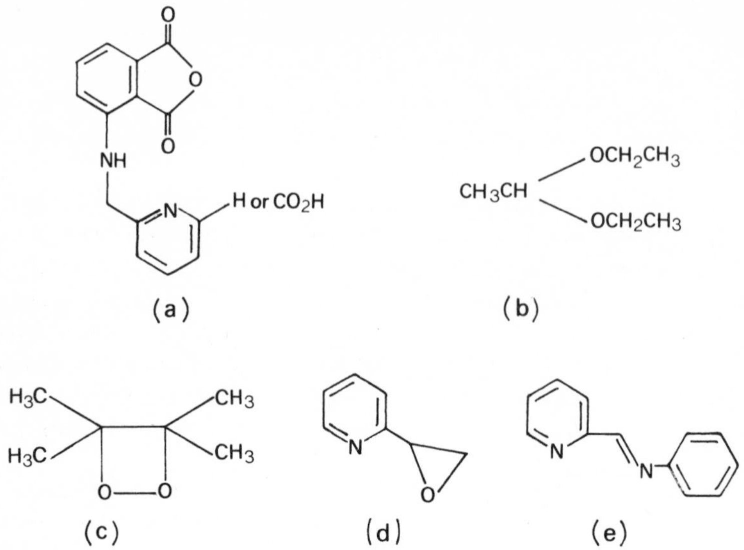

Fig. 12.19

respective M^{2+} ions. The importance of kinetic lability of the metal ions is emphasised in the last (edta) case, because here Fe^{2+}, Co^{2+}, Ni^{2+}, and Cu^{2+} are effective catalysts, but Cr^{3+}, Co^{3+}, and Rh^{3+} are too inert to be effective [276]. In the hydrolysis of diethyl acetal, Fig. 12.19(b), Fe^{3+} is a surprisingly feeble catalyst, being only about as effective as H^+ [270].

The metal ion may do more than merely accelerate the reaction. It may also have some stereochemical effect. This is illustrated by the case of the reactions of 2-pyridyloxiran, Fig. 12.19(d), with chloride, bromide, or methoxide, which become 100% stereospecific in the presence of Cu^{2+} [273]. As in the cases of ester and peptide hydrolysis discussed above, the reactivity of organic molecules coordinated to an inert metal centre is again of interest in this section. An example is provided by substitution by bromide, nitro, or acyl at acetylacetone coordinated to Cr^{3+}, Co^{3+}, or Rh^{3+} [277].

In most of the reactions listed in Table 12.25, the catalytic effect of the metal ion can be ascribed to its positive charge withdrawing electron density from an appropriate part of the organic molecule, thereby facilitating nucleophilic attack (see esters and peptides above). In the iodination of 2-acetylpyridine, the rate-determining step is the enolisation or ionisation of the organic compounds. The actual iodination is of the deprotonated anion. By way of contrast, bromination of aniline is some 10^{10} times *slower* in the complex *trans*-$Co(en)_2(OH_2)(PhNH_2)^{3+}$ than in the free base. Here the rate-determining step is the bromination reaction which, as it involves electrophilic attack, is greatly discouraged by the proximity of the 3+ cobalt centre [278].

Table 12.25 Examples of metal-ion catalysis of reactions of organic compounds.

Substrate	Cation(s)	Reaction	
Acetonitrile[a]	Hg^{2+}		[267]
Acetaldehyde	Zn^{2+}	hydrolysis	[268]
Carboxypeptidase A model, Fig. 12.19(a)	Zn^{2+}		[269]
Diethyl acetal, Fig. 12.19(b)[b]	Fe^{3+}		[270]
2-Acetylpyridine	Ni^{2+}, Cu^{2+}, Zn^{2+}	enolisation/iodination	[271]
Tetramethyl-1,2-dioxetan, Fig. 12.19(c)[c]	several M^{2+}	decomposition	[272]
2-Pyridyloxiran, Fig. 12.19(d)	Cu^{2+}	hydrolysis	[273]
Schiff base, Fig. 12.19(e)	Cu^{2+}		[274]
N-Methyltetraphenylporphinecopper(II)	Cu^{2+}	demethylation	[275]
edta	several M^{2+}, M^{3+}	methylene H/D exchange	[276]

a See Chapter 14 for related reactions of nitriles coordinated to inert metal ions such as cobalt(III); b products are ethanol and acetaldehyde; c each molecule of the dioxetan gives two molecules of acetone.

It is interesting to contrast the $Cu^{2+}aq$ catalysis of Schiff base hydrolysis included in Table 12.25 with the $Fe^{2+}aq$ catalysis of formation of a very similar Schiff base mentioned in the next section.

12.10.4 Organic Compounds: Formation

Whereas in many cases metal ions are effective catalysts for the hydrolytic breakdown of organic compounds (see previous section), on other occasions metal ions are effective catalysts for the construction of complicated organic molecules from simpler species. There are many examples of metal-ion catalysis of the formation of Schiff bases,

$$\text{\large$>$}C=O + H_2N- \xrightarrow{M^{n+}} \text{\large$>$}C=N\text{\large$\diagdown$} + H_2O$$

for instance by Ni^{2+}, by Fe^{2+} [279], and by Zn^{2+} [280]. In this and other cases, it is often not clear whether the metal ion is coordinated to one or other of the reactants, or to both, in the condensation step. In cases where both reactants are coordinated to the same metal ion prior to the condensation step, we have a true **template reaction**. The effectiveness of various metal cations can be assessed either by comparison of rate constants or, less commonly, by establishing product yields after a given reaction time. The latter approach can be illustrated by the condensation of acetone with furan,

With no catalyst present, the yield of product is 1-2% after 48 hours at $20°C$. In the presence of alkali metal halides, the yield after a similar reaction time is between 5 and 10%, while in the presence of a salt of one of several transition metal 2+ cations the yield may be up to 40% [281].

In some systems, a messy condensation reaction leading to a mixture of ill-characterised products is transformed into a clean reaction of synthetic utility by the presence of a metal ion. A good example of this is the reaction of dimethylglyoxal with methylamine [282],

In other cases the nature of the products may depend on the nature of the metal ion employed as catalyst. 1,2-Dicyanobenzene usually condenses to give phthalocyanine when heated with metal salts,

but when treated with UO_2Cl_2 it gives the analogous macrocyclic species derived from five rather than four molecules of 1,2-dicyanobenzene [283]. A similar case of product determination by metal ion is shown in Fig. 12.20 [284].

Extension from here into inorganic biochemistry quickly leads to very complicated systems, for example that of Zn^{2+}aq catalysis of oligonucleotide synthesis [285]. Extension to homogeneous catalysis in organometallic chemistry [286] is outside the scope of this book because the metal centres involved are almost always in low oxidation states stabilised by appropriate ligands. The only important exception is Ag^+, whose role in the catalysis of isomerisation of organic compounds is mentioned below.

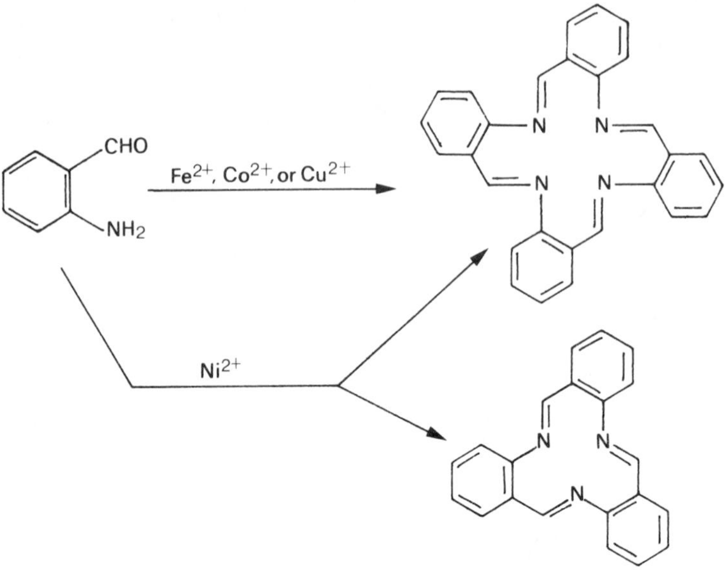

Fig. 12.20 Metal-ion catalysed condensation reactions of o-aminobenzaldehyde.

12.10.5 Organic Compounds: Isomerisation

One specialised area of chemistry which has attracted much interest in the past few years is that of the isomerisation of strained cyclic hydrocarbons in the presence of metal ions and complexes. Several of the catalysts are organometallic species, for example $[Rh(CO)_2Cl]_2$, and so beyond our scope, but the Ag^+ cation is one of the most effective catalysts for these isomerisations. Product, stereochemical, and kinetic studies of its catalysis of such rearrangements have been carried out on numerous compounds [287]. Examples include the isomerisation of a variety of bicyclobutane, Fig. 12.21(a), derivatives, of the cubane of Fig. 12.21(b) to the cuneane of Fig. 12.21(c), of the homocubane of Fig. 12.21(d) to the pentacyclononane of Fig. 12.21(e), and, most recently, of substituted 1,8-bishomocubanes, Fig. 12.21(f), to substituted norsnoutanes, Fig. 12.21(g) [288]. Rate laws and rate constants have been determined for several of these reactions. For isomerisation of cubane the rate law is simply:

$$-d[\text{cubane}]/dt = k[\text{cubane}][Ag^+] .$$

For Ag^+-catalysed isomerisation of the homocubane of Fig. 12.21(h) the kinetic pattern is more complicated. Here there are indications of the intermediacy of an Ag^+ to carbon σ-bonded complex [289].

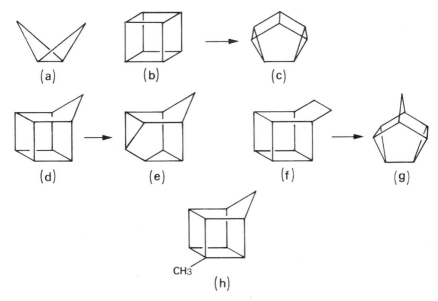

Fig. 12.21   Strained polycyclic hydrocarbons.

Further Reading

Wilkins, R. G. and Eigen, M. in *Advances in Chemistry Series, No. 49,* ed. Gould, R. F., American Chemical Society, Washington, D. C. (1965) Ch. 3.

Sutin, N., *A. Rev. Phys. Chem.,* **17**, 119 (1966).

Basolo, F. and Pearson, R. G., *Mechanisms of Inorganic Reactions, 2nd Edn.,* Wiley/Interscience (1967) p. 107.

Langford, C. H. and Stengle, T. R., *A. Rev. Phys. Chem.,* **19**, 193 (1968).

McAuley, A. and Hill, J., *Q. Rev. Chem. Soc.,* **23**, 18 (1969).

Hewkin, D. J. and Prince, R. H., *Coord. Chem. Rev.,* **5**, 45 (1970).

Kustin, K. and Swinehart, J. H. in *Progress in Inorganic Chemistry,* ed. Lippard, S. J., *Volume 13: Inorganic Reaction Mechanisms,* ed. Edwards, J. O., Wiley/Interscience (1970) p. 107.

Wilkins, R. G., *Accts. Chem. Res.,* **3**, 408 (1970); *Pure Appl. Chem.,* **33**, 583 (1973).

Langford, C. H. in *Ionic Interactions, Volume 2,* ed. Petrucci, S., Academic Press (1971) Ch. 6.

Wilkins, R. G., *The Study of Kinetics and Mechanism of Reactions of Transition Metal Complexes,* Allyn and Bacon (1974).

Hoffmann, H., *Pure Appl. Chem.,* **41**, 327 (1975).

Benson, D., in *Comprehensive Chemical Kinetics, Volume 18: Selected Elementary Reactions,* ed. Bamford, C. H. and Tipper, C. F. H., Elsevier (1976) Ch. 5.

Coetzee, J. F. in *Solute-Solvent Interactions, Volume 2*, ed. Coetzee, J. F. and Ritchie, C. D., Marcel Dekker (1976).

Inorganic Reaction Mechanisms – Chemical Society Specialist Periodical Report, Volumes 1 (1971); 2 (1972); 3 (1974); 4 (1976); 5 (1977).

REFERENCES

[1] Diebler, H. and Eigen, M., *Z. Phys. Chem. Frankf. Ausg.*, **20**, 229 (1959); Eigen, M., *Ber. Bunsenges. Phys. Chem.*, **67**, 753 (1963).

[2] Eigen, M. and Tamm, K., *Z. Elektrochem.*, **66**, 93, 107 (1962).

[3] Winstein, S., Clippinger, E., Fainberg, A. H., and Robinson, G. C., *J. Amer. Chem. Soc.*, **76**, 2597 (1954).

[4] Jackopin, L. G. and Yeager, E., *J. Phys. Chem.*, **74**, 3766 (1970).

[5] Brumm, P. and Rüppel, H., *Ber. Bunsenges. Phys. Chem.*, **75**, 102 (1971)

[6] McCain, D. C. and Myers, R. J., *J. Phys. Chem.*, **72**, 4115 (1968); Berdyev, A. A., Lezhnev, N. B., Nazorova, G. A., and Shubina, M. G., *Russ. J. Phys. Chem.*, **48**, 1644 (1974).

[7] Pearson, R. G. and Buch, T., *J. Chem. Phys.*, **36**, 1277 (1962).

[8] Leung, W. H. and Millero, F. J., *J. Solution Chem.*, **4**, 145 (1975).

[9] Millero, F. J. and Masterton, W. L., *J. Phys. Chem.*, **78**, 1287 (1974).

[10] Davies, A. G. and Smith, W.MacF., *Proc. Chem. Soc.*, 380 (1961).

[11] Brintzinger, H. and Hammes, G. G., *Inorg. Chem.*, **5**, 1286 (1966).

[12] Eigen, M., *Z. Phys. Chem. Frankf. Ausg*, **1**, 176 (1954).

[13] Fuoss, R. M., *J. Amer. Chem. Soc.*, **80**, 5059 (1958); Eigen, M., *Z. Elektrochem.*, **64**, 115 (1960).

[14] E.g. Eigen, M., Kruse, W., Maass, G., and de Maeyer, L., *Prog. React. Kinet.*, **2**, 287 (1964),

[15] E.g. Hammes, G. G. and Steinfeld, J. I., *J. Amer. Chem. Soc.*, **84**, 4639 (1962); *J. Phys. Chem.*, **67**, 528 (1963).

[16] Kurtz, J. and Vriesenga, J. R., *J. Chem. Soc. Chem. Commun.*, 711 (1974).

[17] Lo, S. T. D. and Swaddle, T. W., *Inorg. Chem.*, **14**, 1878 (1975).

[18] Margerum, D. W., Rorabacher, D. B., and Clarke, J. F. G., *Inorg. Chem.* **2**, 667 (1963).

[19] Lin, C. -T. and Rorabacher, D. B., *Inorg. Chem.*, **12**, 2402 (1973).

[20] Letter, J. E. and Jordan, R. B., *Inorg. Chem.*, **10**, 2692 (1971).

[21] Rorabacher, D. B., *Inorg. Chem.*, **5**, 1891 (1966).

[22] Eisenstadt, M., *J. Chem. Phys.*, **51**, 4421 (1969).

[23] Holyer, R. H., Hubbard, C. D., Kettle, S. F. A., and Wilkins, R. G., *Inorg. Chem.*, **4**, 929 (1965).

[24] Cassatt, J. C., Johnson, W. A., Smith, L. M., and Wilkins, R. G., *J. Amer. Chem. Soc.*, **94**, 8399 (1972).

[25] Hoffmann, H., *Ber. Bunsenges. Phys. Chem.*, **73**, 432 (1969).

[26] Nancollas, G. H. and Sutin, N., *Inorg. Chem.*, **3**, 360 (1964).

[27] Saar, D., Macri, G., and Petrucci, S., *J. Inorg. Nucl. Chem.*, **33**, 4227 (1971).

[28] Hammes, G. G. and Morrell, M. L., *J. Amer. Chem. Soc.*, **86**, 1497 (1964).

[29] Bear, J. L. and Lin, C. -T., *J. Phys. Chem.*, **72**, 2026 (1968).

[30] Caldin, E. F., Grant, M. W., and Hasinoff, B. B., *Chem. Commun.*, 1351, (1971); *J. Chem. Soc. Faraday I*, **68**, 2247 (1972).

[31] Grant, M. W., *J. Chem. Soc. Faraday I*, **69**, 560 (1973).

[32] Jost, A., *Ber. Bunsenges. Phys. Chem.*, **79**, 850 (1975).

[33] Kustin, K., Pasternack, R. F., and Weinstock, E. M., *J. Amer. Chem. Soc.*, **88**, 4610 (1966).

[34] Harada, S., Tanabe, H., and Yasunaga, T., *Bull. Chem. Soc. Japan*, **46**, 3125 (1973).

[35] Harada, S., Okuue, Y., Kan, H., and Yasunaga, T., *Bull. Chem. Soc. Japan*, **47**, 769 (1974).

[36] Hoffmann, H. and Nickel, U., *Z. Naturf.*, **26B**, 299 (1971).

[37] Cavasino, F. P., *J. Phys. Chem.*, **69**, 4380 (1965); Calvaruso, G., Cavsino, F. P., and Di Dio, E., *J. Chem. Soc. Dalton*, 2632 (1972); *J. Inorg. Nucl. Chem.*, **36**, 2061 (1974); Nickel, U., Hoffmann, H., and Jaenicke, W. *Ber. Bunsenges. Phys. Chem.*, **72**, 526 (1968); and references [26], [35], and [36].

[38] Cavasino, F. P., Di Dio, E., and Locanto, G., *J. Chem. Soc. Dalton* 2419 (1973).

[39] Kaden, T. A., *Helv. Chim. Acta.*, **53**, 617 (1970).

[40] Lin, C. -T., Rorabacher, D. B., Cayley, G. R., and Margerum, D. W., *Inorg. Chem.*, **14**, 919 (1975).

[41] Steinmann, W. and Kaden, T. A., *Helv. Chim. Acta*, **58**, 1358 (1975).

[42] Hambright, P. and Chock, P. B., *J. Amer. Chem. Soc.*, **96**, 3123 (1974).

[43] Khosropour, R. and Hambright, P., *Chem. Commun.*, 13 (1972).

[44] Taylor, R. W., Stepien, H. K., and Rorabacher, D. B., *Inorg. Chem.*, **13**, 1282 (1974).

[45] Turan, T. S., *Inorg. Chem.*, **13**, 1584 (1974).

[46] Rorabacher, D. B. and Melendez-Cepeda, C. A., *J. Amer. Chem. Soc.*, **93**, 6071 (1971).

[47] Kustin, K. and Liu, S. -T., *J. Chem. Soc. Dalton*, 278 (1973).

[48] Barr, M. L., Kustin, K., and Liu, S. -T., *Inorg. Chem.*, **12**, 1486 (1973).

[49] Bydalek, T. J. and Blomster, M. L., *Inorg. Chem.*, **3**, 667 (1964); Bydalek, T. J. and Constant, A. H., *ibid.*, **4**, 833 (1965).

[50] Kustin, K. and Wolff, M. A., *J. Chem. Soc. Dalton*, 1031 (1973).

[51] Kowalak, A., Kustin, K., Pasternack, R. F., and Petrucci, S., *J. Amer. Soc.* **89**, 3126 (1967).

[52] Rorabacher, D. B. and Moss, D. B., *Inorg. Chem.*, **9**, 1314 (1970).

[53] Hayes, R. G. and Myers, R. J., *J. Chem. Phys.*, **40**, 877 (1964).

[54] Hague, D. N. and Zetter, M. S., *Trans. Faraday Soc.*, **66**, 1176 (1970).

[55] Koryta, J., *Z. Elektrochem.*, **64**, 196 (1960).

[56] Buck, D. M. W. and Moore, P., *J. Chem. Soc. Chem. Commun.*, 60 (1974).

[57] Hague, D. N. and Martin, S. R., *J. Chem. Soc. Dalton*, 254 (1974).

[58] Diebler, H. and Rosen, P., *Ber. Bunsenges. Phys. Chem.*, **76**, 1031 (1972).

[59] Pasternack, R. F., Gibbs, E., and Cassatt, J. C., *J. Phys. Chem.*, **73**, 3814 (1969).

[60] Brubaker, J. W., Pearlmutter, A. F., Stuehr, J. E., and Vu, T. V., *Inorg. Chem.*, **13**, 559 (1974).

[61] Pasternack, R. F., Angwin, M., and Gibbs, E., *J. Amer. Chem. Soc.*, **92**, 5878 (1970).

[62] Karpel, R. L., Kustin, K., Kowalak, A., and Pasternack, R. F., *J. Amer. Chem. Soc.*, **93**, 1085 (1971).

[63] Feltch, S. M., Stuehr, J. E., and Tin, G. W., *Inorg. Chem.*, **14**, 2175 (1975).

[64] Roche, T. S. and Wilkins, R. G., *Chem. Commun.*, 1681 (1970); *J. Amer. Chem. Soc.*, **96**, 5082 (1974).

[65] Kirschenbaum, L. J. and Kustin, K., *J. Chem. Soc. A*, 684 (1970).

[66] Weaver, J. and Hambright, P., *Inorg. Chem.*, **8**, 167 (1969).

[67] Pearson, R. G. and Anderson, O. P., *Inorg. Chem.*, **9**, 39 (1970).

[68] Pasternack, R. F. and Kustin, K., *J. Amer. Chem. Soc.*, **90**, 2295 (1968).

[69] Cabbiness, D. K. and Margerum, D. W., *J. Amer. Chem. Soc.*, **92**, 2151 (1970).

[70] Hambright, P. and Fleischer, E. B., *Inorg. Chem.*, **9**, 1757 (1970).

[71] Jones, T. E., Zimmer, L. L., Diaddario, L. L., Rorabacher, D. B., and Ochrymowycz, L. A., *J. Amer. Chem. Soc.*, **97**, 7163 (1975); Kodama, M. and Kimura, E., *J. Chem. Soc. Dalton*, 116 (1976).

[72] Kaden, T. A., *Helv. Chim. Acta*, **54**, 2307 (1971).

[73] Hertli, L. and Kaden, T. A., *Helv. Chim. Acta*, **57**, 1328 (1974).

[74] Malin, J. M. and Swinehart, J. H., *Inorg. Chem.*, **7**, 250 (1968).

[75] Kruse, W. and Thusius, D., *Inorg. Chem.*, **7**, 1373 (1968).

[76] Pearson, R. G. and Gansow, O. A., *Inorg. Chem.*, **7**, 1373 (1968).

[77] Kustin, K., Taub, I. A., and Weinstock, E. M., *Inorg. Chem.*, **5**, 1079 (1966).

[78] Pagenkopf, G. K. and Margerum, D. W., *Inorg. Chem.*, **7**, 2514 (1968).

[79] Sellers, R. M. and Simic, M. G., *J. Chem. Soc. Chem. Commun.*, 401 (1975).

[80] Diebler, H. *Ber. Bunsenges. Phys. Chem.*, **74**, 268 (1970).

[81] Cohen, H. and Meyerstein, D., *J. Chem. Soc. Chem. Commun.*, 320 (1972); *Inorg. Chem.*, **13**, 2434 (1974).

[82] Elding, L. I., *Inorg. Chim. Acta*, **6**, 683 (1972).

[83] Vargaftik, M. N., German, E. D., Dogonadze, R. R., and Syrkin, Ya. K., *Dokl. Akad. Nauk SSSR*, **206**, 370 (1972).

[84] Pearson, R. G. and Hynes, M. J. in *Coordination Chemistry in Solution*, ed. Högfeldt, E., Swedish Natural Science Research Council, Lund (1972)

p. 461.

[85] Frey, C. M., Banyasz, J. L., and Stuehr, J. E., *J. Amer. Chem. Soc.*, **94**, 9198 (1972).

[86] Hague, D. N. and Eigen, M., *Trans. Faraday Soc.*, **62**, 1236 (1966).

[87] Diebler, H., Eigen, M., Ilgenfritz, G., Maass, G., and Winkler, R., *Pure Appl. Chem.*, **20**, 93 (1969).

[88] Atkinson, G., Emara, M. M., and Fernandez-Prini, R., *J. Phys. Chem.*, **78**, 1913 (1974).

[89] Banyasz, J. L. and Stuehr, J. E., *J. Amer. Chem. Soc.*, **95**, 7226 (1973).

[90] Platz, G. and Hoffmann, H., *Ber. Bunsenges. Phys. Chem.*, **76**, 491 (1972).

[91] Silber, H. and Wehner, P., *J. Inorg. Nucl. Chem.*, **37**, 1025 (1975).

[92] Eigen, M. and Maass, G., *Z. Phys. Chem. Frankf. Ausg.*, **49**, 163 (1966).

[93] Loyola, V. M., Wilkins, R. G., and Pizer, R., *J. Amer. Chem. Soc.*, **97**, 7382 (1975).

[94] Baldwin, W. G. and Stranks, D. R., *Aust. J. Chem.*, **21**, 2161 (1968).

[95] Strehlow, H., *Inorganic Reaction Mechanisms Discussion Group*, Bristol (1974).

[96] Qadeer, A., *Z. Phys. Chem. Frankf. Ausg.*, **88**, 160 (1974).

[97] Wilkins, R. G., *The Study of Kinetics and Mechanism of Reactions of Transition Metal Complexes*, Allyn and Bacon (1974) p. 219; and references therein.

[98] McClellan, B. E. and Freiser, H., *Analyt. Chem.*, **36**, 2262 (1964).

[99] Prince, R. H. and Woolley, P. R., *Angew. Chem., Int. Ed. Engl.*, **11**, 408 (1972).

[100] Stein, T. P. and Plane, R. A., *J. Amer. Chem. Soc.*, **91**, 607 (1969).

[101] Cayley, G. R. and Hague, D. N., *Trans. Faraday Soc.*, **67**, 2896 (1971).

[102] DeSimone, R. E., Penley, M. W., Charconneau, L., Smith, S. G., Wood, J. M., Hill, H. A. O., Pratt, J. M., Ridsdale, S., and Williams, R. J. P., *Biochim. Biophys. Acta*, **304**, 851 (1973); Schrauzer, G. N., Weber, J. H., Beckham, T. M., and Ho, R. K. Y., *Tetrahedron Lett.*, 275 (1971).

[103] Yasanaga, T. and Harada, S., *Bull. Chem. Soc. Japan*, **44**, 848 (1971).

[104] Rabenstein, D. L., *J. Amer. Chem. Soc.*, **93**, 2869 (1971).

[105] Vijh, A. K. and Randin, J. -P., *J. Phys. Chem.*, **79**, 1252 (1975).

[106] Meyerstein, D., *Inorg. Chem.*, **14**, 1716 (1975).

[107] Farrow, M. M., Purdie, N., and Eyring, E. M., *Inorg. Chem.*, **14**, 1584 (1975).

[108] Purdie, N., Farrow, M. M., Steggall, M., and Eyring, E. M., *J. Amer. Chem. Soc.*, **97**, 1078 (1975).

[109] Winkler, R., *Structure and Bonding*, **10**, 1 (1972).

[110] Lehn, J. M., Sauvage, J. P., and Dietrich, B., *J. Amer. Chem. Soc.*, **92**, 2916 (1970).

[111] Cahen, Y. M., Dye, J. L., and Popov, A. I., *J. Phys. Chem.*, **79**, 1292 (1975).

[112] Gouger, S. and Stuehr, J. E., *Inorg. Chem.*, **13**, 379 (1974); and references therein.

[113] Langford, C. H. and Chung, F. M., *J. Amer. Chem. Soc.*, **90**, 4485 (1968).

[114] Antonini, E. and Brunori, M., *Haemoglobin and Myoglobin in their Reactions with Ligands*, Elsevier (1971).

[115] McAuley, A. and Gilmour, A. D., *J. Chem. Soc. A*, 2345 (1969).

[116] Cavasino, F. P., *J. Phys. Chem.*, **72**, 1378 (1968).

[117] Pandey, R. N. and Smith, W.MacF., *Can. J. Chem.*, **50**, 194 (1972).

[118] Mentasti, E. and Pelizzetti, E., *J. Chem. Soc. Dalton*, 2605 (1973).

[119] Funahashi, S., Adachi, S., and Tanaka, M., *Bull. Chem. Soc. Japan*, **46**, 479 (1973).

[120] Diebler, H., *Z. Phys. Chem. Frankf. Ausg.*, **68**, 64 (1969); Birk, J. P., *Inorg. Chem.*, **14**, 1724 (1975).

[121] Patel, R. C. and Diebler, H., *Ber. Bunsenges. Phys. Chem.*, **76**, 1035 (1972).

[122] Baker B. R., Sutin, N., and Welch, T. J., *Inorg. Chem.*, **6**, 1948 (1967).

[123] Espenson, J. H. and Pladziewicz, J. R., *Inorg. Chem.*, **9**, 1380 (1970).

[124] Langford, C. H. and Chung, F. M., *Can. J. Chem.*, **48**, 2969 (1970).

[125] Postmus, C. L. and King, E. L., *J. Phys. Chem.*, **59**, 1216 (1955).

[126] Hamm, R. E., Johnson, R. L., Perkins, R. H., and Davis, R. E., *J. Amer. Chem. Soc.*, **80**, 4469 (1958).

[127] Burgess, J., *Inorganic Reaction Mechanisms – Chemical Society Specialist Periodical Report*, **3**, 163 (1974).

[128] Basolo, F. and Pearson, R. G., *Mechanisms of Inorganic Reactions, 2nd Edn.*, Wiley/Interscience (1967).

[129] Espenson, J. H., *Inorg. Chem.*, **8**, 1554 (1969).

[130] Banerjea, D. and Chaudhuri, S. D., *J. Inorg. Nucl. Chem.*, **30**, 871 (1968)

[131] Schenk, C. and Kelm, H., *J. Coord. Chem.*, **2**, 71 (1972).

[132] Schenk, C., Stieger, H., and Kelm, H., *Z. Anorg. Allg. Chem.*, **391**, 1 (1972).

[133] Clonis, H. B. and King, E. L., *Inorg. Chem.*, **11**, 2933 (1972).

[134] Fleischer, E. B. and Krishnamurthy, M., *J. Amer. Chem. Soc.*, **93**, 3784 (1971).

[135] Thusius, D., *Inorg. Chem.*, **10**, 1106 (1971).

[136] Sasaki, Y. and Sykes, A. G., *J. Chem. Soc. Chem. Commun.*, 767 (1973); *J. Less-common Metals*, **36**, 125 (1974); *J. Chem. Soc. Dalton*, 1048 (1975).

[137] Sandberg, R. G., Auborn, J. J., Eyring, E. M., and Watkins, K. O., *Inorg. Chem.*, **11**, 1952 (1972).

[138] Conocchioli, T. J., Nancollas, G. H., and Sutin, N., *Proc. Chem. Soc.*, 113 (1964); *Inorg. Chem.*, **5**, 1 (1966).

[139] Pavelich, M. J. and Harris, G. M., *Inorg. Chem.*, **12**, 423 (1973); and references therein.

[140] Silber, H. B., *J. Phys. Chem.*, **78**, 1940 (1974).

[141] Wang, H. -C. and Hemmes, P., *J. Phys. Chem.*, **78**, 261 (1974); and references therein.

[142] Qadeer, A., *Z. Phys. Chem. Frankf. Ausg.*, **91**, 301 (1974).

[143] Silber, H. B., Scheinin, N., Atkinson, G., and Grecsek, J. J., *J. Chem. Soc., Faraday I*, **68**, 1200 (1972).

[144] Doyle, M. and Silber, H. B., *Chem. Commun.* 1067 (1972).

[145] E.g. Silber, H. B., *Chem. Commun.*, 731 (1971); Reidler, J. and Silber, H. B., *J. Phys. Chem.*, **77**, 1275 (1973); Farrow, M. M. and Purdie, N., *J. Solution Chem.*, **2**, 513 (1973).

[146] Farrow, M. M., Purdie, N., and Eyring, E. M., *Inorg. Chem.*, **13**, 2024 (1974); Farrow, M. M. and Purdie, N., *ibid*, **13**, 2111 (1974).

[147] Purdie, N. and Vincent, C. A., *Trans. Faraday Soc.*, **63**, 2745 (1967).

[148] Farrow, M. M., Purdie, N., and Eyring, E. M., *J. Phys. Chem.*, **79**, 1995 (1975).

[149] Geier, G., *Ber. Bunsenges. Phys. Chem.*, **69**, 617 (1965); *Chimia*, **23**, 148 (1967).

[150] Silber, H. B., Farina, R. D., and Swinehart, J. H., *Inorg. Chem.*, **8**, 819 (1969).

[151] Onodera, T. and Fujimoto, M., *Bull. Chem. Soc. Japan*, **44**, 2003 (1971).

[152] Stepanov, A. V., *Russ. J. Phys. Chem.*, **49**, 1025 (1975).

[153] Kawai, Y., Takahashi, T., Hayashi, K., Imamura, T., Nakayama, H., and Fujimoto, M., *Bull. Chem. Soc. Japan*, **45**, 1417 (1972).

[154] Kalidas, C., Knoche, W. and Papadopoulos, D., *Ber. Bunsenges. Phys. Chem.*, **75**, 106 (1971).

[155] Mal'kova, T. V. and Ovchinnikova, V. D., *Trudy Ivanov. Khim-tekhnol. Inst.*, 59 (1970) (*Chem. Abstr.*, **73**, 18911q (1970)).

[156] Faraggi, M. and Feder, A., *J. Chem. Phys.*, **56**, 3294 (1972).

[157] Ekstrom, A. and McLaren, A., *J. Inorg. Nucl. Chem.*, **34**, 1009 (1972).

[158] Hanna, S. B. and Hessley, R. K., *Inorg. Nucl. Chem. Lett.*, **7**, 83 (1971).

[159] Steinhaus, R. K. and Margerum, D. W., *J. Amer. Chem. Soc.*, **88**, 441 (1966).

[160] Margerum, D. W. and Rosen, H. M., *J. Amer. Chem. Soc.*, **89**, 1088 (1967); Jones, J. P., Billo, E. J., and Margerum, D. W., *ibid.*, **92**, 1875 (1970).

[161] Holyer, R. H., Hubbard, C. D., Kettle, S. F. A., and Wilkins, R. G., *Inorg. Chem.*, **5**, 622 (1966).

[162] Cobb, M. A. and Hague, D. N., *J. Chem. Soc. Faraday I*, **68**, 932 (1972).

[163] Melvin, W. S., Rablen, D. P., and Gordon, G., *Inorg. Chem.*, **11**, 488 (1972).

[164] Langford, C. H., *Inorg. Nucl. Chem. Lett.*, **9**, 679 (1973).

[165] Jordan, R. B., Dodgen, H. W., and Hunt, J. P., *Inorg. Chem.*, **5**, 1906 (1966).

[166] Rablen, D. and Gordon, G., *Inorg. Chem.*, **8**, 395 (1969).
[167] Cayley, G. R. and Margerum, D. W., *J. Chem. Soc. Chem. Commun.*, 1002 (1974).
[168] Pasternack, R. F. and Sigel, H., *J. Amer. Chem. Soc.*, **92**, 6146 (1970).
[169] Barr, M. L., Baumgartner, E., and Kustin, K., *J. Coord. Chem.*, **2**, 263 (1973).
[170] Farrer, D. T., Stuehr, J. E., Moradi-Araghi, A., Urbach, F. L., and Campbell, T. G., *Inorg. Chem.*, **12**, 1847 (1973).
[171] Hague, D. N. and Kinley, K., *J. Chem. Soc. Dalton*, 249 (1974).
[172] Sharma, V. S. and Leussing, D. L., *Inorg. Chem.*, **11**, 138, 1955 (1972).
[173] Cobb, M. A. and Hague, D. N., *Chem. Commun.*, 192 (1971).
[174] Pasternack, R. F., Huber, P. R., Huber, U. M., and Sigel, H., *Inorg. Chem.*, **11**, 276 (1972).
[175] Mildvan, A. S. and Scrutton, M. C., *Biochemistry N.Y.*, **6**, 2978 (1967).
[176] Hague, D. N., Martin, S. R., and Zetter, M. S., *J. Chem. Soc. Faraday I*, **68**, 37 (1972).
[177] Cobb, M. A. and Hague, D. N., *Trans. Faraday Soc.*, **67**, 3069 (1971).
[178] Cayley, G. R. and Hague, D. N., *Trans. Faraday Soc.*, **67**, 786 (1971).
[179] Cayley, G. R. and Hague, D. N., *J. Chem. Soc. Faraday I*, **68**, 2259 (1972).
[180] Hammes, G. G. and Hurst, J. K., *Biochemistry N.Y.*, **8**, 1083 (1969).
[181] See pp. 193–207 of reference [128].
[182] Duffy, N. V. and Earley, J. E., *J. Amer. Chem. Soc.*, **89**, 272 (1967); Langford, C. H. and Muir, W. R., *ibid.*, **89**, 3141 (1967).
[183] Joubert, P. R. and van Eldik, R., *Inorg. Chim. Acta*, **14**, 259 (1975).
[184] Joubert, P. R. and van Eldik, R., *J. Inorg. Nucl. Chem.*, **37**, 1817 (1975); *React. Kinet. Catal. Lett.*, **3**, 23 (1975); van Eldik, R., *J. Inorg. Nucl. Chem.*, **38**, 884 (1976).
[185] Van Eldik, R., *Z. Anorg. Allg. Chem.*, **416**, 88 (1975); *React. Kinet. Catal. Lett.*, **2**, 251 (1975).
[186] See p. 233 of reference [127].
[187] Crumbliss, A. L. and Wilmarth, W. K., *J. Amer. Chem. Soc.*, **92**, 2593 (1970).
[188] Costa, G., Mestroni, G., Tauzher, G., Goodall, D. M., Green, M., and Hill H. A. O., *Chem. Commun.*, 34 (1970).
[189] E.g. Thusius, D., *J. Amer. Chem. Soc.*, **93**, 2629 (1971); Hayward, G. C., Hill, H. A. O., Pratt, J. M., and Williams, R. J. P., *J. Chem. Soc. A*, 196 (1971).
[190] Hasinoff, B. B., *Can. J. Chem.*, **52**, 910 (1974).
[191] Ashley, K. R., Berggren, M., and Cheng, M., *J. Amer. Chem. Soc.*, **97**, 1422 (1975); Pasternack, R. F., Cobb, M. A. and Sutin, N., *Inorg. Chem.*, **14**, 866 (1975).
[192] Fleischer, E. B., Jacobs, S., and Mestichelli, L., *J. Amer. Chem. Soc.*, **90**,

2527 (1968).

[193] Toma, H. E. and Malin, J. M., *Inorg. Chem.*, **12**, 2080 (1973).

[194] Caldin, E. F. and Hasinoff, B. B., *J. Chem. Soc. Faraday I*, **71**, 515 (1975).

[195] Hasinoff, B. B., *Biochemistry N.Y.*, **13**, 3111 (1974).

[196] Sarkar, S. and Banerjea, D., *J. Inorg. Nucl. Chem.*, **37**, 547 (1975).

[197] Cohen, H. and Meyerstein, D., *J. Chem. Soc. Chem. Commun.*, 320 (1972); *Inorg. Chem.*, **13**, 2434 (1974); Malik, S. K., Schmidt, W., and Spreer, L. O., *ibid.*, **13**, 2986 (1974).

[198] Matthews, R. J. and Moore, J. W., *Inorg. Chim. Acta*, **6**, 359 (1972).

[199] Luz, Z. and Meiboom, S., *J. Chem. Phys.*, **40**, 2686 (1964).

[200] Dickert, F., Fischer, P., Hoffmann, H., and Platz, G., *Chem. Commun.*, 106 (1972); Fischer, P., Hoffmann, H., and Platz, G., *Ber. Bunsenges. Phys. Chem.*, **76**, 1060 (1972).

[201] Dickert, F. and Wank, R., *Ber. Bunsenges. Phys. Chem.*, **76**, 1028 (1972).

[202] Hoffmann, H., Janjic, T., and Sperati, R., *Ber. Bunsenges. Phys. Chem.*, **78**, 223 (1974).

[203] Dickert, F. and Hoffmann, H., *Ber. Bunsenges. Phys. Chem.*, **75**, 1320 (1971).

[204] Williams, J., Petrucci, S., Sesta, B., and Battistini, M., *Inorg. Chem.*, **13**, 1968 (1974).

[205] Coetzee, J. F. and Hsu, E., *J. Solution Chem.*, **4**, 45 (1975); and references therein.

[206] Diamond, A., Fanelli, A., and Petrucci, S., *Inorg. Chem.*, **12**, 611 (1973).

[207] Carle, D. L. and Swaddle, T. W., *Can. J. Chem.*, **51**, 3795 (1973).

[208] Devia, D. H. and Watts, D. W., *Inorg. Chim. Acta*, **7**, 691 (1973).

[209] Bennetto, H. P. and Caldin, E. F., *J. Chem. Soc. A*, 2191, 2198 (1971).

[210] Bennetto, H. P and Caldin, E. F., *J. Chem. Soc. A*, 2207 (1971).

[211] Moore, P. and Buck, D. M. W., *J. Chem. Soc. Dalton*, 1602 (1973).

[212] Chattopadhyay, P. K. and Coetzee, J. F., *Inorg. Chem.*, **12**, 113 (1973).

[213] Cock, P. A., Cottrell, C. E., and Boyd, R. K., *Can. J. Chem.*, **50**, 402 (1972).

[214] Sanduja, M. L. and Smith, W.MacF., *Can. J. Chem.*, **50**, 3861 (1972); **51**, 3975 (1973).

[215] Benton, D. J. and Moore, P., *J. Chem. Soc. Chem. Commun.*, 717 (1972); *J. Chem. Soc. Dalton*, 399 (1973).

[216] Scharfe, R. R., Sastri, V. S., Chakrabarti, C. L., and Langford, C. H., *Can. J. Chem.*, **51**, 67 (1973).

[217] Choi, E. I. and Fleischer, E. B., *Inorg. Chem.*, **2**, 94 (1963).

[218] Brisbin, D. A. and Balahura, R. J., *Can. J. Chem.*, **46**, 3431 (1968).

[219] Kingham, D. J. and Brisbin, D. A., *Inorg. Chem.*, **9**, 2034 (1970).

[220] Brisbin, D. A. and Richards, G. D., *Inorg. Chem.*, **11**, 2849 (1972); Paquette, G. and Zador, M., *Can. J. Chem.*, **53**, 2375 (1975).

[221] Shchori, E., Jagur-Grodzinski, J., Luz, Z., and Shporer, M., *J. Amer. Chem. Soc.*, **93**, 7133 (1971); Shchori, E., Jagur-Grodzinski, J., and Shporer, M., *ibid.*, **95**, 3842 (1973).

[222] Chock, P. B., *Proc. Natn. Acad. Sci. U.S.A.*, **69**, 1939 (1972).

[233] Bear, J. L. and Lin, C, -T., *J. Inorg. Nucl. Chem.*, **34**, 2368 (1972).

[224] Dickert, F., Hoffmann, H., and Janjic, T., *Ber. Bunsenges. Phys. Chem.*, **78**, 712 (1974).

[225] Caldin, E. F. and Bennetto, H. P., *J. Solution Chem.*, **2**, 217 (1973); Chattopadhyay, P. K. and Coetzee, J. F., *Inorg. Chem.*, **15**, 400 (1976); Coetzee, J. F. and Gilles, D. M., *ibid.*, **15**, 405 (1976).

[226] Caldin, E. F. and Grant, M. W., *J. Chem. Soc. Faraday I*, **69**, 1648 (1973).

[227] MacKellar, W. J. and Rorabacher, D. B., *J. Amer. Chem. Soc.*, **93**, 4379 (1971); Shu, F. R. and Rorabacher, D. B., *Inorg. Chem.*, **11**, 1496 (1972).

[228] Baltisberger, R. J., Knudson, C. L., and Anderson, M. F., *Inorg. Chem.*, **13**, 2354 (1974).

[229] Lin, C. -T. and Bear, J. L., *J. Inorg. Nucl. Chem.*, **31**, 263 (1969).

[230] Biryukov, A. A., Shlenskaya, V. I., and Rabinovitch, B. S., *Russ. J. Inorg. Chem.*, **14**, 413 (1969).

[231] Fittipaldi, F. and Petrucci, S., *J. Phys. Chem.*, **71**, 3414 (1967).

[232] Atkinson, G. and Kor, S. K., *J. Phys. Chem.*, **69**, 128 (1965).

[233] Bear, J. L. and Lins, C. -T., *J. Inorg. Nucl. Chem.*, **34**, 2368 (1972).

[234] Reidler, J. and Silber, H. B., *J. Phys. Chem.*, **78**, 424 (1974).

[235] Caldin, E. F. and Godfrey, P., *J. Chem. Soc. Faraday I*, **70**, 2260 (1974).

[236] Lin, C. -T. and Bear, J. L., *J. Phys. Chem.*, **75**, 3705 (1971).

[237] Bennetto, H. P., *J. Chem. Soc. A*, 2211 (1971).

[238] Hughes, M. N., *The Inorganic Chemistry of Biological Processes*, Wiley/Interscience (1972; revised reprint 1975).

[239] Williams, D. R., *The Metals of Life*, Van Nostrand Reinhold (1971).

[240] E.g. Burgess, J., *Inorganic Reaction Mechanisms — Chemical Society Specialist Periodical Report*, **2**, 153 (1972); **3**, 184, 203 (1974); Moore, P., *ibid.*, **4**, 152 (1976).

[241] Bifano, C. and Linck, R. G., *Inorg. Chem.*, **7**, 908 (1968).

[242] Armor, J. N. and Haim, A., *J. Amer. Chem. Soc.*, **93**, 867 (1971).

[243] Espenson, J. H., Bushey, W. R., and Chmielewski, M. E., *Inorg. Chem.*, **14**, 1302 (1975), and references therein.

[244] Jones, M. M. and Clark, H. R., *J. Inorg. Nucl. Chem.*, **33**, 413 (1971).

[245] E.g. Dash, A. C. and Nanda, R. K., *J. Indian Chem. Soc.*, **52**, 289 (1975), and references therein.

[246] Rudakov, E. S. and Kozhevnikov, I. V., *Tetrahedron Lett.*, 1333 (1971).

[247] E.g. Bach, R. D. and Willis, C. L., *J. Amer. Chem. Soc.*, **97**, 3844 (1975).

[248] Johnson, G. L. and Angelici, R. J., *J. Amer. Chem. Soc.*, **93**, 1106 (1971).

[249] Angelici, R. J. and Leslie, D. B., *Inorg. Chem.*, **12**, 431 (1973).

[250] Hix, J. E. and Jones, M. M., *Inorg. Chem.*, **5**, 1863 (1966).

[251] Hay, R. W. and Nolan, K. B., *J. Chem. Soc. Dalton,* 2542 (1974).

[252] Hay, R. W., Jansen, M. L., and Cropp, P. L., *Chem. Commun.,* 621 (1967).

[253] Burgess, J., *Inorganic Reaction Mechanisms – Chemical Society Specialist Peridocial Report,* **1**, 198 (1971); **2**, 219 (1972); **3**, 297 (1974).

[254] Steffens, J. J., Sampson, E. J., Siewers, I. J., and Benkovic, S. J., *J. Amer. Chem. Soc.,* **95**, 936 (1973).

[255] Synge, R. L. M., *Biochem. J.,* **39**, 351 (1945).

[256] Lawrence, L. and Moore, W. J., *J. Amer. Chem. Soc.,* **73**, 3973 (1951).

[257] Grant, I. J. and Hay, R. W., *Aust. J. Chem.,* **18**, 1189 (1965).

[258] Jones, M. M., Cook, T. J., and Brammer, S., *J. Inorg. Nucl. Chem.,* **28**, 1265 (1966).

[259] Long, D. A., Truscott, T. G., Cronin, J. R., and Lee, R. G., *Trans. Faraday Soc.,* **67**, 1094 (1971); Nakata, T., Tasumi, M., and Miyazawa, T., *Bull. Chem. Soc. Japan,* **48**, 1599 (1975).

[260] Cronin, J. R., Long, D. A., and Truscott, T. G. *Trans. Faraday Soc.,* **67**, 2096 (1971).

[261] See pp. 121-2 of reference [238].

[262] Hay, R. W. and Morris, P. J., *Chem. Commun.,* 1208 (1969).

[263] Bender, M. L. in *Advances in Chemistry Series 37: Reactions of Co-ordinated Ligands,* ed. Gould, R. F., American Chemical Society, Washington, D.C. (1963) p. 19.

[264] Krebs, H. A., *Biochem. J.,* **36**, 303 (1942).

[265] Raghavan, N. V. and Leussing, D. L., *J. Amer. Chem. Soc.,* **96**, 7147 (1974).

[266] Prue, J. E., *J. Chem. Soc.,* 2331 (1952).

[267] Yu-Keung Sze and Irish, D. E., *Can. J. Chem.,* **53**, 427 (1975).

[268] Pocker, Y. and Meany, J. E., *J. Phys. Chem.,* **71**, 3113 (1967); and references therein.

[269] Breslow, R., McClure, D. E., Brown, R. S., and Eisenach, J., *J. Amer. Chem. Soc.,* **97**, 194 (1975).

[270] Wada, G. and Sakamoto, M., *Bull. Chem. Soc. Japan,* **46**, 3378 (1973).

[271] Cox, B. G., *J. Amer. Chem. Soc.,* **96**, 6823 (1974).

[272] Bartlett, P. D., Baumstark, A. L., and Landis, M. E., *J. Amer. Chem. Soc.,* **96**, 5557 (1974).

[273] Hanzlik, R. P. and Michaely, W. J., *J. Chem. Soc. Chem. Commun.,* 113 (1975).

[274] Hay, R. W. and Nolan, K. B., *J. Chem. Soc. Dalton,* 548 (1976).

[275] Lavallee, D. K., *Inorg. Chem.,* **15**, 691 (1976).

[276] Norman, P. R. and Phipps, D. A., *Inorg. Chim. Acta,* **17**, L19 (1976).

[277] E.g. Collman, J. P. in reference [263], Ch. 5.

[278] Chawla, N. K., Lambert, D. G., and Jones, M. M., *J. Amer. Chem. Soc.,* **89**, 557 (1967).

[279] E.g. Farrington, D. J. and Jones, J. G., *Inorg. Chim. Acta,* **6**, 575 (1972).

[280] E.g. McQuate, R. S. and Leussing, D. L., *J. Amer. Chem. Soc.,* **97**, 5117 (1975).

[281] Rest, A. J., Smith, S. A., and Tyler, I. D., *Inorg. Chim. Acta,* **16**, L1 (1976).

[282] Busch, D. H. in reference [263], p. 1.

[283] Day, V. W., Marks, T. J., and Wachter, W. A., *J. Amer. Chem. Soc.,* **97**, 4519 (1975).

[284] Hawley, G. and Blinn, E. L., *Inorg. Chem.,* **14**, 2865 (1975).

[285] Sawai, H. and Orgel, L. E., *J. Amer. Chem. Soc.,* **97**, 3532 (1975).

[286] Burgess, J., *Inorganic Reaction Mechanisms – Specialist Periodical Report,* **1**, 275 (1971); Kemmitt, R. D. W. and Burgess, J., *ibid.,* **2**, 277 (1972), Kemmitt, R. D. W. and Smith, M. A. R., *ibid.,* **3**, 384 (1974); **4**, 305 (1976).

[287] See pp. 276-8 of Volume 1 and pp. 278-286 of Volume 2 of reference [286].

[288] Paquette, L. A. and Beckley, R. S., *J. Amer. Chem. Soc.,* **97**, 1084 (1975); Paquette, L. A., Beckley, R. S., and Farnham, W. B., *ibid.,* p. 1089; Paquette, L. A., Ward, J. S., Boggs, R. A., and Farnham, W. B., *ibid.,* p. 1101.

[289] Paquette, L. A. and Ward, J. S., *Tetrahedron Lett.,* 4909 (1972).

Chapter **13**

KINETICS AND MECHANISMS: REDOX REACTIONS

13.1   INTRODUCTION

Aquo-cations can act as oxidants or reductants. In practice examples of the latter are more common, because metals in their higher oxidation states (four and above) tend to occur in the form of oxocations or, more usually, oxoanions rather than as the simple aquo-cations with which we are primarily concerned here. Inspection of tables of redox potentials (Chapter 8) will indicate which aquo-cations can in principle act as oxidants or reductants. The kineticist should not feel entirely constrained by thermodynamic considerations. Many reactions with negative free energy changes go so slowly that oxidation or reduction of water by aquo-cations is often sufficiently protracted for the kinetics of their reactions with other species still to be investigated. Chromium(II), uranium(III), and manganese(III) provide examples of this.

The simplest reaction of aquo-cations is that of reduction by hydrated electrons. This topic is dealt with first. Then follows the main bulk of the Chapter, in which the inner- and outer- sphere mechanisms for redox reactions are discussed and their relevance to aquo-cations described. This Chapter is almost exclusively concerned with reactions in aqueous solution, an unbalanced state of affairs dictated by the great scarcity of kinetic data pertaining to redox reactions of solvated metal cations in mixed and non-aqueous media.

13.2   REDUCTIONS BY HYDRATED ELECTRONS

The reduction of an aquo-cation by a hydrated electron is formally a simple process,

$$M^{n+}aq + e^-aq = M^{(n-1)+}aq .$$

The hydrated electron is a powerful reducing agent, with an estimated redox potential of $-2.7V$. It is an ephemeral species, with a half-life of less than a millisecond even under the most favourable conditions. Nevertheless pulse radiolysis can generate sufficiently high concentrations (about millimolar) of hydrated electrons for kinetic studies, leading to reasonable estimates of rate constants, to be possible.

Second-order rate constants for the reactions of hydrated electrons with aquo-cations are listed in Table 13.1. All these reactions are very fast, with the rate constant for the reduction of Cr^{2+}aq (as for such complexes as $Co(NH_3)_5Cl^{2+}$, $Rh(NH_3)_6^{3+}$, or $Co(bipy)_3^{3+}$) approximating to the value of about 5×10^{10} M^{-1} sec^{-1} expected for diffusion controlled reactions. Rate constants for the reduction of those lanthanide(III) cations which have a well-characterised 2+ state in aqueous solution, that is Eu^{3+}aq, Sm^{3+}aq, and Yb^{3+}aq, are also close to the diffusion controlled limit. For other lanthanide(III) cations, rates are somewhat slower. There is a rough correlation between reduction rate and second ionisation potential (gas phase) for reductions of transition metal cations in oxidation state 2+, specifically Cr^{2+}aq to Zn^{2+}aq. At least for the slower reactions of aquo-metal cations with hydrated electrons, the possibility of equilibrium preassociation prior to electron transfer into the metal orbitals should be borne in mind (see the Eigen-Wilkins mechanism for complex formation, Chapter 12).

Table 13.1 Second-order rate constants $(M^{-1} sec^{-1})$ for reduction of aquo-cations by hydrated electrons at $25°C$.[a]

Cr^{2+} 4.2×10^{10} [1]	Cd^{2+} 5.2×10^{10} [3]	La^{3+} 3.4×10^8 [2]
Mn^{2+} 7.7×10^7 [2]	Pb^{2+} 3.9×10^{10} [3]	Pr^{3+} 2.9×10^8 [2][b]
Fe^{2+} 3.5×10^8 [2]		Nd^{3+} 5.9×10^8 [2]
Co^{2+} 1.2×10^{10} [2]	Al^{3+} 2.0×10^9 [1]	Sm^{3+} 2.5×10^{10} [2]
Ni^{2+} 2.2×10^{10} [2]		Eu^{3+} 6.1×10^{10} [2]
Cu^{2+} 3.0×10^{10} [2]	Cr^{3+} 6.0×10^{10} [2]	Gd^{3+} 5.5×10^8 [2]
Zn^{2+} 1.5×10^9 [2]		Tm^{3+} 3.0×10^9 [1]
		Yb^{3+} 4.3×10^{10} [2]

[a] Rate constants for reactions of aquo-oxo-cations of actinides (UO_2^{2+}aq, NpO_2^{+}aq, NpO_2^{2+}aq, and PuO_2^{2+}aq) with hydrated electrons are between 1.5 and 6×10^{10} $M^{-1} sec^{-1}$ at $24°C$, (Sullivan, J. C., Gordon, S., Cohen, D., Mulac, W., and Schmidt, K. H., *J. Phys. Chem.*, 80, 1684 (1976)); [b] a later determination gave 6×10^6 M^{-1} sec^{-1} (Faraggi, M. and Feder, A., *J. Chem. Phys.*, 56, 3294 (1972)).

13.3 REDOX MECHANISMS FOR COMPLEXES

Aquo-cations are a particular class of complex ion. In this Chapter it is convenient to set out the general pattern for redox mechanisms for complexes before examining the particular class of aquo-cations. Two distinct mechanisms for redox reactions of complexes are recognised. These are the inner-sphere and outer-sphere mechanisms, which are outlined and contrasted in the following two sections (13.3.1 and 13.3.2) and discussed in more detail in succeeding sections (13.3.3 and 13.3.4 respectively).

13.3.1 The Inner-sphere Mechanism

The essential feature of this mechanism is that substitution at one of the metal centres occurs, to give a binuclear ligand-bridged species, previous to the transfer of an electron (Fig. 13.1). The first reaction for which this redox mechanism was unequivocally demonstrated was the chromium(II) reduction of the $Co(NH_3)_5Cl^{2+}$ cation [4],

$$Co(NH_3)_5Cl^{2+} + Cr^{2+}aq + 5H^+ = CrCl^{2+}aq + Co^{2+}aq + 5NH_4^+.$$

The reaction products immediately indicate the transfer of chloride from the cobalt to the chromium. Consideration of the following evidence *proves* that this ligand transfer is an essential feature of the redox process. A comparison of the rate constant for this redox reaction ($>10^3$ $M^{-1}sec^{-1}$ at 25°C) with that for aquation of the $Co(NH_3)_5Cl^{2+}$ cation ($1.7 \times 10^{-6} sec^{-1}$ at 25°C) shows that the chloride cannot become detached from the cobalt prior to the formation of the transition state. Similarly the rate constant for formation of $CrCl^{2+}$ from $Cr^{3+}aq$ (3×10^{-8} $M^{-1}sec^{-1}$ at 25°C) is very much less than the redox rate constant, so the chloride must be bonded to the chromium before the transition state separates into the products. Logic therefore demands that the chloride be bonded to both metals in the transition state. Chloride transfer and electron transfer are intimately connected. The lability of chromium(II) and cobalt(II) complexes is such that they offer no obstruction to the mechanism outlined. Confirmatory experiments using labelled chloride have shown that, except in solutions containing so much chloride that $CrCl^+$ becomes significant as a reductant, all the chloride in the $CrCl^{2+}$ product comes from the cobalt(III) complex and cannot have been derived from free chloride in solution.

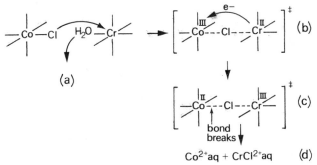

Fig. 13.1 The inner-sphere electron transfer mechanism, for chromium(II) reduction of a chloro-cobalt(III) complex.

The sequence of events can therefore only be described as in the diagrams of Fig. 13.1. As the cobalt(II) to bridging chloride bond will be very much more labile than the chromium(III) to chloride bond (crystal field stabilisation for

$d^3 \gg d^7$), the post-electron transfer transition state must come apart at the position indicated in Fig. 13.1(c). The lability of the cobalt(II) species thus generated is such that it will 'instantly'[†] give $Co^{2+}aq$ and ammonium ions in acidic aqueous solution.

Isotopic labelling experiments on chromium(II) reduction of the $Co(NH_3)_5(OH_2)^{3+}$ cation have confirmed that the chromium is six-coordinate in the transition state, as shown in Fig. 13.1 (b) and (c). If the chromium were coordinated to seven oxygens in the transition state, then only 6/7 of the labelled oxygen would be transferred to the chromium. However the transfer of labelled oxygen was quantitative, and this indicates that the chromium atom in the transition state is bonded to just those six oxygen atoms that will be bonded to it in the final chromium(III) product [5]. A word of caution is needed here, in that this work is the only instance in which evidence has been presented for water acting as a bridging ligand in inner-sphere electron transfer (see Chapter 13.3.3).

The above experimental demonstration of the operation of an inner-sphere redox mechanism in the chromium(II) reduction of $Co(NH_3)_5Cl^{2+}$ represents one of the most convincing elucidations of a mechanism in inorganic chemistry. Its definitiveness arises from the particularly favourable combination of inert and labile metal centres in the starting materials and products. Such favourable combinations are not, as will emerge later, common.

13.3.2 The Outer-sphere Mechanism

In this mechanism the redox step postulated is simply electron transfer[††] between two reactants whose primary coordination spheres remain intact throughout (Fig. 13.2). The very simplicity of this mechanism makes positive proof almost impossible to come by. However there are a considerable number of inorganic

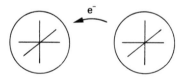

Fig. 13.2   The outer-sphere electron transfer mechanism.

redox reactions which cannot conceivably go by an inner-sphere route and are therefore thought to be outer-sphere processes. These are reactions where two complexes which are both substitution inert undergo rapid electron transfer. Examples include electron exchange between manganate and permanganate, or

[†]Rates of loss of ammonia from cobalt(II) ammines have recently been estimated by conductometric pulse radiolysis techniques (Simic, M. and Lilie, J. *J. Amer. Chem. Soc.* 96, 291 (1974)).

[††]The hydrated electron, e^-aq, is not an intermediate in any known electron transfer reaction.

between ferrocene and the ferrocinium cation, and electron transfer reactions such as those of phenanthroline complexes of iron(II), ruthenium(II), or osmium(II) with such oxidants as hexachloroiridate(IV) [6]. In all these reactions electron transfer is complete within a tiny fraction of a second, indeed the hexachloroiridate(IV) oxidations have rates approaching the diffusion controlled limiting rate for bimolecular processes. Yet in all these reactions, the half-life for the liberation of a ligand from the complex is of the order of minutes, hours, or even days. The inner-sphere path of Fig. 13.1 (a)-(d) is thus not possible. The only conceivable way of generating a bridged transition state would be to increase the coordination number of the metal in one of the reactants (for example, Fig. 13.3). Such a subterfuge seems unlikely for the manganate–permanganate reaction, and less likely still for the ferrocene–ferrocinium reaction or for the reactions of phenanthroline complexes, in view of the steric constraints imposed by the bulky ligands. The generation of a bridged transition state of this type has been proposed in only one reaction, that of neptunium(III) with ruthenium(III) [7], where the coordination number of the large Np^{3+} can readily be increased to seven.

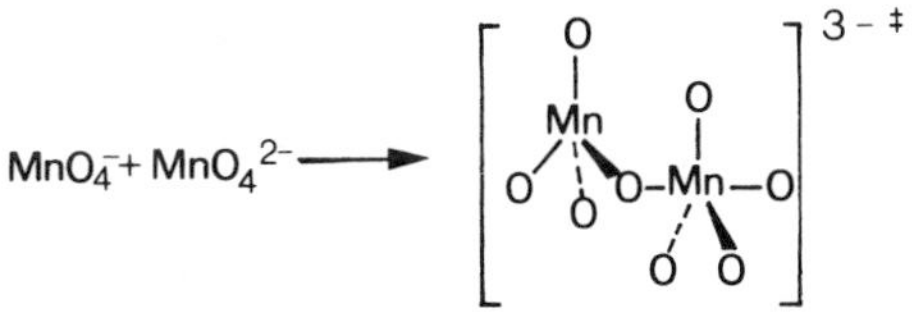

Fig. 13.3 Implausible transition state for the MnO_4^-/MnO_4^{2-} electron exchange reaction.

13.3.3 Inner-sphere Redox Reactions

The basic steps in this type of reaction may be summarised by the sequence below [8, 9]. For illustration, the reaction between a metal cation M in oxidation state $m+$ and a metal cation N in oxidation state $n+$ is considered, with a bridging anionic ligand X^-,

$$MX^{(m-1)+} + N^{n+} \rightarrow [M^{m+} \cdots X^- \cdots N^{n+}]^{\ddagger} \rightarrow [M^{(m-1)+} \cdots X^- \cdots N^{(n+1)+}]^{\ddagger} \rightarrow \text{products.}$$

The bridging group X^- (or X, X^{n-} in general) can be one of a wide variety of ligands. The one essential feature for a bridging ligand is that it should have at least one lone pair of electrons available for interaction with N^{n+}, that is at least one lone pair beyond the electron pair needed to bond to M^{m+} in the first place. Ligands such as ammonia, and most other Group V bases, have only one electron pair available for bonding to a cation. They use this to coordinate to M^{m+}, and therefore cannot act as bridging ligands in inner-sphere redox reactions. On the other hand, Group VI and Group VII ligands have one or more spare electron pairs after bonding to a cation, and so can act as bridging ligands. Bridging possibilities also obviously exist for multi-atomic ligands, for example cyanide, thiocyanate, or the multitude of carboxylic acids and their derivatives. It has been suggested that the methyl group can act as a bridge, despite its lack of a suitable electron pair to form a $Co-CH_3-Cr$ bridge (but cf. S_N2 at carbon, e.g. $HO\cdots CH_3\cdots I!$). However the reaction cited to support this suggestion can also be interpreted as an S_H2 substitution rather than an inner-sphere redox process [10]. The postulate of a methylene bridge, $Co-CH_2-Cr$, in the reaction of $Cr(OH_2)_5(CH_2I)^{2+}$ with $Cr^{2+}aq$ [11] seems more plausible. Water is the one common potential bridging ligand which seems extraordinarily reluctant to act in this capacity, though hydroxide ions can act as effective bridges for electron transfer.

In the classical example (Chapter 13.3.1) of the inner-sphere mechanism, it is the unquestionable transfer of chloride from cobalt to chromium which provides evidence for its operation. Such ligand transfer is likely in most inner-sphere reactions, but it is not an essential requirement. In general for the post-electron transfer situation,

$$[M^{(m-1)+} \underset{(a)}{\ldots} X^- \underset{(b)}{\ldots} N^{(n+1)+}]^{\ddagger},$$

whether bond (a) or bond (b) will break depends on their relative strengths and labilities. If reduced M is substitution-labile and oxidised N substitution-inert, bond (a) will break and ligand transfer result, as in chromium(II) reductions of most cobalt(III) complexes. However, if bond (b) breaks there is no ligand transfer, and indeed no easy proof of the operation of an inner-sphere mechanism. The original example cited for this latter situation was chromium(II) reduction of hexachloroiridate(IV) [4], but this reaction has now been shown to involve outer-sphere electron transfer and an inner-sphere path with chloride transfer [12]. A currently acceptable example of an inner-sphere redox reaction not involving ligand transfer is the chromium(II) reduction of a ruthenium(III)-nicotinamide complex. After electron transfer, the nicotinamide is bridging inert (d^3) chromium(III) and inert (low spin d^6) ruthenium(II), Fig. 13.4(a). The inertness of the latter exceeds that of the former, so there is no transfer of the ligand to the chromium. The detection of the appropriate transient binuclear intermediate shows that an inner-sphere mechanism is operative here.

Persistent intermediates. To be honest, the example at the end of the last paragraph involves some cheating. The detection of such a binuclear intermediate signifies the existence of an intermediate rather than just of a transition state, so one is really dealing with a two-stage mechanism rather than with a simple inner-sphere redox reaction. Several other cases are known of redox reactions in which the generation of two inert centres on electron transfer results in an intermediate of significant lifetime en route to the ultimate reaction products. The chromium(II)-hexachloroiridate(IV) reaction mentioned

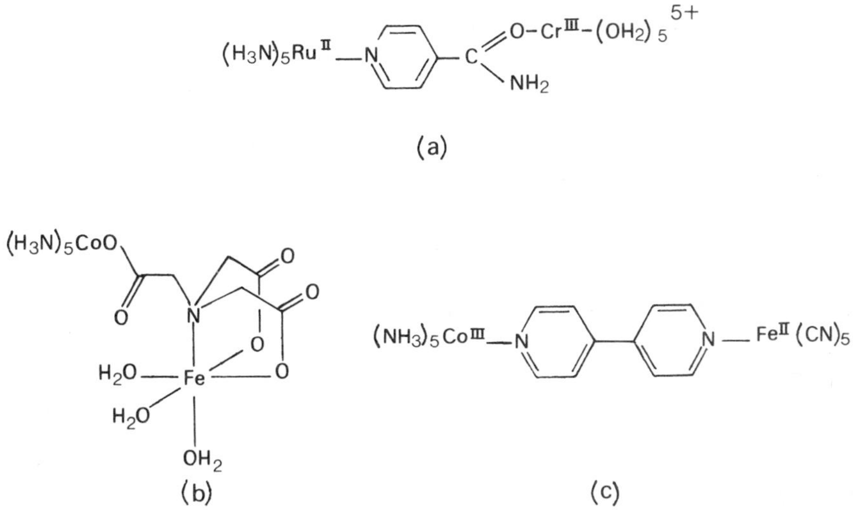

(a)

(b)   (c)

Fig. 13.4 Inner-sphere redox reactions: (a) the ruthenium(III)-nicotinamide-chromium(II) transition state *after* electron transfer; (b) and (c) precursor complexes proposed for inner-sphere redox reactions.

above provides another example, because the coloured chromium(III)-chloride-iridium(III) intermediate, again containing an inert d^3 and an inert d^6 centre, can be seen momentarily on mixing the reactants. In the case of the reaction of hexacyanoferrate(III) with pentacyanocobaltate(II), the bridged anion $(NC)_5Fe^{II}-CN-Co^{III}(CN)_5^{6-}$ can be precipitated as its barium or potassium salt [14].

Oxocations of metals in high oxidation states are also sufficiently inert for the isolation of binuclear intermediates in redox reactions to be possible. In consequence, the reduction of NpO_2^{2+}aq or of UO_2^{2+}aq with chromium(II) leads, after the electron transfer step, to isolable cations of the type $(H_2O)_5Cr^{III}-O-U^V-O^{4+}$ [15]. For the case of UO_2^{2+}aq, ^{18}O-tracer studies have shown that oxygen is ultimately transferred from the uranium to the chromium. The limiting example is provided by vanadium redox chemistry. The reduction by V^{3+}aq of $Ru(NH_3)_5Cl^{2+}$, which hydrolyses during this reaction, produces the $(H_3N)_5Ru-O-V(OH_2)_n^{4+}$ cation. This presumably contains ruthenium(II) and vanadium(IV). It is a very stable species which does not separate into mononuclear complexes. Indeed it is readily preparable from $Ru(NH_3)_5(OH_2)^{3+}$ and VO^{2+}aq [16].

There are thus several examples of binuclear intermediates containing the two cations in their post-electron-transfer oxidation states. Evidence for binuclear intermediates containing two metal ions in their pre-electron-transfer oxidation states is more recent and less direct. The kinetics of reduction of *cis*-$Ru(NH_3)_4Cl_2^+$ by chromium(II) hint at a transient $Ru^{III}-Cl-Cr^{II\ 3+}$ intermediate

[17], while those for the reduction of Co(tet-a)(OH$_2$)$_2^{3+}$ or of its *trans*-[14]-diene analogue by chromium(II) suggest that a Co^{III}-bridge-Cr^{II} species persists long enough to choose between undergoing proton or electron transfer [18]. More direct evidence comes from studies of the reduction of Co(NH$_3$)$_5$(nta) by iron(II). Here it has proved possible to evaluate the stability constant for the formation of the intermediate, whose proposed structure is shown in Fig. 13.4(b). The kinetic parameters for electron transfer within this binuclear complex were also estimated. At 25°C its half-life is approximately five seconds; the binuclear species breaks down quickly once electron transfer has taken place. The activation parameters for this electron transfer are $\Delta H^{\ddagger} = 18.7$ kcal mol^{-1} and $\Delta S^{\ddagger} = 0$ cal deg^{-1} mol^{-1} [9, 19]. For the reaction of Co(NH$_3$)$_5$-(4,4'-bipyridyl)$^{3+}$ with Fe(CN)$_5$(OH$_2$)$^{3-}$ both the rate of formation of the precursor complex, Fig. 13.4(c), and the rate of electron transfer within this binuclear species ($k_{25} = 2.6 \times 10^{-3}$ sec^{-1}) can be measured [20]. Other systems for which there is good evidence for a precursor binuclear complex include the reduction of the Co(NH$_3$)$_5$(OH$_2$)$^{3+}$ cation by hexacyanoferrate(II) [21], where presumably preassociation is helped by large opposite charges ($z_A z_B = -12$) on the reactants, and copper(I) reduction of cobalt(III) complexes containing alkene-carboxylate ligands [22] (see copper(I), Chapter 13.4.4).

Site of reductant attack. Many potential bridging ligands offer alternative sites for attack by a reducing (or oxidising) metal ion, as illustrated in Fig. 13.5. Often the reductant's choice of site of attack is affected by two opposing factors, the steric ease of approach to a site remote from the central metal atom of the complex being reduced as against more favourable bonding interaction with a ligand atom adjacent to the central metal. The ease of electron transfer from the site of attack to the central metal atom may also be an important factor. So in the chromium(II) reduction of a cyano-cobalt(III) complex, ease of access favours chromium(II) approach to the cyano-nitrogen. Moreover an electron could easily be transferred from this to the cobalt. However, the thermodynamically favoured product is CrCN^{2+} rather than CrNC^{2+}. In practice a full kinetic study of the chromium(II) reduction of the Co(NH$_3$)$_5$(CN)$^{2+}$ cation has revealed the sequence. There are three kinetically distinct steps, with half-lives of the order of seconds, minutes, and hours respectively. The known rates of isomerisation of CrNC^{2+} to CrCN^{2+}, and of aquation of the latter, permit

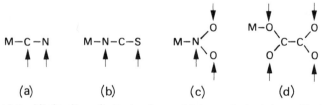

Fig. 13.5   Likely sites of attack of a metal ion reductant (or oxidant) at coordinated ligands.

the last two stages of the redox reaction sequence to be assigned to these processes. Therefore the first, rapid, step must be the actual redox reaction, and must involve interaction of the chromium(II) with the cyano-nitrogen. Hence the full sequence is [23]:

$$Co(NH_3)_5(CN)^{2+} + Cr^{2+} \xrightarrow{5H^+} CrNC^{2+} + Co^{2+}aq + 5NH_4^+$$

$$CrNC^{2+} \longrightarrow CrCN^{2+}$$

$$CrCN^{2+} \xrightarrow{H^+} Cr^{3+}aq + HCN .$$

Similar investigations, evidence, and reasoning suggest that remote attack also predominates in other cases, for example the reduction of $Co(NH_3)_5(NO_2)^{2+}$ by $Co(CN)_5^{3-}$ [24]. However in the chromium(II) reduction of $Co(NH_3)_5(SCN)^{2+}$, both remote (at N) and adjacent (at S) attack takes place [25]. Reduction of this same complex by $Co(CN)_5^{3-}$ takes place almost entirely by attack at the 'adjacent' sulphur atom [26]. In view of the 'soft' nature of the $Co(CN)_5^{3-}$ anion, this result is hardly surprising.

There does not seem to be direct evidence concerning the point of attack at (mono)carboxylato-complexes (Fig. 13.6), though indirect indications favour approach at the uncoordinated oxygen (a) rather than at the coordinated oxygen (b) [27, 28]. Interestingly, steric hindrance to the approach of the reductant has a marked effect. The chromium(II) reduction of $Et_3CCO_2Co(NH_3)_5^{2+}$ is 150 times slower than that of $H_3CCO_2Co(NH_3)_5^{2+}$ [29]. For complexes containing a coordinated dicarboxylato-ligand, a choice of four places for reductant attack exists, Fig. 13.5(d). There is evidence to suggest that the reductant prefers to attack at the carboxylate moiety remote from the cobalt(III), because rate constants for reduction vary greatly with the nature of that part of the ligand molecule between the two carboxylate groups. For instance reduction of succinato-cobalt(III) complexes ($Co^{III}O_2CCH_2CH_2CO_2^-$) takes place at comparable rates to reduction of acetato-complexes. However, when the intervening ligand has a delocalised π-electron pathway for facile electron transfer, as in maleate or

fumarate $(^-O_2CCH=CHCO_2^-)$ or p-phthalate $(^-O_2C-\langle\!\!\!\!\bigcirc\!\!\!\!\rangle-CO_2^-)$ complexes,

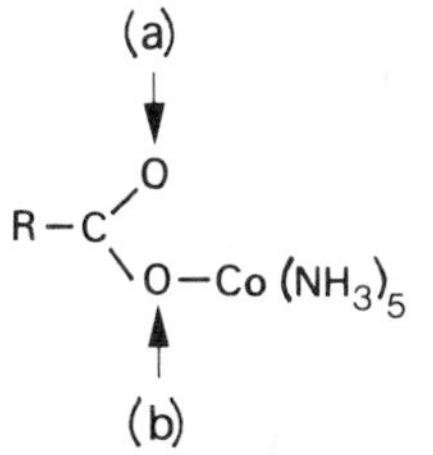

(a)

$$R-C\overset{\displaystyle O}{\underset{\displaystyle O-Co\,(NH_3)_5}{\diagdown}}$$

(b)

Fig. 13.6 Potential points of attack of reductant ions at a carboxylato-cobalt(III) complex.

reduction takes place about a hundred times more rapidly [30]. The situation is less simple than this though, with factors such as the ease of reduction of the ligand itself and the possibility of ligand chelation also needing to be borne in mind. The most effective organic bridging group so far discovered is pyridine-2,4-dicarboxylate. Chromium(II) reducton of its cobalt(III) complex (Fig. 13.7)

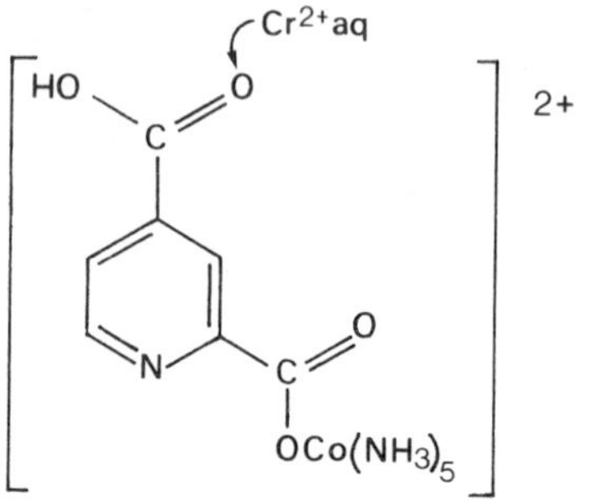

Fig. 13.7 Chromium(II) reduction of a cobalt(III)-pyridine-2,4-dicarboxylate complex.

is nearly 10^5 times faster than that of 'normal' cobalt(III) complexes of carboxylato-ligands. As in previously mentioned cases, the chromium(II) is thought to attack at the remote aldehydic oxygen atom, as indicated in Fig. 13.7 [31]. Remote attack is also thought to be the major route in the reduction of the $Co(NH_3)_5(\text{nicotinamide})^{3+}$ cation and its isonicotinamide analogue

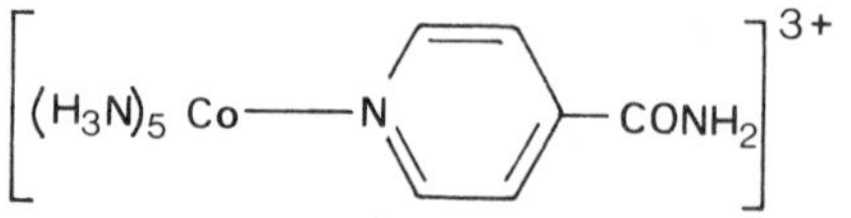

by chromium(II) [32].

Two-stage electron transfer. The foregoing discussion has tacitly assumed that, no matter what the stabilities of the various transition states and intermediates may be, the electron is transferred in one step from one metal ion to the other. There is an alternative to this scheme, in which the electron lingers on the bridging ligand giving a coordinated radical intermediate. About a decade ago it was pointed out that the reducibility of a series of penta-ammine-organic ligand-cobalt(III) complexes correlated with the reducibility of the organic ligands themselves, except for pyridine carboxylate ligands. It was suggested that this deviation could be ascribed to the intermediacy of a tolerably stable radical intermediate [33]. Subsequently, spectroscopic evidence was obtained for a coordinated radical by-product in the chromium(II) reduction of the $Co(NH_3)_5$-$(p\text{-}NO_2C_6H_4CO_2)^{2+}$ cation [29]. The rate constant for the transfer of the electron from the coordinated p-nitrobenzoate radical to the cobalt atom has been measured [34]. Further rate comparisons suggested that chromium(II)

reduction of cobalt(III)-isonicotinamide complexes also involved radical intermediates [35]. Recently the results of a full kinetic and spectroscopic investigation of similar redox reactions involving pyrazine carboxylate and 4-cinnoline
carboxylate ligands have been published, giving full documentation of the coordinated radical intermediates involved [36]. Lest the false impression is given
that chromium(II) is the only reductant which can generate this type of intermediate, the analogous situation in europium(II) reduction of the $Co(NH_3)_5$-
(isonicotinamide)$^{3+}$ cation should be mentioned [37].

The reduction of the nitroprusside anion, $Fe(CN)_5(NO)^{2-}$, to $Fe(CN)_5(NO)^{3-}$
is a one step electron transfer process in solution, but e.s.r. evidence suggests
that this is a two stage reaction when conducted in a methanol glass at $-196°C$.
The e.s.r. results are consistent with the initial transfer of an electron to a
nitrosyl ligand, with subsequent migration of the electron to the iron on an-
nealing the glass [38].

Double bridging in the transition state. When the oxidant (or reductant) contains
two potential bridging ligands, then it is reasonable to enquire whether a double-
bridged transition state or intermediate is feasible. Geometrical considerations
suggest that double bridging could occur for two potential bridging ligands in
cis-positions, but would be unlikely for the analogous *trans*-situation. When
cis-$Co(en)_2(N_3)_2^+$ or its tetra-ammine analogue are reduced by $Cr^{2+}aq$, approximately 1.2 and 1.4 azide ligands per chromium(III) are found in the respective
products. This indicates that double bridging and the consequent transfer of
two azide ligands occurs in a significant proportion of the redox acts [39].
Double bridging also occurs in the similar situation of $Cr^{2+}aq$-catalysed aquation
of *cis*-$Cr(OH_2)_4(N_3)_2^+$ [40].

13.3.4 Outer-sphere Redox Reactions

The simplicity of the outer-sphere mechanism makes its positive identification difficult (Chapter 13.3.2) but permits its theoretical treatment in some
detail. The most convenient feature is the small degree of orbital overlap
between oxidant and reductant. Several chemists have developed theoretical
treatments of outer-sphere redox reactions. Probably the best-known theory is
the version proposed and developed by R. A. Marcus [41]. This theory is fully
discussed, and other contributors acknowledged, in reference [42]. The most
interesting chemical result of Marcus's treatment is the connection between free
energies of activation of redox reactions and their overall free energy changes.
For a series of related reactions, a plot of free energies of activation against free
energies of reaction should be linear (Fig. 13.8) if the reactions are outer-sphere
in character. The other parameters which, in Marcus's theory, determine rates of
outer-sphere redox reactions are the rates of electron transfer between the oxidised and reduced forms of each of the couples which make up the redox
reaction with nett chemical change. Marcus's theory has stimulated much

thought in this area of chemistry, and proved to be of considerable predictive value. However, murmurings of doubt about some of its aspects [43, 44] have been followed by attempts to supersede it by a more rigorous and quantum mechanically based model [45]. Several references to the application of Marcus's theory will be found in subsequent sections of this Chapter (for example pp. 432, 439, 442).

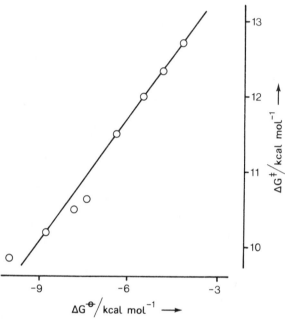

Fig. 13.8 Correlation of Gibbs free energies of activation, $\Delta G^{\ddagger}$, with Gibbs free energies of reaction, $\Delta G^{\ominus}$, for outer-sphere oxidation of substituted tris-1,10-phenanthroline-iron(II) complexes by cerium(IV) in aqueous solution; data taken from Dulz, G. and Sutin, N., *Inorg. Chem.*, **2**, 917 (1963) and Miller, J. D. and Prince, R. H., *J. Chem. Soc. A.* 1370 (1966).

Rates of electron transfer by the outer-sphere mechanism are greatly influenced by several factors. The electron transfer takes place in a much shorter time than that required for atoms to move (Franck-Condon principle). As cation to solvent or ligand distances are different in different cation oxidation states, the transition state geometry must be a compromise between reactant and product geometries. The bond lengthening and shortening involved contribute several kilocalories per mole to the activation energy. A redox reaction involving a change in electron spin angular momentum is formally forbidden, and so will be slow. Outer-sphere redox reactions are particularly sensitive to the presence of other species in the reaction medium. Consequently, cation variation has a marked effect on rates in such reactions as the MnO_4^{2-}/MnO_4^- and $Fe(CN)_6^{4-}/Fe(CN)_6^{3-}$ electron exchange processes.

13.3.5 Inner- vs. Outer- sphere Distinctions

The evidence for the operation of an inner-sphere or outer-sphere mechanism in redox reactions between metal complexes is only definitive in a limited number of systems. When the inertness or lability of the various reactants and products are in precisely the right combination, product characterisation may demonstrate the inner-sphere mechanism. On the other hand, substitution prior to electron transfer may be manifestly implausible and so an outer-sphere mechanism almost certain. However, conditions are not favourable for the direct establishment of mechanism for the majority of redox reactions involving aquo-cations. Resort therefore has to be made to indirect evidence. Several of the lines of approach which have been proposed are mentioned in this section, and all have some more or less unsatisfactory features. Only when several different approaches all lead to the same conclusion can one have a limited degree of confidence in the resulting assignment of mechanism. As will be seen in the section dealing with specific reductants and oxidants, different pieces of indirect evidence are often in conflict. It should also be borne in mind that a simple assignment of mechanism may not be possible, for example in the Cr^{2+}aq reduction of the $IrCl_6^{3-}$ anion where there are parallel inner- and outer- sphere pathways [12].

pH-Dependence of rates. This possible means of discriminating between inner- and outer- sphere mechanisms applies specifically to aquo- and mixed ligand aquo- complexes, and is therefore particularly relevant to this book. A marked dependence of the rate of reaction on reciprocal hydrogen ion concentration is generally observed for redox reactions of known inner-sphere mechanism, $Co(NH_3)_5(OH)^{2+} + Cr^{2+}$aq among them. The pH dependence of known outer-sphere processes is very much smaller. The reason for this difference in behaviour lies in the excellent bridging properties of a hydroxo-ligand. The water molecule is a poor bridging ligand whose presence is associated with outer-sphere electron transfer.

The pH dependence of rates can also indicate the route of electron transfer from a mixed ligand complex. The strongly pH dependent term in the rate law for the chromium(II) reduction of tetrakis-(4-N-methylpyridyl)porphinecobalt-(III) suggests inner-sphere reduction via hydroxo-bridges rather than outer-sphere reduction with electron transfer from the porphyrin periphery [46].

Activation volumes. The expected volume of activation for an outer-sphere redox reaction can be estimated with some confidence [47]. For the necessarily outer-sphere electron exchange between $Co(en)_3^{2+}$ and $Co(en)_3^{3+}$, the estimated $\Delta V^{\ddagger}$ is -18.4 cm^3 mol^{-1}, and the measured $\Delta V^{\ddagger}$ (at 1 atmosphere, 25°C) is -19.8 ± 1.5 cm^3 mol^{-1}. Similarly, for the almost certainly outer-sphere electron exchange between $Fe(OH_2)_6^{2+}$ and $Fe(OH_2)_6^{3+}$, $\Delta V^{\ddagger}$ (estimated) is -14.4 cm^3 mol^{-1} and $\Delta V^{\ddagger}$(experimental, 2°C) is -12.2 ± 1.5 cm^3 mol^{-1}. However, for the electron exchange between $Fe(OH_2)_6^{2+}$ and $Fe(OH_2)_5(OH)^{2+}$, the experimentally determined activation volume of $+0.8 \pm 0.9$ cm^3 mol^{-1} is far from the $\Delta V^{\ddagger}$ of -11.4 cm^3 mol^{-1} calculated for an outer-sphere process. Again, the electron exchange between Cr^{2+}aq and Cr^{3+}aq in fact proceeds mainly via a hydroxo-bridged inner-sphere transition state, and has a measured activation volume of $+4.2 \pm 1.1$ cm^3 mol^{-1}, very different from the estimated value of -11.6 for an outer-sphere mechanism here.

The usefulness of this criterion of redox mechanism can be illustrated by its application to the iron(II) reduction of the cobalt(III) cations $Co(NH_3)_5X^{2+}$. Calculated activation volumes for outer-sphere processes lie between -10.6 and -12.8 cm^3 mol^{-1} for X = F, Cl, Br, or N_3. The actual volumes of activation, which are +11, +8, +8, and +14 cm^3 mol^{-1} respectively, strongly suggest an inner-sphere mechanism for each of these reactions [47].

Oxygen isotopic fractionation factor. This is the ratio between the rates of reaction of $^{16}OH_2$ and $^{18}OH_2$ (or $^{16}OH^-$ and $^{18}OH^-$) complexes. Only an inner-sphere mechanism involves some marked change in bonding to the ligating oxygen, so this ratio could be significantly further from unity for an inner-sphere than for an outer-sphere reaction. As usual, the likely viability of this diagnostic technique was first tested by looking at reactions of known mechanism. Values of the oxygen isotopic fractionation factor, $f = $ (rate for ^{16}O system)/ (rate for ^{18}O system), for such reactions are given in Table 13.2. Though the differences between f values for inner- and outer- sphere reactions look small, the proposers of this method believe them to be real and significant. Values of f of 1.020 for V^{2+}aq reduction and of 1.019 for Eu^{2+} reduction of $Co(NH_3)_5$-$(OH_2)^{3+}$ are similar to the f values for established outer-sphere reactions (Table 13.2). An outer-sphere mechanism was therefore suggested for these V^{2+}aq and Eu^{2+}aq reductions [48].

Table 13.2 Oxygen isotopic fractionation factors, f, for inner- and outer- sphere redox reactions [48].

Mechanism	Reaction	f
Outer-sphere	$Co(NH_3)_5(OH_2)^{3+} + Ru(NH_3)_6^{2+}$	1.021
	$Co(NH_3)_5(OH)^{2+} + Ru(NH_3)_6^{2+}$	1.017
Inner-sphere	$Co(NH_3)_5(OH)^{2+} + Cr^{2+}$aq	1.046

Azide–thiocyanate comparisons. If the predominant route for inner-sphere redox reactions is through reductant attack at the remote end of the potentially bridging ligand, then such reductant approach may well generate the thermodynamically less favoured linkage isomer of a product (see p. 424),

$$M-N-C-S \rightarrow [M-N-C-S-M']^{\ddagger} \rightarrow M + N-C-S-M' \rightarrow M'-N-C-S .$$
$$\underset{M'}{\overset{\uparrow}{}}$$

In the special case of the azide ligand as bridging group there is no alternative product, so inner-sphere reactions may be appreciably easier for azide complexes than for their thiocyanate analogues. Outer-sphere electron transfer rates should be much less sensitive to azide–thiocyanate variation. The ratio of the rate constant for reduction of an azide complex to that for reduction of the analogous thiocyanate complex, $k(N_3^-)/k(NCS^-)$, is therefore potentially diagnostic of mechanism [8]. Table 13.3 shows how different this ratio is for reactions of established inner- or outer- sphere mechanism. This pattern only applies to 'hard' reductants which prefer to bond to nitrogen rather than to sulphur. $k(N_3^-)/k(NCS^-)$ for the reduction of the $Co(NH_3)_5(N_3)^{2+}/Co(NH_3)_5$-$(NCS)^{2+}$ pair of complexes by the 'soft' sulphur-liking anion $Co(CN)_5^{3-}$ is only 1.6 despite the undoubtedly inner-sphere nature of this reaction.

Table 13.3 **Ratios of rate constants for reduction of pairs of azide and thiocyanate complexes (at 25°C; [8]).**

Mechanism	Reaction	$k(N_3^-)/k(NCS^-)$
Outer-sphere	$Co(NH_3)_5X^{2+} + Cr(bipy)_3^{2+}$	4
	$FeX^{2+} + V^{2+}aq$	~1
Inner-sphere	$Co(NH_3)_5X^{2+} + Cr^{2+}aq$	2×10^4
	$CrX^{2+} + Cr^{2+}aq$	4×10^4

This method for assigning mechanisms often gives equivocal answers. Values of $k(N_3^-)/k(NCS^-)$ of about 30 to 40 for vanadium(II) reductions of $Co(NH_3)_5X^{2+}$ and for uranium(III) reductions of $Cr(OH_2)_5X^{2+}$ require explanation rather than define mechanisms [8]. On the other hand, $k(N_3^-)/k(NCS^-)$ ratios for some redox reactions do give a firmer indication of mechanism. For example, ratios of about 6×10^4 for $U^{3+}aq$ reduction [49], $> 3 \times 10^3$ for $Fe^{2+}aq$ reduction [50], and 300 for $Eu^{2+}aq$ reduction [51] of the $Co(NH_3)_5X^{2+}$ pair of complexes strongly suggest an inner-sphere mechanism in all these cases.

The use of the ratio $k(SCN^-)/k(NCS^-)$ in assigning mechanisms of redox reactions has recently been suggested as an alternative to $k(N_3^-)/k(NCS^-)$ [52]. The thiocyanate:isothiocyanate ratio would be of less general applicability in view of the difficulties of preparing both isomers in many instances.

Other methods. Marked catalysis by added ions has been reported for a variety of reactions known, or at least confidently believed, to have an outer-sphere mechanism. Effects of added anions on known inner-sphere redox reactions are much smaller. Hence an examination of the dependence of redox rate on added anions, such as chloride, has been used as an indication of mechanism,[†] for instance in the cases of Yb^{2+}aq [53] and U^{3+}aq [49] reductions.

The ratio of rate constants for the reduction of a given species by Cr^{2+}aq and by V^{2+}aq has also been proposed as a way of distinguishing between the operation of inner- and outer- sphere mechanisms. Values of $k(Cr^{2+}aq)/k(V^{2+}aq)$ of around 0.02 have been reported for a number of outer-sphere reductions [54]. Plots of rates of reduction by V^{2+}aq against the respective rates of reduction by Cr^{2+}aq are linear for a series of outer-sphere processes and for a series of inner-sphere processes. The slope of the correlation line is 1.10 for the former (the Marcus theory predicts 1.00), 0.40 for the latter [55]. This approach has also been applied to reductions of bridged dicobalt complexes, where values of 0.021 to 0.024 for $k(Cr^{2+}aq)/k(V^{2+}aq)$ suggest outer-sphere reactions [56].

Finally one should mention the time-honoured method of comparisons of rate trends. The trend of increasing reactivity for reduction of a series of complexes by Cr^{2+}aq is often the opposite to that for reduction of the same series of complexes by $Cr(bipy)_3^{2+}$. If this is the case, one only has to establish the direction of the reactivity trend for the same series of complexes reacting with a reducing agent of unknown mechanism of reduction to guess this unknown mechanism. If the trend parallels that for reduction by Cr^{2+}aq, an inner-sphere mechanism is indicated, while if the trend parallels that for reduction by $Cr(bipy)_3^{2+}$ an outer-sphere mechanism is assumed. This has proved a popular method for many years; examples of its use will be quoted in following sections. There are sometimes drawbacks and difficulties with the use of this diagnostic method, so the indications should be treated with considerable caution if there is no supporting evidence from some other independent source [57].

The question of inner- as opposed to outer- sphere mechanism distinction also crops up in considering the transfer of electrons between transition metal complexes and electrodes [58]. Inner-sphere electron transfer is suggested in electrodic oxidation of Cr^{2+}aq, and between certain isothiocyanato-transition metal complexes and a mercury electrode.

13.3.6 Two-electron Transfers

The majority of electron transfer reactions between complexes involve the transfer of one electron at a time. For many years evidence has been sought for the simultaneous transfer of two (or more) electrons from reductant to oxi-

---

†However the addition of chloride can lead to serious complications in the interpretation of observed rates and kinetics, as has been illustrated recently in a series of studies of thallium(III) oxidations by Thakuria, B. J. and Gupta, Y. K. *J. Chem. Soc. Dalton* 77, 2541 (1975), and by Gupta, K. S. and Gupta, Y. K. *Inorg. Chem.* 14, 2000 (1975).

dant [59]. Obvious areas in which to conduct this search are in the chemistry of the non-transition elements, or of platinum, where stable oxidation states are two units apart. A much investigated reaction has been the Tl^+/Tl^{3+} electron exchange, some details of whose mechanism still await clarification. The determined activation volume [47] is said to indicate an outer-sphere one electron transfer rate-determining step. However, a recent comparison of this electron exchange under thermal and photochemical conditions suggested that while Tl^{2+} may be involved in the latter, simultaneous two electron transfer is indicated in the former [60, 61]. Simultaneous two electron transfer steps have also been suggested in the Cr^{2+}aq reduction of the $Pt(NH_3)_5Cl^{3+}$ cation [62] and the V^{2+}aq reduction of Hg^{2+}aq [63]. The latter system involves a one-electron transfer path in parallel with the two-electron transfer path.

13.3.7 Non-complementary Redox Reactions

Many sp-block elements have stable cations differing by two units of charge, while many transition elements have stable cations differing by only one unit of charge. An example of the former is thallium, whose stable oxidation states in aqueous media are thallium(I) and thallium(III). An example of the latter is iron, with Fe^{2+}aq and Fe^{3+}aq as its water stable cations. The redox reaction between thallium and iron in aqueous solution,

$$2Fe^{2+}aq + Tl^{3+}aq \longrightarrow 2Fe^{3+}aq + Tl^+aq,$$

can proceed by an initial one-electron or two-electron transfer. Either way one unstable cation is produced, Tl^{2+}aq or Fe^{4+}aq respectively. In this and analogous cases there is no way of avoiding the generation of at least one unstable intermediate. In the case of the iron(II)–thallium(III) reaction, the observation that the addition of Fe^{3+}aq to reaction mixtures reduces the rate of the redox reaction suggests [64] an initial reversible one-electron transfer,

$$Fe^{2+}aq + Tl^{3+}aq \rightleftharpoons Fe^{3+}aq + Tl^{2+}aq,$$

followed by the rapid reaction of the Tl^{2+}aq generated,

$$Fe^{2+}aq + Tl^{2+}aq \longrightarrow Fe^{3+}aq + Tl^+aq.$$

The rate constant for this latter step has recently been measured directly [61]. The addition of Tl^+aq to reaction mixtures does not retard the redox reaction, which rules out the analogous reversible two-electron transfer as initial process,

$$Fe^{2+}aq + Tl^{3+}aq \rightleftharpoons Fe^{4+}aq + Tl^+aq.$$

Such a sequence of one-electron steps is common for thallium(III) oxidations of transition metal cations, though there is evidence for two-electron transfer in its reaction with chromium(II). In this system the chromium(III) product is a hydroxo-bridged binuclear species, which is likely to have been formed by reaction of $Cr^{2+}aq$ with a chromium(IV) species such as $Cr(OH)^{3+}$ [65].

The converse type of reaction, where thallium(I) is oxidised by a transition metal cation, is also believed to involve the intermediacy of thallium(II). An example of this is afforded by its oxidation in cobalt(III) sulphate solution [66].

Consecutive one-electron transfers have also been demonstrated in, for example, the reactions of tin(II) and copper(II) [67] and of antimony(III) and cerium(IV) [68]. The three-electron reductions of chromate by iron(II) or by vanadium(III) are thought to proceed by three one-electron steps [69]. Many more non-complementary redox reactions have had their mechanisms elucidated. A number of early studies are described in reference [70], and later examples can be traced through reference [71].

13.3.8 Isotopic Exchange and Electron Exchange

These, first qualitatively characterised in 1920 [72], form a special class of redox reactions [73]. Isotopic exchange involves only electron transfer between different oxidation states of the same metal in a constant environment [74], for example

$$Fe^{2+}aq + *Fe^{3+}aq \longrightarrow Fe^{3+}aq + *Fe^{2+}aq .$$

Water molecules are reluctant to act as bridging ligands for inner-sphere electron transfer [54], so reactions involving 1+ and 2+ cations only will go by the outer-sphere mechanism, for instance the Cu^+/Cu^{2+} electron exchange, monitored by ^{63}Cu n.m.r. [75]. However, if a 3+ or 4+ aquo-cation is involved hydroxo-species will be in equilibrium with the aquo-cation, and inner-sphere electron transfer by way of a hydroxo-bridged transition state is a likely mechanism. Its operation would be suggested by the appearance of a term in $[H^+]^{-1}$ in the rate law for the exchange reaction.

Kinetic parameters for some isotopic exchange reactions [73, 76–82] between aquo-cations are listed in Table 13.4. The majority of isotopic exchange reactions involve the transfer of a single electron. The only much-studied two-electron transfer is the thallium(I)/(III) reaction,† already mentioned in Chapter 13.3.6. Present opinion [60, 61] is that, at least under thermal conditions, this is a true two-electron transfer. This thallium(I)/(III) reaction has been a popular subject of study for many years. The effects of added anions, which affect rates via interacting with the $Tl^{3+}aq$, have been established. Added transition metal cations,

---

†However, it should be recalled that the first qualitative study of this type of reaction dealt with lead(II)/lead(IV) exchange, in water, pyridine, alcohols, and acetic acid [72].

such as $Fe^{2+}aq$ [83] or $Ce^{4+}aq$ [84], affect rates by acting as one-electron transfer mediators.[†] Similar one-electron transfer mediation has been reported when $Fe^{3+}aq$ is added to the uranium(IV)/(VI) exchange [85]. The implied change of mechanism from two-electron to one-electron transfer in the presence of such transition metal cations is not without precedent in other areas of chemistry. For example, in reactions of Grignard reagents with benzophenone the presence of traces of transition metal impurities in the magnesium metal used for Grignard reagent preparation can alter the reaction course from a polar (two electron) to a single electron transfer pathway [86].

As mentioned at the start of this Chapter, kinetic studies of redox reactions in solvents other than water are rare. Indeed, it is only in this small area of simple isotopic exchange reactions that there has been any real effort to investigate the effects of solvent composition on rates. Reactions such as the above-mentioned thallium(I)/(III) [87] and uranium(IV)/(VI) [88] have been studied in mixed aqueous solvents, as has the iron(II)/(III) reaction [89]. Electron transfer seems to grow markedly slower when water of solvation is replaced by alcohols or by dimethyl sulphoxide. The iron(II)/(III) exchange is about 10^8 times slower in isopropanol than in water. Presumably electron travel across these bulky, saturated organic molecules is more difficult than across solvating water molecules. The last three sentences apply to reactions between fully solvated metal cations, but not necessarily to reactions of the type $Fe(NCS)^{2+}$ + Fe^{2+}, where there is a potential bridging ligand.

Isotopic exchange reactions involving hydroxo-cations (Table 13.4, right-hand side) and those involving oxocations (Table 13.5) [90–94] form special cases. Here the ligating atoms are the same in the two oxidation states, but there are differences in the number of protons attached to them. Electron exchange reactions involving two complexes of the same metal in different oxidation states (and different ligand environments) form a class intermediate between the above isotopic exchange reactions and the bulk of redox reactions discussed in the present Chapter. Their kinetics and mechanisms have been extensively investigated [71]. Examples will be included where relevant in the following sections.

While the majority of studies of electron exchange kinetics have been primarily of academic interest, the uranium(IV)/(VI) reaction has been examined with a view to its possible utilisation as a means of $^{235}U/^{238}U$ separation. However, slow rates and small single stage separation factors are thought to prevent this becoming an economically viable alternative to the established diffusion and centrifugation methods [95].

[†] Similar electron transfer mediation, by the $Ta_6Cl_{12}^{3+}$ cluster cation, has recently been reported for the non-complementary redox reaction between iron(III) and tin(II) (Haynes, D. M. and Higginson, W. C. E. *J. Chem. Soc. Dalton* 309 (1976)).

Table 13.4 Kinetic parameters for isotopic exchange reactions between pairs of aquo-cations in aqueous solution.

	k_{25} /M^{-1}sec^{-1}	$\Delta H^{\ddagger}$ /kcal mol^{-1}	$\Delta S^{\ddagger}$ /cal deg^{-1}mol^{-1}	k_{25} /M^{-1}sec^{-1}	$\Delta H^{\ddagger}$ /kcal mol^{-1}	$\Delta S^{\ddagger}$ /cal deg^{-1}mol^{-1}	
	— M^{n+} + $M^{(n+1)+}$ —			— M^{n+} + MOH^{n+} —			
V^{2+}/V^{3+}	1.0×10^{-2}	12.6	−25	~1.8			[73]
Cr^{2+}/Cr^{3+}	2×10^{-5}	(21)	(−8)	0.7			[73]
Mn^{2+}/Mn^{3+}	~10^{-4}						[76]
Fe^{2+}/Fe^{3+}	3.3(21.6°C)	9.3	−25	2700(21.6°C)	6.8^b	−18^b	[77]
Co^{2+}/Co^{3+}	2.38 (18.35°C)	10.4	−21	325(18.35°C)	8.0	−24	[79]
Ce^{3+}/Ce^{4+}	a			a			[80]
Pu^{3+}/Pu^{4+}				2.0×10^4	2.2	−32	[81]
	— M^+ + M^{3+} —			— M^+ + MOH^{2+} —			
Tl^+/Tl^{3+}	7.0×10^{-5}	17.4	−21	2.5×10^{-5}			[82]

a It was not possible to report second-order rate constants, since the cerium(IV) species involved included both binuclear and hydroxo- species, even in perchloric acid; b values of 11.5 kcal mol^{-1} and −4.0 cal deg^{-1} mol^{-1} have been reported for $\Delta H^{\ddagger}$ and $\Delta S^{\ddagger}$ for the electron exchange reaction between Fe^{2+} and $Fe(OD)^{2+}$ in D_2O [78].

Table 13.5 Kinetic parameters for electron exchange reactions between oxo-cations and aquo-cations in aqueous solution.

	k_{25} /M^{-1}sec^{-1}	$\Delta H^{\ddagger}$ /kcal mol^{-1}	$\Delta S^{\ddagger}$ /cal deg^{-1}mol^{-1}	
$VO^{2+} + V^{2+}$	a	12.3	-16.5	[90]
$VO^{2+} + V(OH)^{2+}$	1.0	10.7	-24	[73]
$VO_2^+ + V^{2+\,b}$	2.58	1.9	-36.8	[91]
$VO_2^+ + V^{3+\,b}$	30.1	8.3	$+24.3$	[73]
$VO^{2+} + VO_2^+$	fast			[73]
$UO_2^{2+} + U^{3+}$	5.5×10^4	4.3	-22.3	[92]
$NpO_2^{2+} + Np^{3+}$	1.05×10^5	1.0	-32	[92]
$PuO_2^{2+} + Pu^{3+}$	2.7	4.8	-40.4	[93]
$PuO_2^{2+} + Pu^{4+}$	37	13.6	-6	[93]
$NpO_2^{2+} + NpO_2^+$		10.2	-12.6	[94]

a $k_{25} = 6.7 \times 10^{-2}$ M^{-1}sec^{-1} for the inner-sphere path and 3.9×10^{-2} M^{-1}sec^{-1} for the outer-sphere path (at $0°$C); b these reactions also go by other paths in parallel (see cited refs. for details).

Table 13.6 Kinetic parameters for chromium(II) reductions of complexes: from reference [99] except where otherwise stated.

	k_{25} /M^{-1}sec^{-1}	$\Delta H^{\ddagger}$ /kcal mol^{-1}	$\Delta S^{\ddagger}$ /cal deg^{-1}mol^{-1}	
$Fe(OH_2)_6^{3+}$	2.3×10^3	5.2	-28	[100]
$Fe(OH_2)_5(OH)^{2+}$	3.3×10^6	4.6	-13	[100]
$Co(OH_2)_6^{3+}$	1.25×10^4	9.5	-7.8	[101]
$Co(OH_2)_5(OH)^{2+}$	6.59×10^3	12.7	$+1.7$	[101]
$Co(NH_3)_6^{3+}$	8.9×10^{-5}			
$Co(NH_3)_5(OH_2)^{3+}$	0.5^a	2.9	-52	[102]
$Co(NH_3)_5(OH)^{2+}$	1.5×10^6			
$Co(NH_3)_5F^{2+}$	2.5×10^5			
$Co(NH_3)_5Cl^{2+}$	6×10^5			
$Co(NH_3)_5Br^{2+}$	1.4×10^6			
$Co(NH_3)_5I^{2+}$	3×10^6			
$Co(NH_3)_5(OAc)^{2+}$	0.35	8.2	-33	
$Co(NH_3)_5(maleateH)^{2+}$	180			
$Co(NH_3)_5(PO_4)$	4.8×10^9			
UO_2^{2+}aq	1.47×10^4	0.79	-36.8	[103]

a $20°$C.

13.4 REDUCTION OF COMPLEXES BY SPECIFIC AQUO-METAL CATIONS

13.4.1 Chromium(II)

This is the most studied and best documented reductant, for reasons discussed earlier. Reduction by Cr^{2+}aq has already been mentioned several times in this Chapter. Reviews and extensive tabulation of kinetic data can be found in references [29. 71, 96–98]. Selected values are listed in Table 13.6 [99–103] to give an idea of the range of reactivities exhibited. Aquo-chromium(II) is very labile, and acts as an inner-sphere reductant towards almost all reducible complexes containing potentially bridging ligands. For outer-sphere reductions by this cation, one has to turn to complexes containing such ligands as saturated Group V bases. The $Co(NH_3)_6^{3+}$ cation is the classic example here. The reduction of ferricytochrome C by chromium(II) can occur by an outer-sphere mechanism, but it is inner-sphere in the presence of chloride or thiocyanate [104].

13.4.2 Vanadium Cations

Vanadium(II). A selection of kinetic data for the vanadium(II) reduction of a variety of inorganic aquo-cations and complexes is given in Table 13.7 [111–119]. Here one has the interesting situation that many redox rates are of the same order of magnitude as water exchange rates. In some reductions, product identification indicates an inner-sphere mechanism. Examples include reductions of $Co(CN)_5(N_3)^{3-}$ and $Co(CN)_5(SCN)^{3-}$ [105]. In these cases the VN_3^{2+} and $VNCS^{2+}$ cations, and the $Co(CN)_5^{3-}$ anion, have been positively identified amongst the products. Moreover the rate of reduction of the thiocyanato-complexes approximates to the rate of substitution of thiocyanate ion at V^{2+}aq, and so to the rate of water exchange at this cation. Other reductions which are inner-sphere in mechanism include those of α-carbonylcarboxylate cobalt(III) complexes [106], of bridged binuclear cobalt(III) complexes [107] such as,

$$\left[(H_3N)_4Co \underset{SO_4}{\overset{NH_2}{<}} Co(NH_3)_4 \right]^{3+},$$

and of mercury(II). In the latter case two-electron transfer appears to occur in a $V-OH-Hg^{3+}$ transition state [63]. There is kinetic evidence for a transient intermediate in the vanadium(II) reduction of nicotinic acid and isonicotinic acid complexes $Co(NH_3)_5L^{3+}$ at low acidities [108].

Rates of reduction of cobalt(III) complexes $Co(NH_3)_5X^{n+}$ by V^{2+}aq and by $Cr(bipy)_3^{2+}$ correlate linearly (Fig. 13.9) for $X = Cl^-$, Br^-, I^-, NH_3, and OH_2. This suggests an outer-sphere mechanism for all this series of redox reactions. However, the rates of reduction of the $Co(NH_3)_5F^{2+}$ cation do not fit this correlation, which suggests an inner-sphere mechanism for the V^{2+}aq reduction of

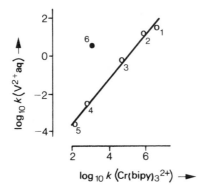

Fig. 13.9 Plot of rate constants for reduction of cobalt(III) complexes by V^{2+}aq and by $Cr(bipy)_3^{2+}$; mechanism assignment ○ outer-sphere and ● inner-sphere. The cobalt(III) complexes are 1 $Co(NH_3)_5Br^{2+}$, 2 $Co(NH_3)_5Cl^{2+}$, 3 $Co(NH_3)_5(OH_2)^{3+}$, 4 $Co(NH_3)_6^{3+}$, 5 $Co(en)_3^{3+}$, 6 $Co(NH_3)_5F^{2+}$ [109].

this complex. It is supported by the marked difference between the activation parameters for the reduction of this complex and for the reduction of the other $Co(NH_3)_5X^{2+}$ listed above [109]. Vanadium(II) reductions of cobaloximes, $Co(dmgH)_2LX$, exhibit an interesting control of mechanism by the ligands present, with the operation of inner- or outer- sphere electron transfer determined by the natures of L and X [110].

The rate of reduction of UO_2^{2+}aq by V^{2+}aq, which is $12 \, M^{-1} sec^{-1}$ at 25°C and zero ionic strength, is similar to that for water exchange at the V^{2+}aq cation. However this seems to be a coincidence, giving a misleading indication of an inner-sphere redox mechanism. The activation parameters for the redox reaction, $\Delta H^{\ddagger} = 7.1 \, kcal \, mol^{-1}$ and $\Delta S^{\ddagger} = -26 \, cal \, deg^{-1} mol^{-1}$, are very different from the activation parameters for water exchange, which are $\Delta H^{\ddagger} = 16.4 \, kcal \, mol^{-1}$ and $\Delta S^{\ddagger} = +5 \, cal \, deg^{-1} mol^{-1}$. The likely mechanism for the redox reaction is therefore outer-sphere [111]. The situation with respect to NpO_2^{2+}aq is very similar [112]. The rate of reduction of iron(III) is considerably greater than the rate of substitution at V^{2+}aq, clearly indicating an outer-sphere mechanism here too [113]. The reduction of $Ru(NH_3)_6^{3+}$ by V^{2+}aq must also be outer-sphere. Again the activation parameters (Table 13.7) are very different from those for water exchange at V^{2+}aq (see above) [114]. Even when there are potential bridging ligands present, such as halides, the V^{2+}aq reduction of most ruthenium(III)–amine–ligand complexes goes by the outer-sphere mechanism. Similarly, the V^{2+}aq reduction of the $PtCl_6^{2-}$ and $Pt(NH_3)_5Cl^{3+}$ ions seems to be outer-sphere in character [120].

Outer-sphere reductions by the V^{2+}aq cation conform well to the predictions of the Marcus theory. A plot of the free energy of activation against free energy of reaction has a slope of 0.56, with an estimated error of ±0.08. This is well within tolerance of the expected slope of 0.5 [44]. The use of

Table 13.7 Kinetic parameters for vanadium(II) reduction of complexes; from [99] except where otherwise stated.

	k_{25} /M^{-1} sec^{-1}	$\Delta H^{\ddagger}$ /kcal mol^{-1}	$\Delta S^{\ddagger}$ /cal deg^{-1} mol^{-1}	
Fe^{3+}aq	1.8×10^4			[113]
$Fe(OH)^{2+}$	$<4 \times 10^5$			[113]
$FeCl^{2+}$	4.6×10^5			
$RuCl^{2+}$	1.9×10^3			
$Ru(NH_3)_5Cl^{2+}$	3.0×10^3	3.0	-30	
$Ru(NH_3)_5Br^{2+}$	5.1×10^3	2.8	-34	
$Ru(NH_3)_5py^{3+}$	1.2×10^5			
$Ru(NH_3)_6^{3+}$	1.3×10^3	0.6	-42	[114]
$Co(NH_3)_6^{3+}$	3.7×10^{-3}			
$Co(NH_3)_5(OH_2)^{3+}$	0.53	8.2	-32	[115]
$Co(NH_3)_5F^{2+}$	2.6			
$Co(NH_3)_5Cl^{2+}$	7.6	7.4	-30	[116]
$Co(NH_3)_5Br^{2+}$	25	9.1	-22	
$Co(NH_3)_5I^{2+}$	120			
$Co(NH_3)_5(N_3)^{2+}$	13	11.7	-14	[51]
$Co(NH_3)_5(OAc)^{2+}$	1.2	11.6	-19	
$Co(NH_3)_5(PO_4)$	1.4×10^7			
$Co(CN)_5(N_3)^{3-}$	110			
$IrCl_6^{2-}$	$>4 \times 10^6$			[117]
Cu^{2+}aq	26.6	11.4	-13.8	[118]
Hg^{2+}aq^a	1.04	15.8	-5.6	[63, 119]
Hg^{2+}aq^b	8.69	14.8	-4.6	[63, 119]
Np^{4+}aq	2.53	9.7	-24.0	[112]
UO_2^{2+}aq	12^c	6.7^c	-34^c	[111]

a One-electron transfer path; *b* two-electron transfer path; *c* these values have been estimated for zero ionic strength from rates determined at various ionic strengths up to approximately 2M.

comparisons between rates of reduction by V^{2+}aq and by Cr^{2+}aq as a means of distinguishing between inner-sphere and outer-sphere electron transfer has been mentioned in an earlier section (Chapter 13.3.5). Polyelectrolyte acceleration of redox reactions in aqueous solution has been demonstrated for reductions both by V^{2+}aq and by Cr^{2+}aq [121].

Vanadium(III) and vanadium(IV). Less information is available concerning the kinetics and mechanisms of redox reactions involving these reductants (Table 13.8 [122-128]). Interestingly, rates of reduction by VO^{2+}aq parallel those for reduction by Fe^{2+}aq. Reaction rates are around 100 times slower for the weaker reductant, VO^{2+}aq [125].

Table 13.8 Kinetic parameters for vanadium(III)^a and vanadium(IV) reductions of complexes.

	$\log_{10} k_{20}{}^b$	$\Delta H^{\ddagger}$ /kcal mol^{-1}	$\Delta S^{\ddagger}$ /cal deg^{-1}mol^{-1}	
Vanadium(III)				
Hg^{2+}aq + V^{3+}aq	-2.2^c			[123]
Cu^{2+}aq + V^{3+}aq	-2.2	21.3	$+5$	[124]
Cu^{2+}aq + $V(OH)^{2+}$aq	-2.2	9.4	-16	
Vanadium(IV)				
Mn^{3+}aq	2.36	11.1^d	-12^d	
Fe^{3+}aq	~-1.2			
Co^{3+}aq	0.32			[125]
Tl^{3+}aq	~-3.9			
Ag^{2+}aq	3.7			
Ce^{4+}aq	3.0^e			[126]
$Ce(OH)^{3+}$aq	1.6	16.8	$+6$	[127]
UO_2^+aq	-3.7^e			[125]
NpO_2^{3+}aq	~3.2	6.7		[128]

a The kinetics of reduction of Fe^{3+}aq and of Co^{3+}aq by V^{3+}aq are complicated [122]; *b* units of k are M^{-1}sec^{-1}; *c* at 35°C; *d* $\Delta H^{\ddagger}$ and $\Delta S^{\ddagger}$ values for $Mn(OH)^{2+}$ + VO^{2+} depend on the values assumed for $\Delta H^{\oplus}$ and $\Delta S^{\oplus}$ for the Mn^{3+}aq $\rightleftharpoons$ $Mn(OH)^{2+}$aq + H$^+$ equilibrium, for details see Rosseinsky, D. R. and Nicol, M. J., *J. Chem. Soc. A*, 1022 (1968); *e* at 25°C.

13.4.3 Iron(II)

Kinetic parameters for some iron(II) reductions are listed in Table 13.9 [129-132]. Iron(III) reaction products are fairly labile, but persist long enough in some cases for their characterisation to be possible. Thus the detection of $FeNCS^{2+}$ as initial product of the reaction between $Co(NH_3)_5(SCN)^{2+}$ and Fe^{2+}aq indicates an inner-sphere mechanism. This reaction is particularly suitable for the demonstration of an inner-sphere mechanism, because the affinity of the Fe^{2+} for the remote nitrogen atom of the –SCN ligand encourages rapid redox reaction so that the redox step is much faster than the subsequent dissociation of the $FeNCS^{2+}$ generated [52]. Inner-sphere mechanisms seem likely for many reactions involving $Fe^{2+}aq$ as reductant, but the reductions of $Co^{3+}aq$ and of $Tl^{2+}aq$ are too fast compared with water exchange rates for the operation of an inner-sphere mechanism here. In these cases the mechanism must therefore be outer-sphere [60, 130].

Table 13.9   Kinetic parameters for iron(II) reductions of complexes.

	k_{25} /M^{-1}sec^{-1}	$\Delta H^{\ddagger}$ /kcal mol^{-1}	$\Delta S^{\ddagger}$ /cal deg^{-1}mol^{-1}	
$Tl(OH)^{2+}aq$	1.4×10^{-2}	17.8	-7	[129]
$Co^{3+}aq$	2.25×10^2	15.2	$+4.7$	} [130]
$Co(OH)^{2+}aq$	3.85×10^2	15.8	$+8.2$	
$Co(NH_3)_5F^{2+}$	6.6×10^{-3}	13.7		
$Co(NH_3)_5Cl^{2+}$	1.35×10^{-3}	12.5	-30	
$Co(NH_3)_5Br^{2+}$	7.26×10^{-4}	13.3		} [99, 131]
$Co(NH_3)_5(N_3)^{2+}$	8.8×10^{-3}			
$Co(NH_3)_5(OAc)^{2+}$	5×10^{-2}			
PuO_2^{2+a}	5×10^3	8.6	-13	} [132]
PuO_2^{2+b}	1×10^3	4.4	-30	

a Inner-sphere pathway; b outer-sphere pathway.

In the iron(II) reduction of the complex $Co(NH_3)_5(nta)$, the binuclear Co-carboxylato-Fe species is not just an ephemeral transition state but an intermediate of half-life about 10 seconds (at $25°C$). The structure suggested for this intermediate has already been illustrated in Fig. 13.4(b) (p. 423) [19].

One would expect that iron(II) reductions of complexes $M(phen)_3^{3+}$, $M(bipy)_3^{3+}$, and $M(terpy)_2^{3+}$, with M = Fe or Ru, would go by the outer-sphere mechanism in view of the known substitution inertness of these Fe(III) and Ru(III) species. These reactions have small or even negative activation enthalpies, tentatively ascribed to pre-association of the reactants prior to the electron transfer step. Normal electrostatic ion association between the reactants is obviously ruled out by the fact that both are positively charged, but association is possible between the $Fe^{2+}aq$ and anions necessarily present. The authors suggest that the reductant may be accommodated within the pockets formed by the large flat heteroaromatic ligands [133]. The inclusion of these redox reactions within the normal Marcus correlations for outer-sphere redox processes has been considered in detail [134].

13.4.4 Other Transition Metal Cations

Kinetic studies of reductions by other transition metal aquo-cations are scattered throughout the chemical literature. Recent work can be traced fairly painlessly through the appropriate sections of references [71] and [99], while older studies are cited in standard texts [2, 3]. Copper(I) seems to have been the most studied cation recently. This may be the result of the relatively recent demonstration that solutions of Cu^+aq, obtained for example by $Cr^{2+}aq$ reduction of $Cu^{2+}aq$ in perchlorate media, are stable enough for use in kinetic studies. Parallels between reactivity trends for Cu^+aq and $Cr^{2+}aq$ reduction, and indirect evidence for hydroxo-bridged transition states in the Cu^+aq reductions of vanadium(III) and of iron(III), suggest inner-sphere mechanisms [135]. Cu^+aq reduction of complexes $Co(NH_3)_5(O_2CH=CR_2)^{2+}$ involves intermediates, where the copper(I) is bonded to the alkene group, of significant lifetime [22]. Ni^+aq, generated by pulse radiolysis, is even more ephemeral than Cu^+aq. The available evidence suggests that Ni^+aq acts as an inner-sphere reagent. Cd^+aq and Zn^+aq reduce cobalt(III) complexes even more rapidly than Ni^+aq. Zn^+aq is thought

Table 13.10   Kinetic parameters for reduction of metal complexes by other transition metal aquo- and hydroxoaquo- cations.

Reductant	oxidant	k_{25} /$M^{-1}s^{-1}$	$\Delta H^{\ddagger}$ /kcal mol^{-1}	$\Delta S^{\ddagger}$ /cal deg^{-1}mol^{-1}	
Cu^+aq	$Fe(OH)^{2+}$	1.6×10^5	12.4	+6.7	[140]
	$Co(NH_3)_5(OH_2)^{3+}$	1×10^{-3}			
	$Co(NH_3)_5(OH)^{2+}$	3.8×10^2			
	$Co(NH_3)_5F^{2+}$	1.11	12.4	−16.6	[141]
	$Co(NH_3)_5Cl^{2+}$	4.9×10^4	5.3	−19.4	
	$Co(NH_3)_5Br^{2+}$	4.46×10^5	3.9	−19.8	
	$Co(NH_3)_5(N_3)^{2+}$	1.5×10^3	5.4	−25.8	
	$Co(NH_3)_5(NCS)^{2+}$	~1 to 3			
	$Co(NH_3)_5(OAc)^{2+}$	$\leqslant 4 \times 10^{-3}$			[135]
	$Co(en)_3^{3+}$	$\leqslant 4 \times 10^{-4}$			[141]
$Ti^{3+}aq^a$	$Fe^{3+}aq$	25			[142]
$Ti(OH)^{2+}aq^a$	VO_2^+aq	5.3×10^5			[143]
$Ti(OH)^{2+}aq$	$Co(NH_3)_5Cl^{2+}$	0.48^b			
$Ti^{3+}aq$	cis-$Co(en)_2Cl_2^+$	$2 \times 10^{-3\,b}$			[144]
$Ti(OH)^{2+}aq$		0.75^b			
$Ti^{3+}aq$	$trans$-$Co(en)_2Cl_2^+$	$9 \times 10^{-3\,b}$			
$Ti(OH)^{2+}aq$		2.8^b			

a Detailed kinetic results for titanium(III) reduction of $Co(NH)_5L^{n+}$, where $L = inter\ alia$ N_3^-, NCS$^-$, or one of a range of carboxylato-ligands, can be found in refs. [138] and [139]; *b* 25.2°C.

to act as an outer-sphere reductant, and Cd^+aq by parallel inner- and outer-sphere paths [136].

Ruthenium(II) is another cation whose strongly reducing properties (it reduces perchlorate, though not trifluoromethylsulphonate) cause kinetic difficulties. The low-spin d^6 electronic configuration of this cation confers great inertness, so it must be expected to act as an outer-sphere reductant towards substitution-inert complexes. Titanium(III) is a less inconveniently powerful reductant, though it does reduce perchlorate at a significant rate. There is evidence for a hydroxo-bridged inner-sphere mechanism, with a transition state $[Ti\cdots OH\cdots VO_2^{3+}]^{\ddagger}$, in its reduction of vanadium(V) [137]. Comparison of rates of reduction of the $Co(NH_3)_5(N_3)^{2+}$ and $Co(NH_3)_5(NCS)^{2+}$ cations indicates an inner-sphere mechanism for reduction by $Ti^{3+}aq$, and evidence has been presented for inner-sphere reduction of carboxylato-cobalt(III) complexes by $Ti(OH)^{2+}aq$ [138]. In general, titanium(III) reductions of cobalt(III) complexes containing 'hard' ligands, such as sulphate, acetate, or fluoride, are inner-sphere in mechanism. Those of complexes containing 'soft' ligands, such as chloride or iodide, are outer-sphere [139].

Kinetic data for some redox reactions coming within the scope of this section are given in Table 13.10 [140–144].

13.4.5 sp-Element Cations

Kinetic and mechanistic information about reductions of complexes by aquo-cations of the sp-block elements is both sparse and disconnected. Examples of studies relating to 1+ cations include those of Tl^+aq reduction of neptunium-(VII) [145], In^+aq reduction of iron(III) (consecutive one-electron transfers) [146], and the Zn^+aq and Cd^+aq reductions of a range of cobalt(III) complexes already mentioned in the previous section in connection with reduction by Ni^+aq [136]. The $Hg_2^{2+}aq$ cation reduces such complexes as $Fe(phen)_3^{3+}$ and $Ru(bipy)_3^{3+}$ by one-electron transfer steps [147]. One-electron transfers have also been proposed for the $Sn^{2+}aq$ reduction of $UO_2^{2+}aq$ [148]. Tin(II) reductions of $Fe^{3+}aq$ and of VO_2^+aq are more complicated [149]. The kinetic pattern has been established for several redox reactions of the transient $Tl^{2+}aq$ cation, including its reduction of $Fe^{3+}aq$ and of $Co^{3+}aq$ [60].

The kinetics of tin(II) reduction of uranium(VI) have been studied in several series of mixed aqueous solvents, including water+ethanol and water+N-methyl-acetamide [150].

13.4.6 Lanthanide Cations

Few of the lanthanide elements give reducing aquo-cations. Only europium, samarium, and ytterbium give 2+ cations of sufficient longevity for kinetic studies to be possible by normal techniques. For example, the Yb^{2+} cation is stable for only up to about half an hour at best. For all three elements the lability of both the 2+ and 3+ oxidation states rules out the possibility of any direct evidence for the operation of an inner-sphere mechanism for redox reactions. Indeed, redox mechanisms in this area are still the subject of much speculation and disagreement.

There are several indications of outer-sphere reductions by Eu^{2+}aq. There is often a lack of variation of redox rate with pH. Redox rates often parallel those for analogous reductions by V^{2+}aq or Cr^{2+}aq of known outer-sphere character [151]. Oxygen isotopic fractionation experiments have also indicated outer-sphere reduction by Eu^{2+}aq [48]. On the other hand some reductions of iron(III) complexes and of vanadium(IV) have been assigned inner-sphere mechanisms, while the reduction of some cobalt(III) complexes is thought to go by parallel inner- and outer-sphere routes [151]! The most recent results in this area suggest that at least the majority of Eu^{2+}aq reductions of carboxylato–cobalt(III) complexes, and of $Co(NH_3)_5(NCS)^{2+}$, $Co(NH_3)_5(SCN)^{2+}$ and $Co(NH_3)_5(N_3)^{2+}$, go by the inner-sphere mechanism [152]. Regardless of these mechanistic doubts, a fair amount of kinetic information exists on Eu^{2+}aq reductions (Table 13.11 [153–156]).

Table 13.11 Kinetic parameters for europium(II) reduction of complexes.

	k_{25} /$M^{-1}sec^{-1}$	$\Delta H^{\ddagger}$ /kcal mol^{-1}	$\Delta S^{\ddagger}$ /cal deg^{-1}mol^{-1}	
$Cr(OH_2)_6^{3+}$	$\sim 1.7 \times 10^{-5}$			
$V(OH_2)_6^{3+}$	9.0×10^{-3}	11.4	-30	[153]
$V(OH_2)_5(OH)^{2+}$	2.0	6.2	-35	
$Fe(OH_2)_6^{3+}$	3.4×10^{3a}			
$Fe(OH_2)_5(OH)^{2+}$	6.0×10^{6a}			[100]
$Fe(OH_2)_5(NCS)^{2+}$	3.2×10^{5a}			
VO^{2+}aq	2.64×10^3	2.8	-33.4	[154]
UO_2^{2+}aq	1.5×10^4	1.5	-34.4	[155]
$Co(NH_3)_6^{3+}$	1.7×10^{-3}	8.8	-42	[151]
$Co(NH_3)_5(OH_2)^{3+}$	7.4×10^{-2}	9.3	-32	
$Co(NH_3)_5 F^{2+}$	1.5 or 2.6×10^4			
$Co(NH_3)_5Cl^{2+}$	3.9 or 4.7×10^2	5.0	-30	
$Co(NH_3)_5Br^{2+}$	2.5×10^2	4.7	-32	[156]
$Co(NH_3)_5I^{2+}$	1.2×10^2			
$Co(NH_3)_5(N_3)^{2+}$	1.9×10^2	5.5	-30	
$Co(NH_3)_5(OAc)^{2+}$	1.8×10^{-1}			
$Ru(NH_3)_5Cl^{2+}$	2.4×10^4			
$Ru(NH_3)_5Br^{2+}$	1.34×10^4	1.0	-37	[28]
$Ru(NH_3)_5(OAc)^{2+}$	1.7×10^5			

a 1.6°C.

The situation is less satisfactory for reductions by Yb^{2+}aq. For a start there is argument about the stability of reagent solutions [53, 156, 157], so that it is hardly surprising that different workers report different rate constants for a given reaction. The indirect methods necessarily used in mechanism assignment here are therefore not soundly based, with consequent confusion in the apparent pattern of mechanisms.

13.4.7 Actinide Cations

Unavoidably, evidence for the assignment of redox mechanisms is again indirect, due to the lability of actinide(III) and actinide(IV) cations. Just as uranium is the most studied element in the group, so U^{3+}aq has proved the most investigated reductant. This interest has been maintained despite the difficulties of working with solutions containing this cation. It is a powerful reducing agent, with $E^{\ominus}(U^{4+}/U^{3+}) = -0.63V$. Thus U^{3+}aq can reduce water, but in practice solutions of this cation can be kept for many hours under suitable conditions. There is fairly convincing evidence that U^{3+}aq reduction of complexes $Cr(OH_2)_5X^{2+}$ occurs by an inner-sphere mechanism, with OH or, where possible, X as bridging ligand in the transition state. The OH bridge appears to be derived from $Cr(OH_2)_4(OH)X^+$ rather than from $U(OH)^{2+}$aq [158]. Uranium(III) reductions of cobalt(III) complexes also generally go by an inner-sphere pathway when this is feasible [49]. Comparison of reduction rates of $Co(NH_3)_5(N_3)^{2+}$ and $Co(NH_3)_5(NCS)^{2+}$ points to inner-sphere reduction [49]. U^{3+}aq reduction both of V^{3+}aq and of $V(OH)^{2+}$aq is inner-sphere [159]. Selected kinetic parameters for uranium(III) reductions of complexes are quoted in Table 13.12 [49, 160].

Table 13.12 **Kinetic parameters for uranium(III) reduction of complexes [49, 160].**

	k_{25} /M^{-1}sec^{-1}	$\Delta H^{\ddagger}$ /kcal mol^{-1}	$\Delta S^{\ddagger}$ /cal deg^{-1}mol^{-1}
$Co(en)_3^{3+}$	0.13		
$Co(NH_3)_6^{3+}$	1.32		
$Co(NH_3)_5F^{2+}$	5.40×10^5		
$Co(NH_3)_5Cl^{2+}$	3.24×10^4	2.7	-29.0
$Co(NH_3)_5Br^{2+}$	1.42×10^4	2.0	-32.7
$Co(NH_3)_5(N_3)^{2+}$	1.08×10^6		
$Co(NH_3)_5(NCS)^{2+}$	18		
$Co(NH_3)_5(OAc)^{2+}$	1.5×10^4		
$Cr(OH_2)_5F^{2+}$	17		
$Cr(OH_2)_5(N_3)^{2+}$	40		
$Cr(OH_2)_5(SCN)^{2+}$	9.0×10^3		

Inner-sphere mechanism have also been proposed for reductions by Np^{3+}aq and Pu^{3+}aq. Activation entropies for the reduction of ruthenium(III) complexes by Np^{3+} are unusually large and negative, with values down to -47 cal $deg^{-1} mol^{-1}$. Perhaps this is due to the generation of transition states containing seven-coordination around the large neptunium (see Chapter 13.3.2) [7].

13.5   OXIDATION OF COMPLEXES BY METAL CATIONS

Two mechanistic questions are usually considered here, first the relative reactivities of M^{n+}aq and $M(OH)^{(n-1)+}$aq, and second whether the redox processes are inner- or outer- sphere in character. A range of recent studies can be tracked down by the use of references [71] and [99]. Redox mechanisms for the oxidation of complexes by cobalt(III) and manganese(III) have been well reviewed in references [161] and [162] respectively. The recent interest in the kinetics of oxidations by silver(II) can be illustrated by reference to a study of its oxidation of vanadium(IV) [163], and of the possible intermediacy of silver(III) in silver(II) oxidations [164]. We shall be returning to oxidation by aquo-metal cations in the next section, where the more extensively investigated area of oxidations of simple inorganic and organic species will be covered.

Some lead(IV) oxidations in acetic acid provide rare examples of kinetic studies of inorganic redox reactions in non-aqueous media. Here the reacting species are probably of the form $M(OAc)_n^{m-}$, and there may also be some dinuclear species involved in solutions containing cobalt(II). The lead(IV) oxidation of cerium(III) is thought to involve one-electron transfers, with a lead(III) intermediate [165], whereas the oxidation of cobalt(II) may involve parallel one-electron and two-electron transfer paths [166].

13.6   REDOX REACTIONS INVOLVING
SIMPLE INORGANIC OR ORGANIC SPECIES

The majority of reactions discussed in this section [167–170] are, by chance, oxidations. The kinetics of only a few reductions, for example by Cr^{2+}aq, V^{2+}aq, Ru^{2+}aq, and Ti^{3+}aq, have been described. In addition, there is one common class of reaction difficult to classify in these black and white terms, that is reaction of an oxidising metal ion with hydrogen peroxide. Here the metal ion is reduced, but the hydrogen peroxide may be both oxidised and reduced, to oxygen and water [171].

Many of the reactions covered in this section proceed by a common mechanism, in which the inorganic or organic reactant, L, forms a complex with the aquo-cation before electron (and atom, where relevant) transfer. This two-stage mechanism was suggested about a century ago. The first hint of a transient intermediate in the iron(III) oxidation of organo-sulphur compounds dates from 1879 [172], while the intermediate complex involved in the iron(III) oxidation of thiosulphate was firmly established in 1930 [173]. The overall kinetic pattern observed for such a reaction sequence,

$$M^{n+} + L \underset{k_{-1}}{\overset{k_1}{\rightleftharpoons}} ML^{n+}$$

$$ML^{n+} \overset{k_2}{\longrightarrow} \text{products},$$

depends on the relative values for the rate constants k_1, k_{-1}, and k_2. Several situations can be distinguished.

(i) k_1, k_{-1} both $\gg k_2$. Here an intermediate ML^{n+} is generated immediately, and its decomposition can readily be monitored. If the equilibrium constant for formation of ML^{n+} is K_1, the kinetic pattern is simply $k_{obs} = K_1 k_2$.

(ii) $k_{-1} \gtrsim k_2$. Here an intermediate may be detectable. If so, its formation and decomposition can be monitored and its stability constant estimated. This state of affairs is exemplified by iron(III) oxidation of catechol [174] and of thiosulphate [175]. In the latter case, the stoichiometry could be determined by Job's method, and the intermediate was found to be of formula $FeS_2O_3^+$ (in aqueous solution).

(iii) $k_2 \gg k_{-1}$. Now the overall reaction rate will be determined by, and equal to, the rate of formation of the intermediate. This situation is common in Co^{3+}aq oxidations.

(iv) When the concentration of the intermediate is very small throughout the reaction, and thus steady-state conditions prevail, the observed rate constant will be given by,

$$k_{obs} = \frac{k_1 k_2}{k_{-1} + k_2}.$$

In this situation the observed kinetic parameters, k_{obs} and $\Delta H^{\ddagger}$ and $\Delta S^{\ddagger}$ derived from the temperature variation of k_{obs}, will be composite quantities. They will only be resolvable if an independent method of determining K_1, the equilibrium constant for the formation of ML^{n+}, is available. In cases where there is no positive evidence for a transient intermediate, it is difficult to tell whether situation (iv) obtains, or whether the redox reaction is a simple one-step process. The consequent difficulties in interpreting the observed kinetic parameters are obvious.

There are two complications commonly encountered in applying the above analysis to redox reactions in this category.

(i) Most of the aquo-cations which act as oxidants, for example $Ce^{4+}aq$, $Fe^{3+}aq$, $Co^{3+}aq$, $Mn^{3+}aq$, and $Tl^{3+}aq$, are acidic (see Chapter 9). Hence one must take into account the fast pre-equilibrium

$$M^{n+}aq \rightleftharpoons M(OH)^{(n-1)+}aq + H^+aq$$

and allow for the likelihood of both $M^{n+}aq$ and $M(OH)^{(n-1)+}aq$ reacting with the substrate.

(ii) Many of the aquo-cation oxidants and reductants are one-electron redox agents, while many organic and inorganic substrates require two electrons. Examples include the oxidation of acetoin to biacetyl [176] and of sulphite to sulphate [177]. In such circumstances the initial reaction products will be reactive free radicals, which may complicate the reaction mechanism and kinetics.[†] Such intermediate radicals have been trapped in the iron(III)–acetoin reaction [176], and proposed (as the thionate, HSO_3, radical) in the iron(III)–sulphite reaction [177]. The effects of added alcohols on the iron(II)–chlorate reaction provide evidence for the intermediacy of Cl and ClO radicals here [178]. There is a further twist in that some species can act as a one-electron or two-electron oxidant or reductant depending on the conditions. Thus sulphite can act as a one-electron reductant, giving dithionate, or a two-electron reductant, giving sulphate. Similarly hydrazine can act initially as a one-electron reagent, ultimately giving nitrogen and ammonia, or as a two-electron reagent, ultimately giving nitrogen after a second two-electron transfer step.

In addition to these kinetic awkwardnesses there may be experimental difficulties. Three sources of these are mentioned here.

(i) Both $Co^{3+}aq$ and $Mn^{3+}aq$ are unstable with respect to oxidising solvent water. Not only does this make kinetic monitoring difficult, indeed impossible for slow reactions, but it also means that there is a dearth of precise physical data, such as pK values, available for these cations. There are also reports that $Co^{3+}aq$ and $Mn^{3+}aq$ undergo dimerisation or polymerisation in aqueous solution if the pH is not sufficiently low.

(ii) Oxidations by cerium(IV) are generally rapid in perchlorate media. Many investigators have therefore succumbed to the temptation to work in sulphate media where reaction is slower. However there is competition in sulphate media between the reductant and the sulphate for the cerium(IV), and a real possibility of forming mixed complexes

---

[†]The formation of free radicals can be circumvented, as in the iron(III) oxidation of α-mercaptocarboxylic acids. This proceeds via dimerisation of intermediates before electron transfer (Ellis, K. J., Lappin, A. G. and McAuley, A. *J. Chem. Soc. Dalton* 1930 (1975)).

$Ce(SO_4)L^{n+}$. The consequent complications in the observed kinetics are obvious.

(iii) There are sometimes technical problems with stopped-flow devices which can lead the unwary to postulate a transient intermediate complex when in fact none exists.

A review of oxidation or reduction of ligands by the metal ion to which they are complexed provides useful background material relevant to the second, redox, stage of these two-stage oxidations and reductions [179].

13.6.1 Oxidations

It is convenient to classify aquo-cation oxidations of simple inorganic and organic species into two groups, those for which there is evidence for an intermediate complex of significant life-time, and those for which there is no such evidence. The latter class can be further subdivided according to their proposed transition states into inner- and outer- sphere processes. For this latter class it is difficult to tell whether lack of evidence for an intermediate indicates the absence of such a species, or merely that it is too short-lived for the detection techniques employed. There is, of course, no sharp dividing line between mechanisms with a transient intermediate and true inner-sphere processes.

For reactions in which there is definite evidence for the existence of a complex between oxidant and reductant, it is possible to distinguish between cases where there is direct evidence, usually spectrophotometric,[†] for such a species, and cases where there is only indirect evidence, usually kinetic. Although an oxidant-substrate complex may be detected, this does not prove unequivocally that this complex is an intermediate in the redox sequence. A pattern of the type shown in Fig. 13.10 may operate [181]. Sometimes, as in cerium(IV) oxidation of alcohols, agreement between spectrophotometric and kinetic estimates for the stability constant of the oxidant-substrate complex provides strong support for the actual intermediacy of this species in the redox reaction.

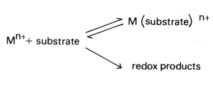

Fig. 13.10

†Normally ultraviolet-visible spectroscopy is used to seek intermediates, but e.s.r. spectroscopy has also proved useful at times. Instances include photo-oxidations of some organic compounds by Ce^{4+}aq and UO_2^{2+}aq [180], and the provision of direct evidence for the intermediacy of $\bullet CH(CO_2H)_2$ radicals in the oxidation of malonic acid by cerium(IV) (Amjad, Z. and McAuley, A. *J. Chem. Soc. Dalton* 304 (1977)).

In the case of the cobalt(III) oxidation of salicylate, the redox process is so slow that it is possible to monitor the kinetics of formation of the intermediate mono- and bis- salicylato–cobalt(III) complexes before significant oxidation takes place [182]. When the redox process is so much slower than the complex formation step, it is possible to isolate and characterise the intermediate complex. A good example of this is provided by thallium(III) oxidation of certain alkenes [183]. An inorganic example is afforded by the isolation, and subsequent redox kinetic study, of the *trans*-$Co(NH_3)_4(OH_2)(SO_3)^+$ cation [184]. In the cases of Co^{3+}aq, Mn^{3+}aq, and Ag^{2+}aq [185] oxidations of water, the 'intermediate' complex is necessarily preformed in the shape of the starting aquo-cation.

In the extreme, the 'intermediate' complex may be so resistant to internal redox reaction that it appears indefinitely stable. For instance, iron(III) oxidises thiomalic acid via a spectrophotometrically characterised blue intermediate [186], but iron(III) reacts with malic acid to produce a malate complex which undergoes no further reaction [187].

Examples of oxidations falling into the categories described above are given, for a range of aquo-metal cation oxidants, in Table 13.13 [188-203]. Many more examples are known, and can be found via references [167] to [170]. The intermediates proposed for the reactions cited in Table 13.13 and in [167] to [170] contain the inorganic or organic substrate in its starting, reduced, form complexed to the metal ion in its starting, oxidising, form. In contrast, the copper(II) oxidation of substituted hydrazines involves intermediates which are thought to consist of the oxidised hydrazine complexed to copper(I). The intermediate derived from copper(II) oxidation of 1,1-dimethylhydrazine, $(Me_2N=N)_2Cu_3Cl_3$, has been isolated and fully characterised [204].

An interesting variant on the theme of the oxidation of simple species by metal ions in solution is the oscillating reaction. The best known example is the Belusov reaction, involving cerium(IV), bromate, and malonate in acid solution. The complicated mechanisms proposed to account for the very unusual periodicity of this and similar reactions, for example those involving manganese(II), are described elsewhere [205].

M^{3+}aq vs. $M(OH)^{2+}$aq as oxidants. For aquocations such as Fe^{3+}aq, the amounts of this reagent in the forms of Fe^{3+}aq and of $Fe(OH)^{2+}$aq can be specified because the pK of Fe^{3+}aq is known fairly accurately (Chapter 9.2.1). However, the pK value may not be known under the precise conditions of the redox kinetic study. For Co^{3+}aq and Mn^{3+}aq the situation is much less satisfactory. These cations oxidise water too rapidly for precise pK determinations to be made by classical physical chemistry techniques. Indeed, most of the estimates of pK values for these aquocations come from kinetic studies. Such an approach is acceptable when the substrate being oxidised is the anion of a strong acid, for example chloride or bromide, because there is no chance

Table 13.13 Examples of aquo-cation oxidations of simple inorganic and organic species, classified by evidence for intermediate species. These examples are drawn from references [167] to [169] except where otherwise indicated; all are in aqueous solution.

Oxidant	Inorganic reductants	Organic reductants
Spectrophotometric evidence for intermediate formation		
Thallium(III)	hypophosphite	alkenols [188]
		oxalic acid
Manganese(III)	hydrazoic acid	hydroquinone
	hydrogen peroxide	
Iron(III)	sulphite [176]	catechol [174]
	thiosulphate	cysteine
		thiols
		thiomalic acid [186]
		thiourea [189]
Cobalt(III)	chloride	cysteine
		malic acid
		thiomalic acid
Cerium(IV)	hypophosphite	acetic acid
		alcohols
		hydroquinone
		malic acid [190]^a
Kinetic evidence for intermediate formation		
Thallium(III)	hypophosphite [191]	alkenes [183]
		alkenols [188]
		cycloalkenes [192]
		o-dihydroxy aromatics [193]
		oxalate
Manganese(III)	bromide	alcohols
		malic acid [194]
Iron(III)	hydrogen peroxide	acetylacetone
	iodide	catechol [174]
	sulphite [195]	
	thiocyanate	
Cobalt(III)	hydrazoic acid	alcohols
	hydrogen peroxide	α-amino-acids
		propionic acid
		pyridine carboxaldehydes [196]
		succinic acid
Cerium(IV)	bromide	acetic acid
	chloride	ethanol
	phosphorous acid [197]	glycerol
		malic acid^a
		oxalic acid
No evidence for intermediate formation		
Silver(I)		ascorbate [198]
Silver(II)	dithionate [199]	
Thallium(III)		catechol [174, 200]^b
Manganese(III)	hydrazine	
	hydroxylamine	
	nitrous acid	
Iron(III)	hydrazine	acetoin [176]
		quinol [201]
		thiols

Table 13.13 *(c'td.)*.

Oxidant	Inorganic reductants	Organic reductants
Cobalt(III)	bromide chlorine dioxide nitrous acid thiocyanate	hydroquinone thiourea [202]
Cerium(IV)		benzaldehyde glycollic acid lactic acid malic acid[c] phenylacetic acids [203]

a In perchloric acid; *b* but see ref. [192] ; *c* in sulphuric acid.

of confusion from parallel $L + H^+ \rightleftharpoons HL^+$ equilibrium. However when HL^+ is a weak acid, a term in the oxidation rate law of the form $k_2[M^{3+}aq][L]$ might arise from the reaction of $M^{3+}aq$ plus L, or equally plausibly from $M(OH)^{2+}$ aq plus HL^+. There is no kinetic way of resolving such a kinetic ambiguity (cf. kinetics of formation of complexes, Chapter 12.6).

Both for cobalt(III) and for manganese(III), $M(OH)^{2+}aq$ seems to be a more reactive oxidant than $M^{3+}aq$. This can be seen clearly in the cobalt(III) oxidation of benzene for example, where the intermediacy of benzeneH$^+$ or of cobalt(III)– benzene complex to complicate the interpretation of the observed kinetics both seem highly implausible. For reactions in which the initial step is the formation of a complex, as for chloride, $Co(OH)^{2+}aq$ is again more reactive than Co^{3+} aq. Obviously electrostatic considerations are not dominant here. It seems that $Fe(OH)^{2+}aq$ reacts more quickly than $Fe^{3+}aq$ with oxidisable substrates for the better documented iron (III) oxidations also. It may be that this difference arises from greater ease of substitution at $Fe(OH)^{2+}aq$, as attested by the relative rates of water exchange at this species and at $Fe^{3+}aq$ (Chapter 11.6.2).

Where there is neither kinetic nor spectroscopic evidence for an intermediate, the oxidations may therefore be simple one-step processes. The question then arises as to whether the mechanism is inner-sphere or outer-sphere in character. Several oxidations by $Mn(OH)^{2+}aq$ have been assigned an inner-sphere mechanism [206]. For reactions of $Co^{3+}aq$ with oxidisable substrates there is such a small range of observed rate constants that one is led to postulate S_N1 substitution at the cobalt. Compelling evidence for outer-sphere reactions is difficult to obtain. There have been tentative suggestions that the oxidation of quinol by manganese(III) or cobalt(III) may be outer-sphere [207].

13.6.2 Reductions

Reductions, like oxidations, of simple inorganic or organic species by aquo-cations may take place via an intermediate complex (see previous section). Chromium(II) reductions of hydrogen peroxide and of alkynes are believed to

take place via the formation of intermediates $Cr(OH_2)_5(OOH)^+$ and $RC \equiv CR$
$$\overset{|}{\underset{Cr^{2+}}{}}$$
respectively. Organo-chromium intermediates have also been proposed for chromium(II) reductions of benzyl chloride, where the intermediacy of the $PhCH_2Cr^{2+}$ cation was demonstrated [208], and of benzaldehyde [209] and salicaldehyde [210]. The chromium(II) reduction of organic compounds was reviewed a few years ago [211], and this review contains many comments on the mechanisms of the reactions discussed. A pulse radiolysis study of the reaction of molecular oxygen with chromium(II) indicated the intermediacy of a complex $CrO_2{}^{2+}$ en route to the di-μ-oxo-dichromium(III) final product [212]. There is kinetic evidence for the intermediacy of a similar complex in the uranium(IV) reduction of molecular oxygen [213].

The much studied reaction of iron(II) with hydrogen peroxide may also involve an intermediate, that is a mixed solvate (cf. chromium(II) above), containing these two components. In the first step in the redox sequence, customarily represented as

$$Fe^{2+} + H_2O_2 \longrightarrow Fe^{3+} + OH^{\cdot} + OH^-$$

the reactants may in fact not be separate but rather combined in the form $Fe(OH_2)_5(O_2H_2)^{2+}$. Evidence to support the participation of such an intermediate comes from experiments which monitored the effects of added anions, for example, chloride, bromide, sulphate, or hexafluorophosphate, on reaction rates. Rates of reaction of the complexes $FeCl^+$, $FeBr^+$, $FeSO_4$, and $FePF_6{}^+$ with hydrogen peroxide are available for comparison. An $FeBrO_3{}^+$ cation seems a likely intermediate in one pathway (see also below) in the reduction of bromate by Fe^{2+}aq [214]. The iron(II) reduction of iodate is more complicated in kinetic pattern and mechanism [215]. The iron(II) reduction of chlorate has yielded neither kinetic nor spectroscopic evidence for any intermediate [178].

Reduction may occur by a simple one-step mechanism, either inner- or outer- sphere with respect to the aquo-cation, rather than through a transient intermediate. There is good evidence for the chromium(II) reduction of alkyl chlorides going via an inner-sphere transition state $[Cr\text{---}Cl\text{---}C]^{\ddagger}$. Also chromium(II) reductions of unsaturated carboxylic acids yield no evidence for intermediate formation. On the other hand, vanadium(II) and titanium(III) reductions of maleate to succinate appear, from the kinetic pattern exhibited, to involve the intermediacy of a reductant-substrate complex [216]. The kinetics of vanadium(II) reduction of I_2, I_3^-, and Br_2 do not indicate an intermediate complex. Moreover the activation energies for these reductions are low, considerably lower than that for water exchange at V^{2+}aq, indicating an outer-sphere mechanism [217]. The iron(II)-bromate reaction mentioned in the previous paragraph proceeds by an outer-sphere electron transfer path in parallel with the intermediate formation path mentioned above [214]. Plutonium(III) reduction of chromium(VI) may also be outer-sphere [218].

Recent studies on the reduction of halogen molecule ions, X_2^- with $X = Cl$, Br, or I, generated by pulse radiolysis, have given information on reduction mechanisms for a series of transition metal aquo-cations. Chromium(II) reductions of Br_2^- and I_2^- are inner-sphere. Chromium(II) and iron(II) reductions of Cl_2^- proceed by parallel inner- and outer- sphere pathways. Manganese(II) and cobalt(II) reductions are again inner-sphere, while only for vanadium(II) does the evidence indicate solely outer-sphere reduction [219].

13.7 NON-AQUEOUS AND MIXED AQUEOUS MEDIA

There is very little information on the kinetics of redox reactions involving solvento-metal cations in non-aqueous and mixed aqueous media. Examples in non-aqueous media include iron(II)/(III) isotopic exchange in iso-propanol [89], and lead(IV) oxidation of cerium(III) [165] and of cobalt(II) [166] in acetic acid. Investigations in binary aqueous mixtures include the thallium(I)/(III) [88], iron(II)/(III) [89], and uranium(IV)/(VI) [88] isotopic exchange reactions in various solvent mixtures, and tin(II) reduction of uranium(VI) in ethanol + water and in N-methylacetamide + water mixtures [150].

FURTHER READING

Stengle, T. R. and Langford, C. H. *Coord. Chem. Rev.* **2**, 349 (1967).

Sutin, N. *Accts. Chem. Res.* **1**, 225 (1968).

Taube, H., *Electron Transfer Reactions of Complex Ions in Solution.* Academic Press (1970).

Proll, P. J. in *Comprehensive Chemical Kinetics,* ed. Bamford, C. H. and Tipper, C. F. H. *Volume 7: Reactions of Metallic Salts and Complexes, and Organometallic Compounds.* Elsevier (1972) chap. 2; Benson, D. *ibid.* chap. 3; Kemp, T. J. *ibid.* chap. 4.

Wilkins, R. G. *The Study of Kinetics and Mechanism of Reactions of Transition Metal Complexes.* Allyn and Bacon (1974).

Haim, A. *Accts. Chem. Res.* **8**, 264 (1975).

Benson, D. *Reaction Mechanisms in Organic Chemistry, 10: Mechanisms of Oxidation by Metal Ions.* Elsevier (1976).

Inorganic Reaction Mechanisms – Chemical Society Specialist Periodical Report, Volumes 1 (1971); 2 (1972); 3 (1974); 4 (1976); 5 (1977) – see Part I in each volume.

REFERENCES

[1]   Tobe, M. L. *Inorganic Reaction Mechanisms,* Nelson (1972), p. 126.
[2]   Basolo, F. and Pearson, R. G. *Mechanisms of Inorganic Reactions, 2nd edn.* Wiley Interscience (1967), p. 157.
[3]   Sykes, A. G. *Kinetics of Inorganic Reactions,* Pergamon (1966), p. 218.
[4]   Taube, H. and Myers, H. *J. Amer. Chem. Soc.* **76,** 2103 (1954).
[5]   Murmann, R. K., Taube, H. and Posey, F. A. *J. Amer. Chem. Soc.* **79,** 262 (1957); Kruse, W. and Taube, H. *ibid.* **82,** 526 (1960).
[6]   See p. 132 of ref. [1].
[7]   Lavallee, D. K., Lavallee, C., Sullivan, J. C. and Deutsch, E. *Inorg. Chem.* **12,** 570 (1973).
[8]   Sutin, N. *Accts. Chem. Res.* **1,** 225 (1968).
[9]   Cannon, R. D. and Gardiner, J. *Inorg. Chem.* **13,** 390 (1974).
[10] Espenson, J. H. and Shveima, J. S. *J. Amer. Chem. Soc.* **95,** 4468 (1973).
[11] Nohr, R. S. and Spreer, L. O. *Inorg. Chem.* **13,** 1239 (1974).
[12] Sykes, A. G. and Thorneley, R. N. F. *J. Chem. Soc. A.* 232 (1970).
[13] See p. 138 of ref. [1].
[14] Haim, A. and Wilmarth, W. K. *J. Amer. Chem. Soc.* **83,** 509 (1961).
[15] Newton, T. W. and Baker, F. B. *Inorg. Chem.* **1,** 368 (1962).
[16] De Smedt, H., Persoons, A. and de Maeyer, L. *Inorg. Chem.* **13,** 90 (1974).
[17] Movius, W. G. and Linck, R. G. *J. Amer. Chem. Soc.* **91,** 5349 (1969).
[18] Liteplo, M. P. and Endicott, J. F. *J. Amer. Chem. Soc.* **91,** 3982 (1969); *Inorg. Chem.* **10,** 1420 (1971).
[19] Cannon, R. D. and Gardiner, J. *J. Amer. Chem. Soc.* **92,** 3800 (1970).
[20] Gaswick, D. and Haim, A. *J. Amer. Chem. Soc.* **96,** 7845 (1974).
[21] Gaswick, D. and Haim, A. *J. Amer. Chem. Soc.* **93,** 7347 (1971).
[22] Hurst, J. K. and Lane, R. H. *J. Amer. Chem. Soc.* **95,** 1703 (1973).
[23] Espenson, J. H. and Birk, J. P. *J. Amer. Chem. Soc.* **87,** 3280 (1965).
[24] Halpern, J. and Nakamura, S. *J. Amer. Chem. Soc.* **87,** 3002 (1965).
[25] Shea, C. and Haim, A. *J. Amer. Chem. Soc.* **93,** 3055 (1971).
[26] Shea, C. J. and Haim, A. *Inorg. Chem.* **12,** 3013 (1973).
[27] E.g. Gould, E. S. *J. Amer. Chem. Soc.* **90,** 1740 (1968).
[28] Stritar, J. A. and Taube, H. *Inorg. Chem.* **8,** 2281 (1969).
[29] Gould, E. S. *J. Amer. Chem. Soc.* **88,** 2983 (1966); Taube, H. and Gould, E. S. *Accts. Chem. Res.* **2,** 321 (1969).
[30] See pp. 502–506 of ref. [2].
[31] Gould, E. S. *J. Amer. Chem. Soc.* **96,** 2373 (1974).
[32] Nordmeyer, F. and Taube, H. *J. Amer. Chem. Soc.* **88,** 4295 (1966).
[33] Gould, E. S. and Taube, H. *J. Amer. Chem. Soc.* **86,** 1318 (1964).
[34] Hoffman, H. Z. and Simic, M. *J. Amer. Chem. Soc.* **94,** 1757 (1972).
[35] Nordmeyer, F. and Taube, H. *J. Amer. Chem. Soc.* **90,** 1162 (1968).

[36] Gould, E. S. *J. Amer. Chem. Soc.* **94**, 4360 (1972).

[37] Norris, C. and Nordmeyer, F. *J. Amer. Chem. Soc.* **93**, 4044 (1971).

[38] Symons, M. C. R., West, D. X. and Wilkinson, J. G. *J. Chem. Soc. Chem. Commun.* 917 (1973).

[39] Haim, A. *J. Amer. Chem. Soc.* **88**, 2324 (1966).

[40] Snellgrove, R. and King, E. L. *J. Amer. Chem. Soc.* **84**, 4609 (1962).

[41] Marcus, R. A. *J. Phys. Chem.* **67**, 853 (1963); **72**, 891 (1968); *A. Rev. Phys. Chem.* **15**, 155 (1964).

[42] See pp. 461–466 of ref. [2].

[43] E.g. Rillema, D. P. and Endicott, J. F. *Inorg. Chem.* **11**, 2361 (1972).

[44] Rillema, D. P., Endicott, J. F. and Patel, R. C. *J. Amer. Chem. Soc.* **94**, 394 (1972).

[45] Schmickler, W. *Ber. Bunsenges. Phys. Chem.* **77**, 991 (1973); Christov, S. G. *ibid.* **79**, 357 (1975), and references therein.

[46] Pasternack, R. F. and Sutin, N. *Inorg. Chem.* **13**, 1956 (1974).

[47] Stranks, D. R. *Pure Appl. Chem., 15th Int. Conf. Coord. Chem.* **38**, 303 (1974).

[48] Diebler, H., Dodel, P. H. and Taube, H. *Inorg. Chem.* **5**, 1688 (1966).

[49] Wang, R. T. and Espenson, J. H. *J. Amer. Chem. Soc.* **93**, 380 (1971).

[50] Espenson, J. H. *Inorg. Chem.* **4**, 121 (1965).

[51] Candlin, J. P., Halpern, J. and Trimm, D. L. *J. Amer. Chem. Soc.* **86**, 1019 (1964).

[52] Fay, D. P. and Sutin, N. *Inorg. Chem.* **9**, 1291 (1970).

[53] Christensen, R. J. and Espenson, J. H. *Chem. Commun.* 756 (1970).

[54] Toppen, D. L. and Linck, R. G. *Inorg. Chem.* **10**, 2635 (1971).

[55] Chen, J. C. and Gould, E. S. *J. Amer. Chem. Soc.* **95**, 5539 (1973).

[56] Wieghardt, K. and Sykes, A. G. *J. Chem. Soc. Dalton* 651 (1974); Hyde, M. R. and Sykes, A. G. *ibid.* 1550 (1974).

[57] Haim, A. *Inorg. Chem.* **7**, 1475 (1968).

[58] Weaver, M. J. and Anson, F. C. *J. Amer. Chem. Soc.* **97**, 4403 (1975).

[59] See pp. 489 et seq. of reference [2].

[60] Falcinella, B., Felgate, P. D. and Laurence, G. S. *J. Chem. Soc. Dalton* 1367 (1974); 1 (1975).

[61] Schwarz, H. A., Comstock, D., Yandell, J. K. and Dodson, R. W. *J. Phys. Chem.* **78**, 488 (1974).

[62] Beattie, J. K. and Basolo, F. *Inorg. Chem.* **10**, 486 (1971).

[63] Green, M. and Sykes, A. G. *J. Chem. Soc. A* 3221 (1970); 3067 (1971).

[64] Ashurst, K. G. and Higginson, W. C. E. *J. Chem. Soc.* 3044 (1953).

[65] Ardon, M. and Plane, R. A. *J. Amer. Chem. Soc.* **81**, 3197 (1959).

[66] Ashurst, K. G. and Higginson, W. C. E. *J. Chem. Soc.* 343 (1956).

[67] Nunes, T. L. *Inorg. Chem.* **9**, 1325 (1970).

[68] Mishra, S. K. and Gupta, Y. K. *J. Chem. Soc. A* 260 (1970).

[69] Davies, K. M. and Espenson, J. H. *J. Amer. Chem. Soc.* **92**, 1884 (1970).

[70] See chap. 8 of reference [3] ; Halpern, J. *Q. Rev. Chem. Soc.* **15**, 207 (1961).

[71] McAuley, A. *Inorganic Reaction Mechanisms – Chemical Society Specialist Periodical Report* see Part I of volumes 1 (1971), 2 (1972), 3 (1974), and 4 (1976).

[72] Hevesy, G. von and Zechmeister, L. *Chem. Ber.* **53**, 410 (1920).

[73] Proll, P. J. in *Comprehensive Chemical Kinetics* ed. Bamford, C. H. and Tipper, C. F. H. *Volume 7: Reactions of Metallic Salts and Complexes, and Organometallic Compounds.* Elsevier (1972), chap. 2; and see chap. 7 of reference [3].

[74] Seaborg, G. T. *Chem. Rev.* **27**, 199 (1940).

[75] McConnell, H. M. and Weaver, H. E. *J. Chem. Phys.* **25**, 307 (1956).

[76] Diebler, H. and Sutin, N. *J. Phys. Chem.* **68**, 174 (1964).

[77] Silverman, J. and Dodson, R. W. *J. Phys. Chem.* **56**, 846 (1952).

[78] Fukushima, S. and Reynolds, W. L. *Talanta* **11**, 283 (1964).

[79] Habib, H. S. and Hunt, J. P. *J. Amer. Chem. Soc.* **88**, 1668 (1966).

[80] Gryder, J. W. and Dodson, R. W. *J. Amer. Chem. Soc.* **71**, 1894 (1949); **73**, 2890 (1951); Duke, F. R. and Parchen, F. R. *ibid.* **78**, 1540 (1956).

[81] Keenan, T. K. *J. Phys. Chem.* **61**, 1117 (1957).

[82] Gilks, S. W. and Waind, G. M. *Disc. Faraday Soc.* **29**, 102 (1960); Roig, E. and Dodson, R. W. *J. Phys. Chem.* **65**, 2175 (1961).

[83] Warnqvist, B. and Dodson, R. W. *Inorg. Chem.* **10**, 2624 (1971).

[84] Wada, G. and Tamaki, K. *Bull. Chem. Soc. Japan* **47**, 1422 (1974); and references therein.

[85] Tomiyasu, H. and Fukutomi, H. *Bull. Chem. Soc. Japan* **48**, 13 (1975).

[86] Ashby, E. C. and Wiesemann, T. L. *J. Amer. Chem. Soc.* **96**, 7117 (1974).

[87] Cecal, A. and Schneider, I. A. *Inorg. Nucl. Chem. Lett.* **10**, 977 (1974).

[88] Melton, S. L., Wear, J. O., and Amis, E. S. *J. Inorg. Nucl. Chem.* **17**, 317 (1961); Melton, S. L., Indelli, A. and Amis, E. S. *ibid.* p. 325.

[89] Sutin, N. *J. Phys. Chem.* **64**, 1766 (1960); Horne, R. H. *Exchange Reactions* Proc. New York 1965 Symposium, IAEA Vienna (1965) p. 67.

[90] Newton, T. W. and Baker, F. B. *J. Phys. Chem.* **68**, 228 (1964); *Inorg. Chem.* **3**, 569 (1964).

[91] Espenson, J. H. and Krug, L. A. *Inorg. Chem.* **8**, 2633 (1969).

[92] Newton, T. W. *J. Phys. Chem.* **74**, 1655 (1970).

[93] Rabideau, S. W. and Kline, R. J. *J. Phys. Chem.* **62**, 617 (1958).

[94] Cohen, D., Sullivan, J. C. and Hindman, J. C. *J. Amer. Chem. Soc.* **77**, 4964 (1955).

[95] Florence, T. M., Batley, G. E., Ekstrom, A., Fardy, J. J. and Garrar, Y. J. *J. Inorg. Nucl. Chem.* **37**, 1961 (1975).

[96] Taube, H. *Pure Appl. Chem.* **24**, 289 (1970).

[97] Taube, H. *Ber. Bunsenges. Phys. Chem.* **76**, 964 (1972).

[98] See pp. 473, 481, 503, and 514 of reference [2].

[99] Linck, R. G. in *MTP Int. Rev. Sci., Inorganic Chemistry, Series 1, vol. 9: Reaction Mechanisms in Inorganic Chemistry,* ed. Tobe, M. L. Butterworths, London and University Park Press, Baltimore (1972) chap. 8; and see p. 136 of reference [1] and p. 481 of reference [2].

[100] Carlyle, D. W. and Espenson, J. H. *J. Amer. Chem. Soc.* **91**, 599 (1969).

[101] Hyde, M. R., Davies, R. and Sykes, A. G. *J. Chem. Soc. Dalton* 1838 (1972).

[102] Zwickel, A. and Taube, H. *J. Amer. Chem. Soc.* **81**, 1288 (1959).

[103] Ekstrom, A. and Farrar, Y. *Inorg. Chem.* **11**, 2610 (1972).

[104] Yandell, J. K., Fay, D. P. and Sutin, N. *J. Amer. Chem. Soc.* **95**, 1131 (1973); Przystas, T. J. and Sutin, N. *Inorg. Chem.* **14**, 2103 (1975).

[105] Davies, K. M. and Espenson, J. H. *Chem. Commun.* 111 (1969); *J. Amer. Chem. Soc.* **91**, 3093 (1969).

[106] Price, H. J. and Taube, H. *Inorg. Chem.* **7**, 1 (1968).

[107] Green, M., Taylor, R. S. and Sykes, A. G. *J. Chem. Soc. A* 509 (1971).

[108] Norris, C. and Nordmeyer, F. R. *Inorg. Chem.* **10**, 1235 (1971).

[109] Hyde, M. R., Taylor, R. S. and Sykes, A. G. *J. Chem. Soc. Dalton* 2730 (1973).

[110] Prince, R. H and Segal, M. G. *J. Chem. Soc. Dalton* 330, 1245 (1975).

[111] Newton, T. W. and Baker, F. B. *J. Phys. Chem.* **69**, 176 (1965).

[112] Burkhart., M. J. and Newton, T. W. *J. Phys. Chem.* **73**, 1741 (1969).

[113] Baker, B. R., Orhanović, M. and Sutin, N. *J. Amer. Chem. Soc.* **89**, 722 (1967).

[114] Jacks, C. A. and Bennett, L. E. *Inorg. Chem.* **13**, 2035 (1974).

[115] Dodel, P. H. and Taube, H. *Z. Phys. Chem. Frankf. Ausg.* **44**, 92 (1965).

[116] Guenther, P. R. and Linck, R. G. *J. Amer. Chem. Soc.* **91**, 3769 (1969).

[117] Thorneley, R. N. F. and Sykes, A. G. *J. Chem. Soc. A* 1036 (1970).

[118] Parker, O. J. and Espenson, J. H. *Inorg. Chem.* **8**, 185 (1969).

[119] Green, M. and Sykes, A. G. *Chem. Commun.* 241 (1970).

[120] Bakač, A., Hand, T. D. and Sykes, A. G. *Inorg. Chem.* **14**, 2540 (1975).

[121] Patel, R. C., Atkinson, G. and Baumgartner, E. *Bioinorg. Chem.* **3**, 1 (1973).

[122] Rosseinsky, D. R. and Higginson, W. C. E. *J. Chem. Soc.* 31 (1960); Higginson, W. C. E. and Sykes, A. G. *ibid.* 2841 (1962).

[123] Green, M., Higginson, W. C. E., Stead, J. B. and Sykes, A. G. *J. Chem. Soc. A* 3068 (1971).

[124] Parker, O. J. and Espenson, J. H. *J. Amer. Chem. Soc.* **91**, 1313 (1969).

[125] Rosseinsky, D. R. *Chem. Rev.* **72**, 215 (1972).

[126] Matusek, M. *Sb. Ved. Praci, Vysoka Skola, Chem.-Technol., Pardubice* **18**, 379 (1969) (*Chem. Abstr.* **72**, 25467g (1970)).

[127] Daugherty, N. A. and Taylor, R. L. *J. Inorg. Nucl. Chem.* **34**, 1756 (1972).

[128] Watkins, K. O., Sullivan, J. C. and Deutsch, E. *Inorg. Chem.* **13**, 1712

(1974).

[129] Johnson, C. E. *J. Amer. Chem. Soc.* **74**, 959 (1952).

[130] Davies, G. *Inorg. Chem.* **10**, 1155 (1971).

[131] Cannon, R. D. and Gardiner, J. *J. Chem. Soc. Dalton* 887 (1972).

[132] Newton, T. W. and Baker, F. B. *J. Phys. Chem.* **67**, 1425 (1963).

[133] Cramer, J. L. and Meyer, T. J. *Inorg. Chem.* **13**, 1250 (1974).

[134] Marcus, R. A. and Sutin, N. *Inorg. Chem.* **14**, 213 (1975).

[135] Dockal, E. R., Everhart, E. T. and Gould, E. S. *J. Amer. Chem. Soc.* **93**, 5661 (1971).

[136] Meyerstein, D. and Mulac, W. A. *J. Phys. Chem.* **72**, 784 (1968); **73**, 1091 (1969).

[137] Ellis, J. D. and Sykes, A. G. *J. Chem. Soc. Dalton* 2553 (1973).

[138] Birk, J. P. *Inorg. Chem.* **14**, 1724 (1975); Martin, A. H. and Gould, E. S. *ibid.* **14**, 873 (1975).

[139] Thompson, G. A. K. and Sykes, A. G. *Inorg. Chem.* **15**, 638 (1976).

[140] Parker, O. J. and Espenson, J. H. *Inorg. Chem.* **8**, 1523 (1969).

[141] Parker, O. J. and Espenson, J. H. *J. Amer. Chem. Soc.* **91**, 1968 (1969).

[142] Tockstein, A. and Matusek, M. *Colln. Czech. Chem. Commun. Engl. Edn.* **34**, 316 (1969).

[143] Birk, J. P. and Logan, T. P. *Inorg. Chem.* **12**, 580 (1973).

[144] Orhanović, M. and Earley, J. E. *Inorg. Chem.* **14**, 1478 (1975).

[145] Thompson, R. C. and Sullivan, J. C. *159th Meeting of the American Chemical Society,* Houston (1970), paper Phys. 114.

[146] Taylor, R. S. and Sykes, A. G. *J. Chem. Soc. A.* 1628 (1971).

[147] Davies, R. , Kipling, B. and Sykes, A. G. *J. Amer. Chem. Soc.* **95**, 7250 (1973).

[148] Mayhew, R. T. and Amis, E. S. *J. Inorg. Nucl. Chem.* **35**, 4245 (1973).

[149] Schiefelbein, B. and Daugherty, N. A. *Inorg. Chem.* **9**, 1716 (1970).

[150] Ho Tan, Z. C. and Amis, E. S. *J. Inorg. Nucl. Chem.* **28**, 2889 (1966); Mayhew, R. T. and Amis, E. S. *ibid.* **35**, 4245 (1973); *J. Phys. Chem.* **79**, 862 (1975).

[151] Doyle, J. and Sykes, A. G. *J. Chem. Soc. A* 2836 (1968); Fan, F.-R. F. and Gould, E. S. *Inorg. Chem.* **13**, 2639, 2647 (1974).

[152] Thamburaj, P. K. and Gould, E. S. *Inorg. Chem.* **14**, 15 (1975); Adegite, A. and Kuku, T. A. *J. Chem. Soc. Dalton* 158 (1976).

[153] Adin, A. and Sykes, A. G. *J. Chem. Soc. A* 1230 (1966).

[154] Espenson, J. H. and Christensen, R. J. *J. Amer. Chem. Soc.* **91**, 7311 (1969).

[155] Ekstrom, A. and Johnson, D. A. *J. Inorg. Nucl. Chem.* **36**, 2557 (1974).

[156] Faraggi, M. and Feder, A. *Inorg. Chem.* **12**, 236 (1973); and see p. 481 of reference [2].

[157] Christensen, R. J., Espenson, J. H. and Butcher, A. B. *Inorg. Chem.* **12**, 564 (1973); Faraggi, M. and Tendler, Y. *J. Chem. Phys.* **56**, 3287 (1972).

[158] Wang, R. T. and Espenson, J. H. *J. Amer. Chem. Soc.* **93**, 1629 (1971).

[159] Ekstrom, A., McLaren, A. B. and Smythe, L. E. *Inorg. Chem.* **14**, 1035 (1975).

[160] Espenson, J. H. and Wang, R. T. *Chem. Commun.* 207 (1970).

[161] Davies, G. and Warnqvist, B. *Coord. Chem. Rev.* **5**, 349 (1970).

[162] Davies, G. *Coord. Chem. Rev.* **4**, 199 (1969).

[163] Baumgartner, E. and Honig, D. S. *J. Inorg. Nucl. Chem.* **36**, 196 (1974).

[164] Wells, C. F. *J. Inorg. Nucl. Chem.* **36**, 3856 (1974); Pelizzetti, E. and Mentasti, E. *J. Chem. Soc. Dalton* 2086 (1975).

[165] Benson, D. and Sutcliffe, L. H. *Trans. Faraday Soc.* **56**, 246 (1960).

[166] Benson, D., Proll, P. J., Sutcliffe, L. H. and Walkley, J. *Disc. Faraday Soc.* **29**, 60 (1960).

[167] McAuley, A. *Coord. Chem. Rev.* **5**, 245 (1970); and see reference [99].

[168] Capon, B., Perkins, M. J. and Rees, C. W. (Editors) *Organic Reaction Mechanisms* Wiley-Intersicence (1965 onwards) see Chapter 14 of each volume.

[169] Kemp, T. J., Chapter 4 of reference [73].

[170] Waters, W. A. *Mechanisms of Oxidation of Organic Compounds* Methuen (1964).

[171] E.g. Jones, T. E. and Hamm, R. E. *Inorg. Chem.* **13**, 1940 (1974); Woods, M., Cain, A. and Sullivan, J. C. *J. Inorg. Nucl. Chem.* **36**, 2605 (1974); Wells, C. F. and Fox, D. *J. Inorg. Nucl. Chem.* **38**, 107 (1976); and references therein.

[172] Andreasch, R. *Chem. Ber.* **12**, 1391 (1879).

[173] Schmid, H. *Z. Phys. Chem.* **148**, 321 (1930).

[174] Mentasti, E., Pelizzetti, E. and Saini, G. *J. Chem. Soc. Dalton* 2609 (1973); Pelizzetti, E., Mentasti, E., Pramauro, E. and Saini, G. *ibid.* 1940 (1974).

[175] Uri, N. *J. Chem. Soc.* 335 (1947).

[176] Thomas, J. K., Trudel, G. and Bywater, S. *J. Phys. Chem.* **64**, 51 (1960).

[177] Karraker, D. G. *J. Phys. Chem.* **67**, 871 (1963).

[178] Higginson, W. C. E. and Simpson, M. E. *J. Chem. Soc. Chem. Commun.* 817 (1974).

[179] Anbar, M. in *Advances in Chemistry, Volume 49* ed. Gould, R. F. American Chemical Society (1965), Chap. 6.

[180] Greatorex, D., Hill, R. J., Kemp, T. J. and Stone, T. J. *J. Chem. Soc. Faraday I* **70**, 216 (1974).

[181] Halpern, J. *J. Chem. Educ.* **45**, 372 (1968).

[182] Sandberg, R. G., Auborn, J. J., Eyring, E. M. and Watkins, K. O. *Inorg. Chem.* **11**, 1952 (1972).

[183] Nadon, L. and Zador, M. *Can. J. Chem.* **52**, 2667 (1974); and references therein.

[184] Thacker, M. A., Scott, K. L., Simpson, M. E., Murray, R. S. and Higgin-

son, W. C. E. *J. Chem. Soc. Dalton* 647 (1974).

[185] Po, H. N., Swinehart, J. H. and Allen, T. L. *Inorg. Chem.* **7**, 244 (1968).

[186] Ellis, K. J. and McAuley, A. *J. Chem. Soc. Dalton* 1533 (1973).

[187] McAuley, A. and McCann, J. P., personal communication.

[188] Byrd, J. E. and Halpern, J. *J. Amer. Chem. Soc.* **95**, 2586 (1973).

[189] Pustelnik, N. and Soloniewicz, R. *Mh. Chem.* **106**, 673 (1975).

[190] Amjad, Z. and McAuley, A. *J. Chem. Soc. Dalton* 2521 (1974).

[191] Gupta, K. S. and Gupta, Y. K. *Inorg. Chem.* **13**, 851 (1974).

[192] Abley, P., Byrd, J. E. and Halpern, J. *J. Amer. Chem. Soc.* **95**, 2591 (1973).

[193] Mentasti, E., Pelizzetti, E. and Pramauro, E. *J. Inorg. Nucl. Chem.* **37**, 1733 (1975).

[194] Levesley, P. and Waters, W. A. *J. Chem. Soc.* 217 (1955).

[195] Carlyle, D. W. and Zeck, O. F. *Inorg. Chem.* **12**, 2978 (1973).

[196] Giuntoli, R. B. and Habib, H. S. *J. Inorg. Nucl. Chem.* **36**, 363 (1974).

[197] Mishra, S. K., Sharma, P. D. and Gupta, Y. K. *J. Inorg. Nucl. Chem.* **36**, 1845 (1974).

[198] Mushran, S. P., Agrawal, M. C., Mehrotra, R. M. and Sanehi, R. *J. Chem. Soc. Dalton* 1460 (1974).

[199] Veith, G., Guthals, E. and Viste, A. *Inorg. Chem.* **6**, 667 (1967).

[200] Pelizzetti, E., Mentasti, E. and Saini, G. *J. Chem. Soc. Dalton* 721 (1974).

[201] Baxendale, J. H., Hardy, H. R. and Sutcliffe, L. H. *Trans. Faraday Soc.* **47**, 963 (1951).

[202] McAuley, A. and Shanker, R. *J. Chem. Soc. Dalton* 2321 (1973).

[203] Trahanovsky, W. S., Cramer, J. and Brixius, D. W. *J. Amer. Chem. Soc.* **96**, 1077 (1974).

[204] Boehm, J. R., Balch, A. L., Bizot, K. F. and Enemark, J. H. *J. Amer. Chem. Soc.* **97**, 501 (1975).

[205] Belusov, B. P. *Sb. Ref. Radiat. Med., Moscow, 1958* 145 (1959); Nicolis, G. and Portnow, J. *Chem. Rev.* **73**, 365 (1973); Noyes, R. M. and Field, R. J. *A. Rev. Phys. Chem.* **25**, 95 (1974); Cooke, D. O. *J. Chem. Soc. Chem. Commun.* 27 (1976).

[206] E.g. Davies, G. *Inorg. Chem.* **11**, 2488 (1972); *Inorg. Chim. Acta* **14**, L13 (1975); Pelizzetti, E., Mentasti, E. and Giraudi, G. *ibid.* **15**, L1 (1975).

[207] Pelizzetti, E., Mentasti, E., Carlotti, M. E. and Giraudi, G. *J. Chem. Soc. Dalton* 794 (1975); and references therin.

[208] Anet, F. A. L. and LeBlanc, E. *J. Amer. Chem. Soc.* **79**, 2649 (1957).

[209] Davis, D. D. and Bigelow, W. B. *J. Amer. Chem. Soc.* **92**, 5127 (1970).

[210] Ševčik, P. *Coll. Czech. Chem. Commun. Engl. Edn.* **40**, 2935 (1975); *Chem. Zvesti* **29**, 9 (1975).

[211] Hanson, J. R. and Premuzic, E. *Agnew. Chem., Int. Edn. Engl.* **7**, 247

(1968).

[212] Sellers, R. M. and Simic, M. G. *J. Chem. Soc. Chem. Commun.* 401 (1975).

[213] Fallab, S. *Angew. Chem., Int. Edn. Engl.* **6**, 496 (1967); Južnič, K. and Fedina, S. *J. Inorg. Nucl. Chem.* **36**, 2609 (1974).

[214] Birk, J. P. *Inorg. Chem.* **12**, 2468 (1973).

[215] Mitzner, R., Fischer, G. and Leupold, P. *Z. Phys. Chem. Frankf. Ausg.* **253**, 161 (1973).

[216] Foust, R. D. and Ford, P. C. *J. Inorg. Nucl. Chem.* **36**, 930 (1974).

[217] Malin, J. M. and Swinehart, J. H. *Inorg. Chem.* **8**, 1407 (1969).

[218] Newton, T. W. *Inorg. Chem.* **14**, 2394 (1975).

[219] Laurence, G. S. and Thornton, A. T. *J. Chem. Soc. Dalton* 1142 (1974).

Chapter **14**

KINETICS AND MECHANISMS:
REACTIONS OF COORDINATED SOLVENTS

14.1 INTRODUCTION

Reactions of coordinated ligands have become a popular subject of study. Reactions of coordinated solvents are a special case that has been relatively little studied. The best documented area is that of the thermodynamics of reversible proton loss from coordinated solvent molecules, especially water (Chapter 9), and of subsequent formation of polynuclear species (Chapter 10). The present Chapter deals mainly with kinetic aspects of these reactions.

Kinetics and mechanisms of other reactions of solvent molecules coordinated to metal ions in solution have been very little examined. Moreover most examples are on the fringes of admissibility, in that they generally involve reactions of mixed solvento-ligand complexes in aqueous solution. Obvious examples here the hydrolyses of coordinated ester or nitrile molecules in such species as $Co(NH_3)_5(ester)^{3+}$ and $Ru(NH_3)_5(nitrile)^{3+}$ complexes [1]. The effect of the central metal ion is always marked, and may be dramatic, as in the rate accelerations of around 10^8 times reported for some coordinated nitriles [2]. The hydrolysis of acetonitrile to acetamide in acetonitrile + water mixtures is catalysed by the Hg^{2+} cation, which coordinates to the acetonitrile [3].

Among those investigations of redox reactions of coordinated solvent molecules which are of similar marginal relevance to this Chapter are the oxidation of coordinated ammonia in $Ru(NH_3)_6^{3+}$ by molecular oxygen to give $Ru(NH_3)_5(NO)^{3+}$ [4] and of coordinated dimethyl sulphoxide in $Co(NH_3)_5$ $(dmso)^{3+}$ [5]. A few further observations on redox processes involving solvents coordinated to metal ions in various guises can be found in reference [6].

14.2 PROTON LOSS

The kinetics of reversible proton loss from several aquo-cations,

$$M^{n+}aq \underset{k_b}{\overset{k_f}{\rightleftharpoons}} M(OH)^{(n-1)+}aq + H^+aq, \qquad (14.1)$$

have been monitored by the electric field jump (or dissociation field effect) relaxation technique, described in reference [7]. Rate constants for the forward and reverse reactions are collected together in Table 14.1 [8–18]. Although values of k_f are approximately the same, around 10^5 sec $^{-1}$, for several aquo-cations, Fe^{3+}aq and Ga^{3+}aq are claimed to lose a proton considerably more rapidly, which is puzzling [13]. For the reverse reaction, values of k_b approach the diffusion-controlled limit. The charge product of the reactants is reflected to a certain extent in the rate constant, with the $Th(OH)^{3+}$aq reacting several times more slowly than the $M(OH)^{2+}$aq cations, which in turn react more slowly than $Cu(OH)^+$aq or $UO_2(OH)^+$aq. The rate constants for the reactions of the two last-named cations with H^+aq resemble those for reaction of other unipositive species, for example HS^-, where $k = 7.5 \times 10^{10}$ sec^{-1} [19]. Similar data for non-aqueous solvents are rare, but there is some information on M^{2+}aq-catalysis of proton exchange with ethanol for a series of metal(II) cations of transition metals and sp-block elements [20].

Table 14.1 Rate constants for the reactions of equation (14.1), in aqueous solution at $25°C$ and an ionic strength close to zero.

Cation	k_f/sec^{-1}	$k_b/M^{-1}sec^{-1}$	
Mn^{2+}aq		1.8×10^{10}	[8]
Ni^{2+}aq		2.2×10^{10}	[8]
Cu^{2+}aq		1×10^{10}	[9]
Zn^{2+}aq		$1.4–1.5 \times 10^{10a}$	[8, 10]
UO_2^{2+}aq	1.7×10^4	1.7×10^{10}	[11]
Al^{3+}aq	1.1×10^5	4.4×10^9	[12]
Ga^{3+}aq	b	b	[13]
In^{3+}aq	1×10^5	9×10^9	[14]
Sc^{3+}aq	2.9×10^5	9.7×10^9	[15]
Fe^{3+}aq	b, c	b, c	[13]
$Co(NH_3)_5(OH_2)^{3+}$	3.6×10^3	5×10^9	[16]
Th^{4+}aq		$\sim 7 \times 10^8$	[17]

a The activation energy of this reaction is 9.5 kcal mol^{-1} [10]; b These reactions are too fast for rate constants to be derived by the dissociation field relaxation techniques; c Variable temperature n.m.r. shift measurements have led to estimates for rate constants for proton exchange with Fe^{3+}aq, and with Cr^{3+}aq [18].

14.3 POLYMERISATION AND DEPOLYMERISATION

14.3.1 Polymerisation

Rates of formation of hydroxo- or oxo- bridged bi- and poly-nuclear species are slower than the rates cited in the previous section. In other words, the reaction of a hydroxo-cation with an aquo-cation is slower than that with a hydrated proton. The rate of formation of the binuclear scandium(III) species is similar to that for the formation of complexes from $Sc^{3+}aq$, which suggests that the rate-determining step in both cases is water loss from $Sc^{3+}aq$ [15]. In the same way, the similarity between the rate of formation of the binuclear iron(III) species and the rate of water exchange at $Fe^{3+}aq$ indicates a common rate-determining step, presumably also dissociative in character [17]. Further polymerisation in iron(III) solutions under mildly alkaline conditions probably involves structural changes in the polynuclear species parallel with stepwise addition of further iron(III) units [21]. Polymerisation of nickel(II) is a more tractable problem, as the product, $Ni_4(OH)_4^{4+}$, is simpler and better characterised than that for iron(III) polymerisation. Linear hydroxo-bridged units are formed immediately on mixing nickel(II) solutions with alkali. The rate-determining step in the formation of cyclic $Ni_4(OH)_4^{4+}$ involves the reaction of these linear polymers with more $Ni^{2+}aq$ [22]. Investigation of the mechanism of formation of polynuclear plutonium(IV) species is complicated by the simultaneous existence of other oxidation states of plutonium in equilibrium with plutonium(IV) (Chapter 8.2.2). Nevertheless the rate law for the disappearance of monomeric plutonium(IV) is simply [23],

$$-d[monomer]/dt = k[Pu^{4+}aq][H^+]^{-2}$$

14.3.2 Depolymerisation

The first step in the depolymerisation of hydroxo-bridged polynuclear cations is often simply first-order in oligomer. This has been shown for $Ni_4(OH)_4^{4+}$ [24] and iron(III) species [22] among others. Often there is no kinetic or other evidence to help in elucidating the rest of the depolymerisation sequence. One of the simpler sequences has been proposed for lead(II) [25].

$$Pb_4(OH)_4^{4+}(cyclic) \xrightarrow{2H^+} Pb_4(OH)_2^{6+}(open) \rightleftharpoons 2\,Pb_2(OH)^{3+} \xrightarrow{2H^+} 4\,Pb^{2+}aq.$$

Dissociation of $Bi_6(OH)_{12}^{6+}$ takes place by parallel direct and proton-catalysed pathways. There is evidence for the intermediacy of a trimeric cation $Bi_3(OH)_5^{4+}$ [26]. Rate data are available for some steps in the above-mentioned depolymerisation schemes, and for the dedimerisation of binuclear titanium(III) [27].

REFERENCES

[1]   E.g. Burgess, J. *Inorganic Reaction Mechanisms – Chemical Society Specialist Periodical Report* **1**, 198 (1972); **2**, 219 (1972); **3**, 297 (1974).

[2] Zanella, A. W. and Ford, P. C. *Inorg. Chem.* **14**, 42, 700 (1975).

[3] Yu-Keung Sze and Irish, D. E. *Can. J. Chem.* **53**, 427 (1975).

[4] Pell, S. D. and Armor, J. N. *J. Amer. Chem. Soc.* **97**, 5012 (1975).

[5] De Oliveira, L. A., Toma, H. E. and Giesbrecht, E. *Inorg. Nucl. Chem. Lett.* **12**, 195 (1976).

[6] Anbar, M. in *Advances in Chemistry, No. 49*, ed. Gould, R. F., American Chemical Society (1965), Chap. 6.

[7] Eigen, M. and De Maeyer, L. in *Techniques of Organic Chemistry, Vol. VIII, Part II*, ed. Friess, S. L., Lewis, E. S and Weissberger, A. Wiley-Interscience (1963), Chap. 18.

[8] Katsman, L. A., Vargaftik, M. N. and Syrkin, Ya. K. *Dokl. Akad. Nauk SSSR* **206**, 645 (1972).

[9] De Maeyer, L. and Kustin, K. *A. Rev. Phys. Chem.* **14**, 5 (1963).

[10] Vargaftik, M. N., Katsman, L. A. and Syrkin, Ya. K. *Izv. Akad. Nauk SSSR, Ser. Khim.* 1890 (1972).

[11] Cole, D. L., Eyring, E. M., Rampton, D. T., Silzars, A. and Jensen, R. P. *J. Phys. Chem.* **71**, 2771 (1967).

[12] Holmes, L. P., Cole, D. L. and Eyring, E. M. *J. Phys. Chem.* **72**, 301 (1968).

[13] Hemmes, P., Rich, L. D., Cole, D. L. and Eyring, E. M. *J. Phys. Chem.* **75**, 929 (1971).

[14] Hemmes, P., Rich, L. D., Cole, D. L. and Eyring, E. M. *J. Phys. Chem.* **74**, 2859 (1970).

[15] Cole, D. L., Rich, L. D., Owen, J. D. and Eyring, E. M. *Inorg. Chem.* **8**, 682 (1969).

[16] Eigen, M. and Kruse, W. *Z. Naturf.* **18B**, 857 (1963).

[17] Eyring, M. and Cole, D. L. in *Fast Reactions and Primary Processes in Chemical Kinetics*, ed. Claesson, S. Wiley-Interscience (1967), p. 255.

[18] Luz, Z. and Shulman, R. G. *J. Chem. Phys.* **43**, 3750 (1965).

[19] Eigen, M. and Kustin, K. *J. Amer. Chem. Soc.* **82**, 5952 (1960).

[20] Hunt, A. H. and Hobbs, M. E. *J. Phys. Chem.* **75**, 1994 (1971).

[21] Sommer, B. A., Margerum, D. W., Renner, J., Saltman, P. and Spiro, T. G. *Bioinorg. Chem.* **2**, 295 (1973).

[22] Clare, B. W. and Kepert, D. L. *Aust. J. Chem.* **28**, 1489 (1975).

[23] Bell, J. T., Costanzo, D. A. and Biggers, R. E. *J. Inorg. Nucl. Chem.* **35**, 623 (1973).

[24] Kolski, G. B., Kildahl, N. K. and Margerum, D. W. *Inorg. Chem.* **8**, 1211 (1969).

[25] Frei, V. and Wendt, H. *Z. Phys. Chem. Frankf. Ausg.* **88**, 59 (1974).

[26] Frei, V., Mages, G. and Wendt, H. *Ber. Bunsenges. Phys. Chem.* **77**, 243 (1973).

[27] Birk, J. P. and Logan, T. P. *Inorg. Chem.* **12**, 580 (1973).

Chapter 15

ENVOI

Readers who have persevered this far may well share the author's impressions regarding the subject of metal ions in solution. A vast amount of information is available on this subject, with some areas understood in fair depth and detail. However there are several surprising and disturbing gaps in knowledge and understanding, sometimes of important aspects. In particular, there is still much ignorance on the fundamental question of solvation numbers and the nature of cation-solvent interactions. The estimation of solvation numbers for cations at which solvent exchange takes place more rapidly than n.m.r. frequencies is still difficult and open to dispute. On the other hand, it has been encouraging to observe how X-ray and neutron diffraction investigations have in recent years led to a greater understanding of the environments of several metal ions in solution. These techniques give especially good estimates of one fundamental quantity, the cation to nearest solvent distance. Secondary solvation is, of course, less well documented and understood than primary solvation. Here there are difficulties of definition that are sometimes as much philosophical as chemical. However, recent n.m.r. work suggests that more information may be forthcoming, probably by way of studies of solvation around kinetically inert transition metal complexes.

On the thermochemical front, the present situation is tolerably satisfactory for aqueous solutions. Improvements in methods of estimating single ion parameters may in the future make small differences to the values listed in Chapter 7, but major changes seem unlikely. Recently there has been much activity on ions in mixed and non-aqueous solvents. Single ion parameters estimated for cations in many non-aqueous media are now available. They are also unlikely to change much, because the basic assumption used in their derivation (now usually $AsPh_4^+ \equiv BPh_4^-$) seems reasonable. The situation is considerably less satisfactory for mixed aqueous solutions, where the several different assumptions used for deriving single ion values give significantly different values for a given ion in a given medium. There seems to be little current activity on redox potentials. Values for aqueous media have, except for a few exotic couples, been firmly established for years or even decades. Data on redox potentials in mixed and in non-aqueous media, on the other hand, are scarce. The technical difficulties involved, especially in respect of comparability between scales in different solvents, will probably continue to impede progress here. Studies on the hydrolysis and polymerisation of metal ions in solution, especially in water, continue in several laboratories. X-Ray diffraction studies of solutions and of model species in crystals, together with a range of spectroscopic techniques, are slowly but steadily improving our knowledge and understanding of the species involved.

The kinetic and mechanistic behaviour of metal ions in solution has been extensively studied and is now well understood in the main. Mechanisms of solvent exchange have proved the most difficult to ascertain with confidence. Now, however, the mechanisms of solvent exchange at the majority of cations have been established with the aid of a variety of approaches, especially the determination of volumes of activation. The mechanisms of formation reactions of solvated metal cations have also been settled, the majority taking place by the Eigen-Wilkins interchange mechanism or by understandable variants of it. Both for solvent exchange and for complex formation, kinetic patterns have been investigated and rationalised as well in non-aqueous media as in water. Only in mixed aqueous media do some areas that are seriously controversial remain. In contrast, redox mechanisms for solvated metal ions have been studied almost exclusively in water. It is a quarter of a century since Taube unequivocally demonstrated the operation of the inner-sphere mechanism for the reduction of a cobalt(III) complex by aquo-chromium(II). The mechanisms of reductions by many aquo-metal cations have now been assigned as inner- or outer-sphere, though redox mechanisms for several other aquo-metal cations remain undecided. The basic patterns are therefore settled for mechanisms of reactions of solvated metal cations. Interest now lies in a few unresolved problems, in reactions at coordinated ligands, and in extensions to biological and industrial problems.

Though a great deal has been achieved in observing and explaining the chemistry of metal ions in solution, still more remains to be done.

List of Symbols

A	coupling constant (n.m.r., e.s.r.)
A	associative (mechanism)
B	Jones-Dole viscosity coefficient
B	Racah parameter for free ion
B'	Racah parameter for complex (solvate)
c	concentration
(c)	solid (crystalline) phase
C	heat capacity

$$C_P \quad \text{heat capacity at constant pressure}$$
$$C_V \quad \text{heat capacity at constant volume}$$
$$C_P^\ominus \quad \text{standard partial molal heat capacity at constant pressure}$$
$$C_P^\ominus(M^{n+}aq) \quad \text{standard partial molal heat capacity for an aquo-metal ion at constant pressure}$$

CFAE	crystal field activation energy
CFSE	crystal field stabilisation energy
d^n	d-electron configuration of a transition metal ion
D	dielectric constant
D	dissociative (mechanism)
DN	Gutmann solvent donor number
Dq	crystal field splitting parameter
e	charge on the electron
e^-	electron
E	redox potential

$$E^\ominus \quad \text{standard redox potential}$$
$$E_{1/2} \quad \text{polarographic half-wave potential}$$

E_T	Reichardt solvent parameter
f	oxygen isotopic fractionation factor
f^n	f-electron configuration of a lanthanide or actinide ion
F	Dubois and Bienvenue solvent parameter
F	Faraday
(g)	gas phase

$g(r)$	radial distribution function	
$g(S)$	Letellier and Gaboriaud solvent parameter	
G	Gibbs free energy (for derivatives see under H below)	
h	Planck's constant	
H	enthalpy	
	H^E	excess enthalpy of mixing
	$\Delta H(\text{abs})$	enthalpy change on absolute scale
	ΔH_{amm}	enthalpy of ammoniation
	$\Delta H(\text{conv})$	enthalpy change relative to conventional (arbitrary) zero
	$\Delta H(\text{hydr})$	enthalpy of hydration
	$\Delta H(\text{solv})$	enthalpy of solvation
	$\Delta H_f^{\ominus}$	standard enthalpy of formation
	$\Delta H_{\text{tr}}^{\ominus}$	standard enthalpy of transfer
	$\Delta H_1, \Delta H_1^{*}$	enthalpies of loss of a proton from an aquo-metal ion
	$\Delta H_{\text{pq}}, \Delta H_{\text{pq}}^{*}$	enthalpies of formation of a polynuclear metal ion
	$\Delta H^{\ddagger}$	activation enthalpy
	$\Delta H_i^{\ddagger}$	activation enthalpy for ligand interchange
HN	hydration number	
I	nuclear spin	
I	ionic strength	
I_a	associative interchange (mechanism)	
I_d	dissociative interchange (mechanism)	
ICB	internal conjugate base (mechanism)	
IP	ionisation potential	
k	rate constant	
	k_b	rate constant for back reaction
	k_d	rate constant for dissociation
	k_{ex}	rate constant for solvent exchange
	k_f	rate constant for forward reaction
	k_i	rate constant for ligand (solvent) interchange
	k_{obs}	observed rate constant
	k_{rc}	rate constant for chelate ring closure
	k_{ro}	rate constant for chelate ring opening
	k_{25}	rate constant at $25°C$
k_B	Boltzmann constant	
K	equilibrium constant	
	K_h	reciprocal $^{*}K_1$
	K_{IP}	equilibrium constant for ion-pair formation
	K_{os}	equilibrium constant for association between

		metal ion and ligand
	K_S	solubility product
	K_W	ionic product for water
	K_1	equilibrium constant for formation of mono-ligand complex
	$K_1, *K_1$	equilibrium constants for hydrolysis of an aquo-metal ion
	pK	$-\log_{10}K$
K		compressibility (for derivatives see under H above)
ln		natural logarithm (base e)
L		ligand
M^{n+}		metal ion
	$M^{n+}aq$	aquo-metal ion
M_S		total spin momentum for a system of electrons
N_A		Avogadro number
P		pressure
r		radius
	r_{eff}	effective radius of an ion
	r_{xtal}	estimated radius of an ion in crystal
R		gas constant
S		solvent molecule
S		Brownstein solvent parameter
S		entropy (for derivatives see under H above)
SCS		sterically controlled substitution (ring closure mechanism)
S_E2		bimolecular electrophilic substitution
S_H2		bimolecular homolytic substitution
S_N1		unimolecular nucleophilic substitution
S_N2		bimolecular nucleophilic substitution
t_+, t_-		transference (transport) numbers for cation, anion
T		temperature
T_1		spin-lattice (longitudinal) relaxation time
T_2		spin-spin (transverse) relaxation time
TA		typically aqueous (solvent mixture)
TNAN		typically non-aqueous negative (solvent mixture)
TNAP		typically non-aqueous positive (solvent mixture)
u_+, u_-		mobilities of cation, anion
V		volume
	$\bar{V}^{\ominus}$	standard partial molar volume
		(for other derivatives see under H above)
w_W		Washburn number
x		mole fraction
	x_n	mole fraction of the nth component
X		Gielen-Nasielski solvent parameter

X	general thermodynamic quantity ($X = H, G, S, V$, etc)
X	halide
Y	Grunwald-Winstein solvent parameter
z	charge on an ion
Z	Kosower solvent parameter
β	compressibility coefficient
β	quotient of Racah parameters, B/B'
β_n	overall stability constant for a complex containing n ligands
β_{pq}	stability constant for a polynuclear metal cation
γ	activity coefficient
δ	chemical shift (n.m.r.)
δ	correction term in Born equation
δ_s	Hildebrand and Scott solvent parameter
ϵ_o	permittivity of free space
η	viscosity
Δ	conductivity (molecule)
λ_+, λ_-	conductivity of cation, anion
λ_+°	conductivity of cation at infinite dilution
λ	wavelength
μ	bridging (ligand)
ν	frequency
ν_n	nth vibrational mode
Ω	Berson, Hamlet and Mueller (Diels-Alder) solvent parameter

Units and Conversion Factors

C.g.s. units have been used throughout this book. Conversion to S.I. units is simple in almost all cases; the two factors

$$1 \text{ cal} = 4.184 \text{ J}$$
$$0°\text{C} = 273.15 \text{ K}$$

take care of the vast majority of conversions. Conversion factors for other units can be found in monographs on S.I. units and in many standard text books.

Solvent Abbreviations

DMA	NN-dimethylacetamide
DMF	NN-dimethylformamide
DMSO	dimethyl sulphoxide
HMPA	hexamethylphosphor(tri)amide
TBP	tributyl phosphate
THF	tetrahydrofuran
TMP	trimethyl phosphate

Ligand Abbreviations

acac	acetylacetonate (pentane-2,4-dionate)

adp	adenosine diphosphate
ala	alanine
aq	water
atp	adenosine triphosphate
bamp	2,6-bis(aminomethyl)pyridine
bic	bicine (NN-dihydroxyethylglycine)
bipy	2,2'-bipyridyl
cr	2,12-dimethyl-3,7,11,17-tetra-azabicyclo[11,3,1]heptadeca-1,2,11,13,15-pentaene [Figure 11.13]
ctp	cytosine triphosphate
cyclam	1,4,8,11-tetra-azacyclotetradecane [Figure 12.3(d)]
diap	2,2-di(aminomethyl)propan-1-ol [Figure 11.12]
dien	diethylenetriamine
dmg	dimethylglyoximate
dmso	dimethyl sulphoxide
dopa	dihydroxyphenylalanine [Figure 12.5(b)]
edda	ethylenediaminediacetate
edta	ethylenediaminetetra-acetate
en	ethylenediamine
gly	glycine
glygly	glycylglycine
hm	histamine
hmpa	hexamethylphosphor(tri)amide
ida	iminodiacetate
mal	malonate
mida	N-methyliminodiacetate
nsa	5-nitrosalicylate
nta	nitrilotriacetate
ox	oxalate
oxine	8-hydroxyquinoline
pada	pyridine-2-azo-p-dimethylaniline [Figure 12.2(a)]
par	4-(2-pyridylazo)resorcinol [Figure 12.10(a)]
phen	1,10-phenanthroline
py	pyridine
sal	salicylate
sb	Schiff base [e.g. Figure 11.7]
ser	serine
sxo	semi-xylenol orange
taab	tetrabenzo[1,5,9,13]tetra-azacyclohexadecine [Figure 11.13]
tach	*cis,cis*-1,3,5-triaminocyclohexane
terpy	2,2',6',2''-terpyridyl
tet-*a,b*	5,7,7,12,14,14-hexamethyl-1,4,8,11-cyclotetra-azatetradecane (isomeric forms) [Figure 12.6(c)]
2,3,2-tet	1,4,8,11-tetra-azaundecane [Figure 12.6(b)]

tetren   tetramethylenepentamine
tmd   trimethylenediamine
tpp   tripolyphosphate
tptz   2,4,6-tripyridyl-*s*-triazine
trans-[14]-diene   5,7,7,12,14,14-hexamethyl-1,4,8,11-tetra-azacyclotetradeca-4,11-diene
            [Figure 12.6(d)]
tren   2,2′,2″-triaminotriethylamine
trenol   2,2′,2″-nitrilotriethanol [Figure 11.12]
triam   2,2-di(aminomethyl)-1-propylamine [Figure 11.12]
trien   triethylenetetramine
triol   2-hydroxymethyl-2-methylpropane-1,3-diol [Figure 11.12]

uda   uramildiacetate [Figure 12.9(a)]

val   valine

Index